UNIVERSITY OF WESTMINSTER

Failure to return or renew overdue books may result in suspension
of borrowing rights at all University of Westminster libraries.

Due for return on:

-7 NOV 2002

1 0 JAN 2003

3 0 SEP 2003

Advances in Intelligent Systems

International Series on
MICROPROCESSOR-BASED AND INTELLIGENT SYSTEMS ENGINEERING

VOLUME 21

Editor

Professor S. G. Tzafestas, *National Technical University of Athens, Greece*

The titles published in this series are listed at the end of this volume.

Advances in Intelligent Systems

Systems

Concepts, Tools and Applications

edited by

SPYROS G. TZAFESTAS

Department of Electrical Engineering and Computer Engineering,
National Technical University of Athens,
Greece

KLUWER ACADEMIC PUBLISHERS

DORDRECHT / BOSTON / LONDON

A C.I.P. Catalogue record for this book is available from the Library of Congress.

ISBN 0-7923-5966-6

Published by Kluwer Academic Publishers,
P.O. Box 17, 3300 AA Dordrecht, The Netherlands.

Sold and distributed in North, Central and South America
by Kluwer Academic Publishers,
101 Philip Drive, Norwell, MA 02061, U.S.A.

In all other countries, sold and distributed
by Kluwer Academic Publishers,
P.O. Box 322, 3300 AH Dordrecht, The Netherlands.

Printed on acid-free paper

CONTENTS

PART I : COMPUTER-AIDED INTELLIGENT SYSTEMS AND TOOLS

1. Sources of Efficiency in Planning : A Survey
A. Tsois and *S.G. Tzafestas*

2. An Interactive Geometric Constraint Solver
I. Fudos

3. An Intelligent Agent Framework in VRML Worlds
T. Panayiotopoulos, G. Katsirelos, S. Vosinakis and *S. Kousidou*

4. Determining the Visual Interpretation of Actions in Interactive Stories
N.M. Sgouros and *S. Sotirchos*

5. An Attribute Grammar Driven High-Level Synthesis Paradigm for Control Applications
G.E. Economakos and *G.K. Papakonstantinou*

6. A Case Study in Specifying the Denotational Semantics of C
N.S. Papaspyrou

10. Intelligent Guidance in a Virtual University
T. Panayiotopoulos, N. Zacharis and *S. Vosinakis*

PART II : INFORMATION EXTRACTION FROM TEXTS, NATURAL LANGUAGE INTERFACES AND INTELLIGENT RETRIEVAL SYSTEMS

11. Question Answering and Information Extraction from Texts
J. Kontos and *I. Malagardi*

12. Named Entity Recognition from Greek Texts : The GIE Project
V. Karkaletsis, C. D. Spyropoulos and *G. Petasis*

21. Content-Based Audio Retrieval Using a Generalized Algorithm
P. Piamsa-Nga, S. R. Subramanya, N. A. Alexandridis,
S. Srakaew, G. Blankenship, G. Papakonstantinou,
P. Tsanakas and *S.G. Tzafestas*

22. Intelligent Retrieval of Temporal and Periodic Data
L. Baxevanaki, E. Ioannidis and *T. Panayiotopoulos*

PART III : IMAGE PROCESSING AND VIDEO-BASED SYSTEMS

23. A Java-Based Image Processing System
P. Androutsos, D. Androutsos, K. N. Plataniotis
and *A. N. Venetsanopoulos*

34. A Color Coordinate Normalizer Chip
I. Andreadis

PART IV : APPLICATIONS

35. Stochastic Modeling in GPS Estimation Algorithms
M. Pachter and *S. Nardi*

36. Digital Image Processing for Weathering Analysis and Planning of Conservation Interventions on Historic Structures and Complexes
A. Moropoulou, M. Koui, Ch. Kourteli, N. Achilleopoulos and *F. Zezza*

37. A Thinning-Based Method for Extracting Peri-Urban Road Network from Panchromatic Images
V. Karathanassi, C. Iossifidis and *D. Rokos*

47. Reliability Cost Assessment in Composite Power Systems using the Monte Carlo Simulation Approach
N. C. Koskolos, S. M. Megalokonomos and *E. N. Dialynas*

PREFACE

Intelligent Systems involve a large class of systems which posses human-like capabilities such as learning, observation, perception, interpretation, reasoning under uncertainty, planning in known and unknown environments, decision making, and control action. The field of intelligent systems is actually a new interdisciplinary field which is the outcome of the interaction, cooperation and synergetic merging of classical fields such as system theory, control theory, artificial intelligence, information theory, operational research, soft computing, communications, linguistic theory, and others. Integrated intelligent decision and control systems involve three primary hierarchical levels, namely organization, coordination and execution levels. As we proceed from the organization to the execution level, the precision about the jobs to be performed increases and accordingly the intelligence required for these jobs decreases. This is in compliance with the principle of *increasing precision with decreasing intelligence* (IPDI) known from the management field and theoretically established by Saridis using information theory concepts.

This book is concerned with intelligent systems and techniques and gives emphasis on the computational and processing issues. Control issues are not included here. The contributions of the book are presented in four parts as follows.

PART I: *Computer-aided intelligent systems and tools*
PART II: *Information extraction from texts, natural language interfaces and intelligent retrieval systems*
PART III: *Image processing and video-based systems*
PART IV: *Applications*

I am deeply indebted to all colleagues who have contributed to the success of *EURISCON'98: The Third European Robotics, Intelligent Systems and Control Conference* (Athens, Greece, June 22-25, 1998) and provided revised and upgraded manuscripts for inclusion in this volume. I am sure that their results, put together in this book, will provide a useful reference pool of knowledge that will inspire and motivate other researchers for further developments in the field. Special thanks are due to the members of the Advisory and Program Committees of the Conference who have helped in the selection of the contributions.

November 1998

Spyros G. Tzafestas

CONTRIBUTORS

ACHILLEOPOULOS N. — *Natl. Tech. Univ. of Athens, Greece*
ALEXANDRIDIS N. — *George Washington Univ., USA*
ALEXOPOULOS V. — *Natl. Tech. Univ. of Athens, Greece*
ANDREADIS I. — *Demokritus Univ. of Thrace, Greece*
ANDROUTSOS D. — *Univ. of Toronto, Canada*
ANDROUTSOS P. — *Univ. of Toronto, Canada*
AVOLIO A. — *Univ. of N.S.Wales, Australia*
AVRADINIS N. — *Univ. of Piraeus, Greece*
BAXEVANAKI L. — *NCSR "Demokritos", Athens, Greece*
BIENVENIDO J.F. — *Univ. of Almeria, Spain*
BITZER B.E. — *Univ. of Paderborn, Soest, Germany*
BLANK I. — *Univ. of Munich, Germany*
BLANKESHIP G. — *George Washington Univ., USA*
CAMACHO F. — *Univ. of N.S. Wales, Australia*
CORRAL A. — *Univ. of Almeria, Spain*
D'ANGELO G. — *CELI: Centro Elab. Ling. Info., Torino, Italy*
DALIAS T. — *Intrasoft S.A., Athens, Greece*
DELOPOULOS A. — *Natl. Tech. Univ. of Athens, Greece*
DIALYNAS E.N. — *Natl. Tech. Univ. of Athens, Greece*
DINI L. — *CELI: Centro Elab. Ling. Info., Torino, Italy*
DI TOMASO V. — *CELI: Centro Elab. Ling. Info., Torino, Italy*
DOULAMIS A.D. — *Natl. Tech. Univ. of Athens, Greece*
DOULAMIS N.D. — *Natl. Tech. Univ. of Athens, Greece*
ECONOMAKOS G.E. — *Natl. Tech. Univ. of Athens, Greece*
FAKOTAKIS N. — *Univ. of Patras, Greece*
FERRI H.C. — *Informatique, CDC/DTA, Arcueil, France*
FRANGAKIS G. — *Univ. of N.S. Wales, Australia*
FRIESEN K. — *California State Univ., USA*
FUDOS I. — *Univ. of Ioannina, Greece*
GARCIA-LAZARO J.R. — *Univ. of Almeria, Spain*
GEORGANTOPOULOS J.R. — *Inst. Lang. & Speech Proc., Athens, Greece*
GERBESSIOTIS A.V. — *Oxford Univ., England*
HALATSIS C. — *Univ. of Athens, Greece*
IKONOMAKIS N. — *Univ. of Toronto, Canada*
IOANNIDIS C. — *NCSR "Demokritos", Athens, Greece*
IOSSIFIDIS C. — *Natl. Tech. Univ. of Athens, Greece*
KARATHANASSI V. — *Natl. Tech. Univ. of Athens, Greece*
KARKALETSIS V. — *NCSR "Demokritos", Athens, Greece*
KATSITRELOS G. — *Univ. of Piraeus, Greece*
KOKKINAKIS G. — *Univ. of Patras, Greece*
KOKKOTOS S. — *Natl. Tech. Univ. of Athens, Greece*
KOLLIAS S. — *Natl. Tech. Univ. of Athens, Greece*
KONTOS J. — *Athens Univ. of Econ. & Business, Greece*
KOSKOLOS N.C. — *Natl. Tech. Univ. of Athens, Greece*

KOUI M.	*Natl. Tech. Univ. of Athens, Greece*
KOULOURIS A.	*Natl. Tech. Univ. of Athens, Greece*
KOURTELI C.	*Natl. Tech. Univ. of Athens, Greece*
KOUSIDOU S.	*Univ. of Piraeus, Greece*
KOZIRIS N.	*Natl. Tech. Univ. of Athens, Greece*
KYRIAKOPOULOS K.J.	*Natl. Tech. Univ. of Athens, Greece*
LALIOTIS L.N.	*Natl. Tech. Univ. of Athens, Greece*
LEE J.S.	*California State Univ., USA*
LEONIDOU M.	*Natl. Tech. Univ. of Athens, Greece*
LERTSUNTIVIT S.	*California State Univ., USA*
LILAS T.	*Natl. Tech. Univ. of Athens, Greece*
LIVADAS P.	*Univ. of Florida, Fl., USA*
MALAGARDI I.	*Athens Univ. of Econ. & Business, Greece*
MARINAGI C.C.	*NCSR "Demokritos", Athens, Greece*
MEGALOKONOMOS S.M.	*Natl. Tech. Univ. of Athens, Greece*
MICHOS S.E.	*Univ. of Patras, Greece*
MOROPOULOU A.	*Natl. Tech. Univ. of Athens, Greece*
NARDI S.	*Airforce Inst. of Tech., OH, USA*
NOVOTNY P.	*Inst. of Radioelectronics Brno, Chech Rep.*
PACHTER M.	*Airforce Inst. of Tech., OH, USA*
PANAGIOTAKOPULOS N.D.	*California State Univ., USA*
PANAYIOTOPOULOS T.	*Univ. of Piraeus, Greece*
PANTAZOPOULOS J.	*Natl. Tech. Univ. of Athens, Greece*
PAPAKONSTANTINOU G.K.	*Natl. Tech. Univ. of Athens, Greece*
PAPASPYROU N.S.	*Natl. Tech. Univ. of Athens, Greece*
PAPAZOGLOU T.M.	*Tech. Educ. Inst., Iraklio, Greece*
PETASIS G.	*NCSR "Demokritos", Athens, Greece*
PIAMSA-NGA P.	*George Washington Univ., USA*
PIPERIDIS S.	*Inst. Lang. & Speech Proc., Athens, Greece*
PLATANIOTIS K.N.	*Ryerson Polyt. Univ., Toronto, Canada*
PLATIS N.	*Inrtasoft S.A., Athens, Greece*
QASEM A.	*Univ. of N.S. Wales, Australia*
RODRIGUEZ F.	*Univ. of Almeria, Spain*
ROKOS D.	*Natl. Tech. Univ. of Athens, Greece*
SGOUROPOULOU C.E.	*Natl. Tech. Univ. of Athens, Greece*
SGOUROS N.	*Univ. of Piraeus, Greece*
SKORDALAKIS E.S.	*Natl. Tech. Univ. of Athens, Greece*
SIGMUND M.	*Inst. of Radioelectronics Brno, Chech Rep.*
SOTIRCHOS S.	*Natl. Tech. Univ. of Athens, Greece*
SPYROPOULOS C.D.	*NCSR "Demokritos", Athens, Greece*
SRAKAEW S.	*George Washington Univ., USA*
STEPHAN T.	*Univ. of N.S. Wales, Australia*
STRATAKIS D.I.	*Tech. Educ. Inst., Iraklio, Greece*
SUBRAMANYA S.R.	*George Washington Univ., USA*
TRIANTAFYLLOS G.	*Athens Univ. of Econ. & Business, Greece*
TRIKKALIDIS D.	*Athens Univ. of Econ. & Business, Greece*

TSANAKAS P. — *Natl. Tech. Univ. of Athens, Greece*
TSAPATSOULIS N. — *Natl. Tech. Univ. of Athens, Greece*
TSOIS A. — *Natl. Tech. Univ. of Athens, Greece*
TZAFESTAS S.G. — *Natl. Tech. Univ. of Athens, Greece*
URBANI D. — *Informatique, CDC/DTA, Arcueil, France*
VENETSANOPOULOS A.N. — *Univ. of Toronto, Canada*
VIBLIS M.K. — *Natl. Tech. Univ. of Athens, Greece*
VICHOT F. — *Informatique, CDC/DTA, Arcueil, France*
VOLIOTIS C.P. — *Natl. Tech. Univ. of Athens, Greece*
VOSINAKIS S. — *Univ. of Piraeus, Greece*
VOTSIS G. — *Natl. Tech. Univ. of Athens, Greece*
XIROUCHAKIS Y. — *Natl. Tech. Univ. of Athens, Greece*
WAN L. — *California State Univ., USA*
WOLISNKI F. — *Informatique, CDC/DTA, Arcueil, France*
ZACHARIS N. — *Univ. of Piraeus, Greece*
ZEZZA F. — *Bari Polytechnic, Italy*

EDITORIAL

The book contains forty seven contributions that present a rich set of theoretical and application developments within the field of intelligent systems and applications.

PART I deals with computer-aided techniques and tools. **Chapter 1**, by *Tsois and Tzafestas*, identifies the principles on which efficient planning is based and explains how various methods affect planning algorithms. The key ideas and concepts in planning are presented accompanied by two specific algorithms that illustrate the results drawn.

Chapter 2, by *Fudos*, presents the development of a user-friendly interactive system for editing and solving geometric configurations that arise in feature-based CAD/CAM systems. The system is built around a powerful graph-constructive constraint solving method, capable of efficiently analyzing certain classes of well-determined, over-determined and under-determined configuarions. To realize the constraint solver a prototype on a SUN workstation running Solaris 2.5.1. was developed. The graphical user interface was build in Java AWT, the core method was implemented in SETL (SET Language), the extensions for detecting under-determined and over-determined configurations were also programmed in SETL, and for numerical solving, MATLAB packages were invoked.

Chapter 3, by *Panayiotopoulos, Katsirelos, Vosinakis and Kousidou*, presents the architecture of an Intelligent Agent which consists of a logical core as well as a virtual representative. The logical structure provides the agent with reasoning capabilities as well as domain knowledge. The virtual world is represented as a 'mental' structure in the agent's knowledge base. The agent is not only perceiving, deciding and acting, but its actions (movement, object manipulation, etc.) are visualized as the actions of a virtual 'avatar' in a VRML world. The applications of such a structure are numerous, including educational software, intelligent VRML micro-worlds, role playing games, interactive movies, etc.

Chapter 4, by *Sgouros and Sotirchos*, describes a novel method for determining parts of the visual interpretation of actions in interactive plots. The method accepts as input a story plot consisting of a tree of possible character actions, their motivations and outcomes. It also accepts as input a 2-D layout of the 3-D space in which the story will play out and a list of possible behaviours for each of the objects in the description. The method uses this input to compose 3-D renditions for each story action by assigning automatically behaviors to the objects in the scene based on perfomance-related descriptions of story actions. These renditions are enriched automatically with appropriate multimedia effects that seek to emphasize their dramtic nature.

Chapter 5, by *Economakos and Papakonstantinou*, deals with the problem of high-level synthesis of control applications in the same AG-driven environment, tuned to handle such cases efficiently. As an illustration, a Kalman filter of a track-while-scan radar system is automatically produced. The basic requirements at the algorithmic level are discussed and proper solutions are given in an AG formalism. The proposed approach raises the level of abstraction for high-level hardware synthesis of control applications. AGs can aid designers implement all high-level synthesis transformations quickly and into a more natural way, thus shortening design space exploration time.

Chapter 6, by *Papaspyrou*, presents some results on specifying a complete and accurate formal semantics of the ANSI C programming language, The denotational approach is followed and use of monads and monad transformers is made, in order to improve the modularity and elegance of the semantics. The specification is divided in three distinct phases: static, typing, and dynamic semantics. Interesting results have been achieved in an attempt to accurately model complex characteristics of C, such as the unspecified order of evaluation and sequence points. An implementation of an abstract interpreter for C programs based on the proposed semantics has also been developed, using Haskell as the implementation language.

Chapter 7, by *Papaspyrou, Sgouropoulou and Skordalakis*, suggests a model for the automation of content-based electronic document filtering. The model is based on multi-agent technology and utilizes a knowledge base organized as a set of logical rules. Implementations of the model that make use of client-server architecture should be able to efficiently access documents distributed over an Intranet or the Internet. The filtering of electronic documents seems to be hard to automate, partly because of document heterogeneity, but mainly because it is difficult to train computers to have an understanding of the contents of these documents and make decisions based on user-subjective criteria.

Chapter 8, by *Koulouris, Koziris, Papakonstantinou and Tsanakas*, presents an array architecture for syntactic pattern recognition. The kernel of all syntactic methods is a parsing algorithm, which is responsible for analyzing the complex object into a sequence of primitive patterns. The complexity of the parsing procedure depends on the class of the underlying grammar. Context-free (CF) grammars combine satisfactorily both the expressive power and the simplicity in their analysis, and they are widely used in syntactic pattern recognition applications. Earley's parsing algorithm is the fastest sequential algorithm that can parse an arbitrary CF grammar in $O(n^{2.81})$. This can be a significant overhead for reasonably large n. Consequently the efficient parallelization of Earley's algorithm is of particular importance to the above areas. The proposed method implements Earley's parsing algorithm into a 1-D array architecture, consisted of $O(n)$ processing elements, reducing the hardware complexity.

Chapter 9, by *Voliotis, Triantafyllos, Dalias and Platis*, presents a dynamic load-balancing (DLB) scheme used to maintain fair load balance on a distributed system that exhibits both I/O bounded and CPU bounded characteristics. The design of the proposed scheme is based on a novel systematic approach. The main characteristics of the method are the automatic adjustment to the workload, the successful manipulation of both fine-grain and coarse-grain scenarios, the scalability, the portability and simplicity of the implementation. The DLB decides in a dynamic way if a number of jobs have to be packed and assigned to a processing unit or if each single job will be processed separately to minimize the waiting time and achieve optimum communication network usage.

Part I ends with **Chapter 10**, by *Panayiotopoulos, Zacharis and Vosinakis*, which presents the architecture of a Virtual Reality Application proposing an approach towards intelligent guidance in a Virtual University. The ground plans of the main University Building have been transformed into 3D VRML models and an Intelligent Agent has been designed and implemented as a virtual representation of a walking human being (avatar) in order to guide visitors around the University. Moreover, a spatial graph was developed, which provides routes through 'information-promising' nodes of the University Building, such as Offices, Laboratories, Amphitheaters, the Central Library, etc. Some nodes (e.g. office nodes) are linked to HTML pages, course presentations, documents, video, etc. The guide plans its route through the spatial graph and communicates with the users through a Java applet that manipulates the VRML scene.

PART II deals with *information extraction* (IE) from texts, *natural language* (NL) interfaces and *intelligent retrieval systems*. **Chapter 11**, by *Kontos and Malagardi*, presents a method for the combination of question answering with IE and its application to texts from different domains among which are scientific texts and an ancient greek legal text. A system based on this method and implemented in Prolog that uses a question grammar combined with a text grammar is used for the demonstration of the method. These two grammars use syntax rules and domain dependent lexicons while the semantics of the question grammar provides the means of their combination. The proposed question answering based approach aims at the creation of flexible IE tools which accept natural language questions and generate answers that contain information extracted from text either directly or after applying deductive inference. Another point of the method concerns the direct logical processing of text avoiding any kind of formal representation when inference is required for the extraction of facts not mentioned explicitly in the text.

Chapter 12, by *Karkaletsis, Spyropoulos and Petasis*, presents the results obtained within the framework of the English-Greek project GIE (Greek IE) through the approach of adapting the Sheffield IE system into the Greek language. An IE system involves mainly two tasks: the recognition of the named entities (e.g. persons, organisations, locations, dates) involved in an event and the recognition of the relationships holding between named entities in that event (e.g. personnel joining and leaving companies in management succession events). Named-entity recognition (NERC) involves two sub-tasks: identification of named entities and their classification into different types, such as organizations and person names. The chapter, discusses the significance of the named entity recognition task in IE, providing results from MUC Conferences, as well as information on existing NERC systems in English and in other languages. The prototype NERC system under current development in the context of GIE is presented. Information is provided on the platform, the corpus, as well as on the modules developed or being developed so far.

Chapter 13, by *Michos, Fakotakis and Kokkinakis*, is motivated by the need of modeling functional style (FS) as well as categorising unrestricted texts in terms of FS in order to attain a satisfying outcome in style processing and thus in IE applications. To this end, a three-level description of FS is given that comprises: (a) the basic categories of FS, (b) the main features that characterize each one of the above categories, and (c) the linguistic identifiers that act as style markers in texts for the identification of the above features. Special emphasis is put on the problems that faces a computational implementation of the aforementioned findings as well as on the selection of the most appropriate stylometrics (i.e. stylistic scores) to achieve better results on text categorisation. Finally, it is shown how the results on text categorisation can be used in IE applications to make current IE systems more adaptive to specific user needs.

Chapter 14, by *Blank*, deals with the processing of a multilingual corpus of technical texts. As the relevant knowledge contained in such texts is concentrated in technical terms, the aim of the study is to extract special purpose terminology. A semi-automatic tool is developed to help knowledge engineers, terminologists and professional translators not only to identify terms but also to detect possible translation equivalences and typical contexts of terms. Language-specific terminology is defined by criteria suitable for an automatic procedure. Related studies in multilingual terminology extraction are also considered and the assumptions underlying these studies are examined on the corpus.

Chapter 15, by *Di Tomaso*, describes TradIuta, a system for retrieving text fragments from documents, based on the IE system IUTA. TradIuta exploits "linguistic matching", that is a technology which, given a text or a collection f texts, allows the retrieval of text units on the basis of different kinds of linguistic analysis. TradIuta has been developed as a pilot application for Machine-aided Human Translation (MAHT). The system suggests existing translations of the source text in the target language, which can be reused as such or as a model for the translation of the current language. Given a sentence to translate, TradIuta, besides matching the string literally, can retrieve similar target language text units, using an algorithm based on linguistic similarity. The similarity between two text units is computed considering the lexemes they contain and the similarity between syntactic structures.

Chapter 16, by *Dini*, described a model architecture for a system of IE, showing how it can be exploited to satisfy application needs emerging from a field such as tourism. The system is based on the interaction of a set of monolingual IE systems with a set of natural language generation engines. In order for the system to be truly multilingual it is shown that a strong emphasis should be put on the design of templates rather than on the processing chain itself. For instance, the project, MIETTA, which is described in the chapter, proves that different IE systems can cooperate to reach the goal of multilingual access to monolingual information just by virtue of an agreement on appropriate communication standards and on a strict design of templates.

Chapter 17, by *Vichot, Wolinski, Ferri and Urbani*, presents a decision support system, named SAPE, connected with an IE system. This application is used by Caisse des

Depots et Consignations (CDC) in order to anticipate takeover bids on the European stock markets. It provides ways to manage the highly complex and moving network of European share-holdings (including reciprocal and cross-share-holdings). For instance, it enables to trace, through a graph-oriented representation, the paths of control between companies. SAPE is available on the CDC group's intranet and is used by fund managers as part of their everyday work. The chapter also describes how the natural language processing (NLP) system, *Exoseme*, uses the economic flow of dispatches from the Agence France-Presse (French press agency) in order to extract information on share-holdings and how this information is managed by the user to provide SAPE database with the large amount of figures and company names needed for its computations.

Chapter 18, by *Georgantopoulos and Piperidis*, presents a method aiming at (semi-) automating the process of eliciting domain specific lexical resources, terminological resources, in the framework of IE applications. The method aims at linguistically processing machine-readable text corpora and extracting lists of candidate terms of the domain, that would then be validated by domain experts. The method proceeds in three stages: a) morphosyntactic annotation of the domain corpus, b) corpus parsing based on a pattern grammar endowed with regular expressions and feature-structure unification, c) lemmatisation. Candidate terms are then statistically evaluated with an aim to skim domain valid terms. This hybrid methodology was tested on a text corpus of 90.000 words. The pattern grammar extracted 130 out of 209 manually coded terms, (62% recall), and the best statistical filter further confirmed 40 of the two-word terms (30% recall), reducing the size of the proposed list to 1/15.

Chapter 19, by *Kontos, Malagardi and Trikkalidis*, presents the design and implementation of a natural language (NL) interface for a motion command understanding system. The system accepts Greek and English as the NL of communication of the user with the system for the execution of motion commands. The system is applied to the communication between a user and an artificial agent, which moves in a virtual environment and accepts commands and knowledge about the objectss and the actions possible in this environment. It is supposed that the agent can move around in a room for executing the user's motion commands. These commands may refer directly or indirectly to the movement of specific objects or the change of their state. The agent knows the names of these objects and their position in the room displayed on the screen. The agent also knows how to execute some basic commands. When the user submits a command, the agent, in order to satisfy the constraints of the verb's meaning, may ask for new information and knowledge about objects and verbs, which may be used for the execution of similar commands in the future. The lexicon of the system is created automatically using a machine readable dictionary, while the learning of the correct interpretation of commands with more than one meanings is accomplished using machine learning by supervision based on visual feedback. The system has the ability to learn from its user to understand and execute correctly motion commands that go beyond its initial capabilities as shown by appropriate examples.

Chapter 20, by *Panayiotopoulos, Avradinis and Marinagi*, deals with the process of generating and monitoring a tutoring dialogue aiming to guide users to solve a problem. This is a process that requires a method to diagnose whether user actions are consistent with the problem's solution or not. This means that the tutoring system can analyse the user's final goal (solving the problem) into a sequence of lower-level goals and define what the user's actions should be in order to achieve these goals. The chapter proposes a temporal forward planning system which is based on the TRLi temporal reasoning system for the representation and reasoning of temporal information. The action representation schema adopted has been described in the backward-chaining version called TRL-Planner. Equation solving was selected as a case study. The forward planner generates the sequence of steps the student follows when an answer is given. This plan is used in order to diagnose the student's misconceptions and misbeliefs, so as to provide guidance and clarify the various concepts, as well as the solution process itself.

Chapter 21, by *Piamsa-Nga, Subramanya, Alexandridis, Srakaew, Blankenship, Papakonstantinou, Tsanakas and Tzafestas*, presents a generalized algorithm for content-based audio retrieval using a concept of a "virtual-node". The algorithm is improved from the "partial-matching" algorithm by exploiting multiresolution data structure of a unified k-tree model. The experimental results of audio retrieval using the virtual-node algorithm show that its retrieval time is less than, and its accuracy is better than the results from partial-matching algorithm. The results of a restricted-format query using this generalized virtual-node algorithm do not require a significant longer time than the conventional restricted-format algorithm.

Part II ends with **Chapter 22**, by *Baxevanaki, Ioannidis and Panayiotopoulos*, which presents an intelligent prototype system for retrieving temporal and periodic data based on the advantages of temporal databases and temporal logics. Temporal References Relational Model (TRRM), a temporal extension of the classical relational schema extension, has been used for the storage of temporal and periodic data. In this model, temporal reasoning is performed by a temporal logic interpreter, called Temporal References Language interpreter (TRLi). The communication of the database and the TRLi is mediated by the TRRM-TRLi meta-interpreter. This tool provides an interface to the user for inserting queries in TRL language and for getting the corresponding results. The basic task of the meta-interpreter is to interpret the queries inseted in a suitable form for the TRLi to generate the suitable SQL queries, process them, retrieve the data from the database and transform them to suitable input to the TRLi for reasoning. The system was tested in two application domains with quite satisfactory results.

PART III is devoted to image processing (compression, expansion, segmentation, transformation, reconstruction, etc.), face recognition and video-based systems. **Chapter 23**, by *Androutsos, Androutsos, Plataniotis and Venetsanopoulos*, presents a Java-based image processing system, as a result of the sudden explosion of architecture-independent programming. Primarily for use as a hands-on educational tool for providing novices with experience in the implementation of image processing algorithms, IMAGEnius is an open-ended, platform-independent, freely available program that can also be used as a testbed for the development of new routines. Much of the simplicity associated with this particular package, stems from the fact that Java provides support for the development of user interfaces, as well as a wide variety of routines that facilitate the manipulation and filtering of images.

Chapter 24, by *Friesen, Panagiotacopulos, Lertsuntivit and Lee*, presents a comparative study of image compression methods using five different transform methods; three standard and two wavelet-based transforms. The image under compression is first transformed with each of the five transforms, the transformed image is quantized to zero-out all transform values that correspond to the high-frequency components of the image, and then image encoding is applied to minimize the data storage requirements for the quantized residuals. The criteria used for the comparison of the methods are: (i) compression ratio, (ii) signal to noise ratio (SNR), and (iii) mean squared error (MSE) of the reconstructed image compared to the original. The result of the comparison is that the 2D Daubechies wavelet transform of 4th order (2D D4WT) transform outperforms all other transforms in SNR and MSE, and is the second to the 2D Haar in data compression.

Chapter 25, by *Alexopoulos, Delopoulos and Kollias*, deals with image expansion using subband filterbanks. The work described extends and enhances a previous method in two ways: (i) by introducing a more accurate model for the relation between the original low-and resulting high-resolution images, and (ii) by offering a computationally attractive algorithm for the implementation of the interpolation procedure. The maximization of the likelihood function is performed recursively by a constrained gradient method. Both (i) and (ii) are succeeded via subband filter representations and operations respectively.

Chapter 26, by *Ikonomakis, Plataniotis and Venetsanopoulos*, presents an efficient color image and video segmentation technique that can be used in the low-level analysis of mulitmedia-based still and video images. Multimedia visual data is becoming increasingly important in many scientific and commercial arenas with the advent of applications, such as multimedia databases, color image and video transmission over the Internet, digital broadcasting, interactive TV, video-on-demand, computer-based training, distance education, video-conferencing and tele-medicine, and with the development of the hardware and communications infrastructure to support visual applications. The proposed technique incorporates many features of the image including color, texture, and the motion feature of video, and will help in the tracking of objects in video for the purpose of coding and compression. It will be used in video shot detection and video indexing. The scheme includes region-based, edge-based, pixel-based, and motion-based techniques of color image segmentation.

Chapter 27, by *Sigmund and Novotny*, presents a simple program for three dimensional (3D) engraving machines controlled by personal computer. A source picture in grey scale is used for the 3D engraving. An engraving depth is determined by a pixel brightness. The program loads a picture in format Windows Bitmap (BMP). The z co-ordinate is assigned to every pixel. A white pixel responds to a product surface and a black pixel is in an entered depth. Other pixels in grey scale are between these levels. It is possible to change an orientation of outer points. The program converts a picture from 2D to 3D and send controlling commands to an engraving machine. The final program was developed with machines Mimaki.

Chapter 28, by *Lilas and Kollias*, proposes an approach to active 3D object reconstruction in which several cameras survey the object such that all sides of it are visible by at least two cameras. A laser beam scans the object and output signals are obtained after the processing of the images of the appropriate cameras. This approach possesses the following advantages: (i) it can be applied to large objects, (ii) provides high accuracy by fusing several sensor data and by super sampling, (iii) it does not require position measurements of moving parts, (iv) any defects and distortions are compensated accurately by software calibration, and (v) proper placement of the cameras provides full coverage of the object.

Chapter 29, by *Doulamis, Doulamis and Kollias*, deals with unsupervised video object segmentation based on an iterative maximum-likelihood (ML) scheme. In particular, the proposed method is applied for extracting of humans (head and body) from complex background, mainly in videophone applications. The probability density function (pdf) of the image to be segmented is assumed to be represented as a mixture of pdfs of individual objects (i.e. human and background). Two approaches are examined. The first assumes that the image pixels and the classes, to which they belong, are independent. The second considers that image pixels and their respective classes, can be modeled as Markov Random Fields (MRFs) following Gibbs distribution. In this case improvement of object segmentation can be accomplished since this model better approximates the local correlation of color information which is encountered in an image. Then, an Expectation-Minimization (EM) algorithm is applied to carry out the estimation of model parameters. However in the second case, it is more difficult to obtain simple expressions for the marginal expectation, involved in the EM algorithm. The perfomance of the proposed scheme is evaluated using the MPEG-4 test video sequences. The main goal of the experiments is the human extraction (segmentation) from, even a complex, background.

Chapter 30, by *Xirouhakis, Votsis and Delopoulos*, deals with the estimation problem of 3D motion and structure of human faces. Information regarding the three dimensional shape and motion of human faces is frequently required in various applications such as computer vision, robotics, video coding, 3D modeling for animation, security systems, enhanced man-machine interfacing, etc. Depending on the application, the extraction of either 3D shape parameters or 3D motion is tackled by mechanical sensors, 3D scanners or stereo cameras. The chapter presents a novel procedure that simultaneously estimates the shape and motion of a human face on the basis of a sequence of three arbitrary views of the head. The proposed algorithm involves estimation of 2D motion estimates of the most characteristic face elements, estimation of the 3D global motion and recovery of the 3D head model.

Chapter 31, by *Leonidou, Tsapatsoulis and Kollias*, proposes an innovatve scheme that produces a computerized system suitable for real time face identification and recognition. It involves the creation of a dynamic face storage database coupled with an identification algorithm. The system aims at recognizing faces irrespectively of orientation, scale and texture transformations. This is achieved by using different identification algorithms that mutually solve the exclusive problems that correspond to the different transformations; a decision mechanism is used subsequently to select the most reliable result. The positioning of faces to be recognized within the image is assumed to be *a priori* known; thus the face is considered to be in "head format". Splitting of the error domain due to the above-mentioned transformations is proposed in the chapter, using exclusive descriptors that successfully represent faces under specific conditions. The need to preserve the facial input geometrical space topology leads to the use of two Self-Organized Maps. Each map uses variant moment discriminators (VMD), representing facial texture as its training set, and clusters the input vectors by recognising correlation and hidden similarities among them. Real images, taken out of the face database created at the University of Bern, are used to illustrate the perfomance of the proposed method.

Chapter 32, by *Tzafestas and Pantazopoulos*, presents a simplified algorithm for rendering parametric curves with extension to rendering (wireframe) parametric surfaces. The basic idea is that of subdividing the space with a stopping criterion similar to that of Catmull. The added feature is the avoidance of stack use in the recursive rendering. Also some disadvantages of other algorithms (like crack in scan-like techniques) are eliminated. Catmull's method is known to be the first successful method for rendering parametric curves. His idea was that of subdiving the space to be rendered until it is reduced to the size of a pixel. Generally, the idea of subdiving the space to be rendered was used afterwards by many workers.

Chapter 33, by *Viblis and Kyriakopoulos*, is concerned with the problem of gesture recognition. Gestures comprise a significant communication factor in both man-machine interfaces and in the case of deaf people. Automatic gesture recognition would enhance communication levels in both of these cases because in the first case it accelerates data input while in the second enables the direct, discrete and effective service of a deaf customer, particularly in public services. The chapter describes a scheme for automatic gesture recognition. It is made evident that the whole process is divided into three phases: (i) gesture segmentation, (ii) gesture analysis, and (iii) gesture classification. An approach towards dealing with the first problem is presented, along with some experimental results, and the associated problems are discussed.

Part III ends with **Chapter 34**, by *Andreadis*, which presents the design and VLSI implementation of a new ASIC which performs real-time conversion of the raw RGB data, obtained from a color sensor, into the rgb normalized color co-ordinates. The high speed of operation is achieved by pipelining the data in a vector fashion. Eight-bit color images have been used, since this resolution is adequate for encoding the composite video signal without noticeable degradation. The inputs to the circuit are the RGB digital data obtained from a color sensor and digitized through flash ADCs. The high speed of operation is achieved by pipelining the data in a vector fashion. Signal processing at video rates is a demanding task. The design has been implemented using the CADENCE VLSI CAD tool. The die size dimensions for the core of the chip are 1.86 mm x 1.77 mm = 3.29 mm^2, for a DLM, 0.7 μm, N-well, CMOS technology. The ASIC is intended to be used in real-time pattern recognition applications, such as robotics, military systems, food, printing, pharmaceutical and agricultural industries. Real-time techniques are important not only in terms of improving productivity, but also in reducing operator errors associated with visual feedback delays.

PART IV presents thirteen applications. **Chapter 35**, by *Pachter and Nardi*, provides a GPS (global positioning system) Kalman update solution estimate comparable to that of the conventional iterative least squares (ILS) algorithm. The Kalman update algorithm differs from the standard Kalman algorithm in that the new measurement used for updating the previous estimate is correlated with the previous estimate. The application of concern is that of space-based satellite radio navigation; in particular the NAVSTAR GPS which provides 3-D user positioning by solving a set of nonlinear trilateration equations using pseudorange measurements. The chapter, in contrast to previous works which use only four satellites, treats an overdetermined system making use of all in view n ($\geq$4) satellites. Experimental simulation results are provided that illustrate and validate the proposed algorithm.

Chapter 36, by *Moropoulou, Koui, Kourteli, Achilleopoulos and Zezza*, deals with engineering applications of Image Processing on decay pattern recgnition. The digital image processing of historic architectural surfaces results on the characteristic distribution pattern of the weathering forms, when the microstructural and textural characteristics of weathered stone are used as interpretation criteria. Image classification to the damage level, according to IAEG, is possible in the case of alveolar decay. The same processing distinguishes in general restoration materials as compatible or incompatible to their original ones. Ultra sound measurements accompanied by image processing of the weathered stones (Integrated Computerised Analysis for Weathering) provide insights for the form and the thickness of the decay layers, and result in a reliable, as far as physicochemical processes of decay are concerned, integrated automatic risk mapping method, regarding the weathering of architectural surfaces and the evaluation of compatible conservation interventions.

Chapter 37, by *Karathanassi, Iossifidis and Rokos*, presents a method for recognizing and extracting the road network in peri-urban areas using SPOT panchromatic images. The perfomance of the method on panchromatic very high spatial resolution air-photos is investigated, taking into consideration that the spatial resolution of the expected remotely sensed digital images in the microsatellite era is going to be increased up to one meter. A particular combination of image representation / description algorithms is proposed, which recognizes road features – not clearly defined in remotely sensed images and often confused with other features and extracts them. The method consists of five algorithms – thresholding, morphological, thinning, linking, and gap filling – that are used sequentially. The only human intervention required, is the definition of a threshold. The proposed approach produces a raster road network representation that is highly complete and locationally accurate. Some experimental results are included in the chapter.

Chapter 38, by *García-Lázaro and Bienvenido*, deals with the integration of multiple software tools on a user friendly environment for the design and analysis of structures, applied to the greenhouse design, with reference to the greenhouse situated in the Province of Almeria (Spain). There were general tools for structural analysis, drawing, and cost evalutaion, but no one resolved the full problem, being necessary to manage different descriptions with each tool. Another problem was the fact that the technicians of the constructors were not skilled with these different tools. The proposal made in this chapter to elaborate an integrated tool that facilitates the definition of the proposed structures using a user-oriented windows interface integrates some available tools and knowledge-based subsystems. The tool (complemented with a simulation tool) allows the definition of customized structures, generating automatically project plans, budgets and structural analysis reports. It manages automatically commercial tools for structural analysis (SAP) and drawing plans (AutoCad, CorelDraw), and integrates a budget generator. The techniques applied are: declarative descriptions (at several levels), a 2-level distributed software architecture, and councelor and evaluation modules. The use of distributed soft architectures and declarative descriptions facilitates this integration, generating extremely adaptable tools. DAMOCIA-Design is implemented and working effectively for commercial and educational purposes.

Chapter 39, by *Rodriguez, Corral and Bienvenido*, complements the material presented in chapter 38, by describing the application of new computing techniques in the development of tools for signal validation an data analysis. The readings of the climatic sensor must be validated in order to their later exploitation, DAMOCIA-VAL (Data validation tool) contains twenty fours filters with this objective. A database of abnormal behaviors is created in order to find patterns that permit automatic corrections. DAMOCIA-EXP (Data Exploitation tool) facilitates the agronomic engineers in browsing the validated experimental data; it allows an easy manipulation of a very large volume of data (>10 Gbytes) and the generation of tables, charts, maps of climatic variables along a day obtained by simulation, etc. These tools have been implemented using, among others: a distributed multiagent architecture, a multiagent independent interface, and a distributed data storage. The main results obtained are: a tool set for data and signal analysis easily generalizable to any other application environment, which constitutes a decision support system; an experimental database composed by three main data collections (sensors, validated and analyzed data); and the verification of the proposed theoretical models.

Chapter 40, by *Tzafestas and Laliotis*, presents an overview of the available techniques used in distributed sensor networks (DSN) and their connection to artificial intelligence (AI) and geographic information systems (GIS). The most popular application of DSN seems to be GPS (global positioning systems). The chapter highlights the difficulties in using DSN systems and reports on the potential and the problems of DSN/GPS linkage. It is shown that DSN enhanced with GPS and GIS and other sensor capabilities can be employed to support useful applications such as: water level forecasting, influence of new buildings in water level for determining flood control policies, determination of population distribution patterns and possible sources of crop diseases, power and traffic control systems, determination of police car, fire truck or ambulance nearest to an emergency, etc.

Chapter 41, by *Qasem, Avolio, Camacho, Stephan and Frangakis*, presents a new technique for non-invasive assessment of aortic pressure modulations during treadmill running. The aim of this investigation was to quantify the relation between the beating amplitude of the peripheral pulse and the central aortic pulse during running at different speeds (and at different heart rates). Peripheral pressure was measured continuously and non-invasively in the finger of eight volunteer healthy subjects while running on a treadmill using a finger cuff device with the hand held steady at heart level. This was done to avoid the effect of hand movement *per se* on pressure modulation. Aortic pressure was estimated from the peripheral pressure signal using the non-invasive system. This device utilises an on-line computerised mathematical transfer function for the adult human arm developed previously from invasive measurements. Beating amplitudes for peripheral and aortic pressure were determined as well as pressure amplification of the aortic pulse for speeds of 9, 12, 15 km/hr. With increased heart rate, the peripheral pulse can be more than twice the central aortic pulse. The beating seen in the peripheral pulse is also present in the aortic pulse, but with a lower amplitude. Beating and pulse amplitudes are amplified to the same degree. Non-invasive estimation of the central aortic pressure may provide a more accurate assessment of the pressure-related effects of running on cardiac load. This may have important implications for assessment of patients with heart disease using treadmill exercise protocols.

Chapter 42, by *Marinagi, Spyropoulos, Kokkotos and Halatsis*, discusses the dominant role that planning should have in the process of scheduling patient tests in hospital laboratories. It specifies the requirements that a planning system should satisfy in this domain and demonstrates a particular planning system, called TRL-Planner, which meets these requirements. The paper also presents an improved dynamic distributed planning/scheduling paradigm which is based o patient-wise distribution and aims to increase its perfomance providing incremental scheduling solutions. The architecture of a system which implements this paradigm is described, giving its advantages over previous approaches.

Chapter 43, by *Panagiotacopulos, Lee, Friesen and Wan*, deals with fatigue identification in spinal muscles of individuals with and without using lumbar supports. The assumption upon which the proposed method is based is that after a certain loading task is performed, the entropy of the wavelet transformed surface EMG signals show notable changes. The results obtained by the wavelet-based method are consistent with those obtained via standard frequency spectrum analysis.

Chapter 44, by *Bitzer*, presents a description of the work on intelligent forecasting for refineries and power systems, undertaken in the IFS BRITE-EURAM project. This includes the objectives, tasks and worksteps associated with the subprojects. The general aim of the thematic network is the realization of an information exchange between universities, research centers, and industry to guarantee the knowledge transfer of basic research from universities/research centers and experiences from industry applications. This knowledge transfer includes the exchange of personnel between the institutes and the training of industry personnel. Therefore, the network has to install the support for this knowledge transfer, including: (i) current overview of the state of the art and the main research activities in the field of intelligent forecasting systems, (ii) installation of common working groups between universities and industry, (iii) workshops for personnel training, (iv) project-wide demonstrations to show the advantages of intelligent forecasting systems. The twenty active members of the project work within three task groups: Energy Time Series, Fault and State forecasting, Modelling and Control Strategies. About fifteen national subprojects are part of these task groups. Some first results of one of these cooperation projects between the University of Paderborn and the local distributor of the German town Dortmund are presented.

Chapter 45, by *Papazoglou*, provides a review of the neural network (NN) based load forecasting work carried out for the electric power system of the Greek island Crete. After a look at the classical statistical and Kalman filter-based forecasting techniques, the architectures of the NNs used are outlined, and the acceptance criteria for load forecasting accuracy are presented. Then a short discussion is provided on the software tools and databases used is user-friendly load forecasting.

Chapter 46, by *Stratakis and Papazoglou*, gives a summary about the characteristics, the requirements and the problems of the Cretan Power System. Then the authors provide a list of the factors that must be taken into account in the formation of the data base for the prediction of power demands according to existing data and the system operator's experience. The mechanisms by which the above mentioned factors affect the fluctuation of the power are explained and an effort is made to estimate the importance of the contribution of each one of them. Finally, conclusions are drawn about the required levels of accuracy (in time and predicted power) of the prediction demand for the Electric Power System (EPS) of Crete, so as to be useful, and the approach of the prediction results for the final user-operator of the system is shown. The utility and adoption of a prediction of demand model in an autonomous EPS such as that of Crete is of great interest due to portable consequences it may have for the safety in the system operation, in the quality of the provided electric power and in the general economic state of the region and the whole country.

Chapter 47, by *Koskolos, Megalokonomos and Dialynas*, presents a method for reliability cost assessment in composite power systems using sequential Monte Carlo simulation. The linear programming technique is used to determine a minimum operation and interruption cost for the system customers. The method used for calculating the interruption cost is greatly influenced by the definition of interruption sequences and the curtailment strategies used. Three techniques for calculating the interruption cost are outlined and discussed. The choice of method to be applied depends on the availability of data regarding the individual bus loads. System reliability and cost indices are obtained for each system bus and the overall system and reliability assessment case studies are discussed.

Taken together, the above forty seven contributions of the book provide an up-to-date and well-balanced picture of the developments in the planning, computation, processing, information extraction, data retrieval and estimation issues of intelligent systems and their applications in engineering and non-engineering areas of modern life.

PART I

COMPUTER-AIDED INTELLIGENT SYSTEMS AND TOOLS

PART IV

1

Sources of Efficiency in Planning : A Survey

A. TSOIS and S. G. TZAFESTAS

Division of Computer Science
Department of Electrical and Computer Engineering
National Technical University of Athens
{atsois, tzafesta}@cs.ntua.gr

1. Introduction

Planning is part of the Artificial Intelligence (AI) research field. In the early days of AI, planning was motivated by the needs of robotics. The planning process of a robot was responsible for generating a sequence of actions which, when executed by the robot, would achieve its goals.

The planning problem, in its general definition, is a very complicated problem. Various forms of the planning problem have been proven to be PSPACE hard [1] while other restricted forms are NP hard [2]. Still, today several planners are successful in dealing with real-world applications like building construction, spacecraft assembly, job-shop scheduling and space mission scheduling [1]. Their ability to handle and solve such complex problems is a result of several decades of research in the field.

In this paper the algorithms and fundamental principles on which efficient planning is based are identified. The main approaches taken in planning are presented and an attempt is made to analyze how, and in which cases they achieve efficiency. Short comparisons between various approaches are also presented. We hope that this presentation and analysis will help researchers in other fields to understand and reuse ideas and techniques developed for planning.

The paper is organized as follows: Section 2 presents the planning problem and a short historical background of the field. Section 3 describes the basic representation method used in planning and a total-order planner that uses this representation. Partial-order planning follows in section 4 along with the presentation of the POP algorithm. Section 5 presents some extensions to the basic representation and discuses how these extensions affect efficiency. Section 6 deals with decomposition methods and abstraction techniques. The main effort in this section is not to present the details of these techniques but mainly to identify how and in which cases they can improve the efficiency of planners. Finally, section 7 contains our conclusions regarding the fundamental principles that enhance the efficiency of planners.

2. The Planning Problem

Planning is about finding a sequence of actions which, when executed, are expected to satisfy the given set of goals. The planning algorithm takes as input a description of the initial state of the world, a description of the goals that need to be satisfied and a description of the possible actions that can be performed. The description of possible actions (in some formal language) is often called a *domain theory*. The output of the algorithm is *a plan* - an action graph - that should be executed in the initial state in order to satisfy the goals.

In order to solve the planning problem one can use the general framework for solving problems by searching. This framework is called *problem-solving* by many AI books. In this general framework one needs a modeling of what a world-state is, a function for each possible action that transforms a world-state into a new world-state according to the effects of the action, and a function that checks whether a world state satisfies the goals of the problem. Given the above elements any complete search algorithm can be applied to find a solution. The only type of additional information the search process can use is the heuristic functions. Still, heuristic functions can only be used to compare two generated states and choose one over the other. They can not be used to control which descendants of a state are generated.

So, why is planning examined separately from the general problem-solving framework? The answer is *efficiency*. Planning is a very *important* problem that needs to be solved for quite complicated instances. Using the general purpose searching algorithms without strong guidance would make even the simple planning problems *expensive* to solve.

In order to achieve strong guidance, researchers of the field suggested to "open up" and restrict the representation of states and actions. Allowing the planner to "look inside" the state and reason with the definition of the possible actions improves drastically the efficiency of the planning process. An additional step is to use the "divide and conquer" principle. Allowing only conjunctions in the definition of the states and actions provides the means to invoke *decomposition*. Finally, exploiting the *nonlinear* nature of plans and using *abstraction* can further improve the efficiency of planners and make them usable for real-world problems.

Planning is closely related to acting. Plans are generated in order to be executed. In fact, planning is just one approach to the problem of generating actions. Another approach is called *situated action*. In the case of situated action the correct action can be easily computed from the current state without the need to consider previous states or future implications. This type of actions is similar to reflexes that living organisms have.

The early planners were developed in order to support the operation of some robot. Still, *planning* is done using a theoretical model of the world while *acting* has to be performed (in most cases) in the real world. The planners that we present here deal only with the initial description of the problem while ignoring the chance that something goes wrong. For example, the execution of an operation may fail or have unexpected results. Furthermore, the state of the world may change because of some external reason. In order to cope with such problems, researchers have developed various kinds of planners that generate fail-safe plans or sense the real world during the execution of their plans and make the appropriate corrections.

Conditional planners are used to generate plans that contain conditional operations. An agent that executes such a plan decides which operations to perform and which to skip based on the readings it acquires during plan execution. This kind of plans is suitable for domains where the planner can anticipate what kind of situation may occur.

Integration of the planning and execution processes is another major approach. In this way the planner can obtain information about the real-world state and modify or extend the plan accordingly. *Re-planning* is a very efficient approach in cases where unexpected and unpredictable errors may occur. The re-planning agents modify their original plan in case their sensors detect a deviation from the expected state. Usually these modifications simply intend to restore the world state to the expected state while avoiding to change the structure of the initial plan.

Although conditional planning and re-planning deal with a more realistic domain they do not add to the efficiency of the basic planning algorithms. In fact, they are extensions of these algorithms that use the same key features to achieve efficiency.

2.1　HISTORICAL BACKGROUND

The first major planning system is STRIPS and was developed by Frikes and Nilson in 1971 [3]. STRIPS was designed as the planning component of the software for Shakey robot project. The system used general problem solving techniques in combination with an efficient representation for actions. The STRIPS representation proved to be so efficient for planning that almost all planners developed ever since used some extension or modification of this representation.

In 1975 Sacerdoti [4] introduced partial-order planning as a way to improve planning efficiency by avoiding "premature commitments to a particular order for achieving sub-goals". In his paper [4], Sacerdoti argued that planners should not search in the world state space but in the space of plans. This change of domain proved to be a very important one for the efficiency of planners.

For several years, terminological confusion has regined in the field of planning. Sacerdoti used the term "nonlinear" to describe the partial-order planners and "linear" to describe non-interleaving planners. The non-interleaving planners are the ones that can not interleave the steps for solving one sub-goal with the steps for solving some other sub-goal. This kind of planners has been proved to be incomplete since they suffer from the famous "Sussman Anomaly". In the same time several authors used the term "linear" to describe plans that have all their steps ordered in a serial sequence and the term "nonlinear" to describe partial-order.

What is now becoming a standard terminology is to use the terms "total-order" and "partial-order". A total-order plan is the one that defines a serial sequence (a complete ordering) of all the steps it contains. If a total ordering is not defined and only local orderings between various steps are defined by the plan then this is a partial-order plan.

After Sacerdoti presented NOAH in 1975 many partial-order planners have been developed. Tate in 1977 combined the clean conceptual structures of INTERPLAN with partial-order plans and developed the NONLIN system [5]. TWEAK appeared in 1987 as a formalization of a generic, partial-order planning system [2][1]. The SNLP algorithm was developed in 1991 by Soderland and Weld [6] as an implementation of the planner described by McAllester and Rosenblitt [7]. Two recent applications of planning in the robotics area can be found in [8,9].

3. Basic representation and algorithms

The majority of the planners use some extension of the STRIPS representation. In this section we will describe the simplest STRIPS representation with no variables or other extensions and present a planning algorithm that use this representation. In the following we will call a "STRIPS planning problem" any planning problem that uses the STRIPS representation for its definition.

The STRIPS planning problem can be described with the triple <O, I, G> where O is a set of STRIPS operators, I is a set of positive literals that define the initial state and G is a set of positive literals that define the goals of the problem. In our case literals are ground predicates - predicates that can not contain variables.

The STRIPS operations are represented with preconditions and effects. The preconditions are a conjunction of positive literals. An operation's effect, on the other hand, is a conjunction that may include both positive and negative literals. Operations may be executed only when their preconditions are true. When an operation is executed, it changes the world description in the following way. All the positive literals in the effect conjunction (call the action's add-list) are added into the state description while all the negative literals (called the action's delete-list) are removed. It's illegal for an operation's effect to include a positive literal and its negation, since this would lead to an undefined result. Also, one common restriction is to require that all negative literals in the effect part of an operation (the literals in the delete-list) appear in the precondition part (in their positive form of course). This means that an operation can not delete a literal from the state description if this literal is not already part of the state description.

A solution to a STRIPS planning problem is a *correct* plan. A plan is a set of operators and a sequence in which these operators have to be executed. The plan is correct if it is legal to execute the operators in the defined order starting from the initial state and if by executing them the initial state is transformed in a state that satisfies all goals in G.

---

[1] Some researchers consider TWEAK to be the first truly partial-order planner.

One should note that the STRIPS representation is restricted to goals of attainment. Furthermore, it restricts the type of goal states that may be specified to those matching a conjunction of positive literals.

3.1. ASSUMPTIONS

The STRIPS representation proved to be suitable for various planning domains and allowed the development of efficient planners. Of course, this efficiency comes at a cost. One obvious disadvantage is the expressive power of STRIPS. For example First Order Logic predicates could be used to represent a much richer class of domains. A second disadvantage is the unrealistic assumptions that have to be adopted. All planners based on STRIPS representation or on some extension of it have to make most of the following simplifying assumptions:

- *Closed world assumption*: all literals which are not added by some operation (or are not mentioned in the initial state) are considered not to be true.
- *Atomic actions*: all actions are executed in an indivisible and uninterruptible way.
- *Deterministic effects*: the effect of executing any operation is a deterministic function of the operation and the state of the world when the operation is executed.
- *Omniscience*: the planner has complete knowledge of the initial state of the world and of the nature of its own operations.
- *Sole cause of change*: the only way the world may change is by the operations chosen by the planner.

3.2. PROPERTIES OF PLANNING ALGORITHMS

The planning algorithms take as input a problem description and try to generate one or more solution plans. The algorithms perform some sort of guided search. For the majority of the problems there is no straightforward way to construct the solution without searching. Trying various combinations and permutations of operations is the only way to find a correct plan. So, all planning algorithms need at some point to search. The states of the search space can represent world-states or even candidate plans. In all cases, the branching factor of the search-tree can be as high as the number of operations in the O set while the depth of the tree can not be less than the number of operations contained in the solution.

When dealing with the planning algorithms one would like to "hide" the search process. After all, searching is common to all planners and various search strategies can be easily incorporated in almost all planners without changing their core algorithm. A very convenient technique used when presenting a planning algorithm is to hide the search process with a non-deterministic function that makes only the right choices. The implementation of such a function (on all non-magical computers) would be a search process.

The planning algorithms can be classified in various ways based on their properties. Some of their most important properties are the following:

- *Soundness.* A planning algorithm is sound if every solution plan that it founds is *correct.* Soundness is easily checked so in practice all planners are sound.
- *Completeness.* A planning algorithm is complete if it finds a solution when one exists. Not all planners are complete and some times it is quite hard to prove this property for an algorithm. As mentioned earlier, the non-interleaving planners are not complete and suffer from the "Sussman Anomaly".
- *Systematicity.* A planner is systematic if it generates and checks only once each search state. This property affects the efficiency of the algorithm but does not necessarily increase the overall performance. Some times, the price the algorithm has to pay in order to maintain this property is too high.
- The domain theory. The representation of operations that a planner can accept is obviously a very important property since it defines the class of planning problems for which the planner can be used.
- The type of generated plans. The plans that a planner generates can be total-order plans or they can be partial-order plans. Furthermore, the plan steps to achieve a particular sub-goal could be interleaved with steps for achieving some other sub-goal or they may be strictly non-interleaved. The type of generated plans depends directly on the nature of the planning algorithm.
- The search space. Planners can search in the space of world-states or they can search in the space of plans. Most of the total-order planners search in the space of world-states while most of the partial-order planners search in the space of plans.
- The reasoning direction. Planners can be progressive, choosing an operation based on the current state and trying to satisfy the set of goals or they can be regressive, choosing operations based on the current set of goals while trying to reach a state satisfied by the initial description of the world. A third option would be to use both previous strategies in an interleaved manner but we are not aware of such a planner.

3.3. A TOTAL-ORDER PROGRESSION ALGORITHM

In this section we will present a progression algorithm that searches in the space of world-states and generates total-order plans. The algorithm is taken from Weld's work [10]. It starts from the initial state of the world and by constantly adding actions to the plan it tries to reach a state where all goals are satisfied.

Total-order plans are just a strictly ordered sequence of operations. A plan is a solution to a STRIPS planning problem if it can be applied to the initial state of the problem and by the time the last operation has been applied the state of the world satisfies all problem's goals.

Algorithm:	ProgWS(world_state, G, O, path)
1. *Termination:*	If world_state satisfies each conjunct in G, return path.
2. *Action selection:*	Let Act = Choose from O an action whose precondition is satisfied by world_state. If no such choice was possible, return failure.
3. *State generation:*	Let world_state' = world_state + Add_list(Act) - Delete_list(Act).
4. *Update path:*	Let path' = Concatenate(path, Act).
5. *Recursive invocation:*	Return ProgWS(world_state', G, O, path').

Figure 1

The ProgWS algorithm is presented in figure 1. The algorithm contains the non-deterministic function Choose in order to "hide" the search process. We can consider that Choose always makes the right choices.

Consider a planning problem defined by the vector <O,I,G>. Then a call to ProgWS(I,G,O,{}) will return a totally ordered sequence of operations - a solution plan (if one exists). Step 1 checks if a solution plan has already been found. Step 2 chooses with the non-deterministic function Choose the correct operation from O that must be applied to the current state (world-state). Step 3 computes the new world-state (world-state') by adding to the current state the positive effects of the selected operation and deleting from it the negative effects of this operation. Step 4 updates the path taken in the search space - the sequence of operations performed to reach the current state. Step 5 recursively calls the algorithm with the new world-state and the new path. The algorithm can be proved to be both sound and complete.

3.4. ANALYSIS

So why is the ProgWS algorithm more efficient that a general problem-solving algorithm? In fact ProgWS is quite close to the problem-solving framework. Step 1 is a goal-checking function; step 3 calculates the new state based on the selected operation and the previous state and step 2 in fact implements a searches for all possible valid sequences of operations.

The only significant difference is hidden within step 2. The Choose function does not have to consider all the operations in the O set but only these that have their preconditions satisfied by the current state. Whether or not this pruning of the search space is significant depends on each particular problem. Since the states are represented as a conjunction of literals and since preconditions of operations are represented in the same manner Choose can immediately identify the applicable actions without any complicated and time consuming matching algorithm. So, the ProgWS algorithm can do better that a general problem-solving algorithm because it can look inside the description of operations and select only the applicable ones. In other words, the efficiency of the algorithm is due to the STRIPS representation that allows the immediate selection of possible operations.

4. Partial Order Planning

For the ProgWS algorithm we defined plans to be a totally ordered sequence of operations. Still, as Sacerdoti mentioned in 1975, plans need not define a total ordering for their operations. Plans, almost always, include operations than do not interact with all other plan's operations and could be executed in various points of the plan. So, there is no need for a plan to specify an ordering between two operations if this in not required for the completeness of the plan. This is a "Least Commitment" practice. By minimizing the number of commitments an algorithm minimizes the probability of reaching a dead end in its search process. With less backtracking the planning algorithm is expected to be more efficient and find a plan faster than a total-order planner.

Although partial-order planners exist for quite a long time only recently researchers have analyzed and compared them formally with total-order planners. The work of Minton et. al. [11] is one of the most up to date work and testifies that partial order planning is not essentially different that total-order planning but, almost always, it can be more efficient. In their paper they proved that the search space of a partial order planner is always less or equal to that of an equivalent total-order planner. Still, this does not imply that in all cases partial-order planners are more efficient. Partial-order planners have to pay a cost for their "Least Commitment" strategy. When the algorithm adds a new operation to the plan it has to identify which orderings are required with regard to the rest of the operations already in the plan. This process is called "Conflict Detection and Resolution" and it can be quite expensive to perform if the operations interact strongly with each other.

From a complete partial-order plan one can get one or more linearizations. A linearization defines a total ordering for the steps of a partial-order plan. This means that all linearizations of some partial-order plan are total-order plans. So, a partial-order plan defines a class of total-order plans. The linearization process can also be reversed and from a total-order plan one can construct various partial-order plans. Among these equivalent partial-order plans there is just one "least constrained" plan - a plan that contains the minimum number of orderings. Backstorm in 1993 formalized the problem of removing unnecessary orderings [12][2].

---

[2] Backstorm shows in this paper that the problem of finding a plan with the fewest orderings over a given operator set is NP-hard.

Partial-order planners usually search in the space of plans. Since the plan is not an ordered sequence of operation it can not be represented any more by the path from the initial state to the final state. So, it is mostly natural to search in the space of plans starting from an initial empty (and probably incomplete) plan and stopping when the plan is complete. The initial empty plan contains only two dummy operations: one *start* operation that has no preconditions and its effects are the set of literals known to be true in the initial state of the problem and one *end* operation which has the set of goals as its preconditions.

In order to represent a plan we need at least two components. The first is the set of operations to be performed in the plan and the second is a set of ordering constrains for these operations. In practice, almost all partial-order planners use one additional set when representing plans. This set is called the set of *casual links*[3]. Casual links where defined by Austin Tate for use in the NONLIN planner   [5]   and represent decisions made by the planner about which operations are used to satisfy which preconditions.

A casual link is a structure with three fields: two contain pointers to plan operations (the link's produced, A_p, and its consumer A_c); the third field is a proposition (literal), Q, which is both an effect of A_p and a precondition of A_c.

Casual links are used to detect when a newly introduced operation interferes with past decisions. We call such an operation a *threat*. More formally, an operation A_n is a threat to the casual link $A_p \xrightarrow{\;Q\;} A_c$ if no orderings exist in the plan to ensure that A_n is executed before A_p or after A_c and A_n deletes proposition Q[4].

If a threat remains unresolved the plan is not complete. This means that some (or all) linearizations of this plan are not valid plans. Formally, a plan is *complete* if every precondition of every operation is *achieved* by some other operation. An operation A_p achieves a precondition Q of some operation A_c if the Q is one of the positive effects of A_p, and if no other operation can possibly cancel out Q. This means that there is no operation A_n that deletes Q and which can be executed after A_p and before A_c.

In order to resolve a threat there are at least two possibilities. The first is called *demotion* and adds to the plan the ordering $A_n < A_p$ ensuring that A_n will be executed before A_p achieves Q. The second is called *promotion* and adds to the plan the ordering $A_c < A_n$. This ordering ensures that A_n will be executed after A_c has been executed and the condition Q that A_n deletes is no longer needed. An extension of this definition of a complete plan that accepts literals with variables is called "Modal Truth Criterion" [2].

---

[3] Some authors call them *protection intervals*.

[4] The SNLP algorithm defines threats slightly different: An operation is considered a threat if Q is mentioned in its effect's list (in the negative as well as in the positive form). This definition leads to the *systematicity* property that reduces the overall size of the search space but does not guaranty an increased planning speed.

4.1. A PO ALGORITHM - REGRESSION

In this section we will present POP, a partial-order planning algorithm. The algorithm starts from the set of goals and gradually adds operations trying to satisfy these goals. By adding an operation, the algorithm eliminates some goal(s) from the goal-set. At the same time new goals may have to be added to the goal-set due to the preconditions of the selected operation. When the transformed set of goals is satisfied by the initial state the algorithm has found a solution plan. This means the algorithm uses regression to attack the problem.

The POP algorithm presented in Figure 2 searches in the space of plans. A plan is represented with the vector <A, R, L>. The set A contains the operations chose to be part of the plan, R is a set of orderings constraints on the set A and L is the set of casual links established for the plan. The algorithm starts with an initial empty plan and an agenda that contains the set of goal conditions. The agenda parameter is used in order to keep track of the conditions for which a casual link has not been established. So, agenda is always a superset of the unsatisfied conditions in the plan. The O set contains the description of all operations the plan may consider to use.

The POP algorithm uses two Choose functions (steps 3 and 9) to hide searching and one Select function (step 2) to hide the progress control strategy. The main search process is done at step 3 when the algorithm chooses the operation that will satisfy the selected precondition. The branching factor of this search depends on the number of operations that add this precondition. The depth of the over all search-tree is equal to the total number of preconditions in the final plan.

4.2. ANALYSIS

It is interesting to note that the algorithm uses redundant information. As previously mentioned the set of casual links L can be deduced from the sets A and R. Also, agenda can always be computed by computing the union of all preconditions for all operations and then deleting the preconditions that appear in the set of casual links (L). Still, the above computations are expensive to perform at each step of the algorithm. Maintaining the value of these sets incrementally can make a significant difference in the performance of the algorithm.

The usage of casual links comes at a cost. One can see that even if a plan is complete the POP algorithm has to perform some additional steps in order to establish casual links for all the preconditions. This will be done even if it does not imply adding any operations or ordering. The POP algorithm has no other way of checking the completeness of the plan.

Algorithm:	POP(<A, R, L>, agenda, O)
1. *Termination:*	If agenda is empty, return <A, R, L>.
2. *Goal selection:*	Let <Q, A_{need}> = Select a pair from the agenda (by definition $A_{need} \in A$ and Q is a conjunct of the precondition of A_{need}).
3. *Action selection:*	Let A_{add} = Choose an action that adds Q (either a newly instantiated action from O, or an operation already in A which can be consistently ordered prior to A_{need}). If no such action exists then return failure.
4. *Plan modification:*	Let L' = L + $A_p \xrightarrow{Q} A_c$, R' = R + {A_{add} < A_{need}}, A' = A + {A_{add}}, R' = R + {A_0 < A_{add} < A_∞}.
5. *Update goal set:*	agenda' = agenda - {<Q, A_{need}>}. If A_{need} is newly instantiated, then for each conjunct, Q_i, of its precondition add <Q_i, A_{add}> to the agenda'.
6. *Casual link protection:*	For every threat A_n in the plan <A', R', C'> choose a resolution method: Demotion: add {A_n < A_p} to R'. Promotion: add {A_c < A_n} to R'.
7. *Recursive invocation:*	Return POP(<A', R', L'>, agenda', O).

Figure 2

If we modify POP to check, at each step, the completeness of the plan then the search space of the modified POP would definitely be smaller than the search space of POP. Analyzing the A and R sets can do such a check. Starting from any complete plan, POP has to search a sub-tree that has a height of at most N and a mean branching factor of almost N/2. N is the number of total preconditions for all operations of the plan. The interesting property about this sub-tree is that all leaf nodes are complete plans. So, in fact POP will never have to backtrack when searching this kind of sub-trees. This proves that POP will perform at most N more steps than the modified POP[5] while the algorithm for checking completeness based on the A and R sets has been proved to be NP-complete.

The above discussion justifies the use of casual links as a way to improve the performance of the planning algorithm and in practice almost all planners use them.

The structure of POP is similar to the architecture defined by the problem-solving framework: check for termination, select a path in the search-tree, generate the new state. Furthermore, applying a function to the state checks the termination in both cases. Still, POP differs from the problem-solving framework in the way it chooses the next operation.

Using the way operations, initial states and goals are represented the Choose function can efficiently select the next operation and not pick one at random. At each step POP selects an operation that can satisfy a goal (or sub-goal) that until then was unsatisfied. This is much better than picking up an operation at random and hopping that this operation will satisfy some goal. This *informed* operation selection process results in an exponentially smaller search space.

So, in fact the STRIPS representation is the one that makes POP outperform any algorithm that would use the problem-solving framework. The fact that POP uses regression makes no real difference. Regression can be distinguished from progression only when we know the semantics of operations. In the problem-solving framework we do not care about the semantics of operations but only define the functions that transforms the states. Also, the fact that the states and the solution represent partial-order plans does not make POP any different from some problem-solver that would also generate partial-order plans. This two last properties of POP, the regression property and the fact that generates partial-order plans make it different, and more efficient than ProgWS or any other total-order planner.

Regression is a source of efficiency for the majority of planning domains. Usually the domains contain many operations with similar preconditions but different effects. So, selecting the operations based on their effects reduces the average branching factor making the search space much smaller. It is important to note that regression would not be possible if the representation method did not describe so clearly the effects of operations and if the operations where not reversible. The STRIPS representation and the assumptions that come with it allow the algorithms to use regression.

The fact that POP generates partial-order plans affects the performance only by reducing the size of the search space. As already mentioned, the size of the search space for a partial-order planner is always smaller that the one of a similar total-order planner. This fact becomes of primary interest because this reduction can be exponential to the size of the search-tree.

---

[5] Assuming a deep-first strategy.

5. Extensions to the basic representation

The previously defined STRIPS representation can be extended in various ways in order to facilitate the definition of more complex planning domains. In fact, the original STRIPS representation allowed the usage of variables in the definition of preconditions and effects. This extension complicates the planning algorithms by requiring unification procedures for literals and explicit representation of variable bindings in the description of plans. Furthermore, it creates another type of threat; the *possible threat* for which an additional conflict resolution strategy called *confrontation* can be applied.

Each of the variables is by definition existentially quantified. Allowing universal quantification is an additional extension to the language and allows the elegant definition of complicated operations. Some of the most important extensions to the STRIPS representation are the following:

- disjunctive and negate preconditions
- universally quantified preconditions and effects
- conditional effects

In the case of conditional effects an operation can define an IF <condition> THEN <effect_A> ELSE <effect_B> statement as part of it's effects. Depending on the truth of the <condition> at the time the operation is executed, the operation's effects may include <effect_A> or <effect_B>.

Each extension to the STRIPS representation complicates further the planning algorithms and in almost all cases enlarges the search space that these algorithms have to examine. Still, these extensions do not require modifications to the core of the planning algorithm. Furthermore, we have not found any important "efficiency trick" that planners use in order to handle the extended representation. The way the planners deal with the extensions is, in almost all cases, straightforward.

The conclusion is that it is very important, for the efficiency of the planning procedure, to use a planning algorithm that has as much representation power as it is required to describe the planning domain. If the planner is able to represent domains richer than the ones used for then it probably performs useless and delaying actions.

6. Decomposition and abstraction

One of the well-known strategies to attack a complicated problem is "Divide and conquer". This strategy has been applied to planning quite from the beginning. The planning problem can be decomposed into complementary sub-problems by dividing the set of goals and the predicates that describe the initial state into sub-sets. Then, the sub-problems are solved separately and at the end the solutions have to be combined. Any inconsistencies created by this combination have to be solved in order to obtain the final plan. In most of the cases various plan steps which belong to different sub-plans have to be merged in order to obtain a good plan.

The efficiency of this approach depends greatly on the way the goal set is decomposed. If the generated sub-plans have too many interactions among them then the process of combining them into one plan is of exponential complexity. Still, in many cases it is possible to come up with quite independent sub-plans that can easily be combined. In this case total-order planners gain from the reduction of the mean branching factor and from the reduced depth of the search-tree that each sub-plan generates. The partial-order planners benefit only from the reduced depth of each search-tree [13].

Another approach to decomposition is to start from a partial plan and divide its steps into sub sets. Then, each sub set is solved as a separate planning problem and then the various sub-plans are merged into the final plan. Again, various inconsistencies can be created during the merge process and they have to be solved with the standard conflict resolution methods.

Even the ProgWS and POP algorithms can be considered to do some sort of simple decomposition. At each step they deal with some pre-condition and merge their decision with the rest of the plan.

A more complicated decomposition method is called *hierarchical decomposition*. Some authors call this decomposition *operator reduction, operator expansion, task reduction* or *hierarchical task network* (HTN) planning. Hierarchical decomposition is adopted by most of the practical planners in order to deal with the complexity of their domains. The method appeared as early as 1974 with Sacerdoti's ABSTRIP and the idea has been successfully used in several *industrial strength* classical planners, such as SIPE [14] and O-Plan [15].

Hierarchical decomposition is about defining and using abstract operators that achieve high level tasks. For each of these operators the domain theory should define one or more decomposition methods. Each of these methods (also called *refinement* methods) decomposes an abstract operator in a group of steps that form a plan that implements the operator. The steps may contain and use other abstract operators so that we finally have a hierarchy of abstract operators. At the lowest level all abstract operators are composed only from primitive operators that can be executed.

The hierarchical planners first construct an abstract plan using abstract operators and then gradually refine them by decomposing the abstract operators (using one of their decomposition methods). During the refinement process, conflicts between steps may appear and the planner has (at some point) to resolve them.

There are an exponential number of ways to create a hierarchical abstraction given a set of operators and only a few of them can be used for efficient hierarchical planning. Researchers in the field have identified a set of properties that makes an abstraction hierarchy useful for planning. The *downward solution property* states than an abstract solution plan always have a concrete solution plan and the *upward solution property* states that no inconsistent abstract plan can be refined to a consistent plan. When the planner deals with a hierarchical abstraction that posses these properties it can efficiently prune the search space.

In the general case hierarchical decomposition has a major advantage over simple partial-order planning: it can guide the planning process [16]. So, the user has better control over the type of solutions generated by the planner. This may translate to a smaller search space for the planner or it may mean that the user is able to define intermediate goals to the planner. The gains come at a cost: the user has to properly define the abstraction hierarchy in order to control the planning process. So, by encoding extra information about the nature of the domain one can guide the planning process, prune the search space and control the type of solutions generated.

In all cases of decomposition the performance gained depends greatly on the way the decomposition is done. In order to obtain a good decomposition one usually needs more that the definition of the planning problem. So, decomposition is something link a heuristic that incorporates additional information about the nature of the problem and restricts (to some extend) the search space.

7. Conclusions

All planning algorithms need to search at some point. This is why the efficiency of the planning algorithms is closely related to the size of their search-tree. There are various factors that affect the size of the search-tree. Among them the most important are: the expressive power of the domain theory, the number of predicates involved, the number of operators, the length of the minimal solution, the way plans are represented, and the search direction.

In planning there are four main methods that help in restricting the size of the search-tree. The first is the usage of a suitable representation that allows the efficient selection of operations. The operations are efficiently selected if their selection is based on their preconditions and their effects. By being able to mach properties of states (or goals) with properties of operators and select only the applicable operators, the algorithms are able reduce the branching factor of their search-tree.

The second method is also related to the representation method. Being able to reason backwards from the goals to the initial state can greatly reduce the branching factor for most of the real planning problems. But regression would not be possible if the operators where not appropriately encoded and if several assumptions where not made.

The third method is to search in the space of partial-order plans. By making as fewer commitments as possible the size of the search space can be exponentially reduced. In order to employ this idea the algorithms needed to change the way search states and plans are represented.

The fourth method is to restrict the kind of generated plans by using hierarchical decomposition. This means that the planner can no longer generate any possible legal sequence of operators but it can only generate plans that much the patterns encoded in the given decomposition methods.

Besides the size of the search-tree, the performance of the planning algorithms depends on the resources required to generate and examine each state. In planning there are at least two main methods used to minimize the complexity of generating and examining a state.

The first method is to use a restricted language to model real world states and operators. By using STRIPS the algorithms can easily examine the state, determine if it is a goal state and, if needed, apply some selected operator to generate a new state. If the representation language for states and operators was more general and expressive it would be harder to check for termination and much more complicated to apply some operator and generate a correct new state.

The second method used to diminish the complexity of the algorithms is to incrementally compute results even if this means maintaining redundant data. Partial-order planners use the set of casual links to detect inconsistencies and check for termination. Casual links are redundant data but they speed up the process of conflict detection and identify all possibly unsatisfied preconditions. The important issue is that the set of casual links is incrementally maintained as the algorithm moves from state to state and no expensive computations are required at each state in order to identify conflicts or find unsatisfied preconditions.

One can note that the above methods are just instances of some, well known, general efficiency advises like:
- Commit as less as possible
- Prefer a special case algorithm over a general one
- Encode in the search process as much information about the problem as possible
- Use redundant data to lower complexity
- Avoid re-evaluation - use incremental computations where possible

The same principles are already applied to other research fields but perhaps by identifying their usage in the Planning domain new ideas on how these principles can be exploited may be stimulated.

References

[1]   S. Russell, P. Norvig. Artificial Intelligence: A Modern Approach. Prentice-Hall, 1995.

[2]   D. Chapman. Planning for conjunctive goals. *Artificial Intelligence*, 32(3):333-377, 1987.

[3]   R. Fikes, N. Nilson. STRIPS: A new approach to the application of theorem proving to problem solving. *Artificial Intelligence*, 2(3/4), 1971.

[4]   E. Sacerdoti. The nonlinear nature of plans. *In Proceedings of the Fourth International Joint Conference on Artificial Intelligence*, 1975.

[5]   A. Tate. Generating project networks. *In Proceedings of the 5th International Joint Conference on Artificial Intelligence (IJCAI-77)*, pp. 888-893, Morgan Kaufmann, 1977.

[6]   S. Soderland, D. Weld. Evaluating nonlinear planning. *Technical Report TR-91-02-03*, University of Washington, Department of Computer Science and Engineering, 1991.

[7]   D. McAllester, D. Rosenblitt. Systematic Nonlinear Planning. *In Proceedings of the 9th National Conference on Artificial Intelligence (AAAI-91)*, pp. 634-639, AAAI Press/MIT Press, 1991.

[8]   S. G. Tzafestas, G. B. Stamou. Concerning Automated Assembly: Knowledge Based Issues and a Fuzzy System for Assembly Under Uncertainty. *Computer Integrated Manufacturing Systems*, Vol. 10, No. 3, pp. 183-192, 1997.

[9]   N. I. Katevas, S. G. Tzafestas, C. G.Pnevmatikatos. The approximate Cell Decomposition with Local Node Refinement Global Path Planning Method: Path Nodes Refinement and Curve parametric Interpolation. *Journal of Intelligent and Robotic Systems* 22:289-314, 1998.

[10]   D. Weld. An Introduction to Least Commitment Planning. *AI Magazine*, 1994.

[11]   S. Minton, J. Bersina, M. Drummond. Total-Order and Partial-Order Planning: A Comparative Analysis. *Journal of Artificial Intelligence Research*, Vol. 2, pp. 227-262, 1994.

[12]   C. Backstrom, Finding least constrained plans and optimal parallel executions is harder than we thought, *In Proceedings of the Second European Workshop on Planning*, 1993.

[13]   Q. Yang, Intelligent Planning, A Decomposition and Abstraction Based Approach, *Springer-Verlag*, 1997.

[14]   D. Wilkins, Practical Planning: Extending the classical AI Planning Paradigm, *Morgan Kaufmann Publishers*, San Mateo CA, 1988.

[15]   K. Currie, A. Tate, O-Plan: The Open Planning Architecture, *Artificial Intelligence*, 51(1), 1991.

[16]   S. Kambhampati, Comparative Analysis of Partial Order and HTN Planning, *SIGART Bulletin*, Vol. 6, no.1, pp. 16-25, 1995.

2

An Interactive Geometric Constraint Solver

IOANNIS FUDOS
Department of Computer Science
University of Ioannina
GR 45110 Ioannina
Greece
e-mail: fudos@cs.uoi.gr

1. Introduction

A new generation of CAD systems has become available in which geometric constraints can be defined to determine properties of mechanical parts. The new design concept, often called *constraint-based design* or *design by features* [9, 17], offers users the capability of easily defining and modifying a design, but introduces the problem of solving complicated, not always well defined, constraint problems [3].

In this chapter, we present the development of a user–friendly interactive system for editing and solving geometric configurations that arise in feature–based CAD/CAM systems. The system is built around a powerful graph-constructive constraint solving method presented in [14], capable of efficiently analyzing certain classes of *well-determined, over-determined* and *under-determined* configurations. Minimal systems of geometric constraints that are not solvable by the core constructive method are detected and may either be handled by a numerical method and treated afterwards as rigid bodies, or edited by the user. A main issue pertinent to geometric constraint solving is the solution selection problem. To this end, we have provided an interactive tool for navigating the constraint solver, to the intended solution. Consistent over-determined sub-configurations can be detected, interactively relaxed and solved appropriately. Under-determined subsystems are detected, isolated and subsequently presented to the user annotated with all possible constraint addition choices for interactive editing. To realize the constraint solver we have developed a prototype on a SUN workstation running Solaris 2.5.1. The graphical user interface was built in Java AWT, the core method was implemented in SETL (SET Language)

21

as part of the work described in [5], the extensions for detecting under–determined and over–determined configurations were also programmed in SETL, and for numerical solving MATLAB packages were invoked.

Section 2 provides an overview of methods for geometric constraint solving, and justifies the selection of the graph–constructive method of [14] for developing an interactive sketcher/solver appropriate for use in CAD/CAM systems. Section 3 briefly outlines the core graph–constructive method used. Section 4 presents the design and development of the graphical user interface, and the basic user interaction and system flow. Section 5 describes our experience with methods for treating over–determined and under–determined constraint configurations and their realization as part of our interactive software. Section 6 offers conclusions.

2. Approaches to Geometric Constraint Solving

We present a brief overview of approaches to geometric constraint solving. We outline the most representative methods and evaluate their behavior in terms of the major concerns faced in CAD/CAM systems: solution selection, interactive speed, edit-ability, handling of over and underconstrained configurations and scope. A first version of this overview was presented in [16].

2.1. NUMERICAL CONSTRAINT SOLVERS

In *numerical constraint solvers*, the constraints are translated into a system of algebraic equations and are solved using iterative methods. To handle the exponential number of solutions and the large number of parameters, iterative methods require sharp initial guesses. Also, most iterative methods have difficulties handling overconstrained or underconstrained instances. The advantage of these methods is that they have the potential to solve large nonlinear system that may not be solvable using any of the other methods. All existing solvers more or less switch to iterative methods when the given configuration is not solvable by the native method. This fact emphasizes the need for further research in the area of numerical constraint solving.

Sketchpad [31] was the first system to use the method of relaxation as an alternative to propagation. Relaxation is a slow but quite general method. The Newton-Raphson method has been used in various systems [24, 27], and it proved to be faster that relaxation but it has the problem that it may not converge or it may converge to an unwanted solution after a chaotic behavior. For that reason, Juno [24] uses as initial state the sketch interactively drafted by the user. However, Newton-Raphson is so sensitive to the initial guess [4], that the sketch drafted must almost satisfy all constraints prior to constraint solving. A sophisticated use of the Newton-Raphson method was developed in [22], where an improved way for finding the inverse Jacobian matrix is presented. Furthermore, the idea of dividing the matrix of constraints into submatrices as presented in the same work, has the potential of providing the user with useful information regarding the constraint structure of the sketch. Though this information is usually quantitative and nonspecific, it may help the user in basic modifications. To check whether a constraint problem is well-constrained, Chyz [10] proposes a preprocessing phase where the graph of constraints is analyzed to check whether a necessary condition is satisfied. The method is however quite expensive in time and it cannot detect all the cases of singularity. An alternative method to Newton-Raphson for geometric constraint solving is homotopy or continuation [2], that is argued in [21] to be more satisfactory in typical situations where Newton-Raphson fails. Homotopy, is global, exhaustive and thus slow when compared to the local and fast Newton's method [23], however it may be more appropriate for CAD/CAM systems when constructive methods fail, since it may return all solutions if designed carefully.

2.2. CONSTRUCTIVE CONSTRAINT SOLVERS

This class of constraint solvers is based on the fact that most configurations in an engineering drawing are solvable by ruler, compass and protractor or using other less classical repertoires of construction steps. In these methods the constraints are satisfied in a constructive fashion, which makes the constraint solving process natural for the user and suitable for interactive debugging. There are two main approaches in this direction.

Rule-Constructive Solvers

Rule-constructive solvers use rewrite rules for the discovery and execution of the construction steps. In this approach, complex constraints can be easily handled, and extensions to the scope of the method are straightforward to incorporate [3]. Although it is a good approach for prototyping and experimentation, the extensive computations involved in the exhaustive searching and matching make it inappropriate for real world applications.

A method that guarantees termination, ruler and compass completeness and uniqueness using the Knuth-Bendix critical pair algorithm is presented in [7, 28]. This method can be proved to confirm theorems that are provable under a given system of axioms [6]. A system based on this method was implemented in Prolog. Aldefeld in [1] uses a forward chaining inference mechanism, where the notion of direction of lines is imposed by introducing additional rules, and thus restricting the solution space.

A similar method is presented in [32], where handling of overconstrained and underconstrained problems is given special consideration. Sunde in [30] uses a rule-constructive method but adopts different rules for representing directed and nondirected distances, giving flexibility for dealing with the solution selection problem. In [36], the problem of nonunique solutions is handled by imposing a topological order on three geometric objects. An elaborate description of a complete set of rules for 2D geometric constraint solving can be found in [34]. In their work, the scope of the particular set of rules is characterized. [20] presents an extension of the set of rules of [34], and provides a correctness proof based on the techniques of [13].

Graph-constructive Solvers

The *graph-constructive* approach has two phases. During the first phase the graph of constraints is analyzed and a sequence of construction steps is derived. During the second phase these construction steps are followed to place the geometric elements. These approaches are fast and more methodical. In addition, conclusions characterizing the scope of the method can be easily derived. A major drawback is that as the repertoire of constraints increases the graph-analysis algorithm needs to be modified.

Fitzgerald [12] follows the method of dimensioned trees introduced by Requicha [26]. This method allows only horizontal and vertical distances and it is useful for simple engineering drawings. Todd in [33] first generalized the dimension trees of Requicha. Owen in [25] presents an extension of this principle that includes circularly dimensioned sketches. DCM [11] is a system that uses some extension of Owen's method. [14] presents an elaborative graph-constructive method, with fast analysis and construction algorithms, and extensions for handling classes of nonsolvable, underconstrained and consistently overconstrained configurations.

2.3. PROPAGATION METHODS

Propagation methods follow the approach met in traditional constraint solving systems. In this approach, the constraints are first translated into a system of equations involving variables and constants. The equations are then represented by an undirected graph which has as nodes the equations, the variables and the constants, and whose edges represent whether a variable or a constant appears in an equation. Subsequently, we try to direct the graph so as to satisfy all the equations starting from the constants. To accomplish this, various propagation techniques have been used, but none of them guarantees to derive a solution and at the same time have a reasonable worst case running time. For a review of these methods see [28].

In a sense, the constructive constraint solvers can be thought of as a subcase of the propagation method (fixed geometric elements for constants and variable geometric elements for variables). However, constructive constraint solvers utilize domain specific information to derive more powerful and efficient algorithms.

2.4. SYMBOLIC CONSTRAINT SOLVERS

In symbolic solvers, the constraints are transformed to a system of algebraic equations which is solved using methods from algebraic manipulation, such as Gröbner basis calculation [8] or Wu's method [35]. Although, these methods are interesting from a theoretical viewpoint, their practical significance is limited, since their time and space complexity is typically exponential or even hyperexponential.

3. The Core Graph-Constructive Method

The method is presented in detail in [14]. We provide here a brief outline. A geometric constraint problem is given by a set of points, lines, rays, circles with prescribed radii, line segments and circular arcs, called the *geometric elements*, along with required relationships of incidence, distance, angle, parallelism, concentricity, tangency, and perpendicularity between any two geometric elements, called the *constraints*. The problem can be coded as a *constraint graph* $G = (V, E)$, in which the graph nodes are the geometric elements and the constraints are the graph edges. The edges of the graph are labeled with the values of the distance and angle dimensions.

Our constraint solving method first forms a number of rigid bodies[1] with three degrees of freedom, called *clusters*. For simplicity we will assume that a maximum number of clusters is formed, each cluster consisting of exactly two geometric elements between which there exists a constraint. Three clusters can be combined into a single cluster if they pairwise share a single geometric element. Geometrically, the combination corresponds to placing the associated geometric objects with respect to each other so that the given constraints can be satisfied. The constraint solving method works in two conceptual phases:

- **Phase 1 (analysis phase):** The constraint graph is analyzed and a sequence of constructions is stipulated. Each step in this sequence corresponds to positioning three rigid geometric bodies (clusters) which pairwise share a geometric element (point or line).

---

[1]A rigid body is a set of geometric elements whose position and orientation relative to each other is known.

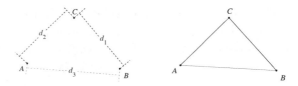

Figure 1. Constraint problem (left), and associated constrain graph (right)

- **Phase 2 (construction phase)**: The actual construction of the geometric elements is carried out, in the order determined by Phase 1, by solving certain standard sets of algebraic equations.

To illustrate the process, consider three points *A*, *B*, and *C* between which distances have been prescribed, as shown in Figure 1 left. The associated constraint graph is shown on the right. In Phase 1 of the constraint solving, we determine first that every pair of points can be constructed separately, resulting in three clusters. Moreover, the three clusters can be combined into a single cluster since they share pairwise a geometric element. The combination merges the three clusters into one. As soon as a single cluster is obtained, Phase 1 considers the constraint problem *solvable*. Phase 1, the analysis phase, consists of two parts:

- the *reduction analysis* that produces a sequence of local cluster merges and handles well-constrained and overconstrained problems, and
- the *decomposition analysis* that produces a sequence of decompositions (that correspond to a reverse sequence of cluster merges) and handles underconstrained cases. The outcome of the reduction analysis is fed as input to the decomposition analysis.

4. User Interaction and System Flow

We have developed a graphical user interface that integrates a geometric constraint solver with an editor for constructing and modifying dimensioned sketches representing CAD cross-sections [9]. For the development of the graphical user interface we used Java's AWT (Advanced Window Toolkit), for three main reasons:

- We were interested in building a rapid prototype to test and tune the effectiveness of our interactive method. Java's AWT provides a relatively easy way for producing graphical user interfaces. Moreover, the new version of AWT has additional flexibility and an improved method for relating user triggered events with actions. These features made AWT a good candidate for developing the GUI of our interactive solver.

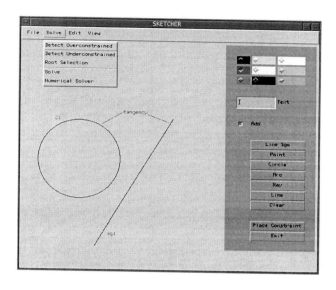

Figure 2. The Graphical User Interface

— Although executing Java code induces a significant overhead in terms of CPU and memory consumption, this overhead is not noticeable in workstations with sufficient memory.
— Producing Java code makes our graphical user interface portable to the vast majority of platforms.

The overall user-system interaction is as follows. The user draws a sketch and then imposes constraints on the sketched elements. The user can then invoke the solver which will try to solve the system of geometric constraints and return a new sketch with the geometric elements placed appropriately. If the solver is incapable of solving completely the sketch the user is notified and certain actions may be taken to correct the source of the problem. After successfully solving a dimensioned sketch the user may add/delete geometric objects and constraints, and re–invoke the solver. We have implemented points, circles, rays, lines, line segments and arcs. The available constraints are distance, tangency, angle, coincidence, concentricity and constraints that can be expressed as a combination of the above. We have also included text, for user annotation, in the form of constrained rectangles.

In Figure 2, a snapshot of the graphical user interface is depicted. The menu for inserting geometric objects and text lays on the right hand side of the main window. At the bottom of this menu above the Exit button there is the the Place Constraint button that transforms the functionality of the right-hand side menu for adding geometric constraints. At the top side

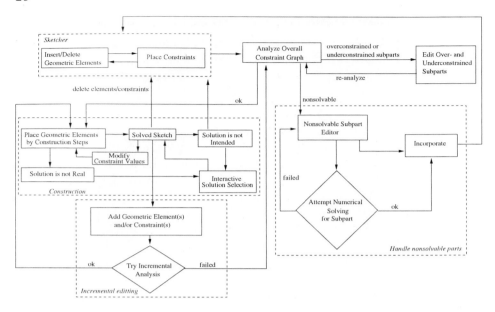

Figure 3. Interaction Flow for Solving a Sketch

there are four pull-down menus. The `File` menu has the `Save/Load` functions, and some auxiliary ones such as `Merge` and `Quit`. The files contain description of geometric objects, constraints, display information and annotation expressed in a language similar to the one presented in [15]. The communication with the main solver program and the numerical solver is realized through files with the same content augmented with solution specific information from the solver to the user and vice versa. The processing of the files is made possible by translating the high level language to an easy to read list of values, by using a compiler developed in C, using the flex and bison tools.

The `View` menu hides or makes visible certain classes of objects such as constraints, geometric elements, and annotation. The `Edit` menu contains the basic `Delete`, `Move` and `Copy` operations which are applicable to both geometric objects and constraints.

The `Solve` menu is illustrated in Figure 2. It contains the basic functions for performing geometric constraint solving of the given dimensioned sketch. Figure 3 shows the overall interaction flow for solving a sketch. The user first attempts to solve the sketch by the extended graph-constructive method outlined in Section 3. This is performed by choosing `Solve` from the `Solve` menu. If the analysis finds a valid sequence of construction steps for placing all geometric elements, this sequence is followed for producing the final Solved Sketch. However, if the solution does not consist of real numbers, or is not the one intended by the user, an interactive tool may be used for navigating the solver to a meaningful solution. The user is presented with the construction sequence and is given the capability of modifying the relative positioning of the objects involved, thus affecting stepwise the overall solution selection.

Changing the values of some constraints (e.g distances, angles), of a solved sketch will result in re evaluating the placement steps (not the graph analysis). When the method cannot solve the configuration, an error message is returned and the user is provided with an enhanced right-hand side menu for interactive intervention. Using this menu, the user may browse the part(s) of the design that correspond(s) to the subgraph(s) that is(are) not solvable. Finding minimal well-constrained subparts is a difficult process that has a worst case time complexity $O(n^4)$ in our implementation. A more sophisticated method $O(n^2)$ has been proposed in [18], but it may need some care to to be applied to rigid bodies without spoiling the quadratic complexity. The nonsolvable parts can be edited by deleting and adding constraints and geometric objects. These modifications are incorporated to the overall design by pressing the `Incorporate` button at the enhanced right–hand side menu. This drives the design process back to the first step. Currently, the `Text`, `Copy` and `Move` buttons are disabled during the process of editing the nonsolvable parts of the design. Another option for handling nonsolvable parts is to invoke the `Numerical Solver` for one or more of the nonsolvable parts. If this succeeds, the parts are subsequently treated as rigid bodies. To support this feature we have defined a rigid body structure in the representation language. If the basic solver detects overconstrained or underconstrained configurations it notifies the user. The process for treating such configurations is described in more detail in the next section.

Finally, the user may choose to incrementally edit the solved sketch by adding constraints and geometric elements. Invoking solve in this case tries to incrementally solve the new design. If incremental solving fails, analyzing the overall system of constraints is attempted. Deleting a geometric element or constraint is not consider an incremental change and analysis of the overall graph is attempted.

5. Handling Over and Under–determined Configurations

Each line or point on the Euclidean plane has two degrees of freedom. Each distance or angle corresponds to one equation. If there are no fixed geometric elements (i.e., geometric elements whose absolute coordinates have been specified explicitly by the user), then we expect that, $|E| = 2|V| - 3$, where $|V|$ is the number of geometric elements and $|E|$ is the number of constraints. This holds for lines and points with distances and angles con-

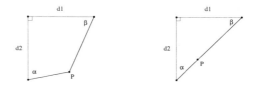

Figure 4. Degenerate Configuration (right) for $\alpha + \beta = 90°$.

straints only. When other objects or constraints exist, we have to substitute with sums for the degrees of freedom and for the constraints. Note that the solution will be a rigid body with three remaining degrees of freedom, because the constraints determine only the relative position of the geometric elements. An example is shown in Figure 4. In the figure, the vertex P of the quadrilateral has a well-defined position when $\alpha + \beta \neq 90°$. But for $\alpha + \beta = 90°$ the position of P is not determined. This "semantic" notion of well-constrained problems can be made specific for the constraint graph analysis, because there the generic problem of constructing a solution is considered independently of dimension values. Intuitively, a dimensioned sketch is considered to be well-constrained, if it has a finite number of solutions for nondegenerate configurations. Similarly, a dimensioned sketch is considered to be underconstrained, if it has an infinite number of solutions for nondegenerate configurations. Finally, a dimensioned sketch is considered to be overconstrained, if it has no solutions for nondegenerate configurations. The intuitive notions above can be made technically precise for the euclidean plane (see e.g., [13]). For the 3D case, however, the necessary and sufficient conditions for specifying whether a configuration is well-constrained are not known.

For an algorithm to test whether a graph is structurally well-constrained in 2D see, e.g. [19, 29]. Note that a structurally well-constrained graph can be overconstrained in a geometric sense, for example if there are three lines with pairwise angle constraints. The core reduction analysis handles structurally well-constrained and overconstrained problems. The decomposition analysis handles structurally underconstrained problems. Thus, the graph analysis may succeed to produce a sequence for placement, even for overconstrained or underconstrained configurations. In any case, the user is notified about the existence of ill-determined configurations. The user may then use the `Detect Underconstrained` and `Detect Overconstrained` buttons to analyze the graph. To detect the overconstrained subgraph, the system executes a reduction analysis and keeps track of cluster merging that states the existence of overconstrained configurations. Then, the user is presented with the corresponding part that is overconstrained, after running a simple algorithm that marks constraints that can be candidates for elimination.

To detect the underconstrained, the result of the decomposition analysis is run, and the plausible constraint additions are highlighted.

6. Conclusions

We have presented an overview of approaches to geometric constraint solving and selected a powerful graph–constructive method for building around it an interactive system for designing and solving dimensioned sketches. We have presented an innovative paradigm of user interaction for using the core method, treating ill–determined configurations, handling nonsolvable cases and performing solution selection. We have developed a prototype to evaluate this paradigm and to fine tune the system–user interaction.

References

1. B. Aldefeld. Variation of geometries based on a geometric-reasoning method. *Computer Aided Design*, 20(3):117–126, April 1988.
2. E. L. Allgower and K. Georg. Continuation and path following. *Acta Numerica*, pages 1–64, 1993.
3. B. Bruderlin and D. Roller (eds). *Geometric Constraint Solving and Applications*. Springer Verlag, 1998.
4. Paula L. Beaty, Patrick A. Fitzhorn, and Gary J. Herron. Extensions in variational geometry that generate and modify object edges composed of rational Bézier curves. *Computer Aided Design*, 26(2):98–107, 1994.
5. W. Bouma, I. Fudos, C. M. Hoffmann, J. Cai, and R. Paige. A Geometric Constraint Solver. *Computer Aided Design*, 27(6):487–501, June 1995. Also available through www, http://www.cs.purdue.edu/people/fudos.
6. B. Bruderlin. Using geometric rewrite rules for solving geometric problems symbolically. *Theoretical Computer Science*, 116:291–303, 1993.
7. Beat Bruderlin. Constructing Three-Dimensional Geometric Objects Defined by Constraints. In *Workshop on Interactive 3D Graphics*, pages 111–129. ACM, October 23-24 1986.
8. B. Buchberger. Grobner Bases : An Algorithmic Method in Polynomial Ideal Theory. In N. K. Bose, editor, *Multidimensional Systems Theory*, pages 184–232. D. Reidel Publishing Company, 1985.
9. X. Chen and C. M. Hoffmann. On Editability of Feature Based Design. *Computer Aided Design*, 27:905–914, 1995.
10. W. Chyz. Constraint management for csg. Master's thesis, MIT, June 1985.
11. D-Cubed Ltd, 68 Castle Street, Cambridge, CB3 0AJ, England. *The Dimensional Constraint Manager*, June 1994. Version 2.7.
12. W. Fitzerland. Using Axial Dimensions to Determine the Proportions of Line Drawings in Computer Graphics. *Computer Aided Design*, 13(6), November 1981.
13. I. Fudos and C. M. Hoffmann. Correctness Proof of a Geometric Constraint solver. *International Journal of Computational Geometry & Applications*, 1995. Also available through www, http://www.cs.purdue.edu/people/fudos.
14. I. Fudos and C. M. Hoffmann. A Graph-constructive Method to Solving systems of Geometric Constraints. *ACM Transactions on Graphics*, 16(2):179–216, 1997.
15. Ioannis Fudos. Editable Representations for 2D Geometric Design. Master's thesis, Dept of Computer Sciences, Purdue University, December 1993. Available through www, http://www.cs.purdue.edu/people/fudos.
16. Ioannis Fudos. *Constraint Solving for Computer Aided Design*. PhD thesis, Depart-

ment of Computer Sciences, Purdue University, August 1995.

17. C. M. Hoffmann and R. Joan. On User-Defined Features. *Computer Aided Design*, 30:321–332, 1998.

18. C. M. Hoffmann, A. Lomonosov, and M. Sitharam. Finding Solvable Subsets of Constraint Graphs. In G. Smolka, editor, *LNCS 1330*, pages 463–477. Springer Verlag, 1997.

19. H. Imai. On combinatorial structures of line drawings of polyhedra. *Discrete and applied Mathematics*, 10:79, 1985.

20. R. Juan-Arinyo and Antoni Soto. A rule-constructive geometric constraint solver. Technical Report LSI-95-25-R, Universitat Politecnica de Catalunya, 1995.

21. H. Lamure and D. Michelucci. Solving geometric constraints by homotopy. In *Proc. Third Symposium on solid Modeling and Applications*, pages 263–269, Salt Lake City, 1995. ACM.

22. Robert Light and David Gossard. Modification of geometric models through variational geometry. *Computer Aided Design*, 14(4):209–214, July 1982.

23. A. Morgan. *Solving polynomial systems using continuation for engineering and scientific problems*. Prentice-Hall, Inc., 1987.

24. G. Nelson. Juno, a costraint-based graphics system. In *SIGGRAPH*, pages 235–243, San Francisco, July 22-26 1985. ACM.

25. J. C. Owen. Algebraic Solution for Geometry from Dimensional Constraints. In *ACM Symp. Found. of Solid Modeling, Austin, TX*, pages 397–407. ACM, 1991.

26. A. Requicha. Dimensionining and tolerancing. Technical report, Production Automation Project, University of Rochester, May 1977. PADL TM-19.

27. D. Serrano and D. Gossard. Combining mathematical models and geometric models in CAE systems. In *Proc. ASME Computers in Eng. Conf.*, pages 277–284, Chicago, July 1986. ASME.

28. Wolfang Sohrt. Interaction with Constraints in three-dimensional Modeling. Master's thesis, Dept of Computer Science, The University of Utah, March 1991.

29. K. Sugihara. Detection of Structural Inconsistencies in Systems of Equations with Degrees of Freedom and its Applications. *Discrete Applied Mathematics*, 10:297–312, 1985.

30. Geir Sunde. Specification of shape by dimensions and other geometric constraints. In M. J. Wozny, H. W. McLaughlin, and J. L. Encarnacao, editors, *Geometric modeling for CAD applications*, pages 199–213. North Holland, IFIP, 1988.

31. I. Sutherland. Sketchpad, a man-machine graphical communication system. In *Proc. of the spring Joint Comp. Conference*, pages 329–345. IFIPS, 1963.

32. Hirimasa Suzuki, Hidetoshi Ando, and Fumihiko Kimura. Variation of geometries based on a geometric-reasoning method. *Comput. & Graphics*, 14(2):211–224, 1990.

33. Philip Todd. A k-tree generalization that characterizes consistency of dimensioned engineering drawings. *SIAM J. DISC. MATH.*, 2(2):255–261, 1989.

34. A. Verroust, F. Schonek, and D. Roller. Rule-oriented method for parameterized computer-aided design. *Computer Aided Design*, 24(3):531–540, October 1992.

35. Wu Wen-Tsun. Basic Principles of Mechanical Theorem Proving in Elementary Geometries. *Journal of Automated Reasoning*, 2:221–252, 1986.

36. Yasushi Yamaguchi and Fumihiko Kimura. A constraint modeling system for variational geometry. In M. J. Wozny, J. U. Turner, and K. Preiss, editors, *Geometric Modeling for Product Engineering*, pages 221–233. Elsevier Science Publishers B.V. (North Holland), 1990.

3

An Intelligent Agent Framework in VRML Worlds

T.Panayiotopoulos, G.Katsirelos, S.Vosinakis, S.Kousidou
Department of Computer Science,
University of Piraeus,
80 Karaoli & Dimitriou str.,
18535 Piraeus, Greece
themisp@unipi.gr,{gkatsi,spyrosv,nova}@erato.cs.unipi.gr

1. Introduction

Agent-based technologies have been rapidly emerging since the beginning of the 1990s [1-5]. J.P.Muller, [4], states that *'Agents are autonomous or semi-autonomous hardware or software systems that perform tasks in complex, dynamically changing environments'*. This means that Intelligent Agents observe the environment, maintain an internal representation of the world, make decisions and perform tasks (executing actions).

Planning is also a key issue in many Intelligent Agent architectures. This is reasonable, as planning provides the means for achieving goals given the definition of certain actions. Planning systems view the problem-solving behavior of agents as a *sense-plan-act* cycle [4]. As B. Hayes-Roth et.al. state, *'An intelligent agent is a versatile and adaptive system that performs diverse behaviors in its efforts to achieve multiple goals in a dynamic, uncertain environment'* [6].

Moreover, Intelligent agents' technology has presented many applications so far, starting from smart e-mail clients to software for robotics applications [7-9].

On the other hand, the area of Virtual Reality seems to be appropriate for the development of Intelligent Agents' applications. VRML, the Virtual Reality Modelling Language, [10], appears very promising as it has abandoned its static modelling style of its first version and moves towards more dynamic modelling techniques with the help of external programming languages like JAVA. As a result, Virtual Reality technology is capable of performing complex behavior [11].

Quite recently an attempt has been made to analyse the role of the underlying model-based semantics for VR modelling and relate it to the behavioural multi-agent architectures [12]. However, this attempt approaches the integration of Virtual Reality with Artificial Intelligence from a theoretical point of view. There have also been remarkable attempts to put an intelligent front-end to Internet applications, such as intelligent interactive systems which are usually implemented with Java applets [13,14].

Therefore it seems that the time has come to integrate the Virtual Reality technology with Intelligent Agents architectures. The experience gained from the areas of temporal reasoning systems [15], temporal planning systems [16] and experimental VRML programming has motivated us towards the area of developing an Intelligent Agent which could observe, decide and act in virtual environments.

In this chapter we present an Intelligent Agent architecture which consists of a logical core, i.e. the Logical Agent, as well as a virtual representative, i.e. its 'Avatar'. The chapter is structured as follows : In Section 2 we present the overall architecture of the system. Section 3 introduces the architecture of the Logical Agent, discusses the categories of knowledge represented by the agent as well as its reasoning capabilities. It also explains how the Virtual Reality Management Unit communicates with the Logical Agent and executes the commands posted by it. In section 4 we discuss some implementation issues and section 5 presents a toy example from the 'maze' world. Finally, concluding remarks summarize the main characteristics of the framework.

2. The overall architecture of the Intelligent Agent

The Intelligent Agent framework consists of two main layers : the *Logical Agent* layer and the *Virtual World* layer. These two modules while they can be viewed as layers at the conceptual level of analysis, they have been implemented as different processes which may be running at two different machines. In fact, the system uses a 2-tier Client/Server architecture, with the Virtual World module on the client side and the Logical Agent on the server side. This has been done to permit the development of network based machine independent applications.

The Logical Agent is a composite module consisting of the *Logical Core*, the *Knowledge Bases* supporting the logical core and the *Agent-to-Environment Communication unit, (AEC)*. On the other hand, the Virtual World module is also composite consisting of the *Static Virtual World*, the *Dynamic Virtual Objects Library*, containing also the agent's *Virtual Representative*, the *Virtual Reality Management Unit, (VRMU)*, as well as an *Environment-to-Agent Communication unit, (EAC)*.

The Logical Core provides the agent with reasoning capabilities. Reasoning is supported by a number of knowledge bases which store various types of knowledge such as static and dynamic knowledge, domain knowledge, knowledge about the agent's capabilities, spatial knowledge about the virtual space, etc. The virtual world is represented as a 'mental' structure in the agent's knowledge base, which is in fact an abstraction maintaining only the important information about it. The Logical Core reasons about the current world situation and according to its goals it sets some abstract actions which must be immediately executed at the virtual layer.

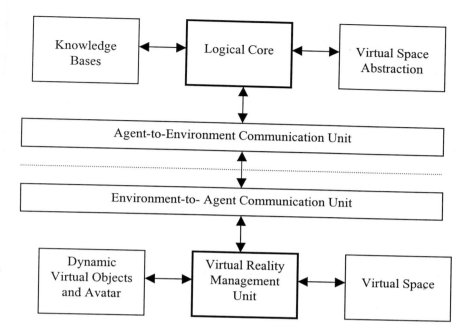

Figure 1. The Overall Virtual Intelligent Agent Architecture

The AEC module serves a twofold purpose : Its bottom level takes care of the process communication between the Logical Agent and the Virtual World module. Its top level takes care of information transfer between the Logical Core and the EAC module. Abstract actions, e.g. move to next room, are received by it and consequently send to the EAC. Finally, the abstract action arrives at the Virtual Reality Management Unit that specifies in detail the received actions. It provides specific values concerning the orientation and position of the avatar, e.g. it specifies the coordinates, orientation and path so that it can successfully move to the next room, and sends them as commands to the Virtual Reality World Browser. The browser executes the command by altering the virtual environment appropriately. When changes have been performed the AEC unit notifies the logical core that the action has been successfully executed and the logical core goes on by updating its internal and external state. Consequently, the agent looks around into the virtual space, gathers any additional information and decides the next step it should take to satisfy its goals.

3. The Structure and Functionality of the Logical Core

The logic part of the agent is a partial implementation of a BDI (*Belief-Desire-Intention*) intelligent agent architecture. In the heart of the logic part is a *look-decide-act loop*, which manipulates the internal and external states of the agent and produces the actions of the agent as a result. The Logical Core maintains many Knowledge Bases which are used to describe the environment, the agent itself, its goals, its 'mental' capabilities, its behavioural capabilities, etc.

3.1. LOOK : LOGICAL REPRESENTATION OF THE VIRTUAL WORLD AND OBSERVATION

Observation is an advanced functionality which in a real situation would require sensors and image understanding. In the situation of Virtual Reality it would require parsing and understanding of Virtual Reality files, i.e. the implementation of a subset of a VR browser. Since this was not our goal we have followed another approach : For each experiment with the Virtual Agent we define the virtual world (objects, positions, orientations, etc.) and using a preprocessor we produce two files, the VR file to be processed by the VR browser and the *World Description* Knowledge Base, *(WD)*, which is an abstarct but complete description of the virtual world in terms of logical assertions. In this sense WD is not part of the agent's knowledge.

Given WD, observation is defined as a '*look*' function which gathers the basic elements of WD observed from a position in it. Therefore, observation has been defined in terms of a '*visible*' *spatial accessibility relation* that has been also asserted into WD. Reasoning on WD results on acquiring all the entities and their observable attributes which compose the *Virtual Reality* Knowledge Base, *(VR)*, which is part of the agent's knowledge. In other words, the 'look' function is a transfer of knowledge from the world to the agent.

For example, if the agent stands in the middle of a room *r* he can observe the surroundings, say a table *t*, a computer *co* on the table, a chair *ch*, and two hallways leading to rooms s and t. In this case the VR would be updated by the following elements :

> *{rooms([r,s,t]), table(t), computer(co), chair(ch), is_at(t,r), is_on(co,t), is_at(ch,r), hallway(r,s), hallway(r,t)}*

While the agent moves around the virtual world VR is updated with new information concerning an abstract spatial description which will be further used for planning the agent's actions.

3.2. AGENT'S KNOWLEDGE

The agent maintains knowledge about its position, e.g. *is_at(myself, r)*, and possessions, e.g. *holding(myself, [key, pocket_computer])*, which is dynamic as it changes over time. It also maintains static knowledge of its *behavioural abilities* in the form of actions :

able(
action(unlock(L),
precond(locked(L), is_at(P,L), lockskey(L,K), is_at(myself,P), ihold(K)),
effects(unlocked(L)))

Beliefs of an agent express its expectations about the current state of the world and its change according to the agents' actions [4]. In particular, B is the set that is known to the agent, either through observation, embedded knowledge, or inference. Given that the architecture is so far limited to a single agent and all agent's actions are successful, B is an objective but not complete representation of the world, i.e. Belief is partial Knowledge. In a multi agent system, 'belief' would include assumptions and knowledge about other agents as well and an agent's beliefs might be subjective and sometimes erroneous.

Finally, the agent keeps track of all actions performed by it. The architecture can use this history to decide based on past actions.

3.3. DESIRES, GOALS AND INTENTIONS

Desire is an abstract notion that specifies preferences over future world states or courses of action. A consistent subset of desires that an agent might pursue compose its *Goals*. *Selection* of one of these goals and *commitment* to it is a process that is called the formation of *Intentions*. *Plans*, on the other hand, are very important for the pragmatic implementation of intentions. It is possible to structure intentions into larger plans and define an agent's intentions as the plans it has currently adopted [4].

In the Virtual Agent Architecture all these are part of the *Mental Structure* of the agent. There are two basic *desires* of the agent : First, to *satisfy its top level goals* and second to *explore* the virtual world. In fact, the agent knows nothing of the virtual world when it first enters it. It therefore does some exploration first, trying always to learn something new about it, and when the time comes and has selected sufficient knowledge to satisfy the goals it starts doing so. Exploration is also tried later on when some goals have been satisfied but lack of knowldege prohibits the agent from satisfying the rest of its goals.

A *top level goal* is defined by the action that must be executed to achieve it and the assumptions that must hold before the action can be executed. Moreover, top goals are partially ordered, so that the prerequisites of each goal are provided in terms of other goals, in a kind of a higher level assumption expressed as an *abstract temporal description*. This further means that some top goals take priority to become *top level intentions*. Commitment to a top level intention, i.e. the current 'active' goal, activates a *planning mechanism* which produces a plan to achieve the goal. Of course, a plan contains intermediate goals and intentions and defines actions to be executed. These actions are part of the behavioural abilities previously mentioned. Each time the first action is selected and sent to the Virtual module for execution as a *Virtual Act*.

3.4. PLAN AND DECIDE

The '*decide*' function manipulates the state of the agent. Since the 'decide' part of the loop is in the middle of it, it has to check the information that has been presented by the last act and look steps of the loop. This is what the 'check_previous_info' function does. The 'check_goal_removal' function is one that checks if the course of actions taken by the agent has lead to the achievement of its current_goal. If so, it removes the goal from its Top Goal list.

decide:
        check_previous_info;
        deep_thinking;

check_previous_info:
        check_goal_removal;

deep_thinking:
        if(validate_previous_plan)
                execute_previous_plan;
        else
                actual_decide;

actual_decide:
        choose_goal;
        make_plan;

Figure 2. A sketch of the decide algorithm.

The produced plans are based on partial knowledge. As the agent executes a virtual act it may visit unexplored places of the virtual world and acquire in this way new pieces of information. Each time new information arrives, the current plan goes under *re-evaluation*, in order to check its suitability to achieve the goal, given the new data. This enables the virtual intelligent agent to be capable of adjusting to a dynamic world.

The 'deep_thinking' function is the one that does this work. First, it validates the existing plan. It has to check whether the information that was collected by act and look affects the previous plan. If not, it just keeps that plan and goes on. Otherwise, it starts the plan-making process all over again.

This consist of two high-level steps. First, it chooses one of the available goals to be the current goal. This process involves estimation of the goals that are feasible at the current point in time. In the case where more than one such goals are available, the estimation concerns the selection of a goal which is more efficient to pursue. This check usually involves some domain-specific estimation. After a goal has been chosen, a plan has to be figured out. The planning mechanism provides it, the agent commits to the first action of the plan and sends it as a command to the Virtual Reality module.

3.5. ACT : ACTION – ABILITY – VIRTUAL ACT

The '*act*' function is a transformation of the World. Based on the agent's last decision, it manipulates the world, in such a way that the effects of the action are carried out. 'act', as 'look', determines whether this action is feasible by making use of the function 'able' previously mentioned.

This function is the second place where a designer might choose to implement properties of the world or the agent, such as difficulties on applying an action in a given world, or modelling disabilities of the agent. Notice that the roles of the 'visible', accessibility relation mentioned above, and 'able' are quite alike. 'visible' mainly refers to the information that the world allows to reach the agent's sensors. Of course one could easily model some kind of sensor disabilities by reducing the accessibility of the function 'visible'.

During communication, the Virtual Reality Management Unit (client) receives commands to perform certain actions. We call these actions *Virtual Acts* as they must be executed on the Virtual World. Virtual acts can be of any kind: from changing the position and the orientation of the objects in the world, to deleting them or replacing them with other objects. They may also take parameters that define them (i.e. move object, destination).

4. Implementation Issues

The system uses a 2-tier Client/Server architecture, with Java/VRML on the client side and C++/Prolog on the server side. The C++ program is used as a gateway to a Prolog meta-interpreter which reasons and determines the actions of the agent. Subsequently these actions are sent as commands to a Java applet. Finally, the Java applet hosts a VRML world where the agent exists as an avatar.

Communication between these two processes takes place by using a standard stream socket over TCP/IP. This arrangement also takes care of the need to be able to run an application on multiple machines. Since TCP/IP sockets are used to decouple the logic component from the VR component, they can also be used for the communication of a multiple agent architecture.

The client side of the architecture consists of a VRML GUI as well as a Java applet that acts as an intermediate medium between the server and the VRML world. The communication between VRML and Java is achieved with the use of EAI (External Authoring Interface). EAI acts as an interface between the two of them and provides a number of functions on the VRML browser that Java can perform in order to affect it. This interaction is interpreted as a change of attributes in the VRML world geometry.

Each time a command is executed, the Java applet sends a verifying message to the C++ server, and the server sends the next string to process. At the end of the logical process, the server sends a signal to stop data transmission.

5. A Maze Example

In this example, the intelligent agent finds his way out of a maze after having to discover some key-information in the maze, and its virtual representative (avatar) is directed through the VRML world following the agent' s orders.

The maze consists of several rooms, which are connected to each other through hallways. Some rooms contain objects like locks and keys, used by the agent in order to exit the maze. The sequence of the agent' s actions should be to *find the blue key, unlock the blue lock, find the red key, unlock the red lock and go to the exit.*

The initial scene consists of the maze and by pressing the "Start" button, the agent, the locks and the keys are instantly loaded. The button "Continue" sets the agent in motion to start its quest for the exit. During its travel from room to room, the agent *perceives* these objects but does not interact with them unless the particular action follows the correct sequence (i.e. although it finds itself in the same room with the red key, it cannot pick it up unless it unlocks the blue lock). The agent stores the information it collects about the environment in its knowledge base. It then uses this gained knowledge in order to accomplish certain tasks like finding the keys, the locks or the exit. Moreover, the agent' s actions change the environment (i.e. when it picks up a key, it automatically no longer exists in the virtual world). The protocol that is followed in the particular application is as follows:

- **commence** – the server asks for permission to transmit data.
- **ok** – the client declares it is ready for receiving data.
- **move** – this action takes two parameters : the object that is about to be moved (in this particular case, the agent) and the destination. The server decides that the object should be moved to the particular destination.
- **delete** – this action takes one parameter : the object that is about to be deleted. The server decides that the particular object should be deleted (either if the agent finds the correct key or if he unlocks a door).
- **end** – this message is sent by the server to the client to declare the end of data transmission.

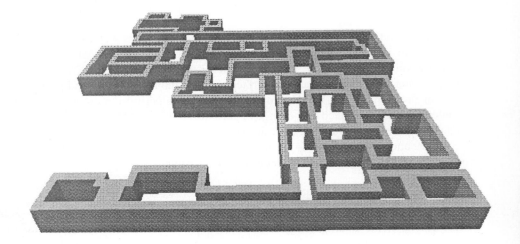

Figure 3. The maze

When the agent is ordered to change room, Java calculates the new avatar orientation according to the target position. Then it starts a thread that rotates the avatar until it reaches the expected orientation and moves it with constant speed to the target room. The delete action is implemented by removing the VRML nodes that represent the objects. These nodes were loaded by pressing the "Start" button at the beginning, with use of the Inline command in VRML.

Figure 4. The Virtual representative finds the blue key

6. Conclusion and future work.

Our aim in this chapter was to present an intelligent agent architecture that combines reasoning abilities, effective GUI and networking extensions. In order to bring such kind if applications closer to reality, we have made use of VRML (Virtual Reality Modelling Language).

In our future plans we intend to put the agent in a more complex VRML world, where he would be able to interact with additional objects. We soon expect to extend the reasoning abilities of the agent and the Virtual Reality metaphor will help towards this direction. It is also in our plans to develop an inter-agent protocol that will enable the agents to communicate with each other and exchange information and knowledge bases, in order to accomplish given tasks.

Both in the industry and in the academic society there is a growing demand for GroupWare and co-operative distributed-work applications, such as video conferencing, remote work, remote people training and education, the needs of which are far beyond simple document exchange as they should be supported by intelligent systems. An intelligent virtual agent could provide its services in such applications.

Acknowledgement

This work was partially funded by the Greek General Secretariat of Research and Technology under the project 'TimeLogic' of ΠΕΝΕΔ'95, contract no 1134.

References

[1]. Y. Shoham, 'Agent-oriented programming', Artificial Intelligence, Vol. 60, pp.51-92, 1993

[2]. M. Wooldridge, N. Jennings, 'Intelligent agents: Theory and practice', Knowledge Engineering Review, Vol. 10(2), pp.115-152, 1995.

[3]. M. Wooldridge, J. Muller, and M. Tambe, editors. 'Intelligent Agents II, Agent Theories, Architectures and Languages', volume 1037 of Lecture Notes in Artificial Intelligence, Springer-Verlag, 1996.

[4]. J. Muller, 'The Design of Intelligent Agents, A Layered Approach', volume 1177 of Lecture Notes in Artificial Intelligence, Springer-Verlag, 1996.

[5]. J. Muller, M. Wooldridge, and N. Jennings, editors. 'Intelligent Agents III, Agent Theories, Architectures and Languages', volume 1193 of Lecture Notes in Artificial Intelligence, Springer-Verlag, 1997.

[6]. B. Hayes-Roth, K. Pfleger, P. Morinot, P. Lalanda, 'Plans and Behavior in Intelligent Agents', ftp://ksl.stanford.edu/pub/KSL-Reports/KSL-95-35.ps

[7]. L. Cavedon, A Rao, and W. Wobcke, editors. 'Intelligent Agent Systems, Theoretical and Practical Issues', volume 1209 of Lecture Notes in Artificial Intelligence, Springer-Verlag, 1997.

[8]. J. Bradshaw, editor, 'Software Agents', AAAI Press / The MIT Press, 1997.

[9]. B.A.Nardi, J.R.Miller, D.J.Wright, 'Collaborative, Programmable Intelligent Agents', Communications of the ACM, Vol.41, no.3, pp.96-104, March 1998.

[10]. 'ISO/IEC 14772-1:1997, VRML97 International Standard', The VRML Consortium, http://www.vrml.org

[11]. B. Roehl, 'Some thoughts on Behavior in VR Systems' (Second Draft : August 1995), URL: http://sunee.uwaterloo.ca/~broehl/behav.html

[12]. M. Prokopenko, V. Jauregui, 'Reasoning about Actions in Virtual Reality', IJCAI-97, Workshop on Nonmonotonic Reasoning, Action and Change, 1997

[13]. E. Denti, A. Natali , A. Omicini. 'Merging Logic Programming into Web-based technology : a Coordination-based approach.' Proceedings of 2nd International Workshop on Logic Programming Tools for Internet Applications, Leuven, Belgium, July 1997.

[14]. S.R. El-Beltagy, M. Rafea, and A. Rafea, 'Practical Development of Internet Prolog Applications using a Java Front End', appearing in [13].

[15]. T.Panayiotopoulos, M.Gergatsoulis, 'Intelligent Information Processing using TRLi', 6th International Conference and Workshop on Database and Expert Systems Applications, DEXA'95, London, U.K., September, 4-8, 1995, appears in Proceedings, N.Revell, A. M. Tjoa (Eds.), pp.494-501, 1995.

[16]. C.C.Marinagi, T.Panayiotopoulos, G.A.Vouros, C.D.Spyropoulos, 'Advisor : A knowledge-based planning system', International Journal of Expert Systems, Research and Applications, Vol.9, No.3, pp.319-355, 1996.

4

Determining the Visual Interpretation of Actions in Interactive Stories

Nikitas M. Sgouros

Dept. of Informatics

University of Piraeus

sgouros@unipi.gr

Stavros Sotirchos

EECS Dept.

Nat. Tech. Univ. of Athens

ssotir@dsclab.ece.ntua.gr

1. Introduction

Recent advances in AI and Web-based multimedia technologies encourage the development of interactive entertainment forms, such as interactive stories, that seek to create engaging stories that are meaningfully interactive. However, these systems can become successful, only if they manage to deliver engaging stories. Consequently, research in this area must develop effective visual interpretation techniques that communicate the informational, aesthetic and dramatic aspects of each story equally well. These techniques should exploit the ability of the computer to dynamically combine different media and create interesting dramatization effects during story delivery.

This paper describes a novel method for determining parts of the visual interpretation of actions in interactive plots. An interactive plot describes a story in which the user takes part as one of the characters and interacts with the rest of the cast to influence the development of the plot. The method accepts as input an interactive plot consisting of a tree of possible character actions along with their motivations and outcomes. It also accepts as input a 2-D layout of the 3-D space in which the story will play out and a list of possible behaviors for each of the objects in the description. The method uses this input to compose 3-D renditions for each story action by assigning behaviors to the objects in the scene based on performance-related descriptions of story actions. These renditions are enriched with appropriate multimedia effects that seek to emphasize their dramatic nature. The effectiveness of this method is currently being tested in extensive user trials of an interactive adventure story that has been deployed on the Web. Possible uses for this research include the development of intelligent tools for the design or real-time direction of interactive stories or games.

The rest of the paper is organized as follows. Section 2 describes the input to the visual interpretation method. Section 3 presents the description and interpretation of actions during story performance. We give an example of the application of this method in section 4 and provide an analysis of preliminary user trials. Finally, section 5 presents some related work, while section 6 is a conclusions and future work section.

2. Input Description

Our method accepts as input a symbolic description of the role (e.g., king, priest etc) that each cast member plays in the story, along with a tree of possible interactions between these characters. The method assumes that the input has been computed by a plot generation system similar to the one described in [1,2]. In the type of stories supported by this description each character is allowed to intervene with the behavior of any other cast member based on his/her goals and the norms associated with the roles that s/he plays. Intervention can be either favorable or unfavorable, depending on whether the initiating character wants to harm or benefit another cast member. Intervention gives rise to the execution of one of possible *action attempts* for achieving it. These can be either actions for satisfying a goal or norm or counter-actions for impeding it. Finally, action attempts and their associated interventions can either succeed or fail in the story. The input plot imposes a partial ordering on the presentation of action outcomes in the story. This ordering ensures that causal relations between story actions are preserved [2].

For example, a character playing the priest role in the story, will seek to intervene with the behavior of the rest of the cast to impose all the norms associated with worshipping and obeying the gods. Furthermore, a character (e.g., X) with a goal of acquiring a valuable resource (e.g., gold) from some other character (e.g. Y) will seek to intervene favorably or unfavorably with Y to satisfy this goal. This intervention will initiate a set of appropriate actions attempts for materializing it. For example, a stealing attempt for the resource in question would materialize an unfavorable intervention, while an exchange of this resource would implement a favorable one. These action attempts can either succeed or fail.

Finally, the method accepts as input a 2-D description of the 3-D space in which the story takes place. We refer to this description as the *story space*. This is a grid-based, top-level view of the 3-D space that describes and stores the position of the set (e.g,, various buildings, trees etc) and the characters participating in the performance. We refer to the 3-D instantiations of these characters as the *actors* of the story.

3. Description & Interpretation of Character Actions

3.1 A FRAMEWORK FOR DESCRIBING CHARACTER ACTIONS

The framework assumes that each character action has a *subject* and (optionally) a *target* associated with it. The subject is the character initiating the action, while the target is the cast member against which the action is directed. Subjects and targets for each action are given in the plot supplied as input to the method.

Visual interpretation uses the following performance-related features in describing each action:

1. **Place.** This feature describes the location in which the action attempt will take place. Place is defined relative to the location of any of the objects in story space. For example, Punishment is currently described as an action that has a subject (the punisher), a target (the character being punished) and a place that is equal to

the location of the target in the story. This means that a punishment attempt occurs whenever the location of the subject coincides with the one for the target of the action. `Sacrifice`, on the other hand, is an action that requires the subject of the action to arrive at the place of the sacrifice (e.g., the temple of the god) in order to take place.

2. **Duration.** This feature defines the minimum duration of the interpretation for each action in the story. Minimum duration can be expressed as an absolute number of user moves in the story from the initiation of the action. Alternatively, an action can last at least as long as it takes for certain conditions to become true in the story. For example, `sacrifice` is an action that lasts at least as long as it takes the subject of the action to go from its current location in the story to the place where the sacrifice will take place. The visual interpretation method does not allow the initiation of any new action attempts while the minimum duration for the current action has not expired. This is to allow the user to sufficiently interact with the interpretation of each action before developing the plot further.

3. **Manner.** Manner refers to the ways in which an action can be realized in a story environment. Currently, an action can have a *violent* or *non-violent* realization. In addition, an action may require physical encounter between its participants or not. For example, `Punishment` or `Confrontation` are realized by a violent encounter of the subject and the target of these actions. `Forbidding`, on the other hand, is an action that uses non-violent realization methods (either verbal or written).

4. **Outcome.** This feature describes the preferred outcome for each action attempt in the story. An attempt can either succeed or fail. The preferred outcome for each attempt is given by the plot. As the next section explains, visual interpretation uses the preferred outcome to decide on the outcome of each action attempt during story performance.

3.2 VISUAL INTERPRETATION

The method accepts as input the story plot that consists of a tree of possible action attempts and their outcome in the story. The method presents each action attempt in the order given by the branch in the plot that the user has decided to follow. In addition, interpretation computes an outcome for each action that is the same with the one described in the plot branch.

The interpretation of each action attempt follows a sequence of four steps:

1. Announce action attempt.
2. Decide on the visual interpretation strategy for the attempt.
3. Determine attempt outcome.
4. Report outcome.

3.2.1 Announce Action Attempt.

During this step, the system informs the user on the initiation of a new action attempt in the story. The user is informed with a text clip on the subject of the action and its motives, on the target of the action, on the place constraints that must hold and on the manner in which the action will be realized. A new action attempt is possible, only if the duration of the previous action attempt in the plot has been exceeded.

Along with the text description of the action attempt the system executes a set of multimedia effects for emphasizing the dramatic character of the attempt. More specifically, the method uses a set of *dramatic situations,* i.e., patterns of character intervention that reflect "interesting" plot structure. One of these situations is *reversal of fortune* which becomes true for a character (e.g. *X*) when X is the target of an action attempt that either materializes the first unfavorable intervention against X in the story or follows an action attempt that implemented a favorable intervention towards X. Another such situation is *rising complication* in which X becomes the target of an action attempt that materializes an unfavorable intervention against X following another unfavorable intervention and action attempt against X by some other character. The method assumes that both situations can induce suspense, therefore during this step it retrieves and executes from a library of multimedia effects appropriate audio clips that can accompany suspense scenes.

3.2.2 Decide on the Visual Interpretation Strategy.

During this step, the system decides on the visual interpretation strategy for this attempt. Currently, this step is performed only for actions that involve the user either as subject or as target of the action. This happens because interpretation assumes that there is only one viewpoint from which story performance is observed. This viewpoint always shows what the user character sees in story space. Visual interpretation uses the following rules in the order given to assign behaviors to the actors in story space:

1. *Satisfy Place Constraints.* The subject of an action must satisfy the location constraints before executing the action. Consequently, if the subject is the user then s/he will be notified that s/he should establish this precondition. If the subject is different from the user, then it will start executing a behavior (e.g., approach) that seeks to move the actor to the location indicated in the Place slot of the action attempt. For example, Punishment or Confrontation require the subject to have the same location as the target of the action. If the user is the subject of this action then s/he will be notified that s/he has to meet with the target. If the subject is some other character then it will start executing a behavior that will lead it at the current location of the target.

2. *Portray the Manner of the Action Attempt.* This step assigns behaviors to the characters involved in each action that relate to the manner of its realization in the story. For example, if the action specifies a violent physical encounter between its participants then whenever the location constraints of the action attempt are satisfied the method schedules a series of multimedia effects that convey the nature of this engagement (e.g., the camera rocks back and forth giving the impression of an engagement etc). Furthermore, any character different from the user will automatically initiate a defensive behavior against any violent action attempt in which s/he is the target. The details of this behavior are determined by the role that this character plays in the story. For example, defensive behavior may range from trying to hide from the subject of the action, to directly confronting him/her or to summoning the guard, in case the target of the action plays the king role. In the case of hiding, if the place constraints for the action require collocation between the

subject and target of the action then this defensive behavior will seek to move the target in positions in the story space where the Place constraints are not satisfied. The method does not assign automatically a defensive behavior to the user character. Instead it assumes that the user is free to choose whether to defend or not against any aggressor. The details of this choice are described by the branching conditions of the plot that is accepted as input by the method.

3.2.3 Determine Attempt Outcome.

This step determines whether an action attempt will either succeed or fail. The final outcome for each attempt is specified in the story plot. Depending on the manner of each action, this step may introduce delays in determining the final outcome that seek to prolong suspense, maximize the dramatic effect of each outcome and ensure that the presentation sequence for all action outcomes during story performance corresponds with the one given in the plot.

For example, actions that are realized as violent physical encounters between their participants are resolved only after a number of such intermediate encounters occur. If the plot determines that the outcome of the action attempt will be favorable for the subject (target) of the action then these intermediate encounters seek to convey the idea that the target (subject) will finally prevail. Consequently, the final outcome cannot be predicted by the user, thus leading to an increase in suspense.

The number of intermediate encounters for each action attempt is determined by the order of presentation of its outcome in the plot. More specifically, the story system will execute intermediate encounters as needed until all the action outcomes that precede the outcome of the current action attempt in the plot sequence are presented.

3.2.4 Report Outcome.

This step informs the user on the outcome of an action attempt. Apart from displaying text clips that describe the result of the attempt, this step executes a number of multimedia effects that seek to convey the dramatic nature of this event. For example, if the outcome is favorable (unfavorable) for the user then the method executes joyful (sad) audio clips. In addition, the outcome report for each action is accompanied by appropriate 3-D animations either of the camera (for results that affect the physical state of the user character) or of the other actors in the story. For example, if an action has a fatal outcome for the user character then the camera moves so as to create the impression that the user character collapses on the ground.

4. An Example

The effectiveness of the visual interpretation of an interative story based on the method described above is currently being tested in a Web-based interactive story. The system consists of a Java applet that interfaces with a VRML browser to produce 3-D renditions of the character actions in the story. The demo resides in: http://www.dsclab.ece.ntua.gr/~defacto/demo/DefactoDemo.htm. Visitors at this site are encouraged to play the part of the story protagonist and then give feedback on the effectiveness of the presentation.

The input plot for this example takes place in ancient Greece where the user character seeks to fulfill his father's last wish and sacrifice to Poseidon, the sea god, in

Corinthos, his home town. To this end, he arrives in Corinthos and tries to perform the sacrifice. However, Eumeneas, the local king and a long-time enemy of his father, attempts to forbid the execution of this sacrifice. The user can then choose whether to obey the royal decree or confront the king. If he chooses to obey the king the story ends with the protagonist failing to fulfill his father's wish. If he tries to confront the king, he motivates Dikosthenis, the local judge, to intervene and seek to punish the protagonist for disobeying the king. The user then has the chance to either surrender to the judge or try to hide from him. If he surrenders then he is send back into exile and the story ends with the protagonist failing to fulfill his father's wish. If he decides to hide then Anacleoussa, the priestess of Poseidon in Corinthos steps in and tries to help the protagonist by asking for Poseidon's help in disobeying the royal decree. In retaliation, Eumeneas tries to send Anacleoussa in exile. The plot indicates that the user should succeed in his attempt to hide from Dikosthenis and confront the king, while Anacleoussa should fail in helping the user. Finally, the input specifies that the outcomes of the sacrifice and the forbidding actions must be presented after the outcome of the confrontation between the user and the king and that the outcome of the punishment attempt by Dikosthenis must precede the outcome of the confrontation. Table 1 describes all the story actions that occur in the plot branch in which the user has decided to confront the king and hide from the judge.

Visual interpretation begins by announcing the execution of the `sacrifice` action attempt by the user. The Place constraints for `sacrifice` indicate that in order for this action to take place the location of the hero must become the same as the location of the temple for Poseidon. Furthermore, the duration of the action indicate that the action must last at least as long as it takes the hero to arrive at the temple. The interactive story system informs the user with a text clip on the need to find the temple to perform the sacrifice and waits till the user gets there. A 2-D map of the story space enables the user to perceive its location more accurately.

Table 1: Example input for the interactive story.

#	Description of Story Actions
1	You seek to sacrifice for Poseidon in order to worship Poseidon.
2	Eumeneas forbids the performance of the sacrifice to Poseidon
3	You decide to confront the king
4	Dikosthenis seeks to punish you for confronting the king.
5	You try to hide from Dikosthenis.
6	Anacleoussa asks for Poseidon's help in overcoming the royal order.
7	Eumeneas seeks to send Anacleoussa in exile.

When the user reaches the temple the system attempts to present the outcome of `sacrifice`. However, the plot indicates that this outcome will not be presented until the outcome of the confrontation between the user and the king is presented, therefore the execution of the sacrifice action pauses till the later outcome is presented. Since `sacrifice` has now exceeded its minimum duration the system proceeds to initiate the next action attempt in the story. This is a `forbidding` action that has no place constraints and a duration of 0 steps. The manner of this action indicate that this is a non-violent action that does not require physical encounter between its subject and target. Therefore, the system presents the user with a text clip informing him of the decision of the king. Furthermore, this attempt constitutes a reversal of fortune for the user since this is the first unfavorable intervention against him in the story, therefore the system accompanies this text clip with an ominous audio clip. The system then proceeds and presents the user with a choice of either confronting or obeying the king as described in the plot. If we assume that the user decides to confront the king the system initiates a confront action attempt with the user as subject and Eumeneas as the target of this action.

The Place constraints for confront require that the subject and the target must be in the same location in order for the action to take place. Since the plot determines that Eumeneas has a fixed location in the story in his palace, the user is informed that he should enter the king's palace in order for the confrontation to happen. At the same time, the system assigns a defensive behavior to Eumeneas, since he is the target of an unfavorable intervention by the user. Because Eumeneas plays the king role, he can summon the guard to protect him from the protagonist. The system introduces a new actor in the story (the Guard) that sits at the palace entrance and seeks to stop the user from entering the palace. The guard becomes the new target for the confront attempt. The Manner for confront indicates that it should be implemented as a violent encounter between the subject and the target. Since the plot indicates that the user should finally prevail in this confrontation at each of these intermediate encounters the user is thwarted by the guard. Each encounter is enriched with appropriate camera moves and audio clips that convey the character of a fight.

Finally, the duration for confront indicates that the attempt should last at least 35 user moves. This number was established during preliminary user trials and was the result of a compromise between the need of the users to sufficiently interact with the interpretation of each action in the story and the need to develop the plot further during performance (i.e., to initiate new action attempts and step-up the action). When the user finally succeeds in confronting the guard then the system presents him with a text clip that informs him on the final action outcome which in this case is successful for the protagonist. This text clip is enriched with a joyful audio clip that seeks to emphasize the favorable character of this event for the user.

For the rest of the story the reader is referred to the demo at the Web site given above. Extensive user trials are currently in progress through this Web site and we expect to have a significant body of results in the next two months.

5. Related Work

Recently, there has been considerable research on the creation of intelligent multimedia interfaces. Our work complements this research, focusing on the development of dynamic presentation techniques for interactive story and game environments. Furthermore, this research complements efforts in creating animated presentation agents for user interfaces [3] by showing how to assign to them appropriate behaviors in interactive narrative environments.

There has been a tremendous amount of work on story dramatization techniques for the performing arts. Our research incorporates part of this work for the creation of multimedia user interfaces, that base story presentation on a content analysis of their material. Another line of research that investigates the relation between user interface design and theater, has resulted in the creation of interface agents, that dramatize user interaction with computer resources [4]. Finally, there has been significant research for the creation of interfaces that feature believable interactive characters (see [5] for an overview). Most of the work in this area has concentrated on reactive control architectures, supporting meaningful interaction patterns between the user and these characters [6, 7]. Our work complements this research, providing higher-level associations between plot developments and directing techniques during story performance.

6. Conclusions & Future Work

This paper describes a novel method for determining parts of the visual interpretation of actions in interactive plots. The method accepts as input an interactive plot and composes 3-D renditions for each action in the story by assigning automatically behaviors to the objects in the scene based on performance-related descriptions of story actions.

Currently, the method assumes that there is only one camera in the story space and this shows what the user character sees. Consequently, the method produces visual interpretations for actions in which the user is one of the participants. Future work will seek to generate visual interpretations for actions in which the user is not participating and automatically compute appropriate positions for the camera in order to observe these actions and convey their dramatic significance effectively. Furthermore, future work will seek to control more parameters that affect story performance. For example, the method currently assumes that story space has constant lighting. Future versions of the method will seek to control this parameter during story performance. Furthermore, the method in its present form does not deal with affective aspects of actor performance in the story or with the believability of their 3-D instantiation. Future work will seek to apply the work in [3,5,6] to create more believable animation and portray more effectively the emotional state of the actors.

References

[1] Sgouros, N. M., Papakonstantinou, G., Tsanakas, P., A Framework for Plot Control in Interactive Story Systems, in *Proc. AAAI-96*, Portland, OR, AAAI Press, 1996.

[2] Sgouros, N. M., Dynamic, User-Centered Resolution in Interactive Stories, *15th International Joint Conference on AI (IJCAI-97)*, Nagoya, Japan, Morgan Kaufmann, 1997.

[3] Rist, T., André, E., Müller, J., Adding Animated Presentation Agents to the Interface, 1997 International Conference on Intelligent User Interfaces, Orlando, FL, ACM Press.

[4] Laurel, B., *Computers as Theater*, Addison-Wesley, 1993.

[5] Maes, P., *Artificial Life Meets Entertainment: Lifelike Autonomous Agents*, Communications of the ACM, vol. 38, no. 11, November 1995.

[6] Hayes-Roth, B., Brownston, L., Sincoff, E., Directed Improvisation by Computer Characters, Tech Rep., KSL-95-04, Dept. of Comp. Sci., Stanford University, 1995.

5

An Attribute Grammar Driven High-Level Synthesis Paradigm for Control Applications

G. E. ECONOMAKOS AND G. K. PAPAKONSTANTINOU
National Technical University of Athens
Department of Electrical and Computer Engineering
Zographou Campus, GR-15773 Athens, GREECE

1. Introduction

Attribute grammars (AGs) were devised by Knuth [8] as a tool for the formal specification of programming languages. In the general case, an AG can be seen as a mapping from the language described by a *context free grammar (CFG)* into a user defined domain. Since their introduction, AGs have been a subject of intensive research [15], both from a conceptual and from a practical point of view producing a large number of automated AG based systems. These systems, usually called *compiler-compilers*, generate different kinds of language processors from their high-level specifications. The development of such systems is the main advantage of AGs over other formal specification methods, since they can also be used as an executable method. This advantage has made AGs one of the most widely applied semantic formalisms.

In the field of electronic design automation, *high-level synthesis* of special purpose architectures [6, 10, 12, 14, 16, 17] presents many similarities with the work undertaken by a traditional compiler. It is defined as the transformation of *behavioral* circuit descriptions into *register-transfer level (RTL)* structural descriptions that implement the given behavior while satisfying user defined constraints, and can be seen as either a compilation process, or as a dataflow computation over a loosely defined (constrained) hardware architecture. The result of this transformation is the exact definition of the optimal (or suboptimal) architecture for each given behavior, with respect to timing constraints, area constraints and more recently, power consumption constraints and test resource constraints.

Recently, a unifying formal framework for high-level synthesis [2, 4] has been proposed based on AGs. Its main functionality is to transform algorithms into architectures. The implementation of a hardware compiler, producing VHDL [1, 13] descriptions of the synthesized circuits and based on the ideas of [2], has been presented in [3, 5].

This chapter faces the problem of high-level synthesis of control applications in the same AG-driven environment, tuned to handle such cases efficiently. As an example, a Kalman filter of a track-while-scan radar system [11] is automatically produced. The basic requirements at the algorithmic level are discussed (like matrix multiplication, parallel or serial input feeding, e.t.c.) and special solutions are given in an AG formalism. The proposed approach raises the level of abstraction for high-level hardware synthesis of control applications. AGs are used as a meta-tool, supporting transformations driven by attribute dependencies.

2. High-Level Synthesis

High-level synthesis [6] takes a specification of the behavior of a digital system along with a set of constraints and goals on the resulting hardware to be satisfied, and finds a structure that realizes the given behavior while satisfying the given goals and constraints. The behavior is usually described as an *algorithm*, similar to programming language descriptions. The structure is a register-transfer level implementation that includes a *data-path* portion and a *control* portion. The data-path contains a network of functional units, registers, and their interconnection. The control activates components of the data-path to realize the required behavior. One of the objectives of synthesis is to find a structure that satisfies the constraints, such as requirements on area, latency or cycle-time, while minimizing other costs. For example, the goal might be to minimize the area while meeting timing requirements.

Due to its complexity, high-level synthesis is divided into a number of distinct yet inter-dependent tasks [14]. First, a hardware description language specification is parsed into an internal representation that models both the control-flow and data-flow of the input behavior. The internal representation is optimized by compiler-like transformations such as dead-code elimination, common subexpression elimination and constant propagation. *Scheduling* and *binding* are then performed to map the behavior into structure. Scheduling assigns operations to control steps, where a control step is the fundamental unit of sequencing in synchronous systems and corresponds to a clock cycle. Binding assigns operations to specific allocated hardware resources. They are closely related and inter-dependent. For example, scheduling attempts to minimize the number of required control steps subject to the amount of available hardware which depends on the results of the binding. Likewise, binding exploit concurrency among operations to allow sharing of hardware resources, where the degree of concurrency is determined by scheduling.

For a given schedule and data-path, control logic must be synthesized to activate components in the data-path according to this schedule. There are many different ways of synthesizing the control, ranging from hardwired control, such as finite-state machines, to microcoded control, where each control step in the schedule corresponds to a microprogram instruction.

The major difficulty in high-level synthesis is the large number of design alternatives that must be examined in order to select the design that best meets the design goals and still satisfies the constraints. The computational complexity of even one of the above subtasks can be intractable in the presence of constraints. For example, the problem of scheduling under resource constraints is provably NP-complete [7]. Compounding the problem is the tight inter-dependencies that exist between scheduling and binding that make the *design space* of possibilities multi-dimensional and irregular. For high-level synthesis to be practical, it is necessary to find an acceptable compromise between the degrees of freedom in exercising design choices and the complexity of the synthesis computation.

Also, high-level synthesis still lacks a formalism of the type already developed for synthesis at lower abstraction levels, that is layout and logic synthesis. Layout models have a good formal foundation in set and graph theory since the algorithms deal with placement and connectivity of two-dimensional objects. Logic synthesis is based on well-known formalisms of Boolean algebra because the algorithms deal predominantly with minimization and transformation of Boolean expressions. High-level synthesis uses formalisms based on several different areas, varying from the programming-language paradigm for the behavioral description to layout models for cost and performance estimation. However, it does not have an identifiable formal theory of its own. Most frequently, a mixture of finite-state machine and register-transfer notations serves as a high-level synthesis formalism.

3. AG-driven High-Level Synthesis

Attempting to overcome the inefficiencies of high-level synthesis, a unifying formal framework based on AGs was proposed in [2]. These early results have been integrated and extended for the development of the AGENDA (Attribute Grammar driven ENvironment for the Design Automation of digital systems) behavioral compiler [5], which performs synthesis by decorating the parse tree of a behavioral circuit description with appropriate attributes. Data-path synthesis is performed using well known heuristics, implemented through attribute evaluation rules. Control-path synthesis is performed driven by the parsing process of each control construct. Overall, the main advantage of AGENDA is that it uses uniform, modular and declarative formalisms to represent all synthesis transformations. This way it defines a higher level of abstraction for designers, where they can describe the primitive actions that they take when they design a chip manually.

Most of the times, high-level synthesis starts with the design captured as an algorithmic description written in a conventional or special purpose *hardware description language (HDL)*. In AGENDA, the HDL used plays a dual and crucial role. On one hand, as in all high-level synthesis systems, it is used to express design functionality. On the other hand, its syntax is used as the underlying CFG that is decorated by a synthesis AG, which implements synthesis heuristics through attribute evaluation rules. It must support a high level of specification abstraction with a strict and well-defined syntax. Such a language, used to describe hardware specifications is HardwareC [9]. A subset of HardwareC is the design capture method in AGENDA.

After design entry, AGENDA transforms the input specification into a *Control/Data Flow Graph (CDFG)* type internal representation. A modified internal representation of the one presented in [16] has been adopted. All operator nodes of the CDFG are described by the set $X = \{x_a\}$ and all operator inputs and outputs are described by the sets $I = \{i_{a,b}\}$ and $O = \{o_{a,c}\}$ respectively. Each operator input is indexed by the operator index a and a second index b that numerically identifies the inputs of the operator. The same applies for outputs. All three sets are implemented as single linked lists with additional links from each operator in X to its corresponding inputs in I and outputs in O. These links represent the edges of the CDFG.

The construction of the CDFG is performed by *attribute evaluation (semantic) rules*, at the time that *syntactic rules*, corresponding to primitive operations of the behavioral description, are used to parse the input string. The order of evaluation is determined by the relations between *synthesized attributes* (whose values depend on attribute instances of their child nodes) and *inherited attributes* (whose values depend on attribute instances of their parent nodes) inserted into the semantic rules. Generally, operator parsing syntactic rules are like the following.

$$operation \Rightarrow operand_1 \; operator \; operand_2 \qquad (1)$$

In each syntactic rule of the type of (1), a standard set of semantic rules is attached which inserts *operator* into X, *operand*$_1$ and *operand*$_2$ into I and the output of *operator* into O.

After CDFG construction, AGENDA performs scheduling using semantic rules and attribute relations to implement different heuristics. For *As Soon As Possible (ASAP)* scheduling, each output must be scheduled in the next control step after all its inputs have been scheduled. This can be accomplished by using a synthesized attribute to pass scheduling information from inputs to outputs, with the following semantic rule attached to (1).

$$operation.s\_ASAP = MAX(operand_1.s\_ASAP, operand_2.s\_ASAP) + 1$$

When the inputs are temporary results of other operations, this input to output connection is enough to perform scheduling. When they are permanent variables of the behavioral description, the control step in which they have been generated is kept in the symbol table and can be obtained easily. Since variables are first assigned and then used, a single pass of the input string is enough to perform ASAP scheduling.

For *As Late As Possible (ALAP)* scheduling, each output must be scheduled in the last control step before its first use. This can be accomplished by using an inherited attribute to pass scheduling information from outputs to inputs, with the following semantic rule attached to (1).

$$operand_i.i\_ALAP = operation.i\_ALAP - 1, \; i = 1, 2$$

When the outputs are temporary results of other operations, this input to output connection is enough to perform scheduling. When they are permanent variables of the behavioral description, the earliest control step in which they will be needed is kept in the symbol table and can be obtained easily. Since variables are first assigned and then used, a single pass of the input string is not enough to perform ALAP scheduling. In this case, a first pass is required to collect dependencies for each output, and a second to perform scheduling. This technique is described in detail in [4].

For resource constrained scheduling, the same ideas hold with the only difference that, for each operator type, a resource table (implemented like symbol tables) is kept with entries for every control step. If an operator is to be scheduled in a control step and no resources are available, it is moved to the next.

The final step of AGENDA is the generation of VHDL descriptions that map the scheduled CDFG into a *finite state machine with data-path (FSMD)* architecture. These VHDL descriptions can be synthesized using modern EDA tools and implemented using different technologies (FPGA, CPLD, ASIC).

4. Control Application Paradigm - Kalman Filter for Track-While-Scan Radar System

The Kalman filter is the most powerful linear estimator for continuous random variables. It has been applied to solve problems in many different domains. In this chapter a Kalman filter to estimate the position of a track-while-scan (TWS) radar system is considered.

The TWS radar system can be modeled using a system equation:

$$X(k+1) = AX(k) + W(k) \qquad (2)$$

where

$$X(k) = [x_1(k), x_2(k), x_3(k), x_4(k)]^T, \ W(k) = [0, u_1(k), 0, u_2(k)]^T$$

$$A = \begin{bmatrix} 1 & C & 0 & 0 \\ 0 & 1 & 0 & 0 \\ 0 & 0 & 1 & C \\ 0 & 0 & 0 & 1 \end{bmatrix}$$

The four state variables are $x_1(k) = \rho(k)$, which represents the range of the radar system, $x_2(k) = \rho'(k)$, which represents the range rate, $x_3(k) = \theta(k)$, which represents the angle and $x_4(k) = \theta'(k)$, which represents the angle rate. The term $W(k)$ is a white noise process and includes $u_1(k)$, which represents the change in radial velocity over interval C and $u_2(k)$, which represents the change in angular velocity over the same interval. C is the sampling cycle time.

What we want from the system is to acquire information about its range and angle. This can be modeled with the measurement equation:

$$Y(k) = BX(k) + V(k) \qquad (3)$$

where

$$Y(k) = [y_1(k), y_2(k)]^T, \ V(k) = [v_1(k), v_2(k)]^T$$

$$B = \begin{bmatrix} 1 & 0 & 0 & 0 \\ 0 & 0 & 1 & 0 \end{bmatrix}$$

Two sensors measure $y_1(k)$, which represents the range and $y_2(k)$, which represents the angle. The term $V(k)$ is a white noise process and includes $v_1(k)$ and $v_2(k)$ which represent range and angle measurement noise respectively.

A Kalman filter can be implemented to estimate the range and angle of the radar system using a set of measurements. The equations for the filter are:

Time update equations (effect of system dynamics):

$$P1(k) = AP(k-1)A^T + Q(k-1) \qquad (4)$$

$$\hat{X}1(k) = A\hat{X}(k-1) \qquad (5)$$

Measurement update equations (effect of new measurement):

$$G(k) = P1(k)B^T[BP1(k)B^T + R(k)]^{-1} \tag{6}$$

$$P(k) = P1(k) - G(k)BP1(k) \tag{7}$$

$$\hat{Y}(k) = B\hat{X}1(k) \tag{8}$$

$$\hat{X}(k) = \hat{X}1(k) + G(k)[Y(k) - \hat{Y}(k)] \tag{9}$$

where

$$Q(k) = E[W(k)W^T(k)] = \begin{bmatrix} 0 & 0 & 0 & 0 \\ 0 & \sigma_1^2(k) & 0 & 0 \\ 0 & 0 & 0 & 0 \\ 0 & 0 & 0 & \sigma_2^2(k) \end{bmatrix}$$

$$R(k) = E[V(k)V^T(k)] = \begin{bmatrix} \sigma_\rho^2(k) & 0 \\ 0 & \sigma_\theta^2(k) \end{bmatrix}$$

$\hat{X}1(k)$ is the *a priori* state estimate, that is the estimate calculated without taking under consideration the new inputs. $\hat{X}(k)$ is the *a posteriori* state estimate, that is after the new inputs are considered. Similarly, $P1(k)$ is the *a priori* error covariance estimate and $P(k)$ is the *a posteriori* error covariance estimate. $G(k)$ is the Kalman gain and $\hat{Y}(k)$ is the output estimate. $Q(k)$ is the system noise covariance matrix, where $\sigma_1^2(k) = E[u_1^2(k)]$ and $\sigma_2^2(k) = E[u_2^2(k)]$. Finally, $R(k)$ is the measurement noise covariance matrix, where $\sigma_\rho^2(k) = E[v_1^2(k)]$ and $\sigma_\theta^2(k) = E[v_2^2(k)]$.

The functionality of the TWS radar Kalman filter is a loop over all $Y(k)$ measurements (infinite if there are infinite measurements). At each iteration, an new estimation is calculated using equations (4)-(9). The initial values can be derived using the first two measurements, $Y(1)$ and $Y(2)$.

What is important in order to design a hardware component for this Kalman filter is to decompose equations (4)-(9) and produce relations between scalars, which can be straightforward mapped into the inputs, outputs and internal registers of the component. Using the system properties (independence of noise sources $W(k)$ and $V(k)$) it can be shown that in the general case the filter equations are:

$$\begin{aligned} P1(k) \quad &= \quad AP(k-1)A^T + Q(k-1) = \\ &= \begin{bmatrix} P1_{11}(k) & P1_{12}(k) & 0 & 0 \\ P1_{21}(k) & P1_{22}(k) & 0 & 0 \\ 0 & 0 & P1_{33}(k) & P1_{34}(k) \\ 0 & 0 & P1_{43}(k) & P1_{44}(k) \end{bmatrix} \end{aligned} \tag{10}$$

where

$$P1_{11}(k) = P_{11}(k-1) + CP_{12}(k-1) + CP1_{12}(k)$$

$$P1_{12}(k) = P_{12}(k-1) + CP_{22}(k-1)$$

$$P1_{21}(k) = P_{21}(k-1) + CP_{22}(k-1), \ P1_{22}(k) = P_{22}(k-1) + \sigma_1^2$$

$$P1_{33}(k) = P_{33}(k-1) + CP_{34}(k-1) + CP1_{34}(k)$$

$$P1_{34}(k) = P_{34}(k-1) + CP_{44}(k-1)$$

$$P1_{43}(k) = P_{43}(k-1) + CP_{44}(k-1), \ P1_{44}(k) = P_{44}(k-1) + \sigma_2^2$$

$$\hat{X}1(k) = A\hat{X}(k-1) = [\hat{x1}_1(k), \hat{x1}_2(k), \hat{x1}_3(k), \hat{x1}_4(k)]^T \qquad (11)$$

where

$$\hat{x1}_1(k) = \hat{x}_1(k-1) + C\hat{x}_2(k-1), \ \hat{x1}_2(k) = \hat{x}_2(k-1)$$

$$\hat{x1}_3(k) = \hat{x}_3(k-1) + C\hat{x}_4(k-1), \ \hat{x1}_4(k) = \hat{x}_4(k-1)$$

$$G(k) = P1(k)B^T[BP1(k)B^T + R(k)]^{-1} = \begin{bmatrix} G_{11}(k) & 0 \\ G_{21}(k) & 0 \\ 0 & G_{32}(k) \\ 0 & G_{42}(k) \end{bmatrix} \qquad (12)$$

where

$$G_{11}(k) = P1_{11}(k)/(P1_{11}(k) + \sigma_\rho^2), \ G_{21}(k) = P1_{21}(k)/(P1_{11}(k) + \sigma_\rho^2)$$

$$G_{32}(k) = P1_{33}(k)/(P1_{33}(k) + \sigma_\theta^2), \ G_{42}(k) = P1_{43}(k)/(P1_{33}(k) + \sigma_\theta^2)$$

$$\begin{aligned} P(k) &= P1(k) - G(k)BP1(k) = \\ &= \begin{bmatrix} P_{11}(k) & P_{12}(k) & 0 & 0 \\ P_{21}(k) & P_{22}(k) & 0 & 0 \\ 0 & 0 & P_{33}(k) & P_{34}(k) \\ 0 & 0 & P_{43}(k) & P_{44}(k) \end{bmatrix} \end{aligned} \qquad (13)$$

where

$$P_{11}(k) = P1_{11}(k) - P1_{11}(k)G_{11}(k), \ P_{12}(k) = P1_{12}(k) - P1_{12}(k)G_{11}(k)$$

$$P_{21}(k) = P1_{21}(k) - P1_{11}(k)G_{21}(k), \ P_{22}(k) = P1_{22}(k) - P1_{12}(k)G_{21}(k)$$

$$P_{33}(k) = P1_{33}(k) - P1_{33}(k)G_{32}(k), \ P_{34}(k) = P1_{34}(k) - P1_{34}(k)G_{32}(k)$$

$$P_{43}(k) = P1_{43}(k) - P1_{33}(k)G_{42}(k), \ P_{44}(k) = P1_{44}(k) - P1_{34}(k)G_{42}(k)$$

$$\hat{Y}(k) = B\hat{X}1(k) = [\hat{y}_1(k), \hat{y}_2(k)]^T \tag{14}$$

where

$$\hat{y}_1(k) = \hat{x1}_1(k)$$
$$\hat{y}_2(k) = \hat{x1}_3(k)$$

$$\hat{X}(k) = \hat{X}1(k) + G(k)[Y(k) - \hat{Y}(k)] = [\hat{x}_1(k), \hat{x}_2(k), \hat{x}_3(k), \hat{x}_4(k)]^T \tag{15}$$

where

$$\hat{x}_1(k) = \hat{x1}_1(k) + G_{11}(k)(y_1(k) - \hat{y}_1(k)), \quad \hat{x}_2(k) = \hat{x1}_2(k) + G_{21}(k)(y_1(k) - \hat{y}_1(k))$$

$$\hat{x}_3(k) = \hat{x1}_3(k) + G_{32}(k)(y_2(k) - \hat{y}_2(k)), \quad \hat{x}_4(k) = \hat{x1}_4(k) + G_{42}(k)(y_2(k) - \hat{y}_2(k))$$

5. High-Level Synthesis of Kalman Filter for Track-While-Scan Radar System

From equations (10)-(15) a behavioral description in HardwareC can be constructed for the TWS radar Kalman filter. This description will act upon the inputs and outputs of a real hardware component. To allow easy parameterization of the component, except from the measurement input and estimation output, inputs for all constant values of the system can be inserted. Such a component declaration is:

```
block kalman(in port yin[16],c[16],s1[16],s2[16],sr[16],sth[16];
             in port reset[1];
             out port xout[16])
```

where c corresponds to C, s1 to σ_1^2, s2 to σ_2^2, sr to σ_ρ^2 and sth to σ_θ^2, reset is a control input used to denote when all measurements have been read and the estimation is to be acquired, yin is the measurement input and xout is the estimation output and the dimensions between brackets are used to denote the number of bits used to represent the corresponding numbers (and not declaration of array objects as in most programming languages).

With this component declaration, a behavioral description for the Kalman filter can be written as:

```
begin
  while reset/=1 do
    begin
      y_1:=read(yin); y_2:=read(yin);
      p1_12:=p_12+c*p_22; p1_11:=p_11+c*p_12+c*p1_12;
      p1_21:=p_21+c*p_22; p1_22:=p_22+s1;
```

```
        p1_34:=p_34+c*p_44; p1_33:=p_33+c*p_34+c*p1_34;
        p1_43:=p_43+c*p_44; p1_44:=p_44+s2;
        x1_1:=x_1+c*x_2; x1_2:=x_2;
        x1_3:=x_3+c*x_4; x1_4:=x_4;
        g_11:=p1_11/(p1_11+sr); g_21:=p1_21/(p1_11+sr);
        g_32:=p1_33/(p1_33+sth);g_42:=p1_43/(p1_33+sth);
        p_11:=p1_11-p1_11*g_11; p_12:=p1_12-p1_12*g_11;
        p_21:=p1_21-p1_11*g_21; p_22:=p1_22-p1_12*g_21;
        p_33:=p1_33-p1_33*g_32; p_34:=p1_34-p1_34*g_32;
        p_43:=p1_43-p1_33*g_42; p_44:=p1_44-p1_34*g_42;
        yh_1:=x1_1; yh_2:=x1_3;
        x_1:=x1_1+g_11*(y_1-yh_1); x_2:=x1_2+g_21*(y_1-yh_1);
        x_3:=x1_3+g_32*(y_2-yh_2); x_4:=x1_4+g_42*(y_2-yh_2)
      end;
    write(xout:=x_1); write(xout:=x_2);
    write(xout:=x_3); write(xout:=x_4)
  end.
```

The description is composed of a main loop that reads a new measurement and updates all system equations. Since the equations have been decomposed from matrix multiplications, as in the general form of the Kalman filter, into scalar arithmetic operations, for the specific filter, they can be straightforward coded into HardwareC. When the **reset** signal gets high, the loop exits and the current state estimation is written to the corresponding output. The user is responsible for setting **reset** high when all measurements have been read.

Measurements and estimations are read and written in and out of the component through single I/O ports. Even though at each sampling point two measurements have to be read, they are serialized through the same line. The same applies for the estimation output. This has been chosen because real hardware components can contain large numbers of functional components (since VLSI densities are continuously growing higher) but limited number of I/O pins.

In HardwareC, serialization is supported through the **read** and **write** primitive functions. These calls lock an I/O port for one control step and do not allow any further interaction before the next control step. For example, the call **read(x)** locks input **x** at some control step i. If a subsequent **read(x)** call is encountered, it will be placed at control step $i + 1$.

In an AG formalism, this can be performed by attaching attributes to keep the last control step each I/O port has been read or written in a special purpose symbol table. Any subsequent I/O activity call can get this value and schedule itself one control step later. The construction of symbol tables using AGs can be found in [4].

Using the above behavioral description AGENDA provided the following output for an ASAP (maximally parallel) implementation:

```
ASAP scheduled data path.
------------------------------
CS=1:   T1:=reset/=1,if T1 {x1_4:=x_4,x1_2:=x_2,y_1:=yin}
CS=2:   if T1 {T20:=c*x_4,T18:=c*x_2,p1_44:=p_44+s2,T12:=c*p_34,
               T10:=c*p_44,p1_22:=s1+p_22,T4:=c*p_12,T2:=c*p_22,
               y_2:=yin}
CS=3:   if T1 {yh_2:=x1_3,yh_1:=x1_1,x1_3:=x_3+T20,x1_1:=x_1+T18,
               p1_43:=p_43+T10,T14:=p_33+T12,p1_34:=p_34+T10,
               p1_21:=p_21+T2,T6:=p_11+T4,p1_12:=p_12+T2}
CS=4:   if T1 {T49:=y_2-yh_2,T44:=y_1-yh_1,T13:=c*p1_34,T5:=c*p1_12}
CS=5:   if T1 {p1_33:=T14+T13,p1_11:=T6+T5}
CS=6:   if T1 {T25:=p1_33+sth,T22:=p1_11+sr}
CS=7:   if T1 {g_42:=p1_43/T25,g_32:=p1_33/T25,g_21:=p1_21/T22,
               g_11:=p1_11/T22}
CS=8:   if T1 {T52:=g_42*T49,T50:=g_32*T49,T47:=g_21*T44,
               T45:=g_11*T44,T42:=p1_34*g_42,T40:=p1_33*g_42,
               T38:=p1_34*g_32,T36:=p1_33*g_32,T34:=p1_12*g_21,
               T32:=p1_11*g_21,T30:=p1_12*g_11,T28:=p1_11*g_11}
CS=9:   xout:=x_1, if T1 {x_4:=x1_4+T52,x_3:=x1_3+T50,x_2:=x1_2+T47,
                          x_1:=x1_1+T45,p_44:=p1_44-T42,
                          p_43:=p1_43-T40,p_34:=p1_34-T38,
                          p_33:=p1_33-T36,p_22:=p1_22-T34,
                          p_21:=p1_21-T32,p_12:=p1_12-T30,
                          p_11:=p1_11-T28}
CS=10:  xout:=x_2
CS=11:  xout:=x_3
CS=12:  xout:=x_4
```

As it can be verified, the data-path contains all primitive operations that are required for the implementation of the Kalman filter, ordered in control steps. If enough hardware resources are available, this schedule can be mapped into an FSMD architecture in a straightforward manner. Since an architectural description of the filter can be generated, it can be manufactured using various design technologies (FPGA, CPLD, ASIC).

6. Conclusions

In this chapter we have shown that high-level synthesis of hardware components can handle control applications efficiently provided a decomposition of the system equations from matrix into scalar arithmetic. AGs can aid designers implement all high-level synthesis transformations quick and into a more natural way, thus shortening design space exploration time. Currently we are evaluating the automated produced results with others produced manually, in order to fine-tune the AGENDA environment to produce more cost effective designs.

62

References

1. J. Bhasker. *A VHDL Primer*. Prentice Hall, 1992.
2. G. Economakos, G. Papakonstantinou, and P. Tsanakas. An attribute grammar approach to high-level automated hardware synthesis. *Information and Software Technology*, 37(9):493–502, 1995.
3. G. Economakos, G. Papakonstantinou, K. Pekmestzi, and P. Tsanakas. Hardware compilation using attribute grammars. In *Advanced Research Working Conference on Correct Hardware Design and Verification Methods*, pages 273–290. IFIP WG 10.5, 1997.
4. G. Economakos, G. Papakonstantinou, and P. Tsanakas. Incorporating multi-pass attribute grammars for the high-level synthesis of ASICs. In *Symposium on Applied Computing*, pages 45–49. ACM, 1998.
5. G. Economakos, G. Papakonstantinou, and P. Tsanakas. AGENDA: An attribute grammar driven environment for the design automation of digital systems. In *Design Automation and Test in Europe Conference and Exhibition*, pages 933–934. ACM/IEEE, 1998.
6. D. Gajski, N. Dutt, A. Wu, and S. Lin. *High-Level Synthesis*. Kluwer Academic Publishers, 1992.
7. M. R. Garey and D. S. Johnson. *Computers and Intractability: A Guide to the Theory of NP-Completeness*. W. H. Freeman and Company, 1979.
8. D. E. Knuth. Semantics of context-free languages. *Mathematical Systems Theory*, 2(2):127–145, 1968.
9. D. Ku and G. De Micheli. HardwareC: A language for hardware design. Technical Report CSL-TR-90-419, Stanford University, 1990. Version 2.0.
10. D. C. Ku and G. De Micheli. *High Level Synthesis of ASICs Under Timing and Synchronization Constraints*. Kluwer Academic Publishers, 1992.
11. C. R. Lee and Z. Salcic. High-performance FPGA-based implementation of kalman filter. *Microprocessors and Microsystems*, 21(4):257–265, 1997.
12. Y-L. Lin. Recent development in high level synthesis. *ACM Transactions on Design Automation of Electronic Systems*, 2(1):2–21, 1997.
13. R. Lipsett, C. F. Schaefer, and C. Ussery. *VHDL: Hardware Description and Design*. Kluwer Academic Publishers, 1993.
14. M. C. McFarland, A. C. Parker, and R. Camposano. The high-level synthesis of digital systems. *Proceedings of the IEEE*, 78(2):301–318, 1990.
15. J. Paaki. Attribute grammar paradigms - a high-level methodology in language implementation. *ACM Computing Surveys*, 27(2):196–255, 1995.
16. D. E. Thomas, E. D. Lagnese, R.A. Walker, J. A. Nestor, J. V. Rajan, and R. L. Blackburn. *Algorithmic and Register-Transfer Level Synthesis: The System Architect's Workbench*. Kluwer Academic Publishers, 1990.
17. R. A. Walker and S. Chaudhuri. High-level synthesis: Introduction to the scheduling problem. *IEEE Design & Test of Computers*, 12(2):60–69, 1995.

6

A Case Study in Specifying the Denotational Semantics of C

N. S. PAPASPYROU
National Technical University of Athens
Department of Electrical and Computer Engineering
Division of Computer Science, Software Engineering Laboratory
Polytechnioupoli, 15780 Zografou, Athens, Greece.
Tel. +30-1-7722486, Fax. +30-1-7722519.

1. Introduction

C is a well known and very popular general purpose programming language which represents, together with its descendants, a strong and indisputable status quo in the current software industry. Its semantics is informally defined in the ISO/IEC 9899:1990 standard [1] using natural language. This causes a number of ambiguities and problems of interpretation, clearly manifested in numerous discussions taking place in the newsgroup comp.std.c. It is worthwhile noticing that members of the standardization committee and other distinguished researchers participating in the discussions often give contradictory answers when asked about the intended semantics of surprisingly small programs, and that their answers are usually based on different possible interpretations of the standard. With all this in mind, the necessity for a formal description of the semantics of C becomes apparent. Such a description would serve as a precise standard for compiler implementation and would provide a basis for reasoning about properties of C programs.

The semantics of many popular programming languages have been formally specified in literature using various formalisms. However, in most cases these specifications are incomplete, inaccurate or both, in varying degrees. By *incomplete* we mean that they do not specify the semantics of the whole language but that of a subset, often leaving out the most complicated features. By *inaccurate* we mean that the formal descriptions are not entirely correct, either because of intended simplifications or by mistake.

Significant research has been conducted recently concerning semantic aspects of C. In what seems to be the earliest formal approach, Sethi addresses mainly the semantics of pre-ANSI C declarations, using the denotational approach and making several simplifications, e.g. requiring left-to-right evaluation [2]. In a different paper, Sethi addresses the semantics of C's control structures using again a denotational framework [3]. In the work of Gurevich and Higgins a formal semantics for C is given using the formalism of evolving algebras [4]. Again, a number of simplifications are made, e.g. no interleaving is possible in expression evaluation and side effects are assumed to take place at the same time that they are generated. In the work of Cook and Subramanian an incomplete semantics for C is developed in the theorem prover Nqthm [5]. Cook et al. have also developed a denotational semantics for C based on temporal logic, which again makes a number of simplifying assumptions, mainly concerning evaluation order [6]. An operational semantics for C has been sketched, in terms of a random access machine, as a part of the MATHS project in California State University. Finally, in the work of Norrish a complete operational semantics for C is given using small-step reductions [7]. To the best of our knowledge, this is the only approach that formalizes correctly C's unspecified order of evaluation and sequence points. No similar denotational approach is known to us.

This chapter summarizes the results of our research, aiming at the development of an accurate and complete formal description for the semantics of the C programming language [8]. For this purpose, we have chosen the denotational approach.[1] As a remedy for the most important drawback of classic denotational semantics, its lack of modularity, we have used a number of monads which represent different aspects of computations.[2] In this way, the developed semantics was significantly improved in terms of modularity and elegance and its development was greatly facilitated.

2. Overview of the semantics

Our denotational specification for the semantics of C can be best understood as part of the *abstract interpreter* depicted in Fig. 1. The left part of the figure is a module diagram of the interpreter, showing the chain of actions performed and the processed data. Each action takes as input the result of previous actions. The initial piece of data is a *source program* and the final result is a representation of this program's *meaning*, i.e. a description of the program's behaviour when it is executed. The right part of the figure presents parts of a small example that will be discussed gradually until the end of this section. It should be noted that the example is simplified and does not correctly specify the semantics of C.

The interpreter consists of three layers, each containing a series of actions. *Syntactic analysis* aims at checking the syntactic validity of the source program and signalling syntax errors. Syntactically correct programs are transformed to abstract parse trees, which represent their structure in detail. *Semantic analysis* is used to define aspects of C that cannot be defined by a context-free grammar. It aims at checking the semantic validity of the program and signalling semantic errors, e.g. use of undeclared identifiers or type mismatches. Finally, *execution* aims at describing the meaning of programs. Our research mainly focuses on the last two layers, containing a total of three actions: *static semantics*, *typing semantics* and *dynamic semantics*. A brief overview of these actions and their collaboration is given in this section. The following three sections present the basics of each action in more detail.

Static semantics keeps track of identifiers that are defined in the source program. It aims at detecting static semantic errors, such as the redefinition of an identifier in the same scope, as well as associating identifiers with appropriate types or values. For each

---

[1] Introductions to denotational semantics, including useful bibliography, can be found in [9] and [10].

[2] Introductions to monads and their use in denotational semantics can be found in [11] and [12].

Figure 1: An abstract interpreter for C.

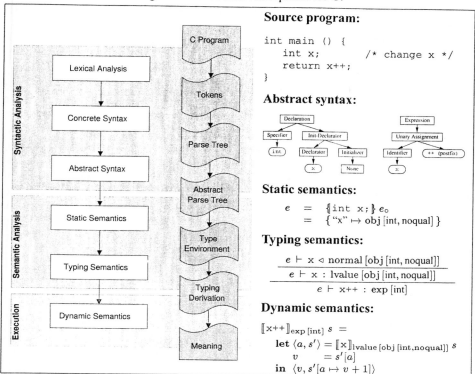

Source program:

```
int main () {
    int x;        /* change x */
    return x++;
}
```

Abstract syntax:

Static semantics:

$$e = \{\!\{ \text{int } x; \}\!\} e_0$$
$$= \{ \text{``x''} \mapsto \text{obj}[\text{int, noqual}] \}$$

Typing semantics:

$$\frac{\dfrac{e \vdash x \lhd \text{normal}[\text{obj}[\text{int, noqual}]]}{e \vdash x : \text{lvalue}[\text{obj}[\text{int, noqual}]]}}{e \vdash x{+}{+} : \text{exp}[\text{int}]}$$

Dynamic semantics:

$$[\![x{+}{+}]\!]_{\text{exp}[\text{int}]} \, s =$$
$$\textbf{let } \langle a, s' \rangle = [\![x]\!]_{\text{lvalue}[\text{obj}[\text{int,noqual}]]} \, s$$
$$v = s'[a]$$
$$\textbf{in } \langle v, s'[a \mapsto v + 1] \rangle$$

syntactically well-formed program phrase P, its static semantic meaning is denoted by $\{\!\{P\}\!\}$. Such meanings are typically types, type environments, i.e. associations of identifiers to types, or functions handling these two. Considering the simple C program that is used as an example in Fig. 1, let us isolate the declaration "int x;". The static semantic meaning of this declaration is a function that updates the type environment by declaring an integer variable "x". If e_0 is the empty type environment, containing no declarations, then the result of $\{\!\{\text{int } x; \}\!\}$ applied to e_0 is the updated environment e shown in Fig. 1, which contains a declaration for "x".

Typing semantics focuses on program phrases. It aims at the detection of type-mismatch errors, such as assignment to a constant value, and at associating syntactically well-formed phrases with appropriate phrase types. Such associations are given by means of *typing derivations*, i.e. formal proofs that phrases are well-typed. Typing derivations use inference rules to prove *typing judgements*, such as "$e \vdash P : \theta$", which states that phrase P has type θ in environment e. In the same example program of Fig. 1, let us now consider the expression "x++". Assuming that the static semantic analysis has resulted in type environment e, it is possible to derive that "x++" is an expression that computes an integer value, in other words "$e \vdash x{+}{+} : \text{exp}[\text{int}]$". A sketch of such a typing derivation is shown in Fig. 1. The derivation makes use of two inference rules, stating in short that variables are l-values and how l-values can be used as operands of postfix "++". The initial assumption means that "x" is declared to be a variable of type int in e, and is not further analyzed here.

Finally, dynamic semantics aims primarily at defining the execution behaviour of well-typed programs. For each well-typed program phrase P of type θ, its dynamic semantic meaning is denoted by $[\![P]\!]_\theta$. Such a meaning is typically a function describing some aspect of the execution of P. The typing derivation for P is important since it determines the way in which the dynamic semantics will be calculated. Going back once more to the same example, let us consider again the expression "x++", together with its typing derivation. Assuming a simple direct semantics for the dynamic meaning of expressions, the dynamic semantics for "x++" is given in Fig. 1. It is a function, taking the initial program state s and returning the result of the expression's evaluation and the final program state. Notice that the typing derivation dictates the types that are used for the dynamic semantics of phrases on both sides of the equation. For the sake of clarity, a represents the address of the object designated by the l-value "x", s' is the program state after evaluating "x" and v is the value stored in a at the program state s'. The result of the evaluation is v, and the final program state is the same as s', with the value stored in a incremented by one.

3. Static semantics

The static semantics of C can be thought of as the symbol table in our abstract interpreter. It calculates the environments containing type information for all identifiers defined in the source program and, for this reason, it mainly deals with the program's declarations. At the same time static semantic errors are detected. Apart from the complicated syntax of declarations that is characteristic of C, the static semantics is further perplexed by the presence of incomplete types in C's type system. Forward declarations of tags used in the recursive definition of structures and unions are also sources of complexity.

The domains that we use in the static semantics of C are summarized in Fig. 2. Most of them are defined as coalesced sums, by enumeration of their elements. The domain ordering relation $\sqsubseteq$ is crucial in the treatment of incomplete types: $x \sqsubseteq y$ denotes that y is a better approximation of a possibly incomplete element x. Notice the number of different types that are dictated by C's type system. Among them, data types provide the basis for the type system, representing types that C programs can manipulate as first class elements. Object types are associated with objects in memory and consist of qualified versions of data types and array types. Denotable types are associated with identifiers in type environments, while identifier types are used for the classification of these identifiers. Member types are associated with identifiers defined as members of structures or unions. Finally, phrase types are associated with program phrases by the typing semantics of §4. Most domains for environments can be taken as functions from identifiers to types; their full definition is omitted in this chapter.

The use of a simple error monad in the definition of the static semantics provides an elegant way of generating and propagating errors. In brief, a monad is a tuple $\langle M, \text{unit}, * \rangle$, where M is a domain constructor, unit : $A \to M(A)$ and $* : M(A) \times (A \to M(B)) \to M(B)$ are polymorphic functions for arbitrary domains A and B, satisfying three monad

Figure 2: Static semantic domains.

Auxiliary domains.

I : **Ide** *(identifiers)*, t : **Tag** *(tags)*, σ : **TagType** *(tag types)*.

Domains for types.

τ : **Type**$_{dat}$	=	void | char | signed-char | unsigned-char	*(data types)*
		| short-int | unsigned-short-int	
		| int | unsigned-int | long-int | unsigned-long-int	
		| float | double | long-double | ptr $[\phi]$	
		| enum $[\epsilon]$ | struct $[t, \pi]$ | union $[t, \pi]$	
q : **Qual**	=	noqual | const | volatile | const-volatile	*(type qualifiers)*
α : **Type**$_{obj}$	=	obj $[\tau, q]$ | array $[\alpha, n]$	*(object types)*
f : **Type**$_{fun}$	=	func $[\tau, p]$	*(function types)*
ϕ : **Type**$_{den}$	=	α | f	*(denotable types)*
m : **Type**$_{mem}$	=	α | bitfield $[\tau, q, n]$	*(member types)*
β : **Type**$_{bit}$	=	int | signed-int | unsigned-int	*(bit-field types)*
v : **Type**$_{val}$	=	τ | f	*(value types)*
δ : **Type**$_{ide}$	=	normal $[\phi]$ | typedef $[\phi]$ | enum-const $[n]$	*(identifier types)*
θ : **Type**$_{phr}$		*(See Fig. 3)*	*(phrase types)*

Domains for environments.

e : **Ent** *(types)*, ϵ : **Enum** *(enumerations)*, π : **Memb** *(members)*, p : **Prot** *(function prototypes)*.

---

laws. Given a domain D, the domain $\mathsf{M}(D)$ is a domain of *computations* resulting in values of type D. Function unit converts values to (trivial) computations and the binary operator $*$ is used to extract the results of computations.[3] Monad E can be defined by taking $\mathsf{E}(D) = D \oplus \mathbf{U}$ where $\mathbf{U}$ is a singleton domain, whose element represents errors.

Using monad E to allow static errors, the static meaning of declarations can be defined as a function of the domain:

▶ $\{declaration\}$: $\mathbf{Ent} \to \mathsf{E}(\mathbf{Ent})$

that is, a function which takes as argument an initial type environment and returns an updated type environment, which contains information about the declared identifiers. The equations defining such functions are often very complex but can be somewhat simplified by the use of several auxiliary operations, defined separately.

The meaning of recursively defined types requires special treatment. In a way similar to the one suggested in [2], a monadic closure operator is used, defined as:

$$\text{mclo } z \ f \ = \ \mathbf{clo} \ z \ (\lambda x. \ x \ * \ f) \ = \ \bigsqcup_{n=0}^{\infty} (\lambda x. \ x \ * \ f)^n \ z$$

on condition that f is continuous and $z \sqsubseteq z * f$. This operator is applied to the initial type environment and the static meaning of a list of declarations, in order to obtain a least upper bound for recursively defined types.

---

[3]This description, although naïve, is sufficient for the purpose of this chapter.

Figure 3: Some of the phrase types.

Phrase type	Description
exp [v]	Expression, whose result is a non-constant r-value of type v.
lvalue [m]	Expression, whose result is an l-value of type m.
val [τ]	Expression, whose result is a constant r-value of type τ.
arg [p]	Actual arguments of a function with prototype p.
stmt [τ]	Statement in a function returning a result of type τ.
decl	Declaration.

Figure 4: Typing judgements.

Main typing relation.

$e \vdash phrase : \theta$ The given *phrase* can be attributed phrase type θ in type environment e.

Predicates as judgements.

P Predicate P is *true*, where P can be any valid predicate over truth values **T**.

$v := z$ The static semantic valuation $z : E(D)$ produces the (non-error) value $v : D$.

Judgements related to environments.

$e \vdash I \vartriangleleft \delta$ Identifier I is associated with identifier type δ in type environment e.

$\pi \vdash I \vartriangleleft m$ Identifier I is associated with member type m in member environment π.

Judgements related to expressions.

$e \vdash E \gg \tau$ Expression E can be assigned to an object of data type τ in type environment e.

$e \vdash E = \text{NULL}$ Expression E is a null pointer constant in type environment e.

$e \vdash T \equiv \phi$ Type name T denotes type ϕ in type environment e.

4. Typing semantics

The primary aim of typing semantics is the association of program phrases with phrase types. Fig. 3 shows all the phrase types that we use. Typing rules are inference rules whose premises and conclusion are typing judgements. Several forms of typing judgements are necessary in order to simplify the typing rules. A summary of the most important ones is given in Fig. 4. Approximately 200 typing rules are used in our approach, in order to specify the typing semantics of C. 70% of those deal with expressions, 10% with statements and the remaining 20% with declarations. In the rest of this section we will present some non-trivial examples of typing rules and we will conclude with a discussion on our typing semantics.

The following rules specify the typing semantics of four types of expressions, in accordance with the ANSI C standard. Rule E1 states that decimal constants with no suffix are attributed type val [τ], where τ is the first integer type that can represent their value. According to E2, identifiers that have been defined as normal variables in an e are l-values of the appropriate type. Similarly, rules E3 and E4 specify the typing semantics of function calls and the indirection operator.

$$\frac{\begin{array}{cc} \text{suffix}(n) = \varnothing & \text{isDecimal(n)} \\ \tau := \text{repType}(n, [\text{int, long-int, unsigned-long-int}]) \end{array}}{e \vdash n \ : \ \text{val}\,[\tau]} \ (E1) \qquad \frac{e \vdash I \lhd \text{normal}\,[\alpha]}{e \vdash I \ : \ \text{lvalue}\,[\alpha]} \ (E2)$$

$$\frac{e \vdash E \ : \ \exp[\text{ptr}\,[\text{func}\,[\tau, p]]] \quad e \vdash \textit{args} \ : \ \arg[p]}{e \vdash E\,(\textit{args}) \ : \ \exp[\tau]} \ (E3) \qquad \frac{e \vdash E \ : \ \exp[\text{ptr}\,[\alpha]]}{e \vdash *E \ : \ \text{lvalue}\,[\alpha]} \ (E4)$$

The following two rules specify in part the semantics of assignments. According to E5, the expression on the left side of the simple assignment operator must be a modifiable l-value, while the expression on the right side must be assignable to the corresponding data type. Rule A1 states that an expression of arithmetic type τ can be assigned to an object of another arithmetic type τ'.

$$\frac{\begin{array}{cc} e \vdash E_1 \ : \ \text{lvalue}\,[m] & \text{isModifiable}(m) \\ \tau := \text{datify } m & e \vdash E_2 \gg \tau \end{array}}{e \vdash E_1 = E_2 \ : \ \exp[\tau]} \ (E5) \qquad \frac{\begin{array}{c} e \vdash E \ : \ \exp[\tau] \\ \text{isArithmetic}(\tau) \quad \text{isArithmetic}(\tau') \end{array}}{e \vdash E \gg \tau'} \ (A1)$$

A small number of typing rules specifies conversions that take place implicitly in the evaluation of expressions. Such conversions are called implicit coercions. For example, rule C1 states that l-values are implicitly converted to the values stored in the designated objects. Even more obviously, according to rule C2 constant values may be treated as normal non-constant expressions.

$$\frac{e \vdash E \ : \ \text{lvalue}\,[\text{obj}\,[\tau, q]] \quad \text{isComplete}(\tau)}{e \vdash E \ : \ \exp[\tau]} \ (C1) \qquad \frac{e \vdash E \ : \ \text{val}\,[\tau]}{e \vdash E \ : \ \exp[\tau]} \ (C2)$$

The typing semantics of statements is specified in a similar way. Rule S1 deals with compound statements. Notice that a compound statement defines a new scope, containing its declarations, and therefore a new environment e' must be calculated, allowing for recursively defined structures or unions. This new environment e' is used for the typing of the compound statement's body. Rules S2 and S3, specifying the semantics of `while` and `return` statements respectively, are relatively easier. In S3, the type of the returned expression must be assignable to the function's returned type.

$$\frac{\begin{array}{c} e' := \textbf{rec}\,\{\!\!\{ \textit{declaration-list} \}\!\!\} \ (\uparrow e) \\ e' \vdash \textit{declaration-list} \ : \ \text{decl} \quad e' \vdash \textit{statement-list} \ : \ \text{stmt}\,[\tau] \end{array}}{e \vdash \{\ \textit{declaration-list statement-list}\ \} \ : \ \text{stmt}\,[\tau]} \ (S1)$$

$$\frac{\begin{array}{c} e \vdash \textit{expr} \ : \ \exp[\tau'] \quad \text{isScalar}(\tau') \\ e \vdash \textit{stmt} \ : \ \text{stmt}\,[\tau] \end{array}}{e \vdash \textbf{while}\ (\ \textit{expr}\ )\ \textit{stmt} \ : \ \text{stmt}\,[\tau]} \ (S2) \qquad \frac{e \vdash \textit{expr} \gg \tau}{e \vdash \textbf{return}\ \textit{expr}\ ; \ : \ \text{stmt}\,[\tau]} \ (S3)$$

The suggested typing semantics for C leads to two forms of ambiguity problems. The first concerns the uniqueness of typing results: the main typing relation does not always provide a unique phrase type for a given program phrase. (Implicit coercion rules such as C1 and C2 are one source of such ambiguities.) This form of ambiguity is in fact useful. A given program phrase can be attributed different phrase types, depending on its role in the program, and a different dynamic semantic meaning may exist for each phrase type. The second form concerns the uniqueness of typing derivations for a given typing judgement.

Figure 5: Dynamic semantic domains.

Auxiliary domains and environments.

$\mathbf{Addr} = \mathbf{Obj} \times \mathbf{Offset}, \; \mathbf{Obj}, \; \mathbf{Fun}, \; \mathbf{Offset}, \; \mathbf{BitOfs}, \; [\![e]\!]_{Ent}, \; [\![p]\!]_{Prot}, \; \mathbf{Cod}, \; \mathbf{Lab}_\tau$

Domains for types.

$[\![\text{void}]\!]_{dat} = \mathbf{U}, \; [\![\text{int}]\!]_{dat} = \mathbf{N}, \; [\![\text{ptr}\,[\alpha]]\!]_{dat} = \mathbf{Addr} \oplus \mathbf{U}, \; [\![\text{ptr}\,[f]]\!]_{dat} = \mathbf{Fun} \oplus \mathbf{U}$

$[\![\text{obj}\,[\tau, q]]\!]_{obj} = \mathbf{Addr}, \; [\![\text{array}\,[\alpha, n]]\!]_{obj} = \mathbf{N} \to [\![\alpha]\!]_{obj}$

$[\![\text{func}\,[\tau, p]]\!]_{fun} = [\![p]\!]_{Prot} \to \mathsf{G}([\![\tau]\!]_{dat})$

$[\![\alpha]\!]_{mem} = [\![\alpha]\!]_{obj}, \; [\![\text{bitfield}\,[\tau, q, n]]\!]_{mem} = \mathbf{Addr} \times \mathbf{BitOfs}$

For example, there are two different derivations concluding with the fact that the sum of two integer constants is an integer expression: the first adds the constants and coerces the constant sum using C3, while the second coerces the summands separately using C3 and adds the resulting expressions. It is required that all different possible derivations for a given typing judgement result in the same dynamic semantics for the program phrase.

5. Dynamic semantics

The dynamic semantics of C specify the execution behaviour of well-typed programs. As a useful side-effect, run-time errors and other sources of undefined behaviour are detected. The most important source of complexity in an accurate definition of C's dynamic semantics is the unspecified evaluation order, combined with the fact that expressions generate side effects. In order to disallow undesired ambiguities, the ANSI C standard has introduced restrictions imposed on expression evaluation with the mechanism of sequence points. Additional restrictions are imposed on the access of objects between consecutive sequence points; however, according to our interpretation of the standard this mechanism does not always prevent non-determinism.[4] The dynamic semantics is further perplexed by pointer arithmetic, complex control statements like for and switch, and the presence of goto in combination with block scopes containing variable declarations.

For each static type, a dynamic semantic domain is defined for representing the dynamic meaning of values of this type. The definitions of some dynamic semantic domains are shown in Fig. 5. Among other things, the domain of integer numbers $\mathbf{N}$ is used to represent values of type int, pointers to objects are represented by the objects' address or a special null value, and addresses of objects are offsets in the biggest (possibly aggregate) objects containing them, in order to correctly model pointer arithmetic. The domain for type environments $[\![e]\!]_{Ent}$ contains the dynamic meanings of all identifiers defined in e, while that for function prototypes contains the values of all parameters. Domain $\mathbf{Cod}$ contains the code of all defined functions and $\mathbf{Lab}_\tau$ contains the meanings of labeled statements in a function returning a result of type τ and is used in the semantics of jumps.

A number of monads is used in order to represent various aspects of the computations that are related to the execution of C programs. Brief descriptions of these monads are

---

[4]This issue has been discussed a lot in comp.std.c. Several opinions have been expressed but no conclusion has been reached.

Figure 6: Monads used in the dynamic semantics.

Auxiliary monads and monad transformers.

$P(D)$ Powerdomain monad, allowing for non-determinism in values of domain D.

$R(M)(T)$ Resumption monad transformer, allowing interleaving in computations of type $M(T)$. Defined as a solution of the equation $R(M)(T) = T \oplus M(R(M)(T))$.

Monads for computations.

$V(T)$ Computation of a constant value of type T, not accessing the memory (value monad).

$C(T)$ Computation of a non-constant value of type T, possibly accessing the memory (continuation monad). Defined as $C(T) = (T \to \mathbf{C}) \to \mathbf{C}$, where $\mathbf{C} = \mathbf{S} \to P(\mathbf{A})$ is the domain of non-deterministic continuations, $\mathbf{S}$ is the domain of program states and $\mathbf{A}$ the domain of final program answers.

$G(T)$ Computation of a non-constant value of type T in the evaluation of an expression, possibly accessing the memory and allowing for interleaving. Defined as $G(T) = R(C)(T)$.

$K_\tau(T)$ Computation of a non-constant value of type T in the execution of a statement, possibly accessing the memory or terminating the function by returning a result of type τ.

---

shown in Fig. 6. The powerdomain monad, based on the convex powerdomain, is used to model non-determinism. The resumption monad transformer is used to model the interleaving in the evaluation of expressions, that is required by C's unspecified order of evaluation.[5]

The definition of dynamic semantics for program phrases is similar to that of static semantics. An important difference, however, is that the typing derivations provide useful information about the semantic meanings of both the defined phrase and its components. Thus, typing derivations control the definition of dynamic semantics, instead of abstract syntax, and there is one dynamic equation for each typing rule. In the rest of this section we will illustrate the definition of dynamic semantics by presenting some small examples.

Let us consider the simple case of typing rule E4 in §4. The dynamic semantics for an expression of the form "$*E$" under the typing given in E4 can be defined as follows:

▶ $[\![\exp[v]]\!] \;\; : \;\; e : \mathbf{Ent} \twoheadrightarrow [\![e]\!]_{Ent} \to \mathbf{Cod} \to G([\![v]\!]_{val})$

$[\![*E]\!]_{\text{lvalue}[\alpha]} \;=\; \lambda e.\, \lambda \rho.\, \lambda \xi.$
     $[\![E]\!]_{\exp[\text{ptr}[\alpha]]} \, e\, \rho\, \xi \; * \; (\lambda\, d_e.\, \mathbf{case}\; d_e\; \mathbf{of}$
       $\mathbf{inl}\; a \qquad\quad \Rightarrow\; \mathbf{unit}\; a$
       $\mathbf{otherwise} \Rightarrow\; \mathbf{error})$

The first line states that the dynamic semantic meaning for phrases of type $\exp[\tau]$ is a function taking as arguments the static and dynamic environments and the code environment and returning an interleaved computation with a result of type $[\![\tau]\!]_{dat}$. Notice the dependent function type, denoted here as $x : A \twoheadrightarrow B(x)$, which dictates a connection between the static and the dynamic environments. The equation that follows defines $[\![*E]\!]_{\text{lvalue}[\alpha]}$ in terms of $[\![E]\!]_{\exp[\text{ptr}[\alpha]]}$. If the pointer contains an object's address, this address is used in the resulting l-value. An error occurs if the pointer is null. Dynamic semantic meanings for phrases of type lvalue $[m]$ are functions of the form:

▶ $[\![\text{lvalue}[m]]\!] \;\; : \;\; e : \mathbf{Ent} \twoheadrightarrow [\![e]\!]_{Ent} \to \mathbf{Cod} \to G([\![m]\!]_{mem})$

---

[5]The resumption monad transformer is defined in detail in a paper that will be submitted for publication in the near future. For an introduction to monad transformers, the reader is referred to [13].

In a similar way, the following equations define the dynamic semantics that correspond to typing rules E3 and C1 of §4 respectively.

$$[\![E \, (args)]\!]_{\exp[\tau]} = \lambda e. \lambda \rho. \lambda \xi.$$
$$[\![E]\!]_{\exp[\text{ptr}\,[\text{func}\,[\tau,p]]]} e \, \rho \, \xi \bowtie [\![args]\!]_{\arg[p]} e \, \rho \, \xi \; * \; (\lambda \langle d_e, d_p \rangle.$$
$$\text{seqpt} \; * \; (\lambda u. \, \textbf{case} \; d_e \; \textbf{of}$$
$$\textbf{inl} \; d_f \qquad \Rightarrow \textbf{let} \; \langle f, b_f \rangle = \xi[d_f] \; \textbf{in} \; \text{isCompatible}\,(f, \text{func}\,[\tau, p]) \to b_f \, d_p \, , \, \text{error}$$
$$\textbf{otherwise} \Rightarrow \text{error}))$$
$$[\![E]\!]_{\exp[\tau]} = \lambda e. \lambda \rho. \lambda \xi. [\![E]\!]_{\text{lvalue}\,[\text{obj}\,[\tau,\varsigma]]} e \, \rho \, \xi \; * \; \text{getValue}_{\text{obj}\,[\tau,q] \mapsto \tau}$$

The first thing to notice is the use of operator $\cdot \bowtie \cdot : G(A) \times G(B) \to G(A \times B)$ for the interleaving of two computations, in order to model the unspecified order in which the function's designator and its arguments are evaluated. The second is the use of seqpt : $G(\mathbf{U})$ to introduce a sequence point just before the function is actually called. The function's dynamic meaning is looked up in the code environment. Furthermore, the function's actual type must be compatible with the type of the designator that is used. In the second equation, the important point is the use of getValue$_{m \mapsto \tau} : [\![m]\!]_{mem} \to G([\![\tau]\!]_{dat})$ which retrieves a stored value from memory.

The dynamic semantics of statements is defined in a similar way. We give here two relatively simple examples, corresponding to the typing rules S2 and S3 of §4. Notice that the dynamic meaning of statements uses also a label environment of type $\mathbf{Lab}_\tau$. The calculation of this environment requires a least fixed point, to allow infinite loops implemented by goto statements.

▶ $[\![\text{stmt}\,[\tau]]\!] \; : \; e : \mathbf{Ent} \twoheadrightarrow [\![e]\!]_{Ent} \to \mathbf{Cod} \to \mathbf{Lab}_\tau \to \mathsf{K}_\tau(\mathbf{U})$

$$[\![\texttt{while} \; (expression) \; statement]\!]_{\text{stmt}\,[\tau]} = \lambda e. \lambda \rho. \lambda \xi. \lambda \ell. \, \mathbf{fix} \; (\lambda g.$$
$$\text{lift}_{\mathsf{G} \to \mathsf{K}} \, ([\![expression]\!]_{\exp[\tau']} e \, \rho \, \xi) \; * \; (\lambda d.$$
$$\text{checkBoolean}_{\tau'}(d) \to$$
$$\text{setBreakContinue} \; \langle \text{unit} \; \mathbf{u}, g \rangle \; ([\![statement]\!]_{\text{stmt}\,[\tau]} e \, \rho \, \xi \, \ell) \; * \; (\lambda u. \, g) \, , \, \text{unit} \; \mathbf{u}))$$
$$[\![\texttt{return} \; expression ;]\!]_{\text{stmt}\,[\tau]} = \lambda e. \lambda \rho. \lambda \xi. \lambda \ell. \, \text{lift}_{\mathsf{G} \to \mathsf{K}} \, (\mathcal{A}[\![expression]\!]_{\exp[\tau']} e \, \rho \, \xi) \; * \; \text{result}$$

The first equation uses the least fixed point operator **fix** to model the semantics of the while statement. The correct continuations that will be used in case of a break or continue statement are passed to the body of the loop by using setBreakContinue. The second equation defines the semantics of the return statement. The expression is evaluated and converted as if by assignment to the function's return type. Notice the use of result : $[\![\tau]\!]_{dat} \to \mathsf{K}_\tau(\mathbf{U})$ which signals the termination of this function and specifies the returned result.

6. Evaluation

A significant effort has been made to evaluate our approach in defining a denotational semantics for the C programming language. In this task, the major issue was to assess how complete and accurate the developed semantics is. Unfortunately, there is no systematic way to evaluate our approach and be absolutely certain that the results are correct: there is simply no way to compare a formal system of this complexity against an informal specification. For this reason, we have resorted in testing an interpreter that directly implements our semantics, by using some test suites for C implementations that were available.

An earlier version of our semantics was first implemented using SML as the implementation language. Later, SML was abandoned and Haskell was used instead, mainly because it has a richer type system, more flexible syntax, elegant support for monads and also because lazy evaluation avoids a number of non-termination problems. The current implementation consists of approximately 15,000 lines of Haskell code, which are distributed roughly as follows: 3,000 lines for the static semantics, 3,000 lines for the typing semantics, 5,000 lines for the dynamic semantics, 3,000 lines for parsing and pretty-printing and 1,000 more lines of general code and code related to testing. As it was expected, the implementation is very slow and this presents a serious handicap in our yet unfinished evaluation process, significantly limiting the size of test programs.

Although the evaluation of our semantics is still under way and minor bugs are waiting to be fixed, the results indicate that the developed semantics is complete and accurate to a great extent, with respect to the ANSI C standard. The most important deviations from the standard are that the developed semantics requires function prototypes to exist for all called functions, something already favoured by the current standard, and that storage specifiers other than `typedef` are currently ignored. Static variables may be preprocessed out, but a solution integrated in the semantics is currently investigated. As less important deviations, the developed semantics requires fully bracketed initializations and forbids the declaration of identifiers, other than labels, in expressions or statements.

7. Conclusion and future work

In this chapter, we have presented a summary of our work in developing a formal semantics for the ANSI C programming language, following the denotational approach. The developed semantics is satisfactorily complete and accurate, with respect to the standard. A significant contribution of our research, besides the developed semantics itself, is the application of monads and monad transformers for the specification of a real programming language. Furthermore, interesting results have been achieved in our attempt to model the interleaving of computations and non-determinism using monads and monad transformers, which may be useful in specifying the semantics of programming languages supporting parallelism.

Our research in the near future will focus on the process of evaluating and improving the developed semantics. Beyond that, we would like to study the practical applications that a formal semantics for C may have in the software industry, especially in tools for program transformation, debugging and understanding. The implementation of the developed semantics also gave rise to an interesting question: what are the characteristics of a programming language that make it suitable for implementing denotational specifications, especially using monadic notation? Finally, another direction for future research aims at studying and specifying the semantics of C's object-oriented descendants, C++ and Java.

References

[1] American National Standards Institute, New York, NY. *ANSI/ISO 9899-1990, American National Standard for Programming Languages: C*, 1990. Revision and redesignation of ANSI X3.159-1989.

74

[2] R. Sethi. A case study in specifying the semantics of a programming language. In *Proceedings of the 7th Annual ACM Symposium on Principles of Programming Languages*, pages 117–130, January 1980.

[3] R. Sethi. Control flow aspects of semantics-directed compiling. *ACM Transactions on Programming Languages and Systems*, 5(4):554–595, October 1983.

[4] Y. Gurevich and J. K. Huggins. The semantics of the C programming language. In E. Börger et al., editors, *Selected Papers from CSL '92 (Computer Science Logic)*, volume 702 of *Lecture Notes in Computer Science*, pages 274–308. Springer Verlag, New York, NY, 1993.

[5] J. Cook and S. Subramanian. A formal semantics for C in Nqthm. Technical Report 517D, Trusted Information Systems, October 1994.

[6] J. Cook, E. Cohen, and T. Redmond. A formal denotational semantics for C. Technical Report 409D, Trusted Information Systems, September 1994.

[7] M. Norrish. An abstract dynamic semantics for C. Technical Report TR-421, University of Cambridge, Computer Laboratory, May 1997.

[8] N. S. Papaspyrou. *A Formal Semantics for the C Programming Language*. PhD thesis, National Technical University of Athens, Software Engineering Laboratory, February 1998.

[9] R. D. Tennent. The denotational semantics of programming languages. *Communications of the ACM*, 19(8):437–453, August 1976.

[10] P. D. Mosses. Denotational semantics. In J. van Leeuwen, editor, *Handbook of Theoretical Computer Science*, volume B, chapter 11, pages 577–631. Elsevier Science Publishers B.V., 1990.

[11] E. Moggi. An abstract view of programming languages. Technical Report ECS-LFCS-90-113, University of Edinburgh, Laboratory for Foundations of Computer Science, 1990.

[12] P. Wadler. The essence of functional programming. In *Proceedings of the 19th Annual Symposium on Principles of Programming Languages (POPL '92)*, January 1992.

[13] S. Liang, P. Hudak, and M. Jones. Monad transformers and modular interpreters. In *Conference Record of the 22nd ACM SIGPLAN-SIGACT Symposium on Principles of Programming Languages (POPL '95)*. San Francisco, CA, January 1995.

7

A Multi-Agent Model for Content-Based Electronic Document Filtering

N. S. PAPASPYROU, C. E. SGOUROPOULOU AND E. S. SKORDALAKIS
National Technical University of Athens
Department of Electrical and Computer Engineering
Division of Computer Science, Software Engineering Laboratory
Polytechnioupoli, 15780 Zografou, Athens, Greece.
Tel. +30-1-7722486, Fax. +30-1-7722519.

A. V. GERBESSIOTIS
Oxford University Computing Laboratory
Parks Rd., Oxford OX1 3QD, UK.

P. LIVADAS
University of Florida, Computer and Information Sciences Department
Gainesville, FL 32611, USA.

1. Introduction

One of the most important tasks in every office is *document management* which includes the subtasks of creating, archiving, retrieving, dispatching, updating and processing documents. In the last decades and in many different ways, there have been attempts to automate these tasks with positive results. Technological advances allow these tasks to be fully automated through the use of computers and specialized software.

Document management is considered as one of the principal applications of computers. Documents comprise 80% of information that is today available in electronic form, in contrast to classic structured databases [1]. Electronic documents are no more a simple analogue of paper documents; they are more dynamic entities in multiple forms and media. They can include information related to their origin and executable code that undertakes their management, something which is not possible in any piece of paper. Documents become the focus of attention when it comes to software development; their easy and user-friendly management is extremely important and new tools are required.

Automation of tasks related to document management has never ceased to be the primary aim of computer science in office environments. The long-term aim is the *paperless office*, whose implementation in the near future still seems unrealizable. In the paperless office, documents will exist in electronic form and be accessed in a way that is easy and "personal" to each user. Many pilot systems have been implemented, studied and evaluated by researchers without significant positive results so far. The most difficult problem today seems to be that of the integration and collaboration of different applications that manage heterogeneous documents.

One solution to the problem of document management that has been proposed recently is the use of *intelligent agents*. An intelligent agent in this context can be perceived as a software entity that mediates between a user and a software system and undertakes tasks that the software system cannot fulfill on its own. The use of an agent as a mediator facilitates and simplifies a user's job and therefore increases his productivity. Agents provide an elegant solution to the problem of integrating heterogeneous forms of information and incompatible applications for document management and dispatching [2, 3]. Related research has proved that they can be used as a means towards the automation of various tasks that are performed in an office (e.g. management of electronic mail [4] and electronic news [5], scheduling of meetings [6], database and library management, etc.) and in the field of education [7]. Furthermore, a significant part of related literature is concerned with the specification of protocols for communication between intelligent agents, as well as their implementation [8].

One particular subproblem of electronic document management is that of *content-based document filtering*. A category of intelligent agents that is strongly related to this subproblem is that of *filtering agents*. Such agents filter information arriving in the form of electronic documents and present to the user only those that the user considers as interesting, while rejecting useless to the user information. Filtering agents usually co-operate with knowledge bases that contain the user's filtering criteria. The intelligence of filtering agents is characterized by the fact that users do not need to explicitly specify the filtering criteria. During a training period, a filtering agent automatically learns the filtering criteria of a particular user under his indirect guidance. Given the rapidly increasing quantity of information that is circulated in electronic form, as well as the increasingly specialized content, filtering agents are becoming an extremely useful tool for efficient access to sources of information. In this chapter we suggest a model for content-based electronic document filtering that is based on a multi-agent system.

2. Definition of the problem

The filtering of conventional printed documents is based on their *contents* and not on external characteristics, such as color, texture or number of pages. The problem of *content-based electronic document filtering* is interesting from a researcher's point of view. It is desirable to develop a system that will automate this process and act as a mediator between the source of information and its human targets. The system should possess adequate intelligence and allow the coding, maintenance and update of the filtering criteria of each of its users. The model that we suggest in the following section attempts to solve this problem. At the same time, we see fit to deal with other related problems, such as *locating* and *collecting* information, its *archiving* and the *exploitation of existing archives*.

One of the basic requirements that a system for automatic filtering of electronic documents is document form *independence*. Total independence may not be feasible; it is at least desirable, however, that users perceive the filtering process in a unified way.

The need for *personalizing* the system for each user results in a model that is very different from those of typical information retrieval systems. The difference lies mainly in the fact that filtering criteria must not stem from an isolated query event, but must result from a period of interaction with the system. The filtering system must be able to cope with long-term changes, based on chains of such events. The issues of *training* and *adaptability* are therefore of great importance also. Besides, the system must be able to selectively forget acquired knowledge, in case the user's interests change. For all these reasons, intelligent agents are an appropriate solution for dealing with all these peculiarities.

In the model's definition, emphasis must be given on the issues of agent *competence* to correctly automate the required tasks and on their *believability*, taking into consideration user psychology and giving the users a feeling of control over the filtering process. The issue of *communication* between agents and users is also very important. The existence of a pleasant and flexible working environment with a user-friendly interface, capable of dynamically adapting to the personal requirements of a user, is an important require-ment. So is an agent requirement for *automatic training*, which can be realized in various ways, such as observations of the user's actions, use of examples and counterexamples, evaluation of user feedback, or through direct user guidance.

3.   Description of the model

As a solution to the problem of content-based electronic document filtering, we suggest a multi-agent model that mainly consists of four intelligent agents and a knowledge base. A schematic overview of the model is shown in Fig. 1. The part of the figure that is enclosed in the dotted line deals exclusively with the filtering of documents. The rest of the figure deals with auxiliary functions that implement necessary parts of the desired automation.

At the two ends of the suggested model, in the upper-left and lower-right corners, are located respectively a source of information and the user. The model, as has already been mentioned, specifies a system that mediates between these two ends. Key positions in the suggested model are occupied by the four intelligent agents, which are shown in the figure as light bulbs. The agents are collaborating software entities that implement the system's functionalities. They are described in detail in the following paragraphs.

In the description of the suggested model, emphasis has been given on its simplicity, so that the model is well defined and can be checked against its specifications. For this reason, the following two simplifications have been made:

(A1) The model supports a single user and a single category of interests for this user.
(A2) The process of header construction, that will be described in the next section, is not
       a user-specific process and does not depend on the contents of the knowledge base.

Directions on how to remove these simplifications are discussed in section 5.

3.1.   HEADER CONSTRUCTION

In order to deal with the heterogeneity of electronic documents and to simplify the fil-tering process, it is necessary to introduce an intermediate stage of processing, during

Figure 1: Overview of the model.

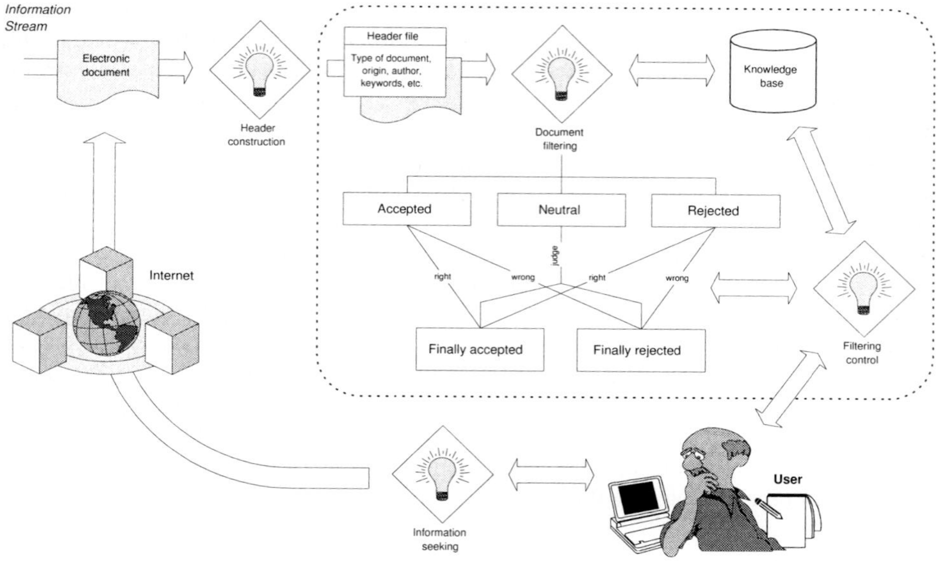

which a *header file* is constructed for every electronic document (header files will be also called *headers* for brevity). The format of headers is common for all forms of electronic documents. The task of header construction is similar to the extraction of appropriate *metadata* from electronic documents [9]. In the proposed model, it is performed by the *header construction agent*, which needs not interact with the user. Following this stage, the filtering process is based solely on headers.

The format of the headers must be general enough to describe accurately the key elements of the documents which are useful for filtering. A header contains a set of entries, each one of which consists of a field and its associated value. A *field* can be an arbitrary sequence of letters and numbers, whereas a *value* can be an arbitrary sequence of characters, surrounded by double quotes. An example of a header describing an electronic announcement is given below:

origin	=	"Reuter News Agency"
author	=	"Alan Smith"
title	=	"New explosion in London underground"
date	=	"December 12, 1996, 17:55 GMT"
category	=	"Foreign current events"
keywords	=	"IRA, explosion, underground, terrorism, casualties, statements"
language	=	"English"
wordcount	=	"1340"

The way in which headers are constructed depends a lot on the application area. It is not possible to predefine the names of fields or their values in a unified way that is appropriate for all application areas and document forms.

The automatic construction of headers from heterogeneous electronic documents is a very hard problem and is not expected to be solved in its general form in the near future. Before this happens, computers must be made capable of understanding all forms and media of electronic documents (text, images, sound, video), which is not feasible with the current state of technology. However, satisfactory approximations have been developed for the case of text documents, which comprise a significant percentage of information available in electronic form. Furthermore, some electronic documents are accompanied by descriptive information in textual form. With all this in mind, automatic construction of headers should not be seen as a utopian aim. If the header construction agent is not able to perform its duty, for one reason or another, human intervention is necessary.

Currently, significant research is conducted in the field of *semi-structured data*, aiming at the extraction of information and the automatic classification of electronic documents. So far, promising results have been achieved and the reader is referred to [10, 11, 12, 13] for a summary. The present work differs mainly in two respects: it allows more freedom of form and content in the extraction of information from electronic documents and it uses an intelligent agent with automatic learning capabilities in the extraction process.

3.2. DOCUMENT FILTERING

The filtering of documents in our model is based on a set of subjective criteria that are specified (mostly indirectly) by the user. We assume that the already constructed headers for the new documents that are about to be filtered have been placed in an *arrival area*. Each header is examined by the *filtering agent*, which finally assigns to it a grade of interest, which will subsequently be called *score*.

The score is a real value in the interval $[-1, +1]$. The value $+1$ corresponds to documents that are considered interesting without any doubt and are therefore accepted for presentation to the user, whereas the value -1 corresponds to documents that are rejected with no doubt. Intermediate values represent the agent's degree of confidence, as far as the filtering process is concerned. The value zero corresponds to documents for which the agent cannot make a confident decision.

After the filtering agent comes up with a score for a particular document, based on the document's header and following an algorithm that will be described in detail in §3.4, the agent classifies the document in one of three possible ways, that determine the destination area of each document. *Accepted documents* are those with score near $+1$, *neutral documents* are those with score near 0 and *rejected documents* are those with score near -1. The values for the two thresholds that separate the three areas must be specified directly by the users, according to their needs and to the degree of fidelity that is sought for the classification process.

A necessary property of the filtering agent is the knowledge of its competence. The agent must be able to estimate its critical capabilities, based on the degree of its training and, mostly, on the results of its previous judgements, as is described in detail in §3.5. This property can be reflected in the model by a real value in the interval $[0, 1]$, representing the degree of the agent's *confidence* in itself. A value of 0 means that the agent does not trust itself at all and should therefore not attempt any classification, whereas a value of 1 means that the agent has total confidence in its decisions. In order to take into account

80

this "hesitation" that is displayed by the agent, our model multiplies the score of each document by the degree of confidence. Following that, the filtering agent classifies the documents based on the product of these two values.

3.3. KNOWLEDGE BASE

The knowledge base contains the criteria for the filtering of documents, coded in an appropriate form. These criteria are specified by the user gradually and mostly in an indirect way. The coding of the criteria must be invisible to the user, if we expect the system to be user-friendly. Furthermore, the form of the criteria must be at least as general as the form of the headers, since they will be applied to them.

In the suggested model, filtering criteria are stored in the knowledge base in the form of *rules*. Each rule consists of a condition and a value of interest. Before a rule is applied to a document's header, it is first examined whether its condition is true. If it is, the score of the document is updated by including the rule's value of interest, otherwise nothing is done. This process is described in detail in §3.4.

The *condition* in a rule is a logical expression that contains field names, which may appear in the document's header, as well as constant values. Useful cases of simple conditions aim at checking whther the value of a specific field in a document's header is equal to (or contains) a specified value. It is also useful to be able to combine simple conditions in more complex ones. The result of applying a rule to a document's header (if there is one) is to attribute a *value of interest* to the document. Such values are elements of a set which will be called $\mathcal{E}$ in the rest of the chapter. The form and properties of this set's elements will be described in detail in §3.4.

The description of the representation of the rules in the knowledge base is outside the scope of this model discussion. It is worth mentioning, however, that realistic implementations of the model would give special emphasis on efficiency issues related to the access of the knowledge base. This becomes critical as the volume of information that is stored in the knowledge base increases.

3.4. SCORE CALCULATION

The calculation of the score that corresponds to a document uses as input the document's header and the knowledge base that contains the required filtering criteria. The score comes up from the application of all rules that are contained in the knowledge base. Each rule whose condition is true for the given document contributes a corresponding value of interest to the total score. The application of all rules may result in a sequence of different interest values, that must be appropriately combined. Our model discriminates between two different kinds of combining the contributed values, which finally lead to a classification of rules in two categories.

The obvious way of combining contributed values is to take all of them into account. Let us assume that the application of all rules to a given document results in the sequence $\langle a_1, a_2, \ldots, a_m \rangle$, where each a_i is a contributed value of interest and an element of the set $\mathcal{E}$. Then, the formula for the calculation of the total score is the following:

$$score \;\; = \;\; \| \, a_1 \oplus a_2 \oplus \ldots \oplus a_m \, \|$$

Figure 2: Properties of operators used in the combination of values of interest.

$a \oplus (b \oplus c) = (a \oplus b) \oplus c$	$\|a\| = \|b\| \Rightarrow \|a \oplus b\| = \|a\| = \|b\|$
$a \oplus b = b \oplus a$	$\|a \oplus \bar{a}\| = 0$
$a \oplus \mathbf{O} = a$	$\bar{\mathbf{O}} = \mathbf{O}$
$-1 \leq \|a\| \leq +1$	$\overline{a \oplus b} = \bar{a} \oplus \bar{b}$
$\|\mathbf{O}\| = 0$	$\|\bar{a}\| = -\|a\|$
$\|a_1\| \leq \|a_2\| \Rightarrow \|a_1 \oplus b\| \leq \|a_2 \oplus b\|$	

In this formula, operator $\oplus$ represents the combination of two values of interest, resulting in a third value, whereas operator $\|\cdot\|$ translates a value of interest (i.e. an element of $\mathcal{E}$) to a value in the interval $[-1, +1]$ of real numbers. The presence of an operator that negates values of interest is also necessary. This operator will map value a to its *negative* value, denoted by $\bar{a}$. The choice of operators $\oplus$ and $\|\cdot\|$, as well as of the negation operator must be made in an appropriate way, so that some properties are satisfied. The desired properties that these operators must have are shown in Fig. 2.

An obvious choice for the definition of set $\mathcal{E}$ is the interval $[-1, +1]$ of real numbers, which renders the presence of operator $\|\cdot\|$ unnecessary. Unfortunately, this choice makes it impossible to define the other operators in such a way as to satisfy all the desired properties.[1] For this reason, $\mathcal{E}$ must be defined in a different way. Each element a of $\mathcal{E}$ is taken to be a pair of the form $\langle x, n \rangle$, where x is a real number in the interval $[-1, +1]$ and n is a natural number. The value of x represents a value of interest. The value of n expresses the *multiplicity* of the value, i.e. the weight that this value carries. It is easy to confirm that the operators defined as follows satisfy the desired properties.

$$\langle x_1, n_1 \rangle \oplus \langle x_2, n_2 \rangle = \begin{cases} \left\langle \dfrac{n_1 x_1 + n_2 x_2}{n_1 + n_2}, n_1 + n_2 \right\rangle & \text{, if } n_1 + n_2 \neq 0 \\ \langle 0, 0 \rangle & \text{, if } n_1 + n_2 = 0 \end{cases}$$

$$\|\langle x, n \rangle\| = x$$

$$\overline{\langle x, n \rangle} = \langle -x, n \rangle$$

$$\mathbf{O} = \langle 0, 0 \rangle$$

In other words, operator $\oplus$ is the weighted mean of the two values of interest.

It is sometimes useful to have certain values of interest that prevail over others. We can thus classify values in two categories: the "privileged" ones and the "common" ones. If the sequence of contributed values for a given document contains only common values, the formula that was given above can still be used. The same is true if only privileged values are present. However, if the sequence contains values from both categories, the total score is computed based only on the privileged values using the same formula; common values are ignored. An example of using such privileged values of interest is a hypothetical user's request to always reject documents written by a certain author, e.g. because the

---

[1] This choice makes it necessary to define $\bar{a}$ as $-a$, and then the associativity property for $\oplus$ cannot be satisfied, if one takes $a = b = +1$ and $c = -1$.

Figure 3: Algorithm for content-based filtering.

```
for each header H in the area of arrival do
    score := unspecified
    for each rule K in the base of authoritative rules do
        if the condition of K is true for H then
            score := max{ score, interest value of K }
    if the score is still unspecified then
        score := O (neutral)
        for each rule K in the base of additive rules do
            if the condition of K is true for H then
                score := score ⊕ interest value of K
    if ‖ score ‖ > threshold for acceptance then
        place the document in the area of accepted documents
    else if ‖ score ‖ < threshold for rejection then
        place the document in the area of rejected documents
    else
        place the document in the area of neutral documents
```

author is not reliable, even if there are many other reasons to accept them.

It would be rather difficult to include this requirement in the mathematical model for values of interest. For this reason, it is not the values that are classified as privileged or common, but the rules that introduce them are classified in two categories: *additive* rules, which introduce common values, and *authoritative* rules, which introduce privileged values. In order to make this distinction clear, in our model we store the rules in two separate knowledge bases. The complete algorithm for the score calculation that is used in our model is shown in Fig. 3.

3.5. FILTERING CONTROL

The process of filtering control is performed by a specialized agent that interacts heavily with the user. After all, the user is the only one who can finally decide whether a document is interesting or not. For this reason, the user is called to check the decisions made by the filtering agent and to classify the documents that were placed in the three aforementioned areas in two new areas: the area of *finally accepted documents* and the area of *finally rejected documents*. With the new classification every combination of areas for a document transitions is possible. As the knowledge base is enriched, it is expected that transitions between areas of differing degree of interest will not be frequent. In the ideal case, the area of rejected documents will only contain uninteresting documents and the user will not have to check it at all, the area of accepted documents will only contain documents that the user will eventually find interesting, and the neutral area will contain as few documents as possible.

Filtering control is divided in two parallel processes: The *training process* aims at the update and enrichment of the knowledge base with the filtering criteria that the user actually applies. In this phase, the user classifies the documents that the agent has placed in the neutral area, unable to correctly classify them. The user has the option to indirectly indicate the criteria according to which he makes the classification by means of a friendly user interface. In this way, the user interacts naturally with the knowledge base by adding or updating rules, without knowing the form and the coding in which these rules exist in the knowledge base. On the other hand, the *error correction process* aims at the correction of errors that were made by the agent in the classification of documents. During this phase, the user classifies the documents that the agent has placed in the other two areas. This process does not affect the knowledge base, but only the filtering agent's degree of confidence.

The process of filtering control may result in changes in the system's state, that is, the knowledge base and the degree of confidence. Such changes will affect the outcome of future document classification. Implementations may provide the option of re-evaluating documents that have already been classified, according to the new state. Apart from that, our model does not address the issue of consistency between the current classification and the current state. In general, classification of a document is bound to be consistent with some state that has existed in the past, but not necessarily the current one.

3.6. INFORMATION SEEKING

The need for content-based document filtering would not be as imperative if the stream of information contained interesting documents with a high probability. This probability can be increased if the users are able to indicate to the system the sources that generate information that is interesting to them. Subsequently, the system must be able to direct information from these sources to the stream of information that reaches the users.

This process is performed by the *information seeking agent*. This agent interacts with the user and, by means of examples, observation or direct guidance, locates the information sources that the user finds interesting and reliable. Then, the same agent undertakes the collection of information from these sources in frequent or prearranged time intervals. The information seeking agent is not a part of the document filtering system. However, its general description is contained in the model because the problem that it tries to solve is very closely related to that of document filtering.

3.7. DISTRIBUTED DOCUMENTS

Efficient access of documents that are distributed over an intranet or the Internet is a key issue in all realistic implementations of our model. The client-server architecture has been successfully used in similar applications and is therefore considered as the most appropriate for implementing our model. In such implementations, servers would typically be responsible for the collection of documents, header construction and information seeking, whereas clients would request and filter document headers, request documents upon user's guidance and contribute to the information seeking process.

With all this in mind, our model in Fig. 1 can be divided in two parts: the client part, contained in the dotted rectangle, and the server part. In a multi-user implementation of our model, the filtering agent and the knowledge base could be part of a second kind of server.

4. Implementation

The suggested model has been used for the partial implementation of an experimental system for electronic document filtering in the environment of a journalist's office. The system, which is named ALEC, is hosted over a local area network of personal computers running Microsoft Windows with direct connection to the Internet and has been developed using Microsoft Visual C++. Apart from the aforementioned simplifications (A1) and (A2), the following limitations were also imposed on ALEC's implementation:

(A3) The header constructing agent can only manage documents of a given form.
(A4) The types of conditions supported are the simplest possible.
(A5) The information seeking agent has not been implemented.

ALEC's capabilities are still limited and the process of updating the selection criteria during filtering control is not yet as indirect as it is desired. The process of header construction is still premature.

A snapshot from ALEC's use is shown in Fig. 4, where, in the process of filtering control, ALEC shows the user a document that he could not classify. The document received a total score of 0.20. From the list of fields, the user can see that an interest value of 0.70 was attributed to the document because its subject was music (apparently ALEC has decided that the user is interested in music) and a negative value of -0.30 because the artist that it deals with is Johann Sebastian Bach (which ALEC considers not one of the user's favourites). The user can direct ALEC to change his selection criteria through this dialog box.

The evaluation of the experimental implementation is not yet complete and there are no trustworthy and adequate in volume performance results from its use. As far as this implementation is concerned, future research will aim at its improvement, on the one hand, and on its complete evaluation in a real environment, on the other.

5. Future directions

The primary target of our future research is the evaluation of our model with the development of realistic implementations. Useful results for the future improvement of our model may also stem from research in the field of *full-text retrieval*. If the filtering agent is capable of searching the full text of the electronic documents, it is possible to specify conditions that are based in the presence and the relative position of words in the text. The technology of full-text retrieval applications is already widely used by various information-seeking agents in the World Wide Web, with very encouraging results. However, as far as our filtering model is concerned, we do not see fit to abolish the stage of header construction.

In order to overcome limitation (A1), additional work is necessary, aiming at the support of multiple knowledge bases that will either belong to different users, or represent different interests of the same user. A possible direction that will be considered is that of integrating all knowledge bases in a global knowledge base, which may also contain generally accepted selection criteria, as well as common characteristics of the system.

Figure 4: Snapshot from ALEC's dialog for filtering control.

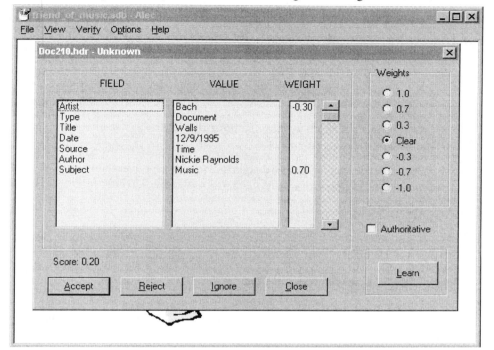

Finally, to overcome limitation (A2), further research will establish an interaction between the header construction agent and the knowledge base, allowing the construction of headers to reflect the user needs. The possibility of creating personalized headers for each user, according to the their individual selection criteria that are contained in the knowledge base, will also be examined.

6. Conclusion

In this chapter we propose a model for automating the tedious and time-consuming process of content-based electronic document filtering. The model is based on a set of collaborating intelligent agents, which mediate between the sources and the targets of information. Emphasis is given the homogeneous treatment of electronic documents, personalization, automatic training, dynamic adaptation to the user needs and believability.

Despite the fact that an implementation of the suggested model is still in a premature stage and that no substantial and trustworthy results from its use are yet available, the first experiences from its application to a journalist's office are encouraging. We believe that, in the future, intelligent agent technology will provide adequate solutions to a lot of problems concerning the automation of tasks that currently make us less productive, including the content-based filtering of electronic documents.

Acknowledgements

This research work is funded by the Greek General Secretariat of Research and Technology, as part of the Programme for Supporting Research (ΠΕΝΕΔ 95) — Project #463.

References

[1] A. Reinhardt. Managing the new document. *BYTE*, pages 91–104, August 1994.

[2] P. Maes. Agents that reduce work and information overload. *Communications of the ACM*, 37(7):31–40, 146, July 1994.

[3] M. R. Genesereth. An agent-based approach to software interoperability. In *Proceedings of the DARPA Software Technology Conference*, 1992.

[4] Y. Lashkari, M. Metral, and P. Maes. *Collaborative Interface Agents*. MIT Press, Cambridge, MA, 1994.

[5] B. Sheth and P. Maes. Evolving agents for personalized information filtering. In *Proceedings of the Ninth Conference on Artificial Intelligence for Applications*, pages 345–352. IEEE Computer Society Press, 1993.

[6] R. Kozierok and P. Maes. A learning interface agent for scheduling meetings. In *Proceedings of ACM SIGCHI International Workshop on Intelligent User Interfaces*, pages 81–88, New York, NY, 1993. ACM Press.

[7] T. Selker. Coach: A teaching agent that learns. *Communications of the ACM*, 37(7):92–99, July 1994.

[8] Y. Shoham. Agent-oriented programming. *Artificial Intelligence*, 60(1):51–92, 1993.

[9] The Dublin Core Metadata Element Set. URL: http://purl.org/metadata/dublin_core/.

[10] The Rufus System. URL: http://www.almaden.ibm.com/cs/showtell/rufus/.

[11] K. Shoens, A. Luniewski, P. Schwarz, J. Stamos, and J. Thomas. The Rufus system: information organization for semi-structured data. In *Proceedings of the 19th Conference on Very Large Databases (VLDB '93)*, pages 97–107, Dublin, Ireland, 1993.

[12] M. Tresch, N. Palmer, and A. Luniewski. Type classification of semi-structured documents. In *Proceedings of the 21st Conference on Very Large Databases (VLDB '95)*, pages 263–274, Zurich, Switzerland, 1995.

[13] D. Smith and M. Lopez. Information extraction for semi-structured documents. In *Proceedings of the Workshop on Management of Semistructured Data, in conjunction with PODS/SIGMOD*, Tucson, AZ, USA, May 1997.

8

An Array Architecture for Syntactic Pattern Recognition

Andreas Koulouris, Nectarios Koziris,
George Papakonstantinou, Panayotis Tsanakas

National Technical University of Athens
Computer Systems Laboratory,
Department of Electrical and Computer Engineering
Zographou Campus, 15773, Athens, Greece

1. Introduction

Syntactic methods are an important tool for tackling recognition and classification problems. In areas such as the analysis of natural languages, or speech recognition, the syntactic approach gains wide interest by the scientific community. The complexity of the patterns and the wide variety of the features that characterize them, makes impractical the use of common decision-theoretic approaches [1]. In many applications, such as the processing of biomedical signals (e.g., ECG, EEG, e.t.a.), syntactic methods have been considered as an alternative way of simultaneous detection and classification of some characteristic complexes of these signals.

The kernel of all such methods is the parsing algorithm, which is responsible for analyzing the complex object into a sequence of primitive patterns. The complexity of the parsing procedure depends on the class of the underlying grammar. Context-free grammars combine satisfactorily both the expressive power and the simplicity in their analysis. They can describe many features of natural languages and are widely used in syntactic pattern recognition applications.

Many efficient parsing algorithms have been developed for this specific class of grammar [1-7]. Earley's algorithm [4] is the fastest sequential algorithm that can parse a CF grammar, without requiring to be in Chomsky or any other normal form. The time complexity of this algorithm, as Graham et al. showed in [6], is $O(n^{2.81})$, which can be a significant overhead for reasonably large n. Consequently, the efficient parallelization of Earley's algorithm is of particular importance to the above areas. One of the most known parallel implementation of Earley's algorithm in the literature is that of Chiang and Fu [8, 9]. Chiang and Fu proposed a 2-D VLSI array having $O(n^2)$ array cells and $O(n)$ processing time.

On the other hand, the feasibility of the automatic parallelization of some special classes of sequential algorithms like nested DO (FOR) – loops has been examined in [10-14]. The minimum parallel time of execution as well as, upper and lower bounds for the number of processors needed to achieve that time, have also been investigated for this special case, providing with optimal methods [10, 11, 16].

We indicate that, Earley's algorithm is very close to the aforementioned form of nested loops, and we implement it into special purpose hardware, using an optimal method proposed by Andronikos and Koziris in [10]. This method makes an efficient mapping of the loops iterations, using the less possible processing elements. We implement Earley's algorithm into an one-dimensional architecture, consisted of O(n) processing elements, reducing significantly the hardware complexity. Furthermore the method can be automatically applied using an appropriate tool like the one presented in [17].

2. Basic Concepts

We give briefly the definitions and the notation, used throughout this paper

Definition 2.1. A Context Free Grammar (CFG) is a quadruple G=(V, N, P, S), where:
- V is the set of symbols of the grammar,
- N is the set of non-terminal symbols (T=V-N is the set of terminal symbols),
- $P \subseteq N \times V^*$ is the set of rules of the grammar, which are of the form A→a, where A∈N and a∈V^*. We call the symbol A, the left symbol of a rule and the symbol(s) α the right part
- S is the start (non-terminal) symbol of the grammar.

Definition 2.2. An input string w=$a_1 a_2 ... a_n$ (possibly empty) is generated from G if, starting from the start symbol S and successively applying any of the rule of the grammar, string w can be derived (i.e., S⇒*w).

Definition 2.3. A recognizer decides whether an input string is generated from a grammar G.

Definition 2.4. Let '•' be a symbol not in V. Then a rule A→α•β (A→αβ is in P) is called "*dotted rule*" and means that the α part of the rule has been found consistent with input string, while the β part still needs to be considered.

Definition 2.5 We call *PREDICT(B)* B⊆V the set: {C→γ•δ | C→γδ is in P, γ⇒*ε, B⇒*Cη for some B in R and some η}

Definition 2.6 We call *predecessor(A)*, A∈V* the set: {B | B⇒*A, B∈N}, that is the set of all the symbols that generate A

Many versions of the initial Earley's algorithm can be found in the literature [1-9). We use the form presented by Fu in [8]. This algorithm builds a recognition matrix T(i,j) whose elements are sets of dotted rules. If at element t_{0n} there is a dotted rule of the form S→α•, then we say that the input string is generated by the corresponding grammar.

Algorithm 2.1 (Earley modified by Chiang & Fu)

```
FOR j=1 to n do
  t_{j-1,j} =Y⊗{a_j}
FOR j=2 to n do
 begin
  FOR i=0 to j-2 do
   t_{ij}=t_{ij}⊗{a_j}
  FOR k=j-1 down to 0 do
   FOR i=k-1 down to 0 do
    t_{ij}=t_{ij}∪t_{ik}⊗t_{kj}
end
IF   there is a rule S→α• in t_{0,n} accept the input string
```

In the previous algorithm Y=PREDICT(N) and the operation $\otimes$ is defined as follows:

- Let Q be a set of dotted rules and R $\subseteq$ V

$$Q\otimes R=\{A\to\alpha U\beta\bullet\gamma \mid A\to\alpha\bullet U\beta\gamma\in Q,\ \beta\Rightarrow^*\lambda,\ \text{and } U\in R\}$$
$$\text{and}$$
$$\{B\to\delta C\xi\bullet\eta \mid \gamma=\lambda,\ B\to\delta\bullet C\xi\eta\in Y,\ \text{and } \xi\Rightarrow^*\lambda, C\Rightarrow^*A\}$$

- Let Q,R be a set of dotted rules

$$Q\otimes R=\{\{A\to\alpha U\beta\bullet\gamma \mid A\to\alpha\bullet U\beta\gamma\in Q,\ \beta\Rightarrow^*\lambda\ \text{and } U\to\delta\bullet\in R\}$$
$$\text{and}$$
$$\{B\to\delta C\xi\bullet\eta \mid \gamma=\lambda,\ B\to\delta\bullet C\xi\eta\in Y,\ \text{and } \xi\Rightarrow^*\lambda, C\Rightarrow^*A\}$$

3. Data Representation

The main characteristics of a VLSI array are first its synchronous and regular flow of constant length data among neighboring cells, and second the same simple and regular internal structure of each cell. In order to fulfilled these requirements we should represent the array elements t_{ij} in a compact form. In addition to this, the internal $\otimes$ operator should be as simple as possible.

Recalling the definition of the $\otimes$ operator, one can see that it operates on a set of rules. Specifically the operation Q$\otimes$R is divided into the following steps:

1.   The set of all the left symbols of the rules in R (in which all the right part has been read) is calculated.

2.   All the rules in set Q, which contain any element of the above set at the right of the dot are found. In these rules, the dot is moved one place to the right.

3.   All the rules in set Q, in which all the right part has been read, are found. The corresponding set   of the predecessors of their left-hand symbol is computed.

4.   Finally,. all the rules in the set PREDICT(N) are computed in the same way as in step 2.

The result is the union of the sets of the rules found in steps 2 and 4. A similar procedure is executed when R is a set of symbols and not a set of rules.

In general, the above steps have different execution times, depending on the index of the elements Q, R. This problem is solved if we represent the grammar with bit-vectors as it was proposed by Chang & Fu in [11].

In the following, a formal description of the implementation of the operator ⊗ in an PASCAL-like algorithm is given, which can be used in a preprocessing level as input to an automatic hardware synthesis tool. Moreover, the proposed implementation allows the grammar to have also ε-productions. (In [11] there was the restriction of an ε-free grammar as input). The removing of the ε-productions may duplicate the rules of the grammar, thus duplicating the length of the data that travel through the VLSI array. Since, in our implementation ε-productions are also allowed, the overall execution time is significantly reduced.

The implementation of the operator ⊗ is illustrated as follows: If the grammar has k symbols, each symbol is encoded into a non-zero k-bit vector. In this encoding every k-bit vector differs only in one bit from the others. The set of the predecessors for each non-terminal is similarly encoded into a k-bit vector. The value of this bit -vector is derived by or-ing all the bit-vectors, that represent the symbols, which belong to this set. For each non-terminal symbol we use its encoding and the encoding of the set of its predecessors. If the grammar has n non-terminal symbols, we store at each cell an array containing $2n$ k-bit vectors. We use the notation *symb[i]*, to point the i-th symbol in this 2n array. Each rule of the grammar is encoded as an array of k-bit vectors. The first bit-vector is the LHS of the rule and the others the RHS. If the grammar has p rules and each rule has at most r symbols at the right part, then the rules of the grammar are stored in a $p(r+1)$ array of k bit–vectors. We use the notation *rule[i]*, to point the position of the i-th rule (i.e. its LHS). Finally we store at each cell three arrays of p (r+1)bit-vectors. Each row in these matrices represents a rule (only the RHS part), and each bit in the (r+1)-bit vector a position of the dot. The first matrix contains the set Y=PREDICT(N). In the second matrix (called M) the bit which denotes the end of the RHS of the corresponding rule is marked in each bit-vector. The third matrix is called *'empty'* and it is symbolized as E. In each row of E, we mark the bit(s), which correspond to symbols that produce the empty string. This matrix is used to transform dynamically the grammar into an equivalent ε-free form, without increasing the size of the main cell. Finally the elements of the parsing matrix T have the same form as the matrices Y,E,M (i.e. a p(r+1) array of bits).

The algorithm is executed in a 5-step procedure. It takes as input the sets Q, R in the form of a p(r+1) bit arrays and calculates the set (matrix) T = Q⊗R. We use the internal variables U, PRED of the form k-bit vector and B of r+1 bit-vector with initial value 0. Finally with A (a k-bit vector) we symbolize the representation of a symbol of the input string. In the following, algorithm the operator ∧ symbolizes the vectorial AND where:

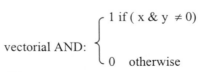

$$\text{vectorial AND:} \begin{cases} 1 \text{ if } (x \,\&\, y \neq 0) \\ 0 \quad \text{otherwise} \end{cases}$$

Algorithm 2 (operator ⊗)

```
/* step1 */
FOR i=1 TO p
   U=U OR rule[i]*(R[i] △ M[i])
U=U OR A
/* step2 */
FOR i=1 TO p
 begin
   B=Q[i]
    FOR j=1 to r
     B[j]=B[j] * (rule[i+j] △ U)
   T(i)=B>>1;
    FOR   j=2 to r+1
    T(i,j)=T(i,j)OR T(i,j-1)*E[i,j]
 end
/* step 3 */
U=0
FOR i=1 to p
 U=U OR rule[i]*(T [i] △ M[i])
/* step 4 */
FOR i=1 to n
   PRED=PRED OR symb[i+1]*(symb[i] △ U)
/*   step 5 */
FOR i=1 TO p
 begin
    T(i,2)=T(i,2) OR (rule[i+1]△PRED)
    FOR   j=2 to r
      T(i,j)=T(i,j) OR T(i,j-1)*E[i,j]
 end
```

Notice, that with the above representation a null bit-vector represents the 'neutral' element for the operator $\otimes$. This gives us the possibility to preserve the regular data flow even if, at some time units we transmit null values. All bit operations can be implemented in parallel, implying even the lowest level inherent parallelism of the algorithm.

4. Dependence Analysis

We rewrite the algorithm 2.1 in the form of a 2-D nested for loop. The new algorithm is:

Algorithm 4.1

```
FOR i=1 TO n
 FOR j=i-1 TO 0
   t(i,j)=t(i-1,j)⊗aᵢ ∪ t(i-1,j)⊗t(i,i-1)⊗ ∪...∪ t(j+1,j)⊗t(i,j+1)
 END
END
```

The equivalence of the two algorithms is obvious. The new algorithm just builds the recognition matrix row by row so the latest element that will be computed is t(n,0] instead of t(0,n].

For the efficient parallelization of the algorithm 4.1 we use the method proposed by Andronikos et al. in [10]. This method partitions the index space of the above double nested loop into the less possible disjoint chains of computations. After the partitioning phase, it assigns each chain to a different processing element, while preserving the dependence relations between the loop iterations. The resulting schedule is proved to be optimal both in terms of time and processor utilization.

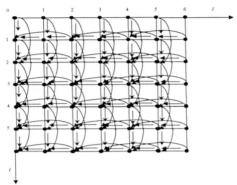

Figure 4.1 The Index Space and the dependence vectors for algorithm 4.1 and for n=6

In our algorithm the set of all possible dependence vectors (called Dependence Set, and symbolized DS), which relate different loop iterations is :

$$DS=\{ \; d_i\,(i\text{-}k,0), \; d_j\,(0,j\text{-}k) \mid 1\le i\text{-}k \le n, \; 1\text{-}n \le j\text{-}k \le 0 \;)$$

From the above set it is clear that satisfying dependencies $d_1(1,0)$ and $d_2(0,-1)$ all other dependencies will automatically be satisfied. This is illustrated in figure 4.1 where the index space and the dependence vectors are presented (We have drawn only the first two dependence vectors in order to avoid a very complex pattern).

5. Mapping into 1-d VLSI array

Unlike the empirical approach by Fu in [8], a systematic way, for mapping algorithm 4.1 to hardware, is used. More specifically, we applied the method [10], which leads to optimal time and space schedules. Table 5.1 presents, the time-schedule for n=6. Processor P_i is responsible for the computation of all the elements in the i-th column of the recognition matrix t.

It is clear that, by applying the above mapping, the proper data flow is ensured.

For example, the computation of element t(4,0) will be done by processor P_1 at the 4-th time step. From algorithm 3 we see, that for the computation of the element t(4,0), the values of the elements t(4, k), t(k, 0) (j $\le$ k $\le$ i-1) are needed. By this time, all these elements (i.e. t(1,0), t(2,0), t(3,0), t(4,1), t(4,2), and t(4,3)) have already been computed and sent to processor P_1.

Time	Processing Elements					
	P_1	P_2	P_3	P_4	P_5	P_6
1	t(1, 0)	t(2, 1)	t(3, 2)	t(4, 3)	t(5, 4)	t(6, 5)
2	t(2, 0)	t(3, 1)	t(4, 2)	t(5, 3)	t(6,4)	-
3	t(3, 0)	t(4, 1)	t(5, 2)	t(6,3)	-	-
4	t(4, 0)	t(5, 1)	t(6,2)	-	-	-
5	t(5, 0)	t(6,1)	-	-	-	-
6	t(6,0)	-	-			

Table 5.1. Time schedule for n=6. Each column of the matrix T is assigned to a different cell

Assigning the computation of each column of the recognition matrix to the same processor, leads to one-dimensional VLSI architecture of n processors. This architecture is illustrated in figure 5.1. Each link is used to transfer both the element t(i,j) and the input symbol a_i, within one time step (Recall from section 3 that both are represented as bit-vectors). The input string $S=a_1a_2...a_n$ is initially loaded, in parallel, to processing elements, so that P_i processor has a_i character (Figure 2).

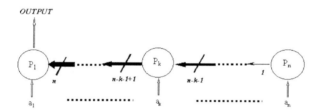

Figure 5.1: One dimensional architecture for the parallel implementation of Earley's algorithm

The number of operators implemented inside each cell is at most equal to the length of the input string as it is shown in figure 3. Most specifically, P_i processor contains i operators $\otimes$. With this architecture, we can obtain the result from the processor 1 at exactly n time steps.

The VLSI cell is illustrated in Fig 5.2. Since we represent the set of rules with an array of bit vectors, it is obvious that the operator $\cup$ can be implemented by OR gates.

Since each cell computes the corresponding t(i,j) within one time instance, we use an additional 1D delay to synchronize the transfer of data from input to output. Thus, the data from *Input$_i$* are forwarded to *Output$_i$* synchronously with the result from the $\cup$ operator. The P_k cell has n-k inputs and n-k+1 outputs, so that neighboring cells have the same number of interconnection links (Figure 5.1).

Finally there are n-k $\geq$ 1 registers which are driven by a control unit. The registers are loaded one by one each time step. Specifically, the output of the $\cup$ operator e is loaded to R_i at the i-th time instance.

Each cell performs the following operation:

$$Cell\ output = operation\ 1 \cup \ldots \cup operation\ k$$

$$where:\ operation\ i = R_i \otimes Input_i$$

The above interconnection strategy, implements the following relations:

$$Output_1 = Cell\ output$$

$$Output_i = Input_{i-1}\ for\ 2 \leq i \leq n- k+1$$

The proposed architecture has the following advantages over the so far presented implementations: First, it has a single time clocking and simple data flow. Second, it only needs one copy of the encoded grammar (symbols, rules, predecessors etc) inside each processing element. Finally, on the contrary with the 2-D architecture of [8], we have at the i-th time step, the result of the parsing of the substring $S^i = a_1a_2 \ldots .a_i$, as the output of the P_1 cell. This could be very useful in cases where, we want to collect information about a sub-string of the entire string (e.g. in pattern recognition problems).

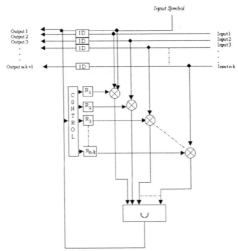

Figure 5.2.: The internal structure of the cell P_k.

6. Conclusion – Further Work

In this paper, we presented a new hardware architecture for the Earley's parsing algorithm by applying the automatic parallelization method presented in [10]. Most possibilities of parallelization of the algorithm were investigated, using the theory of dependence vectors, and the results have been used for the efficient utilization of the hardware. Future work includes the implementation of the presented methodology, in a real FGPA VLSI chip, using the VHDL tool and the environment presented in [17].

References

[1]   K. S .Fu "Syntactic Pattern Recognition and Applications" *Prentice-Hall 1982*

[2]   D. H. Younger "Recognition and parsing of context-free languages in time n^3" *Information and Control* Vol. 10 pp. 189-208 1967

[3]   V. Acho, J. D. Ullman "The theory of Parsing Translating and Compiling" Vol. I *Prentice Hall* Inc.

[4]   J. Earley "An efficient context-free parsing algorithm" *Commun. of the ACM* Vol. 13 pp. 94-102.

[5]   L. Valiant "General context free recognition in less than cubic time" *Journal of Computer and System Science* Vol. 10, pp. 308-315, 1975.

[6]   S. L. Graham, M. A. Harrison and W. L. Ruzzo, "On line context-free languages recognition in less than cubic time" in Proc, *8th Annu. ACM Symp. Theory of Comput.* , May 1976.

[7]   T. Kasami "An efficient recognition and syntax analysis algorithm for context free languages", Science Report AF CRL-65-758, Air Force Cambridge Research Laboratory, Bedford, Mass. 1965

[8]   Y. T. Chiang and K. S. Fu " Parallel parsing algorithms and VLSI implementation for syntactic pattern recognition" *IEEE Trans. Pattern Anal. Mach. Intell.* PAMI-7 (1985).

[9]   H.D.Chen, X..Chen "Shape recognition using VLSI Architecture" *The Int. Journal of Pattern. Recgn. and Art. Intell. 1993.*

[10]   T. Andronikos, N. Koziris, Z. Tsiatsoulis, G. Papakonstantinou, P. Tsanakas "Lower Time and Processor Bounds for Efficient Mapping od Dependence Algorithms into Systolic Arrays" *Journal of Parallel Algorithms and Applications* Vol 10 pp. 177-194, 1997

[11]   Andronikos, T., Koziris, N., Papakonstantinou, G., and Tsanakas, P. "Optimal Scheduling for UET/UET-UCT Generalized n-Dimensional Grid Task Graphs", *Proceedings of 11th IEEE/ACM International Parallel Processing Symposium (IPPS 97), pp. 146-151,* Geneva, Switzerland.

[12]   Bampis, E.,. Delorme, C., and Konig, J.C. Optimal Schedules for d-D Grid Graphs with Communication Delays. *Symposium on Theoretical Aspects of Computer Science (STACS96)*. Grenoble France 1996.

[13]   Koziris, N., Papakonstantinou, G., and Tsanakas, P. Automatic Loop Mapping and Partitioning into Systolic Architectures. *Proceedings of the 5th Panhellenic Conference on Informatics*, Dec. 1995, pp. 777-790, Athens

[14]   Lee, P.-Z. and Kedem, Z.M. Mapping Nested Loop Algorithms into Multidimensional Systolic Arrays. *IEEE Trans. Parallel Distrib. Syst.*, vol. 1, no. 1, pp. 64-76, Jan. 1990.

[15]   Moldovan, D.I. ADVIS: A Software Package for the Design of Systolic Arrrays, *IEEE Trans. Computer Aided Design*, vol CAD-6, no 1, pp. 33-

40, Jan. 1987.

[16]   Shang, W. and Fortes, J.A.B., Time Optimal Linear Schedules for Algorithms with Uniform Dependencies, *IEEE Trans. Comput.*, vol. 40, no. 6, pp. 723-742, June 1991.

[17]   N. Koziris, G.Economakos, A.Pappas, Sylvain de Noyer, Timothee Mangenot, G Papakonstantinou, P.Tsanakas. "Automatic Mapping and Partitioning of Nested Loops into Systolic Arrays Architectures Using VHDL-The MASH environment", Internal Technical Report Digital System and Computer Laboratory 1997

9

A Smart Load Balance Scheme for an Automatic Arbitrage Detection System

Costas P. Voliotis

Computer Systems Lab, Department of Electrical and Computer Engineering,
National Technical University of Athens, Zografou Campus, GR-15773, Athens, Greece,
minimal@softlab.ntua.gr

George Triantafyllos

Department of Computer Science --Athens University of Economics and Business.
Product Development Division -- *Intrasoft S.A.*
triantaf@aueb.gr, triantaf@intrasoft.gr

Tasos Dalias, Nikos Platis

Product Development Division -- Intrasoft S.A.

1. Introduction

Models that detect arbitrage opportunities in capital markets require real-time computing intensive systems that exhibit fast system throughput. A key characteristic of such models is that the complexity and consequently the execution time varies from a few milliseconds to several minutes. The response time of such a system is critical since arbitrage opportunities rarely last for more than a few minutes. An HPCN system is an ideal platform for implementing applications that exhibit the above characteristics. Yet, these requirements cannot be easily met unless special consideration is given to the way jobs (arbitrage models) are scheduled and dispatched in the distributed environment where the application is executed. Thus, the need for a self-adapted load balancing mechanism is arising. Such a load balance scheme must be scalable to accommodate for increased workload requirements by making efficient usage of the available resources. Since the application is executed on a multi-user environment consisting of heterogeneous computing elements, (i.e. personal computers, workstations) with varying availability of processing resources, the load balance scheme must ensure the portability and the interoperability of the application

In this paper we present a smart and efficient Dynamic Load Balancing (DLB) scheme capable of adapting to the workload of the system. This DLB scheme optimizes the application response time by converting the inherent coarse grain parallelism into a fine grain, while maintaining maximum utilization of the communication network.

The method proposed here is an enhancement of the Orchid [1][2][3] dynamic load balancing scheme based on compile-time and run-time techniques [4][5]. Orchid provides a centralized load balancing scheme appropriate for fine-grained parallel programming applications using the farm model. The Orchid load balancing is based on the assumption that each job executed by an active process takes short and constant time. The execution time, common for all the jobs is predefined at compile time. Each time, only a single job is assigned to each process of an Orchid application. This method was designed to support only fine-grain message passing applications.

The load balancing system presented here is much more flexible since it uses job-handling policies guided by the computation load, as well as by the statistical analysis of the job execution times. The statistical analysis is performed on-line during the life of the application. The DLB dynamically decides if a number of jobs must be packed and assigned to a processing unit (achieving this way the maximum CPU usage) or if each single job will processed separately (minimizing this way the wait time and achieving the optimum communication network usage).

A novel five-steps decomposition approach [6] for practical dynamic load balancing schemes has been used in the analysis and the design phase of this project. Special effort has been spent to keep the design general and independent from the details of the application. The presented DLB scheme is implemented as an object oriented module using the MPI (Message Passing Interface) parallel programming platform [7]. Therefore it is portable and can be re-used as stand alone load balancing module in wide range of parallel-platforms and applications.

In the following paragraphs we provide an overview of the application where our load balancing scheme was used. We then present the design process of the proposed LB and its characteristics. Following that we presents experimental results used for the evaluation of the proposed system.

2. An Overview of the application's architecture.

As mentioned in the introduction, the application we consider, is an arbitrage detection system that provides financial advice to traders. This system is modeled as a client-server application which accepts financial information from Reuters TriArch network [8], computes synthetic assets, and publishes information regarding arbitrage violations back into the same network. Client software presents the results to traders upon request. The server is built as a set distributed process that executed on several low cost personal computers and workstations of various architectures interconnected via a high-speed network that form a distributed computing system.

The architecture of the server is depicted in Figure 1. It consists of an input module, an output module, a job dispatcher module and one or more worker modules. The input module serves the system by receiving data from the outside network, checks for integrity, and converts it to format suitable for internal processing. Similarly, the output module receives data produced by the server, converts it to format suitable for publication, and transmits it to the Triarch network in the form of Reuters Information Code (RIC) packets.

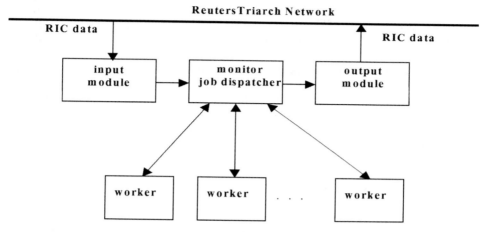

Figure 1: Server architecture

The monitor/dispatcher module is the heart of the system. It maintains a list of models to be evaluated every time their inputs are updated. Thus, when a new datum arrives in the input queue of the monitor, a set of jobs (models that need evaluation) is created. These jobs are entered into one of the job queues available to the system and they are dispatched for execution to the next available worker module. A worker module, thus, executes jobs assigned to it by the monitor, and sends the results back to the calling module.

Each of the previously described modules executes as a process on any of the available processors of the system. Communication between processes takes place using the standard message passing interface, MPI. Thus, the system can be easily ported on any distributed environment that supports MPI. Several hardware configurations are possible depending on the volume of the models being monitored at a given installation. A low performance system can be built by configuring all processes to execute on a single processor. For larger systems, the configuration depends on the number of models to be monitored as well as on the nature of these models, i.e. their computational demands.

The server is a computationally demanding system. The demand rises from the frequency of changes in the input data, as well as from the number of models associated with each input stimuli. Additionally, the computing power required by the system depends on the nature of the models used. The two extremes are : a) a lot of computationally light models (a few milliseconds), resulting in an I/O bounded system, b) a few computationally intensive models (taking several seconds to compute) resulting in a CPU bounded system.

3. The designing process of the Load Balancing module.

A practical solution of the dynamic load balancing problem must deal with the following issues:

Load Evaluation: Some estimation of the possible computer's load must be provided to first determine that a load imbalance could happened.. Estimates of the work loads associated with individual tasks must also be maintained to determine which tasks should be transferred to best balance the computation.

Profitability Determination: Once the loads of the computers have been measured, the presence of a load imbalance can be detected. If the cost of the imbalance exceeds the cost of load balancing, then load balancing should be initiated.

Work Transfer Vector Calculation: Based on the measurements taken in the first phase,
the ideal work transfers necessary to balance the computation are calculated.

Task Selection: Tasks are selected for transfer or exchange to best fulfill the vectors provided by the previous step. Task selection is typically constrained by communication locality and storage requirement considerations.

Workload Adaptability: Both the workloads of a task and the load balancing policy should be adapted to the total workload of the computer. In other words, if the cost of a task migration is more than the profit achieved by the CPU load balancing, this migration should be avoided and the job must be executed locally.

The application of the procedure above results to the baselines of our server load balancing scheme.

- The main task of our server is the real-time evaluation of the arbitrage models which, as explained earlier, exhibit variable amount of computational needs. This heterogeneity of the CPU time requirements for the execution of these models is the first characteristic of the application the DLB module must deal with. A smart load balancing scheme could have a very important impact on the total performance of the server and thus, to the number of the arbitrage opportunities detected.
- The workloads of the available processors are measured indirectly during the lifetime of the application, using the following heuristic. This heuristic calculates the average workload as the average delay involved by the processor's workload. Our scheme consists of workers, statically assigned to processors that execute dynamically assigned tasks.

Processor workload estimation Heuristic.

Workload estimation at time 0, of processor k: $Workload_0 = 0$
Workload estimation at time t of processor k is the average value of the processing overhead

$$Workload_{k,t} \equiv \frac{\sum_{r=0}^{r=t} \left(\sqrt{\left(ActualExecutionTime_r - EstimatedExecutionTime_r \right)^2} \right)}{NumberOf\ Proceeded\ Processes_k}$$

- In order to keep the proposed load balancing scheme as light as possible, in terms of the calculations needed, and to simultaneously preserve the profitability of the load balancing scheme, we assign, to the network processors, packets of jobs instead of single jobs. Each packet contains a variable number of jobs that depends on the sum of their estimated execution times and the workload of the target processor. This way the inherent coarse-grain parallelism of our application is converted to fine-grain.
- A main feature of our design is the usage of a dual job-queue. "Small" jobs (fast in execution time) are stored separately from the "large" jobs. The small jobs grouped in packets and assigned to the first idle worker process. Large jobs are not packed and assigned to processors as is.
- The packet execution time (PET) that determines the number of the jobs grouped as a packet, could be recalculated during the lifetime of the application, keeping this way the assignment of a job profitable.

4. Detailed description of our dynamic load balancing scheme.

As referred previously, the proposed dynamic load balancing scheme deals with packets of jobs that assigned dynamically to worker processes. The distribution of computation load is achieved with the fair distribution of the job packages among the active workers.

An elementary job is represented internally with a simple structure such as the one shown on Figure 2. Each job structure carries a header that identifies the job, two time fields that store the time the job was created and the time the job was completed, the type of the model to be executed, the time this particular job needs to be executed, as well as all the model specific parameters needed for invoking a particular model. It is the responsibility of the load balancing mechanism to package these jobs into packages and send these packages to available workers.

JOB Header	entry time	completion time	Execution Time	Model Class	model specific parameters

Figure 2: The format of a job.

A key feature of the server design and subsequently our load balancing module is the use two job queues (see figure 4). One queue is dedicated to servicing "small" jobs while the other queue is servicing "large" jobs. Each worker processor isassigned two processes dedicated to execution of jobs from each of the above queues, respectively. This way the possibility for a "poverty case", due to a sequence of large jobs, is minimized at the expense of a slightly deteriorating response time.

Slow-queue jobs assigned to the processors one-by-one while fast-queue jobs may be grouped into packets. Each packet carries a variable number of small jobs. The number of these jobs is defined by their total PET. The DLB de-queues jobs from the fast-queue until the sum of their execution times reaches a threshold. This threshold presents the ideal size of a task for the current workload of network. This threshold could be re-estimated when a very large number of jobs have been stored on the fast queue and wait to be executed

The server keeps also, locally the status of each assigned worker. The data included in this structure are the process id (MPI rank) the processor name, the state of the worker (READY, BUSY), the classification as slow or fast and the number of jobs proceeded by this worker so far. Jobs assigned to the workers if they are in state READY.

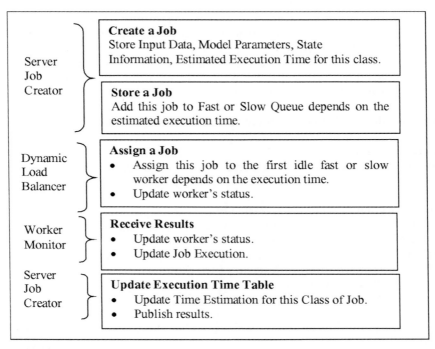

Figure 3: The life cycle of a job

The DLB uses MPI simple messages to assign jobs to the workers. Each such message contains some information for the target worker and a few job descriptions as showed below:

Packet Header	Job 1	Job 2	. . .	Job N

Figure 3 shows the various steps of a job processing and indicates the software modules that perform each step. The server job creator creates and stores the jobs into a job queue depending on their estimated execution time. The DLB assigns the jobs to a worker while it keeps track of the worker's status. The worker receives the message, computes the designated algorithm and returns the results back to the calling process. Finally the main process receives the results of a particular job and publishes them in the TriArch network.

Figure 4 gives a schematic representation of this dynamic load balancing mechanism.

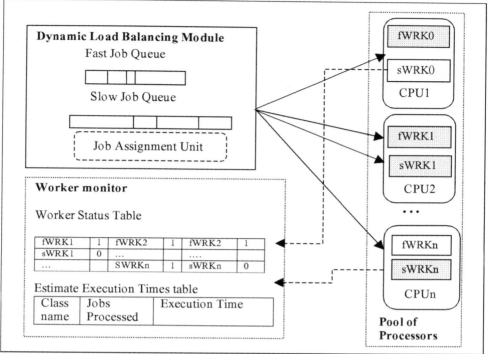

Figure 4: Architecture of the Dynamic Load Balancing module

5. Characteristics of the proposed load balancing method

Many of the published load balancing techniques exhibit some of the following deficiencies:

- They are not scalable. Many load balancing schemes rely on the availability of global knowledge of the load distribution in a system. The technique in paper uses an heuristic for the self-computation of processor workload (see paragraph 3), and doesn't has any restriction for the number of processors, the topology, or communication network.
- They perform complicate calculation and therefore cannot be used on real time-applications. The method presented here uses only simple elementary calculations and could characterized as a "light-load balancing scheme".
- They are application-specific. Techniques that apply to specific data decompositions and problem domains are inherently limited in their usefulness and poorly support the evolution of an application. The framework presented here is very generic, both in terms of its programming interface and its methodology, and could be applied to several problems in concurrent computing.

- They are too complex to reasonably implement. The complexity of many load balancing algorithms in the literature makes errors in their implementation highly likely. Effective implementations of these methods typically involve many details omitted from the description of the algorithms. The methods presented in this paper are quite simple, and any subtleties are illuminated.
- They are not portable. Most of the techniques proposed so far base their operation on machine or architecture specific details and therefore doesn't support the portability of the application. Our method uses simple MPI [7] messages and therefore is fully portable.
- They cannot be automatically adapted on the network workload and the nature of a specific application. Our load balancing scheme calculates both the expected execution time of a class of jobs and the processors workload in real-time during the lifecycle of an application. The former two quantities encapsulate any possible communication overhead. Therefore our method is fully self-adjusted.

6. Load Balance evaluation.

Evaluation of the proposed load-balancing scheme is not a trivial issue due to the fact that the application can exhibit heavy I/O demands, heavy CPU demands, or both simultaneously, depending on the models loaded on the server.

To test the DLB scheme, we created three input sets of models, each resulting to different demands on the server:

A.	I/O Bounded	The server is loaded with many fast models (execution time < 10 msecs)
B.	CPU Bounded	The server is loaded with few slow models (execution time between 1 and 5 seconds)
C.	Mixed case	The server is loaded with a mixture of fast and slow models

We executed these models with three different configurations of the server (where applicable), as follows:

1.	Each of the worker nodes executes one worker process and each worker process executes one job at a time
2.	Each of the worker nodes executes one worker process and each worker process receives a packet of jobs for execution
3.	Each of the worker nodes executes a fast and a slow worker process and each worker process receives a packet of jobs for execution

These configurations were chosen so as to exhibit the benefits of enhancing the basic load-balancing strategy (configuration 1), with job packaging (configuration 2) and additionally with separate queues for fast and slow jobs (configuration 3).

The server was driven with pre-recorded actual financial data taken from several European stock exchanges. The rate of the input data was kept approximately constant for all experiments to a level that created overload conditions for the server. Thus, for all experiments in this paper we were able to measure the maximum throughput of the server as well as the worst-case response times.

Each time, we let the server execute until 10,000 jobs were inserted in the queue(s) and measured the response time (time from the moment the input was fed in the server until the final output was produced) and the throughput (number of jobs completed per second) of the server. Input sets A and B, were tested with server configurations 1 and 2, because configuration 3 is not applicable with these input sets. Input set 3 is the most general one and was tested with all configurations described above. The results of the test cases are summarized in the table that follows:

Test setup	Inputs received	Jobs produced	Run time (seconds)	Jobs completed	Jobs in fast workers	Jobs in slow workers	Throughput (output /sec)	Respon se time
A1	868	10000	244	3428	3428	N/A	14,05	Fig. 5
A2	809	10000	329	9903	9903	N/A	30,1	Fig. 6
B1	4603	10000	1300	2280	N/A	2280	1,75	N/A
C1	2256	10000	515	1773	N/A	N/A	3,44	N/A
C2	2235	10000	510	1665	N/A	N/A	3,26	N/A
C3	2146	10000	675	5898	4783	1115	8,74	N/A

Table 1: Results of the DLB evaluation

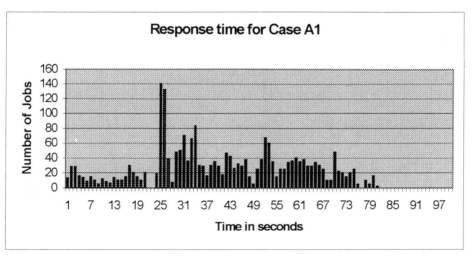

Figure 5: Case A1: Many fast jobs executed on one worker per node without job packing

The results of the tests conducted confirm that the proposed load-balancing scheme is indeed effective. When the server was loaded with fast jobs, which imposed heavy I/O demands on the system, the packaging of the jobs reduced the communication overhead and resulted in doubling the throughput of the server (see table 1). Job packing also reduced the response time of the server as seen from figures 5, 6. Load-balancing had no effect on the throughput of the server when it was loaded only with slow models, because the actual execution time of each model far exceeded the communication overhead and thus any load-balancing strategy would be of limited use. Finally, in the case when the server was fed with a mixture of fast and slow models, its performance greatly increased when the models were piped into different queues according to their (dynamically calculated) execution time (see rows C1, C2, C3, in table 1). This setup allowed fast models to proceed quickly through the server, whereas throughput of slow models again depended mainly on the models themselves. It should be noted that mere packaging did not prove effective in this case: it seems that slow models governed the throughput of the server in this case as well and blocked the fast ones.

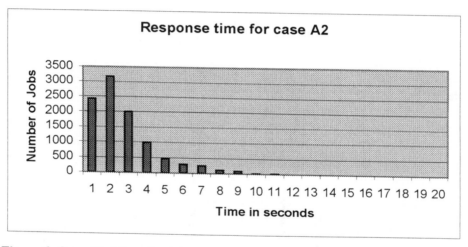

Figure 6: Case A2: Many fast jobs executed on one worker per node using job packing

7. Speedup

This set of experiments attempts to measure the speed up when the number of worker processors is increased. Three different system configurations with one, two and five worker processors were chosen in order to obtain indicative results. The inputs in the system were the same with the ones in the previous experiments.

The number of executed jobs in all cases was expected to have linear increase when adding processors to the system, under the assumption that the workload will be enough to keep all the processors busy while executing the experiment. Figures 7, 8 shows the number of jobs completed as a function of time, in 30 seconds intervals. Figure 7 shows the results for the CPU bounded models, which exhibit a linear relation with the number of workers.

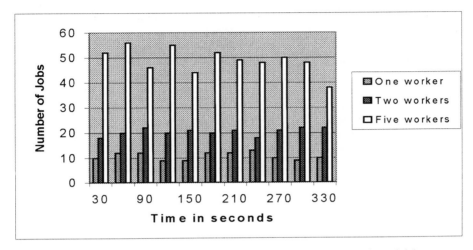

Figure 7: Number of executed Jobs per 30 seconds (CPU bounded models).

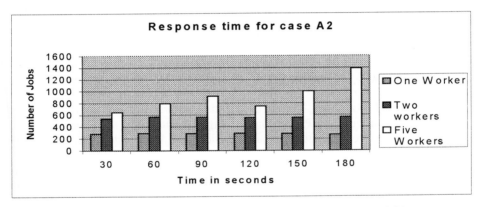

Figure 8: Number of executed Jobs per 30 seconds (I/O bounded models).

Figure 8 shows the results at I/O bounded models. Here the speedup does not always increase linearly. The explanation is merely that some of the workers are idle because there is lack of work. Only after 180 seconds the queue fills with jobs and the number of jobs becomes five times the one with one processor

8. Conclusions

In this paper we presented a load-balancing scheme used to maintain fair load balance on a distributed system that exhibits both I/O bounded and CPU bounded characteristics. The design of the proposed scheme is based on a novel systematic approach. The main characteristics of our method are the automatic adjustment to the workload, the successful manipulation of both fine-grain and coarse-grain scenarios, the scalability, the portability and the simplicity of the implementation.

Our experiment shows that job packaging, a key feature of our DLB, greatly improved the performance of the system by reducing the response times. When the server is loaded with CPU bounded models the speedup obtained increased linearly with the number of available workers.

9. References.

1. K. Voliotis, G. Manis, A. Thanos, G. Papakonstantinou and P. Tsanakas , Facilitating the Development of Portable Parallel Applications on Distributed Memory Systems, Proceedings of Massively Parallel Programming Models MPPM-95 conference, pages 176-184, Berlin, Oct 12-15 1995, IEEE Computer Society Press 1995.
2. K. Voliotis, G. Manis, Ch. Lekatsas, P. Tsanakas, and G. Papakonstantinou, ORCHID: A Portable Platform for Parallel Programming, Journal Of Systems Architectures, The Euromicro Journal, Vol. 43, pages 459-478, Elsevier Science B.V, April 1997.
3. K. Voliotis, An Environment for the development of parallel programming applications, PhD Dissertation, National technical University, Athens, October 1995.
4. H. E. Bal, M. F. Kaashoek, Object Distribution in Orca using Compile-Time and Run-Time Techniques, in: Proceedings of Eight Annual Conf. on Object-Oriented Programming Systems, Languages and Applications, Washington DC, Sept. 1993, pp 162-177.
5. H. Bal, M. Kaashoek, and A. Tannenbaum. Orca: A language for parallel programming of distributed systems. IEEE Transactions on Software Engineering, 18(3):190-205, 1992.
6. Jerrel Watts and S.Taylor, A Practical Approach to Dynamic Load balancing,IEEE Transactions on Parallel and Distributed Systems, vol. 9, No. 3, March 1998.
7. M. Snir, S. Otto, S. Huss-Lederman, D. Walker, J. Dongarra, MPI-The Complete Reference, M.I.T. Press, 1996.
8. Reuters SSL.4.0.2 Triarch Programming Guide (Developers Guide).

10

Intelligent Guidance in a Virtual University

T. Panayiotopoulos, , N. Zacharis, S. Vosinakis
Department of Computer Science,
University of Piraeus,
80 Karaoli & Dimitriou str.
18534 Piraeus, Greece
themisp@unipi.gr, spyrosv@erato.cs.unipi.gr

1. Introduction

Virtual Reality technology [1-3], has introduced a new spatial metaphor with very interesting applications on Intelligent Navigation [4], social behaviour over virtual worlds [5], full body interaction [6], virtual studios [7], etc.

During the past few years, the Virtual Reality Modelling Language (VRML) has emerged as the de facto standard for describing 3-D systems on the World Wide Web. It is platform-independent, easy to use, and gives Web authors the possibility to embed virtual worlds inside their pages. While Java has dramatically altered the way applications are created and distributed, VRML's impact goes beyond, changing the nature of the applications themselves while enriching and deepening the meaning of the data it encapsulates [8]. Recently, a purely theoretical attempt has been made in order to analyse the model-based semantics for Virtual Reality modelling [9]. Another attempt [10] tries to classify the types of behaviour present in Virtual Reality systems.

An interesting application would be the creation of worlds that represent real places or buildings, where the user can be able to access various kinds of information interacting with objects or avatars and travel at the same time in the virtual space. Such applications would require the development of overlying modules which would be capable of communicating with the VRML browsers, providing special control on certain virtual entities, maintaining Information Databases, Interacting with the user, etc. In this system, we have created an interactive application that guides the user inside a virtual university. Visitors can communicate with the program through a command driven system and have a virtual representation of their requests. More specifically, they are able to "walk" through a virtual building with seven different floors that represents the central building of the University of Piraeus and interact with a guide presented as a human-like avatar. The guide can lead visitors to important places inside the building, according to their information needs, and display the appropriate multimedia documents.

2. Overall Architecture and Interface

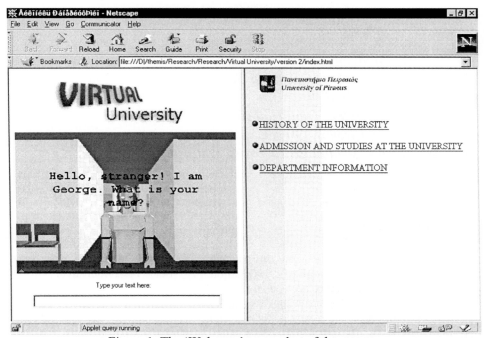

Figure 1. The 'Welcome' screenshot of the program

The system consists of several different parts that communicate with each other according to the user's requests:

- **The User Interface:** It consists of two frames. The first one displays the virtual world that shows the 3-D content as well as a text field where the user can type his/her commands, and the second one is the Information Panel, where the multimedia HTML pages are loaded.
- **The VRML models:** They are subdivided into the static models that represent the main building of the University of Piraeus and the avatar model, which is a virtual representation of a guide.
- **The Multimedia Library:** HTML pages that are displayed in the Information Panel and contain text, image and sound.

- **The Information Database:** It contains information about persons, places, research groups, departments and other entities of the university and also provides links to the multimedia pages.
- **The Spatial Graph:** A graph containing spatial information about the floors of the main building to help the avatar's navigation.
- **The Avatar Control:** The part that is responsible for the avatar's actions inside the virtual world. It uses the spatial graph to implement the movement commands.
- **The Information Control:** The process that searches the Information Database, displays result messages, and loads multimedia pages on the Information Panel.
- **The Command Interpreter:** The part that processes the user's commands, creates a set of actions, and calls the Avatar Control and/or the Information Control to implement them.

When the application starts, it loads a static model representing a floor of the university's main building and the avatar model. It also displays the starting page on the Information panel (Figure 1) and presents a message to the user. According to the user's command it displays multimedia content by calling the information control, or causes an avatar action processed by the avatar control. Furthermore, the user can access information directly by clicking on objects of the virtual world.

The whole architecture of the system can be displayed in the following diagram:

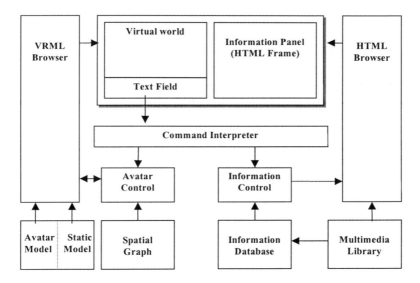

Figure 2. The overall architecture

3. Command Interpreter

When the user enters a command in the text field, the command interpreter tries to recognize it and creates a sequence of actions that are processed by the respective parts.

It repeats calling the Avatar Control or the Information Control using specific parameters until the sequence is over, and then it is ready to read the next user command. The commands recognized by the system are divided into four categories:

Movement Commands: They cause a movement of the avatar, represented as a "virtual walk", that is followed by the user's viewpoint. These commands are: Goto <floor number>, Goto <office number>, Goto <individual>, Goto <room name>. Their meaning is obvious.

Informative Commands: These commands display information to the user according to his request. The result can be a hypermedia HTML page or simple text.

> **TellMeAbout <individual name>:** Displays information about the individual.
> **TellMeAbout <office number>:** Displays the professors that use the office.
> **TellMeAbout <course name>:** Displays the professors that are teaching that course.
> **TellMeAbout <subject>:** Displays the professors that are related to the subject.
> **TellMeAbout <department>:** Displays general information about a department of the University.
> **TellMeAbout <room name>:** Gives information about other interesting rooms, such as the library, laboratories, etc.

Compound Commands: These commands use a combination of movement and information giving to display more complex actions.

> **Tour <department>:** The avatar guides the user through a whole department, providing information about the important places.
> **Tour <floor>:** Guidance in the specified floor
> **Tour Library:** The avatar presents the Main Library of the University

Various Commands: They perform various actions, such as:

> **ReturnToGuide:** The user's viewpoint returns to the position of the guide
> **Help:** Shows on-line help
> **About:** Information about the system
> **Stop:** Terminates instantly and returns control to the user.

4. The Virtual Guide

4.1 THE VIRTUAL UNIVERSITY

The university model has been created using the ground plans of the real building. The walls and the additional objects were placed in the virtual world by translating their real coordinates into the respective 3-D coordinates of the model. There is a basic VRML file that contains the avatar model, the lights and the viewpoints and seven different worlds representing the floors that are called from the basic world using a Switch node. We have used this architecture, instead of having a single world for the whole building, in order to lower the complexity of the VRML models and achieve better performance. Otherwise, the program would require huge memory and increased processor speed and it would be impossible to run in average home computers. Whenever a different child Node is selected in the Switch node, the browser loads and displays the new floor without affecting the main world.

4.2 AVATAR MODEL

The avatar is a virtual human that is presented to the user to help him navigate inside the building. It can display phrases, walk from any place to another, change floors, and load various information on the screen. It consists of two separated parts: the **avatar model** that is responsible for its visual representation and the **avatar control** that is responsible for its behaviour.

The avatar model is a VRML object that presents a human being plus a set of Interpolators used to simulate the movement of its hands and legs during the walking process. The model is divided into seven different parts according to the avatar's limbs. There are six Interpolators to change their orientation and one Time Sensor to control the animation. More specifically, the avatar's body parts are:

➢ the main body and head
➢ the right arm
➢ the left arm
➢ the right leg
➢ the left leg

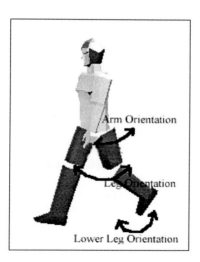

Figure 3. Orientation changes of avatar's body

Each leg is subdivided into its thigh and shin. The position of the shin is relative to that of the whole leg, so changes in its orientation do not affect its position in the body. Furthermore, all the limbs of the avatar maintain their relative positions, therefore a movement of the avatar does not require independent movement of all body parts.

The avatar's limbs have been routed to six independent OrientationInterpolators. An OrientationInterpolator interpolates between two orientations by computing the shortest path on the unit sphere between the two orientations. The interpolation is linear in arc length along this path [2]. The orientation changes in the avatar's body have been depicted in figure 3. There are three interpolators for each side – left and right - controlling the movement of the arm, the leg and the shin.

Finally, all the interpolators are triggered by one TimeSensor, which loops every two seconds. Each time the sensor is enabled, the avatar starts animating by moving its hands and legs. This animation starts whenever the avatar changes its translation to simulate walking.

4.3 THE WALKING PROCESS

The next task that the avatar can perform is "walking". The avatar control uses a set of functions to control the avatar's movement and rotation, and to make sure that it follows the shortest path without colliding to solid objects. The most primitive function is the one that transports it from one point to another using linear movement. First of all, taking into account the current coordinates of the avatar and the coordinates of the "target", it calculates the angle φ that is defined by the horizontal line and the line that passes through the two points, as shown in the following diagram:

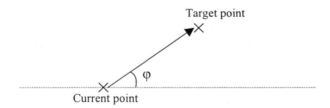

Figure 4. Rotation of the avatar

After that, the program changes the rotation of the avatar, so that its orientation in the two dimensional horizontal field is φ. A new thread is started, that makes the avatar move with constant speed by constantly changing its position using $\sin(\varphi)$ and $\cos(\varphi)$ to preserve its route. For example, if the desired speed is 2 meters per second and the thread changes the avatar's translation every quarter second, then the avatar must be moved by half meter after each time fraction. The new values for x and z coordinates of the avatar are as follows:

$$x = x + 0.5 * \sin(\varphi)$$
$$z = z + 0.5 * \cos(\varphi)$$

Each time the avatar makes a move, the new coordinates are compared to the target coordinates, and when the distance between them is less than the distance covered by a time fraction, the system assumes that the destination is reached and places the avatar in the correct position.

When the thread starts for the first time, the program enables the Time Sensor that creates the animation of "walking" and the avatar's hands and legs are constantly moving during its transportation.

This function is not enough to simulate the avatar's movement. It must also be able to avoid any solid objects that stand in its way. Therefore, the avatar control uses a spatial graph of the floors of the University to create a "safe" path for the avatar to follow. Each floor has been assigned to a complex undirected graph, the nodes of which store the three-dimensional coordinates of the corresponding position in the Virtual world as well as some additional information. The "avatar control part" ensures that the avatar walks only between the nodes of the spatial graph, so that a possible collision between the virtual guide and an object of the static world is avoided. Nodes are divided into two groups:

Intermediate nodes that exist only to provide a path for the avatar's movement.

Termination nodes, that correspond to virtual places with access to informative links. Only these nodes can be requested from the user as a target, and once they are approached, they provide an appropriate link to the multimedia library as well as the information database, so that a relative content is displayed in the information panel.

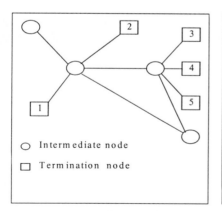 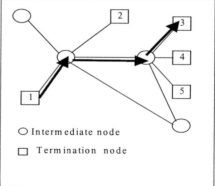

Figure 5: (a) An example of a spatial graph (b) A route between two nodes

Whenever the avatar has to reach a certain destination, the spatial graph is loaded, and the program tries to find the shortest path between the current position and the desired node. The route is planned, and the avatar starts walking through the selected path. For example, if the user requested a walk to the node 3, and he was currently located at node 1, the avatar's route would be as described in figure 5b.

4.4 CAMERA CONTROL

The avatar control is responsible for the movement of the guide in the virtual space, but it has to do more than just changing the translation of the avatar. The user's viewpoint must also be moved to have a constant visual representation of the walking process. Therefore, we have used three different cameras to control the animation and to ensure that the user's view of the scene is correct. The system uses:

> ➢ Two cameras attached to the avatar: they both have the height of the average human eye – at about 1.6 meters. The first one (CAMERA A) is located two meters behind the avatar and the second one (CAMERA C) two meters in front of the avatar.
> ➢ One independent camera: It also has the height of the human eye, and can be placed anywhere in the scene (CAMERA B).

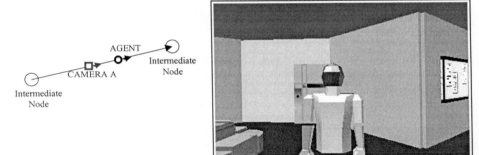

Figure 6. Diagrammatic representation and output of Camera A

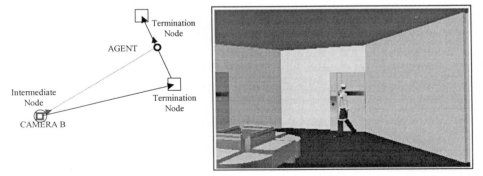

Figure 7. Diagrammatic representation and output of Camera B

When the animation starts, the user views the output of the first camera, which automatically follows the avatar's movement. Whenever the avatar moves from an intermediate node to a termination node or between termination nodes, the independent camera is placed at the current position of the user, maintains its translation, and changes constantly its orientation so that the avatar is always at the center of the screen. After that, when the avatar has reached its destination, it rotates 180° and the third viewpoint (CAMERA C) is enabled, so that the user looks straight at the avatar and is able to read its messages.

AGENT

CAMERA C   Termination
Node

Figure 8. Diagrammatic representation and output of Camera C

5. Implementation

The Virtual University system has been implemented using interactive VRML 2.0 worlds, a Java applet and HTML pages. It runs inside a single page of a simple Web browser with Java capabilities and a VRML 2.0 plug-in. The page consists of two basic frames, one for the VRML world and Java applet, responsible for the 3D content and the user interaction, and one for the HTML pages that display the requested information.

As far as communication between a VRML world and its external environment is concerned, an interface between these two is needed. This interface is called External Authoring Interface (EAI) and it defines the set of functions on the VRML browser that the external environment can perform to affect the VRML world [11]. The EAI allows a currently running Java applet to control a VRML world, just like it would control any other media [12].

6. Conclusions and Future Work

In this chapter we have presented the architecture of an intelligent guidance system inside a virtual environment. It is an effective method for easily accessing the desired information from a huge amount of pages using virtual reality techniques and travelling at the same time in a representation of a real building. It is based on the latest features of the World Wide Web, and therefore it is multi-platform and can be accessed by a number of users while running on a single server.

We are currently working on extending the system's interactive capabilities and making it more attractive to the common users. Moreover, we are planning to add multi-user support using avatars to represent the users, and to put more intelligent agents [13] as well as other interactive objects such as elevators, doors, computers, etc.

Acknowledgement
The system described in this chapter was partially funded by the EPEAEK project (EKT., Subprogram 3., Measure 3.1., Action 3.1.B) entitled "Modernisation of the Central Library of the University of Piraeus", funded by the European Community and the Greek Ministry of Education and Religious Affairs.

References
[1].   J. Vince, "Virtual Reality Systems", ACM Press, 1995.

[2].   The VRML Consortium Incorporated., "VRML97 International Standard", (ISO/IEC 14772-1:1997) http://www.vrml.org/Specifications/VRML97, 1997

[3].   S.Vosinakis, T.Panayiotopoulos, 'State of the Art in Virtual Reality', Internal Report, University of Piraeus, Dpt. Of Computer Science, 1997 (in Greek).

[4].   N. Zacharis, T. Panayiotopoulos, "A Learning Recommendation Agent in – Virtual Environments", International Conference for Artificial Intelligence and Soft Computing, Cancun, Mexico, 1998.

[5].   Y Honda et al, "Virtual Society: Extending the WWW to support a multi-user interactive shared 3D environment", Procs VRML 95, San Diego, 1995.

[6].   P. Maes, et al, "The ALIVE system: full-body interaction with autonomous agents". Proceedings of Computer Animation '95, 1995.

[7].   S. Gibbs, C. Arapis, C. Breiteneder, V. Lalioti, S. Mostafawy, J. Speider, "Virtual Studios: An Overview", IEEE Multimedia, pp.18-35, Jan – Mar 1998.

[8].   C. Marrin, B. McCloskey, K. Sandvik, D. Chin, "Creating Interactive Java Applications with 3D and VRML", http://cosmo.sgi.com/developer.html, Silicon Graphics, 1997.

[9].   M. Prokopenko, V. Jauregui, "Reasoning about actions in Virtual Reality." IJCAI-97 Workshop on Nonmonotonic Reasoning Action and Change, 1997.

[10]. B. Roehl, "Some Thoughts on Behavior in VR Systems (Second draft: August, 1995)", URL: http://sunee.uwaterloo.ca/~broehl/behav.html, 1995

[11]. C. Marrin, "Proposal for a VRML 2.0 Informative Annex. External Authoring Interface Reference", http://cosmo.sgi.com/developer.html, Silicon Graphics, 1997.

[12]. J. Doppke, D. Heimbigner, and A. Wolf, "Software Process Modeling and Execution within Virtual Environments", ACM Transactions on Software Engineering and Methodology, Vol.7, No.1, pp. 1-40, January 1998.

[13]. T. Panayiotopoulos, G. Katsirelos, S. Vosinakis, S. Kousidou, "An Intelligent Agent Framework in VRML worlds", Third European Robotics, Intelligent Systems & Control Conference, EURISCON'98, Athens, June 1998.

PART II

INFORMATION EXTRACTION FROM TEXTS, NATURAL LANGUAGE INTERFACES AND INTELLIGENT RETRIEVAL SYSTEMS

11

Question Answering and Information Extraction from Texts

J. KONTOS AND I. MALAGARDI
Department of Informatics
Athens University of Economics & Business
76 Patission St., 104 34 Athens, Greece
E-mail: jpk@aueb.gr

1. Introduction

The research presented in this chapter is part of a project which aims at the development of a novel method for information extraction and knowledge acquisition from texts combined with question answering. The present state of the art in information extraction [1, 2] is based on the template approach. The template approach relies on a predefined user model which guides the extraction of information and the instantiation of a template as the result of the extraction process.

Our question answering based approach aims at the creation of flexible information extraction tools, which accept natural language questions and generate answers that contain information extracted from text either directly or after applying deductive inference.

Our approach also addresses the problem of implicit semantic relations occurring either in the questions or in the texts from which information is extracted. These relations are made explicit with the use of domain knowledge. Examples of application of our methods are presented in this paper concerning three domains of quite different nature. These domains are: Oceanography, Medical Physiology and Ancient Greek Law.

Another point of our method is to process text directly avoiding any kind of formal representation when inference is required for the elaborated in [6, 7, 8] as the ARISTA method. This is a new method for knowledge extraction of facts not mentioned explicitly in the text. This idea was first proposed in [3, 4] and was applied to simple information extraction from texts in [5] and further acquisition from texts that is based on using natural language itself for knowledge representation.

2. Method

Our method is question answering based and aims at the creation of flexible information extraction tools which accept natural language questions and generate answers that contain information extracted from text either directly or after applying deductive inference. The information extraction task can thus be performed interactively enabling the user to submit natural language questions to the system and therefore allowing for greater flexibility than template based systems.

Our method uses a question grammar combined with a text grammar for the extraction of information. These two grammars use syntax rules and domain dependent lexicons while the semantics of the question grammar provides the means of their combination. An illustrative question grammar fragment is presented below.

2.1. QUESTION GRAMMAR

The semantics of a question grammar may provide the means of the combination of question processing with information extraction. An illustrative fragment of such a question grammar is presented below for the Ancient Greek Law case:

q(Q,TAP):-f(Q,Question_Word,R1),f(R1,Benefit,R2),
f(R2,to,R3),f(R3,TEKNON,R4),f(R4,EleytheraS,R5),onom(EleytheraS,Eleythera),
relative_clause(R5,Act,Agent),
relation(TEKNON,EleytheraS,Relation),
template(S,s,Agent,Act,Eleythera),
template(S,a,Eleythera,Relation,Eleytheron),
template(N,s,Eleythera,Relation,Eleytheron).

q(Q,TAP):-f(Q,Question_Word,R1),f(R1,Benefit,R2),
f(R2,tin,R3),f(R3,Perioysian,R4),f(R4,EleytheraS,R5),onom(EleytheraS,Eleythera),
relative_clause(R5,Act,Agent),
relation(TEKNON,EleytheraS,Relation),
template(S,s,Agent,Act,Eleythera),
template(S,a,Eleythera,Relation,Eleytheron),
template(N,s,Eleythera,Relation,Eleytheron).

relative_clause(Rel_clause,Verb,Sub_Obj):-pronoun(Rel_clause,R1),
f(R1,Verb,R2),morph(Verb,Verb_root),
prepositional_phrase(R2,Sub_Obj).

prepositional_phrase(P,SO):-f(P,pros,SO);f(P,SO,"").

pronoun(P,R):-f(P,pros,R1),f(R1,tin,R2),f(R2,opoian,R);f(P,i,R1),f(R1,opoia,R).

Examples of questions that may be processed with the above grammar are:

qp:-q("ti lamvanei to teknon eleytheras pros tin opoian erhetai doylos").
qw:-q("poios lamvanei tin perioysian eleytheras pros tin opoian erhetai doylos").
qr:-q("ti lamvanei to teknon eleytheras i opoia erhetai pros doylon",P).

2.2. TEXT SYNTACTIC ANALYSIS

The text syntactic analysis performed by the system that was implemented on the computer is based on an original parsing method appropriate for languages with relatively free word order like Greek.

This method consists of the automatic translation of every sentence into a number of logical facts written in Prolog and the recognition of syntactic constituents as logical combinations of these facts. These facts take the form of a logical predicate with three arguments. The first argument specifies the number of the sentence that contains a given word, the second argument specifies the position of the word in the sentence and the third specifies the word itself. In traditional methods of syntactic analysis by computer, which is mainly used, for the analysis of English texts one syntactic rule must be written for every particular sequence of words. This means that if we apply such a method for the syntactic analysis of Greek a plethora of syntactic rules will be needed for the same constituent due to the word order freedom of this language. On the contrary the method followed in the present system allows the statement of a single syntactic rule for the parsing of two or more equivalent syntactic structures that differ only in the relative position of the words involved.

The form of sentences that can be analyzed by the rules developed for the present system consists of one verb and its complements since this was the form of the hypotheses found in the corpus. The case of the missing subject is treated using the valency of the verb for predicting the number of its complements. The complements of the verbs are recognized by syntactic rules that analyze the following basic forms of noun phrases:

- Pronouns
- Nouns
- Article + Noun
- Article+ Participle
- Adjective+ Noun
- Noun Phrase + or+ Noun Phrase
- Noun Phrase + and+ Noun Phrase
- Noun in Nominative+ Noun in Genitive
- Noun+ Pronoun+ Article+ Noun

These forms are recognized by text grammar rules written in Prolog of which illustrative examples follow and where c(X,Y,Z) means that Z is the concatenation of X and Y.

Text Grammar

pr(N,s,M,E2):-s(N,S),f(S,E2,R),f(R,M,_).

pr(N,a,V,E1):-ap(N,A),f(A,E1,R),f(R,V,_).

pr(N,a,E1,V,E2):-ap(N,A),f(A,E1,R),
          f(R,V,R1),f(R1,E2,"").

pr(N,s,E1,V,E2):-s(N,A),f(A,E1,R),
          f(R,V,R1),f(R1,E2,"").

---

p(S,np,NP,P):-w(S,N1,D),w(S,N2,N),N2=N1+1,
          l(D,_,d,Nu,P,G),l(N,_,e,Nu,P,G),
          c(D,N,NP).

p(S,np,NP,P):-w(S,N1,D),w(S,N2,N),N2=N1+1,
          l(D,_,d,Nu,P,G),l(N,_,met,Nu,P,G),
          c(D,N,NP).

p(S,np,NP,P):-w(S,N1,Ad),w(S,N2,N),N1<>N2,%N2=N1+1,
          l(Ad,_,ad,Nu,P,G),l(N,_,e,Nu,P,G),
          c(Ad,N,NP).

p(S,np,NP,P):-w(S,N1,E1),w(S,N2,i),N2=N1+1,
          w(S,N3,E2),N3=N2+1,
          l(E1,_,e,Nu,P,_),l(E2,_,e,Nu,P,_),
          c(E1,i,NP1),c(NP1,E2,NP).

p(S,np,NP,P):-w(S,N1,E1),l(E1,_,e,_,P,_),
          w(S,N2,i),N2=N1+1,
          p(S,npq,NPQ,P,N3),N3=N2+1,
          c(E1,i,NP1),c(NP1,NPQ,NP).

p(S,np,NP,P):-w(S,N1,E1),w(S,N2,kai),N2=N1+1,
          w(S,N3,E2),N3=N2+1,
          l(E1,_,e,Nu,P,_),l(E2,_,e,Nu,P,_),
          c(E1,kai,NP1),c(NP1,E2,NP).

p(S,np,NP,P):-w(S,_,NP),l(NP,_,e,_,P,_).
p(S,np,NP,P):-w(S,_,NP),l(NP,_,pr,_,P,_).

```
p(S,npc,NPc,o):-w(S,_,No),l(No,_,e,_,o,_),
              w(S,_,Na),l(Na,_,e,_,g,_),r(No,Na,R),
              c(No,Na,NPc),write(S,R),nl.

p(S,npc,NPc,o):-w(S,_,No),l(No,_,ad,_,o,_),
              w(S,_,Na),l(Na,_,e,_,g,_),r(No,Na,R),
              c(No,Na,NPc),write(S,R),nl.

p(S,npq,NP,P,N1):-w(S,N1,E1),l(E1,_,e,_,P,_),
              w(S,N2,ek),N2=N1+1,w(S,N3,D),
              l(D,_,d,_,g,_),N3=N2+1,
              w(S,N4,Pr),l(Pr,_,pr,s,g,_),
              N4=N3+1,w(S,N5,E2),l(E2,_,e,_,g,_),
              N5=N4+1,c(E1,ek,S1),c(S1,D,S2), c(S2,Pr,S3),c(S3,E2,NP).

p(S,se,SE,V,NPc,_):-a(S,_,V,NE),l(V,_,iv,_),
                  p(S,npc,NPc,o),c(V,NPc,S1),c(NE,S1,SE).

p(S,se,SE,V,NP,_):-a(S,_,V,Ne),l(V,_,iv,_),
                  p(S,np,NP,o),c(Ne,V,CV),c(CV,NP,SE).

p(S,se,SE,V,_,A):-a(S,_,V,Ne),l(V,_,v,_),w(S,_,A),
                  l(A,_,ad,_,a,_),c(Ne,V,CV),c(CV,A,SE).

a(S,N1,V,den):-w(S,N1,den),w(S,N2,V),N2=N1+1.
a(S,N,V,""):-w(S,N,V).
```

2.3. EXTRACTION OF IMPLICIT SEMANTIC RELATIONS

During the processing of some texts the problem of discovering and extracting by computer implicit semantic relations between concepts occurred. The discovery of such semantic relations requires the codification and processing by computer of the appropriate domain knowledge [9,10,11].

The example of a legal text (the ancient Greek text of the Law Code of Gortys) is used below for the illustration of the kind of knowledge used. This knowledge is divided in two main parts: a) ontology of actions and b) specification of implicit relationships between entities. The actions for the domain of the ancient Greek text are expressed with verbs, which are classified as follows:

1. offences: rape, take
2. existence: live, exist, die
3. general actions: leave, bear, divorce, guarantee, leave, marry

The categories of implicit relationships between nouns are "being relative" and "responsibility" e.g.:

1.   being relative: brother of father
2.   responsibility: responsible of divorce

This knowledge of implicit relationships is used for the analysis of noun phrases of the following forms:

- noun in the nominative + noun in the genitive
- adjective in the nominative + noun in the genitive

These relationships are produced in the answers to the appropriate questions.
The above knowledge is expressed in Prolog.

2.4. KNOWLEDGE EXTRACTION FROM TEXT

The deductive computer analysis of texts is traditionally performed in two stages. In the first stage the text is translated by computer or more commonly by hand into some formal representation. In the second stage reasoning is performed using this formal representation of the content of the text. In our system the translation step is avoided and the analysis is performed directly from the natural language texts following the ARISTA method of text analysis [3, 4, 5, 6, 7, 8].

2.5. THE LEXICA

The texts used as examples of application of our system obviously belong to quite different domains and therefore pose the requirement for specialized lexica. Each specialized lexicon contains all the words from the domain that the particular text processed by the system belongs to. These words are grouped in three categories depending on the number of their characteristic attributes.

The first category consists of words, whose entries have a single attribute that specifies the part of speech. The words in this category are mainly function words. The second category consists of verbs and their entries have two attributes. The first attribute specifies whether the verb is transitive or intransitive and the second attribute specifies the number of the verb. The third category has entries with four attributes and contains the nouns, adjectives and participles found in each text. The four attributes specify the part of speech, the number, the case and the gender of every word of this category.

3. Extraction from the Primary Production Texts

The set of texts on which the combination of question answering with information extraction was first applied [5] were abstracts of research papers from the domain of primary production in the sea. Primary production concerns the growth of phytoplankton and its dependence on environmental factors such as nutrients. These abstracts were taken from the "Deep-Sea Research Oceanographic Literature Review" for the years 1978, 1979 and 1980. The information extracted from these abstracts concerned facts about the causal dependence of biological processes such as growth and photosynthesis on environmental factors such as solar radiation and various chemical elements or compounds. These facts constitute the basic elements of scientific knowledge in this domain and are normally predicated with time and space information. An illustrative question answering example for this application would be:

Question: "What organism depends on what nutrient?"

Answer: "tricornutum depends on nutrient"   from 1
"phytoplankton depends on N"   from 2
"phytoplankton depends on nutrient"   from 3
"phytoplankton depends on nitrate"   from 4

Where the numbers 1-4 correspond to the text sentences from which the answers were extracted and are given below:

1. Growth of tricornutum related to nutrient content.
2. Numbers of phytoplankton correlated with N in the photic zone.
3. Nutrient enrichment in the basin stimulates phytoplankton growth.
4. Spatialdistribution of nitrate correlated with phytoplankton activity.

The system described in [5] that was capable of producing the above results was the first to accomplish a direct attack on the problem of question answering combined with information extraction from unformatted texts. The questions posed to this system were processed using a semantic grammar, which was augmented with some form of ontology of the domain.

The speed of the system was increased when its implementation was based on finite state automata parsing instead of the then traditional grammar based parsing method. It is noticeable that recent work on information extraction has 'resurrected' the finite state method of parsing in order to face the speed problems due to large corpora [2].

4. Extraction from the Lung Mechanics Text

The following example is an extract from a medical physiology book [12] in the domain of lung mechanics. The processing of this text for information extraction was first presented in [7] using a number of scenaria related to causal knowledge chaining. This chaining results from deductive reasoning, which is performed by the system in response to the user's question. This text consists of the following sentences:

1. The alveolar pressure rise forces air out of the lungs.
2. The alveolar pressure rise is caused by elastic forces.
3. Elastic forces include elastic forces caused by surface tension.
4. Elastic forces caused by surface tension increase as the alveoli become smaller.
5. As the alveoli become smaller, the concentration of surfactant increases.
6. The increase of the concentration of surfactant reduces the surface tension.
7. The reduction of the surface tension opposes the collapse of the alveoli.

If the user submits the question "What process of alveoli causes flow of lungs air?" The answer "become smaller" is produced together with the following explanation:

alveoli become smaller causes increase of elastic forces because
surface tension elastic forces is part of elastic forces and
alveoli become smaller causes increase of surface tension elastic forces

alveoli become smaller causes rise of alveolar pressure because
alveoli become smaller causes increase of elastic forces and
elastic forces causes rise of alveolar pressure

alveoli become smaller causes flow of lungs air because
alveoli become smaller causes rise of alveolar pressure and
rise of alveolar pressure causes flow of lungs air

A second question that may be submitted by the user is: "What process of alveoli opposes collapse of alveoli?" which requires the definition of causal polarity for proper treatment of the verb "opposes". After positive and negative causal polarity are defined as "+cause" and "-cause" respectively the system gives the answer again "become smaller" but now the explanatory text generated by the system is as follows:

alveoli become smaller +causes reduces of surface tension because
alveoli become smaller +causes increase of surfactant concentration and
increase of surfactant concentration +causes reduces of surface tension

alveoli become smaller -causes collapse of alveoli because
alveoli become smaller +causes reduction of surface tension and
reduction of surface tension -causes collapse of alveoli

5. Extraction from the Gortys Law Text

The last example concerns the extraction of knowledge from the ancient Greek text of The Law Code of Cortys text is considered as the most ancient legal code of Hellenic Antiquity. This text exists as a stone inscription in the ruins of the ancient city of Gortys, which is situated on the Greek Island of Crete. The inscription is dated at the end of 6th or at the beginning of 5th century B.C.

The whole text amounts to more than 600 lines and about 3000 words. The writing is "boustrophedon", the first line of each column running from right to left and the rest of the lines alternate in direction. The Code is inscribed in the archaic Greek alphabet of eighteen letters including F (digamma). The dialect of the Code was the Cretan Doric Greek dialect and particularly the dialect of Central Crete. Each of its regulations is formulated as a conditional sentence in the third person, with the protasis consisting of the assumed facts and the apodosis consisting of the legal consequences.

The Inscription includes the following items: Property, Marriage and Kinship, Heiress, Rape, Adultery and Divorce, Illegitimate Children, Adoption, The Administration of Justice [13]. The user may submit to our system a question either in Greek or in English that is answered by the system after performing the necessary information and knowledge extraction from the Greek Gortys text. Two illustrative questions are:

1) "What does the child of a free woman inherits to which a slave comes?"
   Whose answer is "the estate of the free woman"

2) "Who inherits the estate of a free woman to whom a slave comes?"
   Whose answer is "the free child"

The part of the ancient legal text from which the information necessary for answering the above questions is extracted consists of two sentences with the following rendering in English:

1. If a slave comes to a free woman then a free woman bears free children.
2. If a free woman bears a free child then the child inherits her estate.

6. Conclusion

A novel approach was introduced in this chapter for the implementation of a question-based tool for the extraction of information from texts. This effort resulted in the computer implementation of a system that answers questions directly from text using Natural Language Processing. Domain knowledge concerning categories of actions and implicit semantic relations was found useful for performing information extraction from the texts treated.

References

[1]. Cowie, J., and W. Lehnert. Information Extraction. *Communications of the ACM*. Vol. 39, No. 1, pp. 80-91, 1996[I1].
[2]. Pazienza, M. T. *Information Extraction*. LNAI Tutorial. Springer, 1997.

[3]. Kontos, J. Syntax-Directed Processing of Texts with Action Semantics. *Cybernetica*, 23, 2 pp. 157-175, 1980.

[4]. Kontos, J. Syntax-Directed Plan Recognition with a Microcomputer. *Microprocessing and Microprogramming*. 9, pp. 227-279, 1982.

[5]. Kontos, J. Syntax-Directed Fact Retrieval from Texts with a Micro-Computer. *Proc. [I2]MELECON '83*, Athens, 1983.

[6]. Kontos, J. Natural Language Processing of Scientific/Technical Data, Knowledge and Text Bases. *Proceedings of ARTINT Workshop. Luxembourg*, 1985.

[7]. Kontos, J. ARISTA: Knowledge Engineering with Scientific Texts. *Information and Software Technology*, Vol. 34, No 9, pp.611-616, 1992.

[8]. Kontos, J. *Artificial Intelligence and Natural Language Processing* (In Greek) E. Benou, Athens, Greece, 1996.

[9]. Malagardi, I. *Comparative Analysis of "na" and "ya na" sentences of the Greek language with the equivalent structures of German language and related problems in their machine translation*. Unpublished Dissertation. University of Athens, 1995a.

[10]. Malagardi, I. The resolution of the subject ambiguity in sentences with "ya na" using domain knowledge, and related problems in machine translation. *Proceedings of 2nd. International Congress on Greek Linguistics*. Salzburg, 1995b.

[11]. Malagardi, I. Computer Determination of Relations between the Elements in Noun Phrases of Sublanguages. *17th annual meeting of the Department of Linguistics*. Aristotle Univ. of Thessaloniki, 1996.

[12]. Guyton, A. C. *Textbook of Medical Physiology*. Eighth Edition, An HBJ International Edition. W.B. Saunders, 1991.

[13]. Willets, R. F. *The Law Code of Gortyn*. Kadmos : Supplement I. Berlin,1967.

12

Named Entity Recognition from Greek Texts : The GIE Project

Vangelis Karkaletsis, Constantine D.Spyropoulos and George Petasis

Software and Knowledge Engineering Laboratory
Institute of Informatics and Telecommunications,
N.C.S.R. «Demokritos»,
Tel: +301-6503196-7, Fax: +301-6532175
e-mail: {vangelis, costass, petasis}@iit.demokritos.gr

1. Introduction

Today's overload of information, particularly through the World Wide Web, makes difficult the user's access to the right information. The situation becomes even more difficult due to the fact that a lot of this information is in different languages. Therefore, it is important to apply an information process that will extract from all that volume of information only the facts that match user's interests, and allow the user to access facts written in a different language. Information Extraction (IE) technology can meet these requirements, since unlike what happens with information retrieval and filtering technology, in IE the user interests are on specific facts extracted from the documents and not on the documents themselves. Some documents may contain the requested keywords but be irrelevant to the user's interests. Working with specific facts instead of documents provides users information more relevant to their domain of interest.

The IE systems developed so far, extract, in most cases, fixed information from documents in a fixed language. However, in order for the IE technology to be truly applicable in real life applications, meeting the above requirements, IE systems need to be easily adaptable (customisable) to new domains and users interests, as well as to multiple languages. During the last decade, substantial progress has been made in developing reliable Information Extraction (IE) technology. IE technology is currently exploited in real applications, such as the extraction of information for companies acquisitions [1],[2],[3], stock exchanges [4], companies profits and losses [5], joint ventures and management succession events [6],[7],[8], as well as for the understanding of military messages [9] and police reports [10],[11],[12].

However, the existing IE technology concerns widely spoken languages and mainly English. So far, according to our knowledge, there is not any IE system for the Greek language, although there is activity in the area of language engineering in Greece. Our laboratory is currently participating in two IE projects, the EU-funded project ECRAN[1],

---

[1] ECRAN (Extraction of Content: Research at Near Market) is a Language Engineering project (LE-2110) funded partially by the European Commission (EC), which involves Thomson (FR), SIS (GE), Univ. of Sheffield (UK), NCSR "Demokritos" (GR), Univ. of Ancona (IT), Univ. of Tor Vergata (IT), and Univ. of Fribourg (SU).

and the bilateral (English-Greek) project GIE[2] (Greek IE). In GIE we cooperate with the University of Sheffield aiming to adapt the Sheffield IE system into the Greek language. An IE task involves mainly two sub-tasks: the **recognition of the named entities** (e.g. persons, organisations, locations, dates) involved in an event and the **recognition of the relationships holding between named entities** in that event (e.g. personnel joining and leaving companies in management succession events). A **named entity** (NE) is a phrase, which serves as a name for something or someone. According to this definition, the phrase in question must by a noun phrase (NP). Clearly, not all NPs are named entities. An important feature of NPs is that they may contain or be contained in other NPs. In general NEs are short NPs, i.e., they contain a small number of words. **Named-entity recognition (NERC)** involves two tasks: recognition of NPs that are NEs, classification of NEs into different types, such as organizations and person names.

In this chapter, we present the prototype named entity recogniser (NERC) we are currently developing for the Greek language, in the context of GIE. The GIE prototype is being developed over the language engineering platform GATE of the University of Sheffield. More specifically, in section 2 of the chapter, we discuss the significance of the named entity recognition task in IE, providing results from MUC Conferences, as well as providing information on existing NERC systems in English and in other languages. Section 3 presents the prototype NERC system we are currently developing in the context of GIE. Information is provided on the platform, the corpus, as well as on the modules developed or being developed so far. Some first evaluation results are presented and the major problems are also discussed. We conclude this chapter discussing the significance of a NERC system for the Greek language, and the need for customisation tools to facilitate the adaptation to new domains.

2. Named Entity Recognition Task in IE

The progress in Information Extraction (IE) technology is due to the increase in available resources such as machine readable dictionaries and text corpora, in computational power and processing volume as well as due to the development of Language Technology techniques that can be applied in practice. This progress is proved from the results of Message Understanding Conferences – MUCs where several IE systems are evaluated (see MUC Website in http://www.muc.saic.com/). Named entity recognition is one of the evaluation tasks, which provides also the better results (see section 2.1), proving that this technology can be applied in practice. The identification of named entities in a corpus along with their classification as persons, organisations, etc., can be useful not only as the first stage of a complete IE system, but also for other tasks, such as indexing of documents, maintenance of data bases containing information for the identified persons, organisations, etc. That's why our laboratory emphasises on the need for the development of a NERC system for the Greek language.

---

[2]   GIE (Greek Information Extraction) is a bilateral project between NCSR "Demokritos" (GR) and Univ. of Sheffield (GB), funded by the Greek General Secretariat of R&T and the British Council.

2.1 NERC in Message Understanding Conferences (MUC)

The systems participating in MUCs should process texts, identify the texts that are relevant to the domain, and fill templates which contain slots for the events to be extracted and the entities involved. Information analysts design the template structure, that is the information that needs to be extracted by domain specific texts. The domain areas examined so far in MUCs are the following:

MUCK	*Navy messages (1987)*
MUCK-II	*Navy messages (1989)*
MUC-3	*News for terrorist attacks (1991)*
MUC-4	*News for terrorist attacks (1992)*
MUC-5	*Company news (joint ventures, micro-electronics products) (1993)*
MUC-6	*Company news (management succession) (1995)*
MUC-7	*Orders of airline companies (1998).*

NERC was one of the tasks evaluated in MUC-6 and MUC-7 (the other tasks in MUC-6 were coreference resolution, template element filling and scenario template filling, whereas in MUC-7 the task of template relationship filling was also added). The main measures used for the evaluation of MUC tasks are **recall** and **precision**. The recall measure counts the number of words/phrases that are assigned the correct tag, out of the total number of words/phrases in the corpus. On the other hand, precision counts the number of words/phrases assigned the correct tag, out of the total number of words/phrases that are assigned a tag (either correct or wrong).

NERC modules represent the most mature IE technology. The best score obtained in the NERC task in MUC-6 was 96% recall and 97% precision. The MUC-6 best overall raw scores are shown in Table 1 [13]:

Task	Recall (%)	Precision (%)
Named Entity	96	97
Coreference (High recall)	63	63
Coreference (High Precision)	59	72
Template Element	74	87
Scenario Template	47	70

Table 1. Best Overall Scores in MUC-6 Tasks

2.2 NERC in the ECRAN Project

ECRAN project is developing a new generation of techniques aiming at bringing IE near to the market, facilitating the customisation to new application domains, as well to the users requirements. The language engineering platform GATE [14] of the University of Sheffield is used in ECRAN for developing and integrating the modules produced (see Fig. 1).

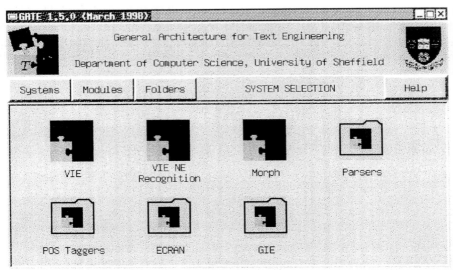

Figure 1. The GATE language engineering platform

During the two first years of ECRAN (the project ends at December 1998), a stable version of the English language system for IE has been established for the company news domain. A stable version of the French system has also been established to handle information about films/videos in French. An Italian IE application has been developed, which performs Named Entity recognition in the domain of financial news. Within the GATE framework, modules have been either ported to new languages (e.g. named entity recognition in French and Italian), imported from already existing resources or redesigned.

During the last year of ECRAN, we examine techniques for the rapid customisation of the English IE system into new domains. One of the customisation tasks currently examined, is NERC. NERC usually exploits a grammar of named entity rules. These rules specify when a sequence of words is named entity and also the type of this entity. Usually the NERC grammar is built manually, by experts in a particular domain. Manual grammar construction is clearly problematic when we want to port the NERC system to a new domain. For this reason, learning from examples can be useful. The aim of our research for this customisation task in ECRAN, is to have a NERC system that is adaptive to new domains exploiting machine learning techniques.

3. NERC in GIE

The named entity recogniser in GIE is based on the relevant English recogniser developed at the University of Sheffield over the GATE language engineering platform. It involves the following modules: tokeniser, sentence splitter, part of speech tagger, gazetteer look up, named-entity parse, name matcher.

The following sub-sections present the work done so far in GIE for the development of the Greek NERC system, starting from the platform, the corpus, and the modules developed or customised so far. Fig. 2 shows the modules of GIE over the platform of GATE.

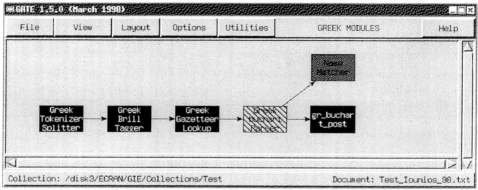

Figure 2. Modules of GIE Named Entity Recogniser

3.1 The GATE Platform

The language engineering platform GATE (v1.5.0) was installed according to the guidelines of Sheffield team. The platform was provided by Sheffield together with the English IE system VIE. The installation was done on two operating systems, first in Linux and then in SunOS and Solaris. The NERC prototype system is implemented in Solaris (v2.5.1).

3.2 The Corpus

We used corpora from three different sources in order to train and then evaluate our system.

– A text corpus on "management succession events" was provided by the company "Advertising Week" (DIAFHMISTIKH EBDOMADA) (http://www.addweek.gr). The corpus contains texts on personnel leaving or joining companies for the period from 1/96 until 6/98. The corpus size is about 50,000 words. A part of this corpus (about 15.000 words) was hand-tagged in order to train our part of speech tagger. The rest part is used for testing our system.

– The second text corpus on "stock market news" was provided by the company "Kapa-TEL" (http://www.kapatel.gr). The corpus contains news for the period from 1/97 until 4/98. The corpus size is about 85.000 words and it has not been used yet in our experiments.

– The third text corpus, is a general theme hand-tagged corpus, which was provided by the WCL[3] of Patras University for training our part-of-speech tagger. The size of that corpus is about 125.000 words.

---

[3] Wired Communications Laboratory, Dept of Electrical and Computer Engineering, University of Patras, Greece

3.3 Tokeniser and Sentence Splitter

This module was implemented in Sicstus Prolog. The tokeniser accepts raw text as input and produces a list of tokens and their boundaries (byte offsets). An identifier is assigned to each token. The sentence splitter uses the tokens produced to generate a list of sentences. Each sentence is described by its span and the list of its constituents (identifiers of tokens). An identifier is also assigned to each sentence.

The tokeniser uses a set of rules in order to identify the tokens. Examples of these rules are presented below:
- a character string represents a token when one of the following characters occurs: <space>, `, ", (,), [,], «, », <new-line>, <tab>.
- a character string represents a token if after «.», either «'» or «"» occurs.

The sentence splitter uses also a set of rules. Examples of such rules are the following:
- the characters «!», «;», «?» mark always the end of a sentence.
- the occurrence of «.» or «:» marks the end of a sentence in certain cases.

3.4 Part-of-Speech Tagger: the Brill tagger

The Brill tagger is a rule-based part-of-speech tagger. It works by first assigning each word its most likely tag, and then changing word taggings based on contextual cues. There are two stages in training (see Brill tagger README file in http://www.cs.jhu.edu/~brill/):
- Rules are learned to predict the most likely tag for unknown words. (Example: if a word ends in "ed", it is probably a past tense verb). These rules operate on word types. If the outcome of applying these rules is that a word should be tagged with a particular tag, this holds for all occurrences of the word in the corpus.
- Rules are learned to use contextual cues to improve tagging accuracy. (Example: change the tag of a word from verb to noun if the previous word is tagged as a determiner). These rules operate on individual word tokens.

A new set of part-of-speech tags for the Greek language was specified, which contains 61 tags (the initial tagset for the English language contains 48 tags). We had to define new tags in order to take into account issues such as the gender for nouns and adjectives, number for adjectives and verbs, etc. We had to decide whether we would use more tags in order to represent more features for the Greek words (e.g. cases for nouns, adjectives and verbs, mood for verbs, etc.). We finally decided to use a rather limited tagset for the Greek language (although larger than the English ones) for efficiency reasons. Our intention is to combine the results of the Brill tagger with a Greek morphological analyser, in a similar way to the one used for the Italian NERC system in ECRAN project, where the Brill tagger was used after the morphological analyser in order to solve any ambiguities produced.

We have also to note that our decision to train the Brill tagger for the Greek language was due to the fact that we couldn't find and use a Greek lexicon. A rich lexicon would give better results without the need for hand tagging.

The Brill tagger was trained into a part of the corpus on management succession events (about 15,000 words) that was hand-tagged with the specified Greek tags, and on the larger hand-tagged corpus provided by WCL (125.000 words). For that second training we had to create a "translator" in order to convert the tags used in the WCL corpus to our tagset. More specifically, the whole tagged corpus (15.000 + 125.000 = 140.000 words) was split in two corpora of equal size. In the first stage of training the lexicon and the lexical rules were learned from the first corpus. The contextual cues were learned during the second stage.

Some first evaluation was performed on a part of the corpus on "management succession events". More specifically, the corpora for May 1998 and June 1998 (1/6 – 12/6/98) were processed by the trained Brill tagger. The number of words tagged correctly by the tagger over the whole number of words in the corpus was computed (i.e. the recall of the tagger). The results for each of the two corpora and in total are presented in Table 2.

Corpus	Size (words)	Words Wrongly tagged	Recall (%)
May 98	1211	54	95,5
June 98	544	29	94,7
Total	1755	83	95,3

Table 2. Evaluation of Greek Brill tagger

The result is not very satisfactory, since the tagger was trained in a rather large corpus. We still have to test whether we would have similar results if we used as training corpus only the 15.000 corpus on "management succession" (i.e. same domain with the test corpus). In other words, we have to examine the behavior of the Greek Brill tagger according to the domain (training and testing).

In any case, these are just our first results. We still have to evaluate the Brill tagger in a larger corpus in order to have more reliable results.

Another test that should also be performed is related to the training with the large general-theme WCL corpus. According to the training strategy followed, the corpus was split into two, without taking into account the structure of the specific corpus. The corpus covers many domains and the documents are grouped according to the domain. A better splitting strategy would be to split into two each domain-specific group of documents, in order for the Brill tagger to take into account all the domains covered.

3.5 Gazetteer Lookup

This module attempts to identify phrases and keywords related to named entities, as defined for the management succession task (persons, organisations, locations, dates). This is done by searching a series of pre-stored lists (gazetteers) of organisations, locations, date forms, currency names, etc.

In order to use that module for the Greek language, we have to create Greek gazetteers. The gazetteers used so far contain some of the proper nouns identified by the Brill tagger which were then classified by hand in the different types of named entities (mainly persons and organisations). More specifically, the current status of gazetteer list is the following (in number of entries): Persons: 842, Organisations: 475, Locations:

154, Titles: 107, Dates: 34, Company Designators: 19. Table 3 contains indicative samples of these lists.

Φραγκίσκος	Σκάι 100,4 FM	Ουκρανίας
Τράγκας	ΣΚΑΪ 100,4 FM	Ουγγαρίας
Τονιά	ΣΚΑΙ	Ουγγαρία
Σόφη	Ποπ-Κορν	Ολλανδίας
Σόνια	Ποπ&Ροκ	Ολλανδία
Σωτήρης	Ποπ Κορν	Νοτίου Ελλάδος
Σίσσυ	Μελωδία FM 100	
Σίμος	Μελωδία	
Σίλια	ΜΕΛΩΔΙΑ FM 100	
Σέργιος	Ι.Γ. Δραγούνης & Υιοί ΑΕ	Σύμβουλος Media
Λία	Ι.Γ. Δραγούνης & Υιοί	Σύμβουλος Marketing
Λένα	Ι.Γ. Δραγούνης	Πρόεδρος Διοικητικού Συμβουλίου
Ιφιγένεια	Ι. Γ. Δραγούνης & Υιοί ΑΕ	Διεύθυνση Στρατηγικού Σχεδιασμού
Ισίδωρος	Ι. Γ. Δραγούνης	Διεύθυνση Επικοινωνίας
Ι.	Flash 96.1	Διεύθυνση Διαφήμισης
Ελισάβετ	Flash 96,1 FM	Διεύθυνση Marketing
Ελεονόρα	Flash 96,1	Senior Product Manager
Ελεάνας	Flash 9,61 FM	Senior Media Planner
Ελεάνα	Flash 9,61	

Table 3. Samples of Gazetteers

The development of such lists for the domain of management succession presented a lot of problems. The names of persons and organisation in the relevant corpus are actually bi-lingual, that is there are several English names especially in the case of organisations (see Table 3). This means that we actually have to maintain two sets of gazetteers: one for Greek and one for English names. Unfortunately for our tests, this is not the only problem. We found out that several of the names are bilingual in their own, that is they are composed from English and Greek characters. Greek characters are used in English names and vice versa. Thus, although a name exists in a gazetteer, it is possible (and this occurs frequently) that it won't be found and subsequently tagged. Another problem concerns the different writings of the same name, such as the name of the radio station "Flash" in Table 3, which in our training corpus occurs with 5 different forms. There is also the problem of Greek proper nouns declension ("Ελεάνα" in nominative and "Ελεάνας" in accusative) and with accented characters, which are not used all the times. We plan to develop a module that will be responsible for identifying these cases and resolve them.

We also hope that we will be able to get soon gazetteers for Greek locations, names and organisations by the WCL laboratory. These gazetteers are the result from the EU-funded project ONOMASTICA, where WCL was participating. The integration of those gazetteers will improve significantly the results of the NERC system.

3.6 Named-Entity Chart Parser

The parser is a modification of the Gazdar and Mellish bottom-up chart parser [15]. It applies a Named Entity grammar to construct proper noun phrases. In the named entity grammar of English IE system, there are 189 rules for organisations, persons, locations, temporal and number expressions. The following information is taken into account by the grammar: part-of-speech tags of the words in the NE and close to it, gazetteer tags for the words in the NE and close to it, punctuation.

We had to create new rules for the Greek language, excluding most of the English rules (see in Fig. 3 the resulting parse tree for the named entities appearing in a sentence from the Greek corpus). This was due to the nature of the Greek language and of the specific corpus. For instance, several of the English rules for organisations are based on the existence of a company designator (i.e. Ltd., Co.), which is not used in the Greek corpus. There was only one case in the test corpus where such a designator was found (A.E. which means S.A., but unfortunately, although this was in the gazetteer list, it was written with English characters in the corpus and thus it couldn't be matched against the list). Another example is the case of person names, where there are English rules based on person title (i.e. Mr., Mrs). However no such titles occur in the Greek corpus.

It seems that in the case of Greek corpus, the named entity rules should be mainly based on the existing gazetteer tags. A rich gazetteer is thus required in order to improve the named entity parser results. However, it is difficult to create and maintain such rich gazetteers. New empirical rules should be included in order to identify unclassified proper nouns as named entities and classify them as persons, organisations, etc. To achieve this we have to take into account the context of the proper nouns. For this purpose we performed some first tests, the results of which are presented below, introducing some new contextual rules. In most of the cases, as it is also shown from these first evaluation results, such rules can prove effective. However, we have to note that these rules represent actually the writing style of the specific technical writer of the corpus. A new technical writer may introduce a new style, canceling in practice those rules or even introducing erroneous results. We need to make a lot of tests implementing different scenaria (actually this is the 2nd year's task of GIE) with different sets of rules.

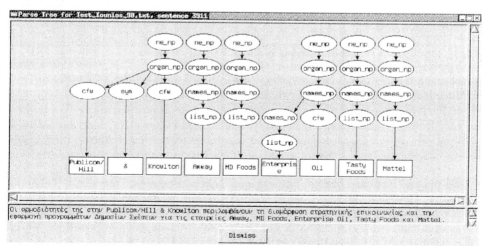

Figure 3. The parse tree for the named entities

We present below the results of two evaluations for two types of named entities (persons, organisations). These evaluations were performed in parts of the corpus on management succession events, which are different from the parts used for training (i.e. evaluation on unseen data).

In the first evaluation (Table 4(a,b,c)), all the named-entity rules are dependent on the gazetteer tags, whereas in the second evaluation (Table 5) a few contextual rules for persons were added in two stages (Rule 1, Rule 2). In both evaluations, we compute the recall and precision figures.

Corpus	Named Entities (words)	Named Entities Correctly Tagged (words)	Recall (%)	Named Entities Tagged (words)	Precision (%)
Jan. 98	187	112	60	117	96
May 98	144	121	84	126	96
June 98	79	68	86	68	100
Total	410	301	73	311	97

Corpus	Persons (words)	Persons Correctly Tagged (words)	Recall (%)	Persons Tagged (words)	Precision (%)
Total	156	109	70	118	92

Corpus	Organisations (words)	Organ. Correctly Tagged (words)	Recall (%)	Organ. Tagged (words)	Precision (%)
Total	254	192	76	193	99

Table 4(a),(b),(c) First Evaluation results of NERC

Corpus	Persons (words)	Persons Correctly Tagged (words)	Recall (%)	Persons Tagged (words)	Precision (%)
Rule 1	156	120	77	129	93
Rule 2	156	136	87	145	94

Table 5. Second evaluation results of NERC

As we can see in Table 4(a), there are differences in recall figures for named entities in the three corpora examined. Especially, concerning January 98 corpus the low recall is due to the large number of names for persons and organisations that are not covered by the gazetteer list. On the other hand, the precision figures are high, since the parser is based mainly on the gazetteer tags and the tags for proper nouns, reducing the number of wrongly tagged noun phrases as named entities. Concerning the second evaluation for persons (Table 5), there is a significant increase with the addition of two sets of rules in two stages. The first set of rules is not corpus dependent, but rather language dependent. However, the larger increase (10%) occurs in the second stage where the added set of rules is corpus dependent. The second set of rules represents a writing style that seems common for this type of company news (management succession events). However, it is possible that a new technical writer would use a different style reducing the positive effect of these rules in practice. In any case, these figures are just our first results. We believe that by the end of GIE we will be able to present more reliable results from evaluations in larger corpora and using more extensive resources (i.e. lexicon, gazetteer lists).

4. Concluding Remarks

The GIE project represents the first effort for the development of an IE system (actually a named-entity recogniser) for the Greek language. We strongly believe that IE technology is important for the Greek market too. This also results from our discussions with people working in the Greek industry. There is a need for reliable Greek IE technology in several domains, such as the stock market and the company news. This small project is actually the first step towards the development of such a technology for the Greek language. Several actions should still be taken concerning the Greek linguistic resources. A rich lexicon is essential part of an IE task, in order to support not only the morphological analysis task but also the parsing and discourse processing tasks in a complete IE system.

Customisation to new domains is another issue. Actually, it is the major issue in IE technology. Research effort is mainly devoted to making IE more adapted to new domains without the huge human labour overhead of constructing new templates by hand for each change of domain. The experience of our laboratory in the development of customisation tools for the needs of the ECRAN project can be useful also for the Greek language [16]. For this purpose we submitted together with the French company CDC-Informatique a proposal in the context of the French-Greek bilateral scientific cooperation programme for a project aiming at the development of customisation tools for IE for the French and the Greek language. The funding of such a project will allow us to better exploit the results of the GIE project towards the development of a more complete IE system and a set of customisation tools for the Greek language.

References

1. Cowie J., Wakao T., Jin W., Pustejovsky J. and Waterman S.. The diderot information extraction system. In Proceedings of the First Conference of the Pacific Association for Computational Linguistics (PACLING 93), Vancouver, Canada, 1993.
2. Jacobs P.S. and Rau L.F.. Scisor: Extracting information from on-line news. Communications of the ACM, 33(11):88-97, 1990.
3. Wilks Y. Diderot: a text extraction system. In DARPA Speech and Natural Language Workshop. Morgan Kaufmann, San Mateo, CA, 1991.
4. Vichot F., Wolinski F., Tomeh J., Guennou S., Dillet B., Aydjian S., High Precision Hypertext Navigation Based on NLP Automatic Extractions, Hypertext, Information Retrieval, Multimedia (HIM'97), Dortmund, Germany, (30):161-174, October, 1997.
5. Andersen P.M., Hayes P.J., Huettner A.K., Nirenburg I.B., Schmandt L.M.and Weinstein S.P. Automatic extraction of facts from press releases to generate news stories. In Proceedings of the Third Conference on Applied Natural Language Processing, pages 170-177. ACL, 1992.
6. ECRAN: Extraction of Content: Research at Near Market, http://www2.echo.lu/langeneg/en/le1/ecran/ecran.html

142

7. MUC5, 1993. Proceedings of the Fifth Message Understanding Conference, San Francisco, Calif.: Morgan Kaufmann.
8. MUC6, 1995. Proceedings of the Sixth Message Understanding Conference, San Francisco, Calif.: Morgan Kaufmann.
9. DARPA Speech and Natural Language Workshop, Harriman, NY, 1992.
10. AVENTINUS: Advanced Information System for Multinational Drug Enforcement. http://www2.echo.lu/langeneg/en/le1/aventinus/aventinus.html
11. Evans R.and Hartley A.F.. The traffic information collator. Expert Systems: The International Journal of Knowledge Engineering, 7(4):209-214, 1990.
12. Gaizauskas R., Evans R., Cahill L.J., Richardson J., and Walker J.. Poetic: A system for gathering and disseminating traffic information. In S.G.Ritchie and G.T.Hendrickson, editors, Conference Preprints of the International Conference on Artificial Intelligence Applications in Transportation Engineering, pages 79-98, San Buenaventura, California, 1992.
13. Gaizauskas, R., Wilks,Y. «Information Extraction beyond Document Retrieval», University of Sheffield, Dept. of Computer Science, CS-97-10, 1997.
14. Cunningham, H., Wilks, Y., Gaizauskas, R., GATE - a General Architecture for Text Engineering, *16th Conference on Computational Linguistics (COLING'96)*, 274-279, 1996.
15. Gazdar G. and Mellish C, 1989. Natural Language Processing in Prolog. Addison-Wesley, 1989.
16. Paliouras G., Karkaletsis V. and Spyropoulos C.D., "Machine Learning for Domain-Adaptive Word Sense Disambiguation". Proceedings of the LREC Workshop on "Adapting Lexical and Corpus Resources to Sublanguages and Applications", Granada, Spain, May 26, 1998.

13

Using Functional Style Features to Enhance Information Extraction from Greek Texts

S. E. Michos, N. Fakotakis, and G. Kokkinakis
Wire Communications Laboratory
Div. of Telecommunications and Information Technology
Dept. of Electrical Engineering and Computer Technology
University of Patras
GR-26500, Patras, GREECE

1. Introduction

Current Information Extraction (IE) systems extract, in most cases, *fixed* information from documents [1,2]. This information pertains only to four distinct tasks: *named entity recognition, coreference identification, template elements filling*, and *scenario-based template elements filling*. Thus, providing these systems with the capability of locating stylistic features in a text and thus detecting its genre, it would be possible to meet specific user interests. For instance, users are often looking for texts on a certain topic with particular, quite narrow generic properties, such as authoritatively written documents, opinion pieces, scientific articles, and so on.

Several attempts have been made for achieving a statistical analysis of style by counting certain words or phrases (i.e., the so-called *style markers*) in texts and comparing the results to a relative norm, in order to decide what type of style the text is [3]. However, these systems are not yet able to interpret the results. Another interesting approach in style variation makes use of a multi-dimensional methodology in order to chart the various ways in which language varies [4]. Although the above methodology is very useful thanks to its cross-linguistic and diachronic orientation, it presents serious technical obstacles and seems time-consuming to implement.

Very recently, they were presented systems based on stylistic information for text categorisation [5,6]. However, all these approaches, though thought provoking and entirely different from a theoretical point of view, are coped only with information retrieval experiments.

In this chapter we present a text categorising computational model for Modern Greek (MG) that is based on stylistic information, namely a three-level stylistic description of functional style (FS). Our work is strongly motivated by the need of categorising unrestricted texts in terms of FS in order to attain a satisfying outcome in style processing and thus in IE applications. In order to achieve this purpose, we have relied on the statistical analysis of large MG text corpora as well as on empirical methods. Finally, with the view of making the required information available, we were based on existing systems for morphological and syntactic analysis.

In the next section the three-level description of FS is briefly outlined. For a more detailed presentation the interested reader can look for [7]. This section ends up with the way unrestricted texts can be categorised in terms of FS as well as the selection of appropriate stylometrics to achieve the intended results on text categorisation. Then, in section 3 we present the computational implementation of our model by giving the model requirements, an overview of the computational model and a clarifying example. An early evaluation of it follows in section 4. In section 5 it is shown and discussed how the results on text categorisation can be used in IE applications to make current IE systems more adaptive to specific user needs. Finally, in section 6 some conclusions are drawn and future work directions are given.

2. Background

2.1. THREE-LEVEL FS DESCRIPTION

FS is the quantitative and qualitative use of language in a specific social relationship for a specific communication aim. It is usually encountered in texts where the personal style of the author is overshadowed by the functional objectives. In order to model FS as better as possible we have adopted a hierarchical description that is composed of the following levels (see Figure 1):

Level 1
This level comprises the five basic categories of FS, that is *public affairs* style, *scientific* style, *journalistic* style, *everyday communication* style and *literary* style. Although the definition of a complete set of FS categories seems to be an unsolved problem, it is stressed here that this classification conforms to what many scholars call a potential and logical set of FS categories [8].

Level 2
This level includes the main features that characterise each one of the above categories, that is *formality*, *elegance*, *syntactic complexity* and *verbal complexity*.

Level 3
This level is composed of the linguistic identifiers that act as style markers in texts for the identification of the above features. These identifiers are divided into verbal and syntactic ones and are given below:

(a) **Verbal identifiers**: idiomatic expressions like "ρίχνω λάδι στη φωτιά" (add fuel to the fire) or "πηγαίνω κατά διαβόλου" (go by the board), "sophisticated" expressions like "επ' άπειρον" (in perpetuity) or "γνήσιο τέκνο" (true-born issue), scientific terminology like "ισοζύγιο" (balance) or "πληκτρολόγιο" (keyboard), "formal" words like "άρση" (lifting) or "μεταστροφή" (swing) or "εμφαντικά" (emphatically), poetic words like "άτι" (steed) or "ξεροβόρι" (icy wind), abbreviations like "ΗΠΑ" (USA) or "ΕΚ" (EC) or "ΟΗΕ" (UN).

(b) *Structural identifiers*: number of words per sentence, number of conjunctions per sentence, number of sentences per paragraph, verbs-nouns ratio, verbs at third person-verbs ratio, nouns at genitive case-nouns ratio, subordinate-main sentences ratio, adjectives-nouns ratio, adverbs-verbs ratio, active-passive voice ratio.

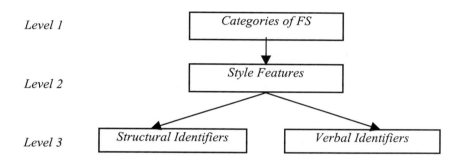

Figure 1. A three-level FS description.

Three points should be mentioned here. First, it is obvious that both a morphological and syntactic analysis of the text at hand must be available. Second, the above description would be more accurate if a semantic and/or pragmatic analysis of texts could also be available. In this case, it could be expanded to include also semantic and/or pragmatic identifiers. Nevertheless, the aim of this work is to deal with unrestricted texts, so such an effort seems unrealistic regarding the excessive computational cost that yields. Third, in order to obtain as language-independent results as possible from such a description, we attempted to build the set of style markers as generally as possible. So, intrinsic elements of MG such as the use of special verbal endings that could be comprised in the third level, have been ruled out. Surely, for getting better results it could be useful to apply the three-level description to a specific language by incorporating such special elements. It must be noted here that each language has its own set of words and expressions that compose the verbal identifiers. Thus, if a word or an expression is characterised as idiomatic in one language, its translation into another language may not be idiomatic at all.

2.2. FS IDENTIFICATION

Generally, by checking the style markers in a text we are able to draw conclusions about the effect they have on the four style features. This fact can facilitate us in making an estimation on the text FS category. So, the linguistic identifiers of the third level act as style markers for the style features of the second level as it is explained below:

Formality
Regarding the verbal identifiers, the large use of "formal" words and "sophisticated" expressions as well as the infrequent presence of abbreviations and idiomatic expressions characterise formal texts. Concerning the structural identifiers, the following style markers have been detected in formal texts: great number of words per sentence, small number of sentences per paragraph, great number of conjunctions per sentence, low verbs-nouns ratio, high nouns at genitive case-nouns ratio, high verbs at third person-verbs ratio, predominance of the passive voice over the active one and high subordinate-main sentences ratio.

Elegance
From the verbal point of view, many idiomatic expressions and poetic words characterise elegant texts. From the structural point of view, these texts have been observed to possess high adjectives-nouns ratio, high adverbs-verbs ratio, low verbs-nouns ratio, high verbs at third person-verbs ratio and predominance of the active voice over the passive one.

Syntactic complexity
Syntactically complex texts are characterised by great number of words per sentence, great number of sentences per paragraph, great number of conjunctions per sentence, low verbs-nouns ratio, high nouns at genitive case-nouns ratio, high verbs at third person-verbs ratio, high adjectives-nouns ratio, high adverbs-verbs ratio and high subordinate-main sentences ratio.

Verbal complexity
Verbally complex texts are characterised by many "sophisticated" expressions, plenty of scientific terminology, many "formal" words, a lot of abbreviations and poetic words and few idiomatic expressions.

Then, after having recognised the degree of effect of the four style features in a given text, the identification of its FS can be based on the following set of estimation rules:

Public affairs style: Formal and syntactically complex to a large extent, elegant and verbally complex to a small extent.

Scientific style: Formal and verbally complex to a large extent, elegant and syntactically complex to a small extent.

Journalistic style: Elegant and syntactically complex to a large extent, verbally complex and formal to a small extent.

Everyday communication style: Formal, elegant, syntactically complex and verbally complex to a small extent.

Literary style: Elegant to a large extent, formal, syntactically complex and verbally complex to a small extent.

The presented approach to text categorisation was based on three main factors: (a) the empirical selection of the style markers and style features, (b) the statistical processing of large MG text corpora of about 100,000 words, and (c) the empirical assessment of the statistical results with the view of identifying FS in unrestricted texts as impartially as possible.

2.3. DETERMINATION OF STYLE MARKERS NORMS

Expressions like "great number of conjunctions per sentence" or "low verbs-nouns ratio" are referred to the comparison of the text's number of conjunctions per sentence and text's verbs-nouns ratio to the corresponding ones of the language norms. It has proved that such linguistic quantities are very similar among languages. For example, for English and French the conjunctions are approximately 4% and 3% of the words respectively, while the verbs-nouns ratio is approximately 0,6 and 0,5 respectively [9].

In Table 1 we give the set of style markers norms for MG as it was derived from statistical analysis of large tagged MG text corpora taken from the ESPRIT-860 project (it should be mentioned here that the texts were selected to belong to all FS categories) [10]. This set can be easily ported to other languages with slight modifications of its values. It has also to be noted that some values, especially those referring to verbal identifiers, are approximate, since it is not yet possible to have an acceptable average for them.

Table 1. Style markers norms for MG.

Style Markers	Norm
number of words per sentence	15
Number of conjunctions per sentence	0,6
Number of sentences per paragraph	5
Verbs-nouns ratio	0,5
verbs at third person-verbs ratio	0,6
Nouns at genitive case-nouns ratio	0,25
Subordinate-main sentences ratio	1,5
Adjectives-nouns ratio	0,3
Adverbs-verbs ratio	0,4
Active-passive voice ratio	1,5
Idiomatic expressions	0,02
"sophisticated" expressions	0,01
Scientific terminology	0,01
"formal" words	0,05
Poetic words	0,01
Abbreviations	0,02

2.4. TEXT CATEGORISATION METHODOLOGY

According to the previous stylistic description, if the detected value of a style marker is different from that of its norm, then this style marker may have a positive or negative effect on a certain style feature. For example, if the active-passive voice ratio has been found to be greater than the norm, then this style marker has a positive effect on the elegance and a negative one on the formality as it can be derived from the descriptions of these two features in section 2.2.

Additionally, a style feature is considered to be "to a small extent", if the percentage of the style markers that have a positive affect on it is lesser than 50% (<50%). Furthermore, a style feature is considered to be "to a large extent", if the corresponding percentage of the style markers that have a positive affect on it is greater than 65% (>65%). If the previous percentage is between 50% and 65% (>50% and <65%), then this percentage is ambiguous and cannot lead to a valid estimation of the feature impact.

Finally, employing the set of the estimation rules of the section 2.2 makes the estimation on the FS category of a given text. Needless to say that every time we have four measured percentages that equal the number of four style features. Therefore, if at least three of the above percentages are unambiguous (i.e., <50% or >65%), we look for the estimation rule that best matches the results. If there are two of them, we do make an estimation but this estimation cannot lead to a definite FS category. In this case, a further analysis of the given text is needed in order to draw a more precise conclusion of its FS category. On the other hand, if at least two of the percentages are ambiguous, an estimation is no longer feasible. Again in this case a further analysis of the given text is needed in order for an estimation to be feasible. Obviously, in several cases the extraction of a valid estimation is a quite difficult process, especially when the size of the text is too small.

3. Implementation

3.1. REQUIREMENTS

As it has been mentioned in the previous section, in order to develop a system that will implement the three-level description of FS, a morphological and a syntactic analyser should be available.

- The former must be able to provide verbal information (e.g., "formal" word, abbreviation, poetic word, etc.) besides the pure morphological information (e.g., part-of-speech, case, number, etc.) for each word of the text.

- The latter must be able to recognise predefined expressions (e.g., "sophisticated" ones, idiomatic ones, etc.), calculate syntactic quantities (e.g., number of words per sentence, number of sentences per paragraph, etc.), and provide syntactic information (e.g., main sentences, subordinate sentences, etc.) for every sentence of the text.

3.2. OVERVIEW OF THE MODEL

The presented model is the integration of three distinct modules as it is shown in Figure 2. These modules are described below:

(a) The Morphological Analyser (MA) is a two-level processor based on a PC-KIMMO description of MG. Its lexicon contains about 30000 words. For a detailed presentation the interested reader can look for [11].

(b) The Syntactic Parser (SP) is a computational model that is able to parse unrestricted texts of 'quasi free' word order languages such as MG. For a detailed presentation the interested reader can look for [12].

(c) The Stylistic Analyser (SA) is the module that implements the presented method for text categorisation based on stylistic information.

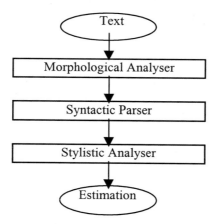

Figure 2. The Overview of the model.

So, when the morphological and syntactic processing of the text have been carried out, all the required style markers values are available, and the SA is able to make an estimation about the FS of the text based on the three-level stylistic description.

It has to be underlined that the two models, MA and SP, already existed and were not designed especially for stylistic analysis. Hence, in order to adopt them on our model several modifications had to be done. So, the MA has been extended to include the required verbal information. On the other hand, the SP has improved with the incorporation of two sub-modules: the first one is able to recognise characteristic expressions, and the second one is endowed with special functions to calculate the required style markers' values.

3.3. A CLARIFYING EXAMPLE

With the view of clarifying further the above methodology to text categorisation in terms of FS, we give in this section a detailed example of identification of the FS category of a text based on it. We have used a text of 3500 words taken from a *newspaper* that has been analysed in the framework of the ESPRIT-860 project. It has to be noted that this analysis provided only a part of the aforementioned set of style markers. All the rest have been calculated manually.

After the morphological and syntactic analysis of the sample text, the set of the values of the style markers was available. The results, the corresponding deviations from the norm values as well as their effect on each style feature are shown in Table 2. Note that the symbols (+) and (-) stand for positive and negative effect on a certain feature respectively.

Table 2. Results of the analysis of the sample text.

Style Markers	Value	Deviation (%)	Formality	Elegance	Syntactic Complexity	Verbal Complexity
Number of words per sentence	27,7	+85	+		+	
Number of conjunctions per sentence	1,17	+95	+		+	
Number of sentences per paragraph	2,74	-45	+		−	
Verbs-nouns ratio	0,59	+18	−	−	−	
Verbs at third person-verbs ratio	0,79	+27	+	+	+	
Nouns at genitive case-nouns ratio	0,27	+8	+		+	
Subordinate-main sentences ratio	2,3	+53	+		+	
Adjectives-nouns ratio	0,62	+101		+	+	
Adverbs-verbs ratio	0,57	+43		+	+	
Active-passive voice ratio	1,53	+2	−	+		
Idiomatic expressions	0,05	+150	−	+		−
"Sophisticated" expressions	0,008	-20	−			−
Scientific terminology	0,01	0				−
"Formal" words	0,04	-20	−			−
Poetic words	0	-100	+	−		−
Abbreviations	0,001	-95	+			−

Taking into account the results of this table we calculated the percentages of the style markers that have a positive effect on each style feature. These can be summarized as follows:

Formality:	8/13 ≈ 62%	(>50% and <65%)
Elegance:	5/7 ≈ 71%	(>65%)
Syntactic Complexity:	7/9 ≈ 78%	(>65%)
Verbal Complexity:	0/6 ≈ 0%	(<50%)

From the observation of these percentages, we can conclude that the sample text is elegant and syntactically complex to a large extent and verbally complex to a small extent. Regarding the formality of this text we cannot make a valid estimation of this feature impact, since its percentage has been found to be ambiguous. Finally, the estimation rule that best matches these results is that of the *journalistic style* since at least three of the above percentages are unambiguous (i.e., elegance, syntactic complexity, and verbal complexity).

4. Evaluation

In Table 3 they are shown the analysis results and the estimations the model produced for five sample texts, each one of them belonging to a different FS category. In spite of the small size of these sample texts (about 210-841 words), the model managed to identify correctly the FS of 3 texts (i.e., public affairs, scientific and everyday communication).

Moreover, the estimation for the literary text led to two FS categories (i.e., literary or journalistic) and only for one sample text (i.e., the journalistic one), the estimation is not correct. This was due to the unusual high percentage of the formality of this text.

Table 3. Analysis results for 5 sample texts.

FS Category	Words	Formality (%)	Elegance (%)	Syntactic Complexity (%)	Verbal Complexity (%)	Estimation
Public Affairs	841	77	57	78	17	*Public Affairs*
Scientific	500	77	29	56	67	*Scientific*
Journalistic	320	77	43	67	17	*Public Affairs*
Everyday Commu/tion	210	23	43	11	0	*Everyday Communication*
Literary	395	31	71	56	0	*Literary or Journalistic*

However, in order to attain the best possible results, it has been estimated that the number of words in a sample text must be at least 500.

5. Discussion

There are some current sophisticated systems that take advantage of FS features in order to attain better results in applications such as text generation (e.g., PAULINE) or machine translation (e.g., STYLISTIQUE) [13,14]. Particularly, STYLISTIQUE can identify the stylistic goals of the writer by choosing a goal from three dimensions, such as *clarity-obscurity, concreteness-abstraction,* and *staticness-dynamism.* On the other hand, PAULINE uses the following stylistic rhetorical goals to generate texts: *formality, simplicity, timidity, partiality, detail, haste, force, floridity, color, personal reference, open-mindedness,* and *respect.*

In our opinion, similar techniques could be used in IE applications. In this case, the results on text categorisation, as they were presented previously, can be used in IE applications to make current IE systems more adaptive to specific user needs. For instance, a user of an IE system may want to look for *journalistic documents* on the ecology topic that are *in favour of* it. Another user may want to read *scientific articles* on the future of information systems that are characterised by *open-mindedness.* Finally, another user may want to search for *letters* to the editor of a magazine whose authors take a *neutral* position in a topic addressed by a past issue of this magazine. In all the above cases, the IE process could be greatly facilitated by the utilisation of FS features.

In conclusion, providing the IE systems with the capability of locating FS features in a text and thus detecting its genre, it would be possible to meet specific user interests. Besides, these systems would be easily customisable to new user interests by incorporating in them new FS features.

6. Conclusions

Stylistic aspects, though necessary in deep understanding of language, have been neglected in computational linguistics research. These problems had been too vague and ill defined to be dealt with by computational systems. However, in this work, we have presented an empirical model based on a formal description of FS that makes the problem of text categorisation more amenable to computational solution. It is hoped that this research will lead to a system sophisticated enough to cope with various applications including grammar and style checking, natural language generation, style verification in real-world texts, and recognition of style shift between adjacent portions of text (e.g., paragraphs).

It can be understood that the more the deviation of a linguistic identifier is from the norm, the more significant its effect is on the estimation process. For instance, a text that has a verbs-nouns ratio equal to 0,2 (i.e., deviation from the norm = 60%) is considered more formal than another one that has 0,3 (i.e., deviation from the norm = 40%). For this reason, it is obvious that the deviation of the style marker value from the norm must be taken into account by the model by means of a *weights mechanism.* However, in those cases, that the percentage of the deviation of a linguistic identifier from its norm is sufficiently small, if not negligible, we are looking for some *threshold values* that will ensure the correct evaluation of our results.

Short-term research is currently focused on some problems that faces the computational implementation of the aforementioned findings as well as the selection of more precise and appropriate stylometrics to achieve better results on text categorization. Towards this direction, the extraction of the most appropriate language norms for all the presented style markers on one hand and the formulation of the most accurate estimation rules on the other hand, are the key points for the successful completion of this research. Long-term research will be concentrated on determining a more thorough set of FS features as well as an updated set of FS categories in order to achieve the best possible results in IE applications that will provide users with user-tailored IE systems.

References

[1] AZZAM S., HUMPHREYS K., GAIZAUSKAS R., CUNNINGHAM H. & WILKS Y. (1997), *"Using a Language Independent Domain Model for Multilingual Information Extraction"*, Proceedings of the 2nd Workshop on Multilinguality in Software Industry, Nagoya.

[2] BOWDEN P., HALSTEAD P. & ROSE T. (1996), *"Knowledge Extraction and Text Analysis Using Conceptual Relation Markers"*, Proceedings of the AISB 1996 Workshop on Language Engineering for Document Analysis and Recognition, Brighton.

[3] CLUETT R. (1990), *"Canadian Literary Prose: A Preliminary Stylistic Atlas"*, ECW Press.

[4] BIBER D. (1995), *"Dimensions of Register Variation: A cross-linguistic comparison"*, Cambridge University Press.

[5] KARLGREN J. (1996), *"Stylistic Variation in an Information Retrieval Experiment"*, Proceedings of the Association for Computational Linguistics.

[6] KESSLER B., NUNBERG G. & SCHUTZE H. (1997), *"Automatic Detection of Text Genre"*, Proceedings of the Association for Computational Linguistics.

[7] MICHOS S. E., STAMATATOS E., FAKOTAKIS N. & KOKKINAKIS G. (1996), *"An Empirical Text Categorizing Computational Model Based on Stylistic Aspects"*, Proceedings of the 8th IEEE International Conference on Tools with Artificial Intelligence, Toulouse.

[8] RIESEL E. (1963), *"Stilistik der deutschen Srache"*, 2nd Edition, Moskau.

[9] DERMATAS E. & KOKKINAKIS G. (1995), *"Automatic Stochastic Tagging of Natural Language Texts"*, Computational Linguistics, vol. 21, no. 2, pp. 137-163.

[10] TECHNICAL ANNEX OF THE ESPRIT-860 PROJECT (1986), *"Linguistic Analysis of the European Languages"*.

[11] SGARBAS K., N. FAKOTAKIS & G. KOKKINAKIS (1995), *"A PC-KIMMO Based Morphological Description of Modern Greek"*, Literary and Linguistic Computing, Vol. 10, No. 3, Oxford University Press, New York.

[12] MICHOS S. E., N. FAKOTAKIS & G. KOKKINAKIS (1995), *"A Novel and Efficient Method for Parsing Unrestricted Texts of Quasi-Free Word Order Languages"*, International Journal on Artificial Intelligence Tools, Vol.4, No. 3, pp. 301-321.

[13] HOVY E.H. (1990), *"Pragmatics and Natural Language Generation"*, Artificial Intelligence, vol. 43, pp. 153-197.

[14] DiMARCO C. & HIRST G. (1993), *"A Computational Theory of Goal-Directed Style in Syntax"*, Computational Linguistics, vol. 19, no. 3, pp. 452-459.

14

Lexical Knowledge Extraction from Technical Texts

Ingeborg Blank
Centre for Information and Language Processing
University of Munich
Oettingenstr. 67, D - 80538 Munich

1. Introduction

This chapter deals with the processing of a multilingual corpus of technical texts, that is a collection of texts translated into several languages. The aim is to elaborate an adequate method for the acquisition of lexical knowledge, especially terminological knowledge, from a text corpus. A method has been developed that combines techniques and tools for symbolic processing with statistic filters. The application of the method to such a corpus resulted in a tool suitable for the semi-automatic extraction of terms and some other lexical and semantic properties, e. g. translation equivalences, semantic relations between terms or typical contexts of terms.

The relevant knowledge contained in technical texts is concentrated in technical terms. Thus, the terminology extraction process can be considered as a particular case of an information extraction task. This task, although language-specific, is domain-independent. Thus, once the method defined, it can be used in many knowledge domains.

Moreover, a multilingual corpus of technical texts provides a major source of linguistic information useful for translation, according to Isabelle: "Given the staggering volume of translations produced year after year, it is quite obvious that existing translations contain more solutions to more translation problems than any other existing resource. Unfortunately translators can currently derive very little benefit of this fact." (Isabelle, 1992: 8)

Translators of technical texts spend a lot of time on terminological work. The acquisition of bilingual lists of terminological expressions is difficult and time consuming. It is, therefore, worthwhile to investigate methods to compile such lists as automatically as possible. Lexical knowledge of this kind is necessary for various monolingual and multilingual NLP systems and tools, such as information extraction systems, machine translation or computer aided translation systems, translation memories, terminological databases, electronic dictionaries etc.

2. Method overview

The prerequisite for the task of terminology extraction is a definition of terms with criteria that can be captured by an automatic procedure, such as morphological, syntactic and statistic criteria. Thus, the first step of this work was the definition of terms by suitable criteria.

Then, the extraction of terminological knowledge is performed through a set of pipelined processes. The first stage, or text handling stage, covers the format analysis of the input text and the identification of textual units, such as document, paragraph and sentence boundaries, abbreviations etc. Then, the handled texts have been aligned on the sentence level using statistical techniques. The following step covers tokenization, lemmatization and tagging, i. e. word forms in the text are identified, reduced to their basis lemma and annotated with part-of-speech. The extraction of likely terminological units and the evaluation of the extracted units follows at the end.

3. The corpus

A trilingual (German-English-French) corpus of technical texts comprising about 12 million words was provided by the European Patent Office in Munich (EPO). The corpus includes two subcorpora each one containing a special type of documents.

- The major one is the DBA subcorpus consisting of about 1000 decisions of the boards of appeal (about 10 million words). Each decision is written in one of the three languages and then translated into the other two.

- The other one is the EPC subcorpus which is the collection of the articles and rules governing the European patent system.

The texts contained in the above corpora have legal value. Therefore, the main part of the terminology included therein is juridical and the remaining part is relating to all technical fields mentioned in the International Patent Classification (IPC) system covering all domains of chemistry, mechanics or physics.

The corpus used is particularly suitable for defining and extracting multilingual terminology, for the following reasons:

- it is structured in a very concise, homogenous and uniform manner,

- it is sufficiently big to be statistically relevant and

- the texts are written in a legal context, i.e. the translations are of good quality.

For ergonomic reasons, the present study was restricted to German and French texts only. Part of the EPC subcorpus (10000 words) was used for the definition of terms and the part of the DBA subcorpus referring to chemistry (40000 words per language) was used for the extraction of terms.

4. Definition of terminology

The basis for the linguistic definition of terms was literature from terminology, translation science, information retrieval, linguistics and computational linguistics. Terminology science, as founded by Wuester, is an interdisciplinary domain that aims at the definition, collection, storage and diffusion of terminology. Terms are usually defined by semantic criteria according to ISO/DIS 1087. Such definitions of terms, however, are not suitable for an automatic procedure. A program for identifying likely terminological units, must take into account the form of terms, i.e. their syntactic and morphological properties. Definitions of that kind can be found in some branches of terminology science and in computational linguistics.

The prescriptive branch of terminology science provides descriptions of the external form of terms as well as "norms" ruling the formation of new terms. In most studies in computational linguistics, technical terms are defined as noun phrases that satisfy a rather restricted set of morpho-syntactic patterns.

Thus, following the above definition, nouns (simple or compound) and noun phrases (built up according to some frequent patterns) are considered to be candidate terms e.g. "Beschwerdeverfahren" in German, "appeals procedure" in English or "procédure de recours" in French.

In order to check the accuracy of this syntactic definition of terminology, parts of the EPC subcorpus were manually parsed in maximal-length noun phrases and, when necessary, segmented in smaller phrases. Such a subcorpus is particularly suited for the detection of terminology. It contains definitions of the basic concepts of the European patent system and the corresponding terms for expressing these concepts. Moreover, in some cases noun phrases are explicitly marked as terms[1].

Once the terms detected, their morpho-syntactic properties have been determined. It was important to adopt a definition of the notion "term" that facilitates the comparison of terms in both languages. Word formation is very different in French and German. French compounds consist of orthographically separated elements (e. g. "chambre de recours"), German compounds are often formed by composition of morphemes resulting in one orthographic word (e.g. "Beschwerdekammer"). The recognition of both types of compounding is not trivial in automatic processing.

The definition of candidate terms finally used for the extraction was elaborated and checked on a part of the EPC subcorpus (5000 words for each language). Candidate terms are defined as noun phrases satisfying a restricted set of **part-of-speech patterns** and are classified by their **length**, i.e. the number of nouns, adjectives, verbs, and participles; e. g. the German term "Beschwerdeverfahren" has length 1 whereas the French term "procédure de recours" has length 2.

4.1. Candidate terms in French

Due to the particular properties of the word formation in French, it is sometimes impossible to establish a clear distinction between a free syntagma and a compound. This problem is discussed more in detail in several studies (Daille, 1994; Jacquemin, 1991; Bourigault, 1994). Two compounds can overlap and build a new compound, for instance "procédure de conversion d'hydrocarbures" (length 3) can be considered as a

---

1. E. g. "Patents granted by virtue of this Convention shall be **called** *European patents*" (Art. 2 (1) EPC).

merge of two compounds of length 2, namely "procédure de conversion" and "conversion d'hydrocarbures", both occurring in the corpus. On the other hand, the whole nominal phrase is translated by one compound in German ("Kohlenwasserstoffumwandlungsverfahren").

The automatic recognition and extraction of French compounds is difficult for the following reasons:

- it is impossible to determine whether a morpho-syntactic structure is a sequence of length 2 before the detection of all compounds of length 3 and

- it impossible to determine whether a morpho-syntactic structure is a sequence of length 3 before the detection of all compounds of length 2.

For this reason we decided to adopt a broad definition of potential terms taking into account all noun phrases of length 2 and the most frequent types of noun phrases of length 3. For the unclear cases the denomination "terminological unit" would be more appropriate than the denomination "term". The types of French terminological units used for the extraction stage are summarized in (Blank, 1998).

4.2. Candidate terms in German

The major part of German technical terms are compounds. In handbooks of terminology almost all other types of formation are very often considered as just a transitional state for the formation of a "real compound", (e. g. "Entscheidung über den Widerruf" with the corresponding compound "Widerrufsentscheidung"). The analysis of the German part of the EPC corpus resulted in a description for German terms reported in (Blank, 1998).

5. Extraction of candidate terms

The extraction of French and German candidate terms from the corresponding corpora is performed through a set of pipelined processes. The processes involved are text handling, sentence alignment, tokenization, lemmatization and POS-tagging. Part of these tasks could be performed with existing tools such as the INTEX system (Silberztein, 1993) or the CISLEX system (Maier-Meyer, 1995), others were carried out by programs that we implemented ourselves.

Text handling is carried out using various programs in C and Perl. These programs analyze the physical appearance of the input text and map these format characteristics in a generalized mark-up language. Afterwards these programs identify textual units, such as document, paragraph and sentence boundaries, abbreviations etc.

5.1. Sentence alignment

The input of the alignment process is a source text and a target text, i. e. the translation of this source text. Alignment exists on several levels. To align a text with its translation on the sentence level is to show which sentences of the source text are translated by which sentences of the target text. In most cases one sentence of the source text is translated by one sentence of the target text. However, there are also harder cases of mappings, such as 1:2, 2:1 or 2:2 mappings.

Two sentence alignment algorithms had been implemented, tested and evaluated in a previous study (Blank, 1995). It turned out from this comparative evaluation that the Church-Gale algorithm performs well on texts that are structured in a manner that sentence and paragraphs boundaries can be automatically identified correctly. For the present work we used an implementation of the Church-Gale method that had yielded an accuracy of about 95% on a test corpus of about 400 000 words.

5.2. French corpus

The INTEX system is a tool for various lexical tasks, e. g. lemmatization, building of concordances, POS-tagging and search of linguistic patterns. It includes several built-in dictionaries (e. g. the dictionary of 800 000 inflected forms of simple words which contains basically all the simple words of the French language) and grammars; both are represented by finite-state automata.

We used INTEX on the French corpus for the tokenization, the lexical analysis (lemmatization and POS-tagging) and the search of candidate terms. Patterns were defined as regular expressions formed up by POS categories, a given word or a list of words. INTEX converted the regular expressions to finite state automata and applied them to the corpus. This resulted in the extraction of sequences that were only considered as candidate terms if they were linguistically correct. INTEX can be tuned for special purposes by means of a user dictionary, a preference lexicon, local grammars etc. (Blank, 1997).

5.3. German corpus

The German subcorpus has been tokenized, lemmatized and annotated with POS categories by means of the CISLEX system. This system is based on a set of electronic dictionaries for German. The dictionary of simple words contains 265 000 base forms and over 1.2 million inflected forms. A complex semantic encoding has been provided for the major part of these simple words (Langer, 1996). The morphological encoding is based on 28 different morphological features and over 270 different morphological feature bundles that are subsets of morphological feature values. Additionally to this, the system recognizes compounds and segments them into simple form components with an accuracy of about 98%. The CISLEX system is implemented by means of a finite-state technology.

Sequences corresponding to the previously defined patterns were extracted by a program written in Perl that takes into account the fine-grained morphological information contained in the CISLEX. German has a great number of agreement features that can be morphologically realized. This concerns especially nouns, adjectives and determiners.

5.4. Results

The extraction task can be considered as a recall and precision problem. The extraction program takes a text and a morpho-syntactic definition of terms as input and provides candidate terms as output. A very restricted definition of terms yields to a high precision, i.e. the probability of getting candidate terms that are really technical terms is growing. With this approach, on the other hand, recall will drop, i.e. a part of the technical terms occurring in the text will not be extracted by the program. The extraction program used in this study promotes completeness using the definition of candidate terms presented in the previous section. This approach is justified because it is easier for the terminologist, or the translator or the knowledge engineer to eliminate some "likely terminological units" than to find "real terminological units" that escaped detection by the program.

As the extraction is carried out on lemmatized texts, terms are grouped according to their lemmas and the "real" frequency can be calculated. The output of the program must be evaluated in two stages. First it must be checked whether the extracted units are linguistically correct (i.e. well-formed noun phrases) and filter out incorrect sequences.

There is no statistical measure to judge wether an extracted noun phrase is a grammatically well-formed noun phrase. This problem can be checked only by human intervention. However, high-frequency noun phrases were, in general, correct.

In a second pass it should be judged whether the linguistically correct noun phrases are really domain-specific terminology. However, this question can only be correctly replied by the skilled person in the specific technical field. The work reported by Daille (1994) deals with the evaluation of about 20 statistical formulas for testing termhood. The conclusion is, that the majority of these formulas do not perform very well except frequency and mutual information that, however, favours rare phenomena. In the present study we used only the frequency formulas and adopted human revision as described by Church & Dagan (1994).

5.4.1. French corpus

83.02% of the extracted sequences were linguistically correct. Incorrect sequences were mainly due to disambiguation errors in POS tagging and to segmentation disambiguations of compounds of length 2 and length 3. The correctly extracted sequences can be divided in two subgroups:

- sequences with a number of occurrences higher than 4: 95% were also semantically correct (i.e. domain- specific terminology) and

- sequences with a number of occurrences between 2 and 4: 60-90% were semantically correct.

5.4.2. German corpus

About 98% of the extracted nouns (either simple or compound) were linguistically correct. The extraction of complex noun phrases reached an accuracy of 65%; for this kind of structures a more complex parsing system would be necessary.

The semantic correctness of the correctly extracted sequences varies according to the formation type and the number of occurrences. Among the different types of extracted sequences, the compounds have the highest probability to represent domain-specific terminology. 60-90% of the correctly extracted sequences with a number of occurrences greater or equal to 3 represented domain-specific terminology.

6. Applications

Studies in multilingual terminology extraction and automatic matching of translation equivalences concern mainly English-French corpora (Church & Dagan, 1994; Gaussier, 1995; Kupiec, 1993). The study of Eijk (1993), based on an English-Dutch corpus, mentioned similar problems in matching translations as those of the present study. This is, probably, due to the fact that word formation in Dutch is similar to German and the word formation in English is similar to French.

The above studies share a similar approach and are based on some assumptions about terminology and translations that are, however, not explicitly stated. We examined whether these assumptions hold on the present German-French corpus (Blank, 1998). The examination of translation equivalences revealed that the candidate terms extracted for French were, in general, translated in German by phrases belonging to the set of candidate terms extracted from the German subcorpus. This observation is less frequent the other way round i.e. from German to French. The reason for this is, probably, that the German structures with length 2 or 3 are not so often domain-specific terminology (except the structure adjective noun). This fits with the considerations about the formation of German terminology found in handbooks. This means that we extracted more German than French candidate terms as it can be shown by a simple numerical evaluation of the extraction. For this reason we did not propose an automatic matching procedure for translations but we investigated other applications based on the extracted data.

The results are presented in the form of a concordance tool that assists translators and terminologists in constructing glossaries. This tool provides, among others, the following information:

(i) candidate terms and associated concordance lines,

(ii) contextual information for candidate terms and

(iii) grouping of candidate terms with common constituents.

6.1. Candidate terms and associated concordance lines

Each candidate term is presented with the sentence of the source text from which it was extracted and the sentence(s) of the target text aligned with said sentence. One should examine the relevant lines of the text in order to decide whether a candidate term is indeed a term, and to identify the multiword terms that are omitted from the candidate term list. The local and the global frequency are also indicated for each candidate term.

Table 1 gives an example of such a concordance[1].

Table 1: Parallel text with candidate terms

French text	German text
II. Le 11 août 1982, la requérante a fait opposition à ce *brevet européen*, et en a demandé la révocation pour **défaut de nouveauté**, en faisant valoir notamment de *nouvelles antériorités*.	II. Gegen diese Erteilung des *europäischen Patents* hat die **Einsprechende** am 11. August 1982 Einspruch eingelegt und den **Widerruf** des Patents wegen *mangelnder Neuheit* beantragt. Die Begründung wurde unter anderem auf *neue Entgegenhaltungen* gestützt.
Rien dans l'**état de la technique** ne permettait d'affirmer que **l'utilisation d'hexaméthylènediamine** dans la **préparation de zéolites** s'imposait à l'évidence.	Es gebe auch im Stand der Technik keine Anhaltspunkte, die die Verwendung von **Hexamethylendiamin** bei der Herstellung von Zeolithen naheliegend erscheinen lassen.
Le procédé revendiqué dans le brevet en litige permettait de préparer directement une zéolite sans alcali, possibilité qui devait être considérée comme inattendue.	Es sei als überraschend anzusehen, daß durch das Verfahren des **Streitpatents** direkt ein *alkalifreier Zeolith* hergestellt werden kann.
Les autres antériorités avaient été publiées durant le **délai de priorité**, et ne pouvaient donc être prises en considération, puisque la priorité avait été revendiquée à juste titre; en effet, le fait que les **résultats d'analyses** ne soient pas identiques dans les exemples du fascicule de brevet d'une part et dans le texte du **document de priorité** d'autre part n'entraînait pas la perte du **droit de priorité**.	Die anderen **Entgegenhaltungen** seien im **Prioritätsintervall** veröffentlicht und daher - da die Priorität zu Recht beansprucht sei - nicht zu berücksichtigen; denn **Prioritätsverlust** trete nicht dadurch ein, daß die **Analysenergebnisse** in den Beispielen der **Patentschrift** einerseits und den **Prioritätsunterlagen** andererseits nicht identisch seien.

1. The following annotations are used: in the French text compounds of type "Noun DE Noun" are written in bold, compounds of type "Noun Adjective" in italics, other types of extracted structures are underlined; in the German text nominal compounds are marked in bold, phrases of type "Adjective Noun" in italics and other types of structures are underlined.

6.2. Contextual information for candidate terms

Verbal, nominal or other contexts in which each term is used are indicated. Language-specific syntagmatic lexical information is very important for translators. Texts that are correct on this level are perceived as fluent and natural.

For example, the French term "procédure orale" appears only with 8 different verbs, such as "la procédure orale s'est tenue"; the German term "mündliche Verhandlung", that is the translation equivalent of "procédure orale", appears with 5 different verbs, such as "die mündlicher Verhandlung hat stattgefunden" (see Blank, 1998).

6.3. Grouping of candidate terms with common constituents

All noun phrase terms that have either the same head or other constituents in common are grouped together in a kind of web. Such a grouping of linguistically related terms makes it easier to judge their validity and gives a lexical and semantic overview of the terms of a certain domain. Figure 1 and 2 in the annex present an example of such a grouping of terms. From this "terminological web" terms like "brevet", "demande de brevet", "titulaire de brevet", "revocation de brevet", "demande de revocation de brevet", "protection par brevet", "brevet en litige", "brevet litigieux" etc. are grouped together. It would be possible to construct from this grouping a kind of terminological hypertext web as described by Bourigault (1994).

7. Conclusion

In this study we developed a method for the extraction of German and French lexical knowledge, especially terminology, terms from a bilingual corpus of patent documentation. The results are used for designing a concordance tool suitable for translators and terminologists. A linguistic definition of German and French terms was elaborated which was the basis for the extraction algorithm. We checked the assumptions underlying the procedures commonly used for the matching of translations. Due to the particularities of the word formation in each language, an automatic matching of translations was not considered. The concordance tool developed seems to be an efficient assistance in terminological work.

Further work will concern a more complex evaluation of the results, semantic encoding, other language couples and new domains. It will be checked whether the extracted terminology matches the French and the German version of the International Patent Classification (IPC) system. We plan also to apply the semantic encoding provided in the CISLEX for the German terms.

Moreover, we will examine whether the results of the present study can be repeated for other language couples with closer morphology than German-French (e.g. English-French). A similar study will concern a corpus in another field than the patent domain.

Acknowledgment

We would like to thank the European Patent Office for providing a machine readable version of the corpus used in the present study. We would also like to thank the stuff

164

of the Language Service and of the Principal Directorate of Chemistry of the EPO for their contribution in this project.

References

Blank, I. (1995). "Sentence alignment: methods and implementations. In *Traitement automatique des langues* Vol. 36, numéro 1-2, (pp. 81--89).

Blank, I. (1997). Computerlinguistische Analyse mehrsprachiger Fachtexte. Doctoral thesis. University of Munich, Centrum fur Informations- und Sprachverarbeitung.

Blank, I. (1998). "Computer-aided analysis of multilingual patent documentation". *In Proc. of the First International Conference on Language Resources & Evaluation* (pp. 765-774), Granada.

Bourigault, D. (1994). LEXTER, un Logiciel d'EXtraction de TERminologie: Application à l'acquisition des connaissances à partir de textes. Thèse de doctorat. Ecole des Hautes Etudes en Sciences Sociales, Paris.

Church, K. W. & Dagan, I. (1994). Termight: Identifying and translating Technical Terminology. *In Proc. of the 4th Conference on Applied Natural Language Processing* (pp. 34--40), Stuttgart.

Daille, B. (1994). Approche mixte pour l'extraction de terminologie: statistique lexicale et filtres linguistiques. Thèse de doctorat en informatique fondamentale. Université Paris VII.

Eijk, P. van der (1993). Automating the acquisition of Bilingual Terminology. *In Proc. of the Meeting of the European Chapter of the Association for Computational Linguistics* (pp. 113--119), Utrecht.

Gaussier, E. (1995). Extraction automatique de lexiques bilingues par des méthodes statistiques. Thèse de doctorat en informatique fondamentale. Université Paris VII.

Isabelle, P. (1992). Bi-textual aids for translators. *In Proc. of the Annual Conference of the UW Center for the New OED and Text Research*.

Jacquemin, C. (1991). Transformation des noms composés. Thèse de doctorat en Informatique Fondamentale. Université Paris VII.

Kupiec, J. (1993). An Algorithm for Finding Noun Phrase Correspondences in Bilingual Corpora. *In Proc. of the 31th Annual Meeting of the Association for Computational Linguistics* (pp. 17--22), Columbus, Ohio.

Langer, S. (1996). Selektionsklassen und Hyponymie im Lexikon. Doctoral thesis. University of Munich, Centrum fur Informations- und Sprachverarbeitung.

Maier-Meyer, P. (1995). Lexikon und automatische Lemmatisierung. Doctoral thesis. University of Munich, Centrum fur Informations- und Sprachverarbeitung.

Picht, H. & Draskau, J. (1985). Terminology: an introduction. Guilford: The University of Surrey.

Silberztein, M. (1993). Dictionnaires electroniques et reconnaissance lexicale automatique, Paris: Masson.

Sta, J.-D. (1995). Comportement statistique des termes et acquisition terminologique à partir de corpus. *In Traitement automatique des langues* (pp. 119--132), Vol. 36.

Annex

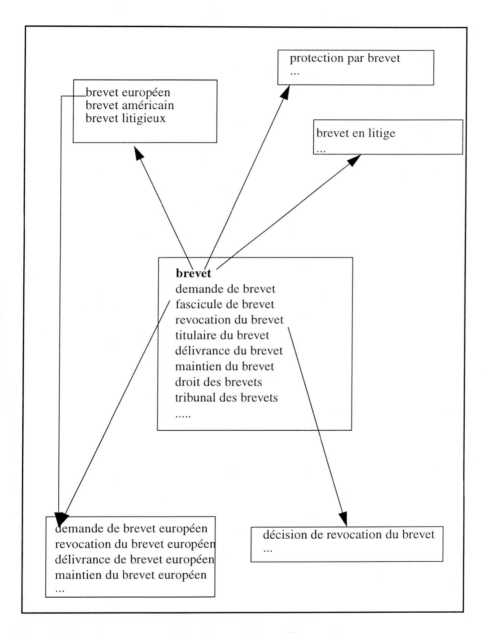

Figure 1: French terms containing the word "brevet" (patent)

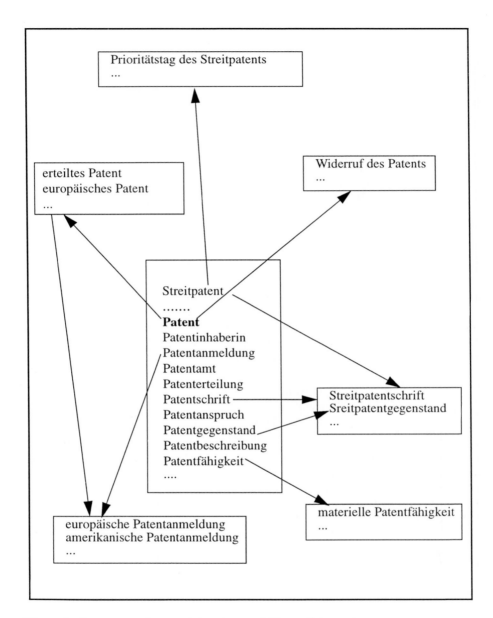

Figure 2: German words containing the word "Patent" (patent)

15

Information Extraction Techniques for Multilevel Text Matching

VITTORIO DI TOMASO
GIANFRANCESCO D'ANGELO
*CELI - Centro per l'Elaborazione del Linguaggio e dell'Informazione
Torino, Italy*

1. Introduction: adapting an information extraction system to a new task

In this Chapter we describe TradIuta, a system for retrieving text fragments from documents, based on the information extraction system IUTA. TradIuta exploits "linguistic matching", that is a technology which, given a text or a collection of texts, allows the retrieval of text units on the basis of different kinds of linguistic analysis.

TradIuta has been developed as a pilot application for Machine-aided Human Translation (MAHT). The system scans a source text and tries to match strings (a sentence or a part thereof) against a translation memory, that is a database of paired source and target language strings. The system suggests existing translations of the source text in the target language, which can be reused as such or as a model for the translation of the current sentence. Given a sentence to translate, TradIuta, besides matching the string literally, can retrieve similar target language text units, using an algorithm based on linguistic similarity. The similarity between two text units is computed considering the lexemes they contain and the similarity between syntactic structures. In fact, the idea behind TradIuta is that successful text matching procedures can be based on linguistic knowledge, which improves the flexibility and robustness of the matching algorithm, allowing for a high degree of granularity: in the envisaged application, text matching is based on strings, lemmas and syntactic chunks. However, the same technology can be enriched adding higher level functional-semantic analysis to match text units on the basis of predicate argument structure. To obtain these results, TradIuta exploits the same finite state technology developed for information extraction in the IUTA system (see [1], [2] and [3]), adding a new set of tools for creating and querying a translation memory database.

2. Machine-aided human translation

Machine-aided Human Translation (MAHT) is usually distinguished from Machine Translation (MT) by the fact that where in MT the translation proper is performed by the computer (even if a human may help in pre-editing and/or post-editing), in MAHT the translation is performed by a human. In the latter case, the role of the computer is to provide tools to help the translation process (see [4]). In recent years, MAHT became a fruitful field, both for research and commercial systems because of the growth of the amount of texts that has to be circulated in audiences often speaking different languages. A particularly promising field is, of course, the field of technical documentation, where the translation task is to render the objective content of a document into another language without addition and omission. The gain obtained from (partial) automation this task is usually said to be an overall gain of 40% to 50% in the time needed to completed the translation (see [4] for details).

In this Chapter we will concentrate on a experimental MAHT system tailored to the needs of an independent professional translator, that is a free-lance professional translator, who is not necessarily working as a member of a larger team. A free lance translator usually doesn't have easy access to real MT systems requiring high computational resources, but s/he can as well benefit from automating part of his workload.

A typical MAHT environment will offer tools to:

- Access a bilingual terminology database;
- Access a translation memory;
- Integrate the translation tools in a standard text processing application.

Commercial stand-alone systems are able, among other functions, to automatically analyse the source text, attaching keyboards shortcuts to the terms and sentences found in the terminology database and in translation memory. This enables the subsequent replacement of portion of the target text with their translation found in the terminology database or in the translation memory.

A translation workbench can be used both in a pre-processing phase and/or as an on-line tool integrated in the text processor. In both cases, the user sets a level of confidence that a candidate match must fulfil in order to be accepted. Acceptance can be fully automated or supervised. Experience shows that, in some cases, 100% confidence (that is exact match) is necessary for automated translation of portion of texts, whereas confidence as low as 60% can still be useful in supervised mode.

TradIuta addresses the issue of accessing translation memory databases using linguistic technology for sentence matching. The source language for TradIuta is Italian and the target language is English. TradIuta does not address many crucial issues for MAHT (neither on the content side, such as lexicon acquisition, nor on the integration side, such as interfacing with text editors). The idea behind the experiment is to adapt an already existing system, that is IUTA, to a new task, without devising a completely new set of tools and applications. The results we obtained with TradIuta show that techniques developed for information extraction can be rather easily adapted to a different domain, which shares with information extraction the goal of reducing the amount of irrelevant information in a text, in order to retrieve its most relevant structure.[1]

---

[1] Notice that if IUTA is extended with the capability of handling other languages, TradIuta will be ready also for other couple of languages, practically without any effort.

The information extraction engine IUTA[2] (Italian Unrestricted Text Analyser) developed at CELI is based on a set of pipelined modules which progressively refine the input texts in order to extract domain dependent templates. Its global architecture follows the "generic information extraction" proposed by Hobbs [5]. The modules of interest for TradIuta are the pre-processor, the morphological analyser and the shallow parser. The pre-processor recognises fixed expressions (dates and currency expressions, numbers, addresses, acronyms, proper names, etc.) producing a SGML mark-up text, which is passed to the morphological module. The morphological module produces couples <*unique lemma, inflectional information*> using a morpho-syntactic dictionary. Part of speech tagging is then performed, using an unsupervised Brill Pos-tagger (see [6]) designed for Italian. Shallow syntactic analysis is conceived as a process of chunking, along the lines of [7] (see also [8] for an analogous application for Italian). Non recursive constituents are recognised and properly labelled by a set of finite state automata, which group together all categories that unambiguously form a constituent.

3. Functions of a translation memory

3.1 DEFINITIONS

The final report of EAGLES (see [9]) proposes the following definition for *translation memory* (TM for short):

> a translation memory is a multilingual text archive containing (segmented, aligned, parsed and classified) multilingual texts, allowing storage and retrieval of aligned multilingual text segments against various search conditions.

Various translation systems based on TMs will, in general, provide tools for two different types of functions:

- Off line functions, such as analysis, import and export
- On line functions, that is in-translation functions, such as retrieval, updating and automatic translation.

TradIuta only addresses issues concerning the off-line analysis of a multilingual text leading to the insertion of new entries into the translation memory and the on-line retrieval of a target text translating (part of) a source text. TradIuta retrieval procedure allows both *exact match* and *fuzzy match*:

- *Exact match* is a perfect character by character match between current source sentence and a stored target sentence
- Everything else is a *fuzzy match*.

In TradIuta fuzzy matches are based on linguistic analysis: a translation memory entry

---

2 IUTA and TradIuta are both developed in Java and have been tested on Unix and Windows platforms. IUTA is currently being used in the European Union project MIETTA (LE4-8343) as the information extraction engine for Italian.

is retrieved when it contains identical lemmas or identical syntactic structures with the current text unit.

The key feature of TradIuta is that both compilation of TM entries and in-translation retrieval exploit the same technology, that is finite state technology, readapted from the information extraction system IUTA. In particular, compilation of TM entries is a *three levels analysis* of input texts, which identifies word forms, lemmas and content words used as syntactic heads on the basis of shallow syntactic parsing,. The retrieval procedure matches source text units against TM in three different ways: string matching, lemma matching and syntactic head matching. As a result, text matching is always based on some meaningful linguistic information.

3.2 OFF-LINE ANALYSIS: PREPROCESSING, LEMMATIZATION, PARSING AND COMPILING

Off-line analysis is the process that starts with a multilingual text in some format and outputs a translation memory entry. In our case, a translation memory entry is couple, containing the analysis of the Italian sentence up to syntactic chunks, and the English counter part of that sentence. The analysed Italian sentence is then compiled into finite state automata, which will be stored in the database, ready to be used at in-translation time to answer a user query.

TradIuta reuses modules of the IUTA system for pre-processing and parsing of input text:

- Textual parsing: the module Pretesto pre-processes the text, recognising punctuation and special texts elements, such as proper names, codes, numbers, dates, currencies. If necessary, pre-editing can introduce mark-up tags into the document.
- Linguistic parsing: morphological analysis (lemmatisation), part of speech tagging and syntactic analysis into chunks.

Chunking is a crucial step, because is the basis of the compilation of translation memory entries. Chunks are the cores of "major" sentences, i.e. NP, VP, PP, AP, AdvP. They are always maximal, non-recursive and contain word tokens mutually linked through unambiguous dependency chains ([7], [8]).

In IUTA, chunks are labeled with a syntactic category (such as NX for nominal chunks and VX for verbal chunks) and contain a list of the syntactic daughters of the chunk. The head is always the last daughter of the chunk. Daughters of a chunk are labeled with the form of the word and contain information on lexical items. The common structure of chunks is depicted in (1) and (2).

$$(1) \quad \text{Chunk} \quad = \quad \begin{bmatrix} \text{S-CAT} \\ \text{Daughter}_1 \\ \text{..............} \\ \text{Daughter}_n \\ \text{Daughter}_{head} \end{bmatrix}$$

$$(2) \quad \text{Daughter} \quad = \quad \begin{bmatrix} \text{FORM} \\ \text{INFLECTIONAL CLASS} \\ \text{CATEGORY} \\ \text{LEXEME - ID} \\ \text{SEMANTICS} \end{bmatrix}$$

For example, the chunking of the Italian sentence *visualizza il nome della stampante corrente*, is reported in (3).

(3) a. Visualizza il nome della stampante corrente

b.
$$\begin{bmatrix} S\text{-}CAT = VX \\ D_{head} = \begin{bmatrix} FORM = visualizza \\ INFL = INDIC\_PRES\_THIRD\_SING \\ CAT = VERBO \\ LEX\text{-}ID = visualizzG78233 \end{bmatrix} \end{bmatrix}$$

$$\begin{bmatrix} S\text{-}CAT = NX \\ D_1 = \begin{bmatrix} FORM = il \\ INFL = MASC\text{-}SING \\ CAT = ARTICOLO \\ LEX\text{-}ID = ART\text{-}DET \end{bmatrix} \\ D_{head} = \begin{bmatrix} FORM = nome \\ INFL = MASC\text{-}SING \\ CAT = NOMEMASCHILE \\ LEX\text{-}ID = nom24918 \end{bmatrix} \end{bmatrix}$$

$$\begin{bmatrix} S\text{-}CAT = PX \\ D_1 = \begin{bmatrix} FORM = de \\ INFL = NO\text{-}MORPH \\ CAT = PREPOSIZIONE \\ LEX\text{-}ID = PRED\text{-}DI \end{bmatrix} \\ D_2 = \begin{bmatrix} FORM = lla \\ INFL = FEM - SING \\ CAT = ARTICOLO \\ LEX\text{-}ID = ART - DET \end{bmatrix} \\ D_{head} = \begin{bmatrix} FORM = stampante \\ INFL = FEM\text{-}SING \\ CAT = NOMEFEMMINILE \\ LEX\text{-}ID = stampantG64496 \end{bmatrix} \end{bmatrix}$$

$$\begin{bmatrix} S\text{-}CAT = AX \\ D_{head} = \begin{bmatrix} FORM = corrente \\ INFL = SING \\ CAT = AGGETTIVO \\ LEX\text{-}ID = ADJ\text{-}CORRENTE \end{bmatrix} \end{bmatrix}$$

After chunking, the process of compilation of source sentences into finite state automata is started, producing the translation memory entry that is stored in a database. The overall workflow of the compilation is reported in fig.1.

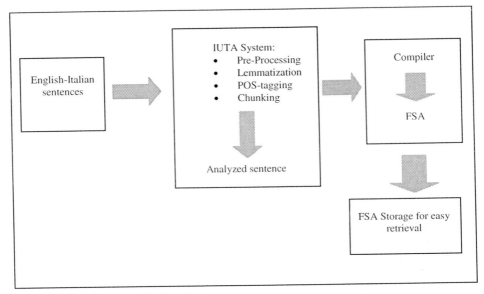

Figure 1: TradIuta workflow when compiling translation memory entries.

3.3   COMPILING THE TRANSLATION MEMORY AS A FINITE STATE AUTOMATON

Compilation of translation memory entries is based on shallow parsing of text units into syntactic chunks. The FSA compiler uses the syntactic chunks to build three finite state automata (cfr. [10] for extensive literature on FSA), which are stored in the database as the translation memory entry for the text unit. When a translation of a given text unit is to be retrieved, the three levels matching procedure will traverse the finite state automata (that is the translation memory entries), trying to recognise the input text.

The three automata, A_{STRING}, A_{LEMMA}, A_{HEAD}, are compiled in such a way that state names are always different in a single automaton and state names are the same across automata: the three automata have the same number of states, and the same state names. In fact, the automata differ only in the strings to be consumed to proceed from one state to the next.[3]

More formally, if Q_{STRING}, Q_{LEMMA} and Q_{HEAD} are finite sets of states of the automata, R_{STRING}, R_{LEMMA} and R_{HEAD} are transition functions, and W_{STRING}, W_{LEMMA} and W_{HEAD} are sets of input strings obtained from chunks for each automata, we have:

---

[3] The requirement of having always the same number of states has the consequence that dummy states with empty transition strings are introduced when IUTA cannot find a certain information. Typically, unknown words will cause empty transition strings in the lemma automaton and non-recognized chunks will cause empty transition strings in the head automaton. In either case, the only possible transition will be along the string automaton, because empty transition strings are always treated as failures, but different solutions are possible.

(4) $Q_{STRING} = Q_{LEMMA} = Q_{HEAD} = Q$

(5) $R_{STRING}: QxW_{STRING} \rightarrow Q$
       $R_{LEMMA}: QxW_{LEMMA} \rightarrow Q$
       $R_{HEAD}: QxW_{HEAD} \rightarrow Q$

(6) $\forall s_i \in Q$
       $R_{STRING} (s_i, w_{STRING}) = s_j$
       $R_{LEMMA} (s_i, w_{LEMMA}) = s_j$
       $R_{HEAD} (s_i, w_{HEAD}) = s_j$

At compile time, each Italian sentence is analysed up to the level of syntactic chunks. At this point, each chunk is used to build a transition of the automata in such a way that from each state, the next state can be reached consuming a different string. Arc labels (that is transition strings) are obtained from the chunk in the following way:

- the arc labels for A_{STRING} are built concatenating word forms of the daughters of the chunk;
- the arc labels for A_{LEMMA} are built concatenating lexeme-id of the daughters of the chunk;
- the arc labels for A_{HEAD} are built taking the lexeme-id of the daughter head of the chunk.

(7)   For each chunk C_i, D_i is the set of daughters of C_i and $h_i \in D_i$ is the head of C_i:
       w_{STRING} = the concatenation of all forms in $d_i \in D_i$
       w_{LEMMA} = the concatenation of all lexeme-id in $d_i \in D_i$
       w_{HEAD} = the lexeme-id of h_i

As a consequence, arcs of A_{STRING} are labelled with sub-strings of the original sentence, arcs on A_{LEMMA} are labelled with lexeme-id of words in the original sentence and arcs on A_{HEAD} are labelled with the lexeme-id of words which are the head of chunks in the original sentence.

Final states (and also some intermediate states) are associated with translations, so that, when a succeeding automaton finds a state associated with a translation, it will display the translation. All automata are stored in a database and the complete database represents our translation memory.

An input text unit can be matched in four cases: (i) because it has exactly the same words (exact match) of an entry in the TM; (ii) because it has the same words with different inflections (the lemma automaton contains only unique lexeme-id, abstracting from inflectional information) of an entry in the TM; (iii) because chunks of input sentence share the head with chunks of an entry in the translation memory; (iv) a combination of (i-iii).

We can consider now an example. Let's start with two sentences, from which an entry of the TM must be created.

(8)   a.   Visualizza il nome della stampante corrente.
       b.   Displays the name of the current printer

When the two sentences are added to the TM, the automaton in fig.2 is created. Notice that in the figure a single automaton is depicted, equivalent to the three automata that are actually compiled. Arc labels contain three strings, separated with a comma, which can be alternatively consumed to reach the final state: for example the transition from q_1 to q_2 is possible when the input is the string *ilnome*, or the lexeme-id *art-det-nome24918*, or the lexeme-id *nome24918*.

In fact, to solve at least some efficiency issue, at compile time the set of automata is first collapsed into a single, non-deterministic automaton, and then simplified into a deterministic FSA. The automaton is then saved into a format that allows an efficient retrieval of an edge, given a starting state and a string to analyse.

Different types of inputs can be fuzzy matched using the TM entry compiled from sentence (8), for example:

a.   <any inflection of *visualizza*> < any inflection of *nome*> <any inflection of *stampante*> <any inflection of *corrente*>

b.   <any chunk with head *visualizza*> < any chunk with head *nome*> < any chunk with head *stampante*> < any chunk with head *corrente*>

For inputs of the previous types, TradIuta will display the translation of the sentence (8) which can be used as a model to translate the current input sentence. In fact, the basic strategy for fuzzy match adopted in TradIuta is the following: if the system can't obtain an exact match with A_{STRING}, it tries to match abstracting from inflection features (A_{LEMMA}). If this abstraction still results in a failure, it looks for a sentence with the same syntactic structure of the input sentence and the same head for each phrase (A_{HEAD}).

The most common case is that, given the input sentence, an exact match against the TM is only partially possible, that is the input sentence share some words with an entry in the TM, but also contains different words. In these cases, TradIuta will adopt a mixture of exact match, fuzzy match and skipping. All identical chunks in source and target sentence will be exactly matched (that is with A_{STRING}, which uses string identity), all other chunks will be fuzzy matched, either with lexeme-id identity or with head identity. TradIuta can also skip an arbitrary number of unrecognised chunks up to maximum number defined by the user. The overall confidence of a translation is a function of the number of exact matches, fuzzy matches and skipped constituents.

ON LINE-ANALYSIS: RETRIEVAL

Retrieval of translation memory entry given a text to be translated comprises two steps:

- Analysis of the query[4] through IUTA
- Matching of the query against the translation memory database and retrieval of the target English sentence

The analysis can be done either sentence by sentence or document by document. We will concentrate on the former (the latter simply consists in analysing a whole document sentence by sentence and matching each sentence onto the TM database).

---

[4] We use the database terminology here because translation memories can be thought as databases and retrieval as a database query.

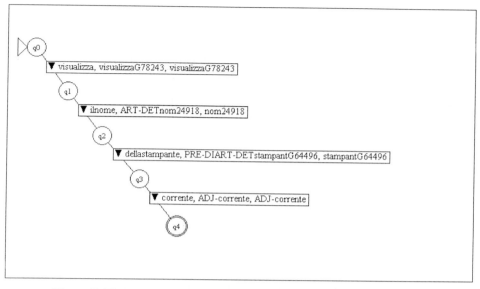

Figure 2: The automaton representing the TM entry obtained from (8)

The overall workflow of retrieval is depicted in fig.3.

The crucial step of in-translation function is the matching of the current text unit against the TM database and the retrieval of the translation. Given the kind of technology we adopted, this step consist in using the FSA to consume the source sentence, until a final state (or eventually a non final state) containing a translation is reached, and the translation is displayed.

To obtain a translation, the three automata try to consume a sentence S_i chunk by chunk in the following way (see fig.4 for a description of the algorithm for text matching):

- First string matching is tried, that is A_{STRING} is traversed. If it yields a translation, it is displayed
- If string matching fails, the element to consume is passed to A_{LEMMA}. If it succeeds (and it is not a final state) the analysis will proceed again with A_{STRING}.
- If lemma matching fails, the element to consume is passed to A_{HEAD}. If it succeeds (and it is not a final state) the analysis will proceed again with A_{STRING}.

4. Conclusion: reusability of linguistic matching

The technology described in this Chapter can be thought as a way to implement a "linguistically-based text-search procedure". The text matcher does not make use of regular expressions, but of linguistic information, so that the user can search a document also on the basis of its linguistic structure.

176

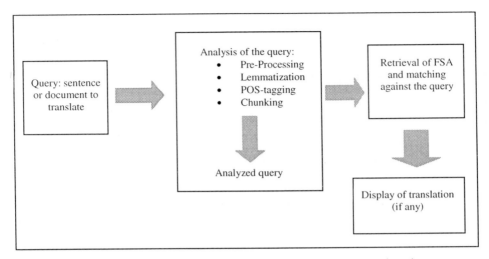

Figure 3: TradIuta workflow when retrieving translation memory entries given a query.

In a limited domain, and given the possibility of performing the described pre-compilation into finite state automata, this technology overcomes the typical problem of text searching (see [11] and [12] for extensive references), that is the inflexibility inherent in any character-matching system.

Whereas string matching with regular expressions (or, at least, wild characters) can match the query string *apple* with the text *apples*, when queries are formulated as multiword character strings, the problem posed by word order become difficult to treat. In fact, a query on *information extraction* does not match a document containing the phrase *extraction of information*. A text matcher exploiting linguistic knowledge can search for phrases with the same syntactic heads (while setting constraints on other lexemes in the phrase), retrieving all documents containing phrases with *extraction* as head (and *information* as a modifier of the head). The problem here is, of course, the effort necessary to parse the query and the documents in order to build the finite state automata. However, if the application can afford this, the retrieval procedure will be both flexible (because it can use syntactic and functional structure) and efficient (because FSA is an efficient technology for text searching)

We are currently using a text matcher based of FSA technology as a fall back solution in an information extraction application. The idea is that significant text units in a certain domain can be identified on the basis of some content words and the syntactic functional relations among them. The application will contain a set of patterns compiled as FSA and will scan documents searching for those patterns.

A similar approach can also be useful in automatic indexing for the task of term-phrase formation. A term-phrase formation process controlled only by word co-occurrence and the document frequency of certain words is usually considered unable to generate a high quality number of phrases, whereas the addition of syntactic criteria can improve the results (see [11]). The use of shallow parsing (chunking) makes the syntactic analysis efficient, because the shallow parser does not try to solve ambiguities.

```
For each chunk Cᵢ of the input sentence do:
            string = form(Cᵢ), lemma = lemma(Cᵢ), head = head(Cᵢ), state = initial
            skipped = 0, maxSkipped = n
            Start StringAutomaton from state with string
            if StringAutomaton succeeds
                        if state contains a translation then display the translation
                        else if state is final then display a translation and stop
                        else state = new state of automata
            else Start LemmaAutomaton from state with lemma
            if LemmaAutomaton succeeds
                        if state contains a translation then display the translation
                        else if state is final then display a translation and stop
                        else state = new state of automata
            else Start HeadAutomaton from state with head
            if HeadAutomaton succeeds
                        if state contains a translation then display the translation
                        else if state is final then display ntranslation and stop
                        else state = new state of automata
            else if skipped <= maxSkipped
                        Skip(Cᵢ)
                        skipped = skipped + 1
            else Fail
```

Figure 4. The algorithm used for text matching

The text matcher can be used to retrieve semi-fixed expressions or collocations from corpora. In the domain of MAHT, a similar application can be useful when building terminological databases. If it is possible, for a particular domain, the identification of a set of core text units, then the text matcher will be able to retrieve all linguistically similar units, ignoring inflectional and syntactic differences. Finally, linguistic matching can be used to scan corpora and to search for dictionary examples as an aid in a word sense disambiguation task.

The future development of the system will add a level of functional-semantic analysis (the same used in the IUTA system and described in [2]), which should improve the text matcher performance. In this case a FSA based text matcher like TradIuta will become very similar to a standard information extraction system.

References

[1]   Bolioli A., Dini L., Di Tomaso V., Goy A. & Sestero D. "MILK: a Hybrid System for Multilingual Indexing and Information Extraction". Proccedings of Recent Advaces in Natural Language *Processing* (*RANLP 1997*), Tzigov Chark, Bulgaria, 11-13 September 1997.

[2]   Dini L. "Parallel Information Extraction System for Multilingual Information Access". Chapter of this book.

[3]   Dini L, Di Tomaso V. & Segond F. "Error Driven Word-Sense

178

Disambiguation". To appear in *Proceedings of ACL 1998*, Montreal, Canada, 10-14 August 1998.

[4] Boitet C. "Machine-aided Human Translation". In Varile G.B & Zampolli A. *Survey of the State of the Art in Human Language Technology*. Press Syndicate of the University of Cambridge, 1997.

[5] Hobbs J. "The Generic Information Extraction System". In Sundheim B. (ed.) *Fourth Message Understanding Conference*. San Mateo, California. Morgan Kaufmann, 1992.

[6] Brill E. "Unsupervised Learning of Disambiguation rules for Part of Speech Tagging". In *Natural Language Processing Using Very Large Corpora*. Kluwer 1997.

[7] Abney S. "Parsing by chunks". In *Principle-Based Parsing: Computation and Psycholinguistics*. Kluwer, 1991.

[8] Federici S, Montemagni S. & Pirrelli V. "Shallow Parsing and Text Chunking: A view on Underspecification in Syntax". In Briscoe T. & Carroll J. (eds.) *Proceedings of the ESLLI-96 Workshop on Robust Parsing*. Prague, 1996.

[9] EAGLES Consortium. "EAGLES Final Report". Eagles Document EAG-EWG-PR2. 1996.

[10] Roche E. & Schabes Y. (eds) *Finite State Language Processing*. MIT Press, 1997.

[11] Salton G. *Automatic Text Processing*. Addison-Wesley, 1989.

[12] Witten I.H, Moffat A. & Bell T.C. *Managing Gigabytes*. Van Nostrand Reinhold, 1994.

16

Parallel Information Extraction System for Multilingual Information Access

LUCA DINI
*CELI-Centro per l'Elaborazione del Linguaggio e dell'Informazione
Torino, Italy*

1. Introduction

Information Extraction[1] is mostly a monolingual technology. This feature is evident when considering the most important worldwide initiative focusing on Information Extraction, i.e. the Message Understanding Conference. In this competition English has always been the unique target language, with the exception of MUC-6 (MET-1) where Spanish and Chinese were considered as well. Even the introduction of these languages, however, has not significantly changed the monolingual status of the competition, as all the systems were assumed to be completely independent. This situation contrasts with Information Retrieval where, almost since the beginning ([1]) multilinguality was one of the important research tracks, as attested by the introduction of a Cross-Language IR track in TREC-6.

In a sense, this situation is induced by the very nature of the technology. Indeed, there is a commonly agreed assumption that IE should produce templates or database records from natural language texts. So, at the end of the processing chain, there is not a human with his/her own language, but a database or a repository of templates. On the contrary, the typical processing chain for Information Retrieval is from human to human (query processing and presentation of the results), passing through information mainly encoded by humans (the document base). Human involvement is what makes multilinguality an issue.

Besides the lack of human involvement on the final edge of the processing chain, the other reason why so little attention was paid to multilinguality in IE, is probably that there is a clear cut way to switch a monolingual system into a multilingual one: as soon as two systems dealing with two different languages share the same template system, the sum of the two give raise to a multilingual system. This shifts the attention from the processing phase to the one of template design. In particular, for a system of templates to be language independent, it must assume that it contains a fixed number of predicates ranging only over named entities, numbers and other predicates. More

---

[1] I would like to thank Andrea Bolioli, Vittorio Di Tomaso, Gregor Erbach and Hans Uszkoreit for important support and advices.

180

formally, assuming that templates are described in terms of typed feature structures, the following definition could be given:

A language independent system of templates is described by
1. A finite set of predicates, defined by the types of the description language.
2. A finite set of attributes
3. A possibly non finite set of values represented by types, integers and strings. Whereas a string appears as a value of an attribute, it has to denote a named entity, i.e its linguistic realization must not change across languages.

This is obviously only one way of formulating constraints on language independent templates. For instance, as an alternative, it is perfectly conceivable that strings rather than types represent predicates. In this case, it is not necessary that all strings denote named entities, but it has still to hold true that the number of strings not denoting named entities has to be finite. In any case, the basic idea is still the same: templates should not contain fields that are string of text "unknown" to the system.

For instance the application of IUTA described in ([2]) *is* language dependent. In that case the task was to analyze job announcements and to store them into a database. However, being the possible range of job qualifications quite wide, the choice was made of not listing all of them internally to the system. For instance, we decided that, in the string *in qualità di X* ("as a X"), the string value of X should be the filler of one of the field **role** of the DB record. This was an easy choice, which simplified the processing on a monolingual side, but had the effect of making the templates intrinsically language dependent. Indeed, the filler of the **role** field is now a fragment of Italian language, with no correspondence in the abstract metalanguage of templates.[2]

Notice that the strategy of handling multilinguality in IE by adopting language independent templates is not the only possibility. In machine translation there are two basic approaches, i.e. the interlingua based approach, and the transfer based approach. What we hinted at so far is the analogous of the interlingua approach in the field of IE. However, there might be cases where the other approach is sensitive as well. In this case different systems would "talk together" by resorting to special domain tailored

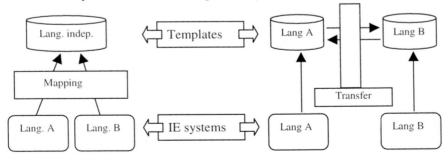

Figure 1: Different views of Multilingual IE

[2] For another example of language dependent templates see for example the description of the MUC-4 templates as emerging from [14].)

bilingual dictionaries and transfer components (cf. Fig. 1). Since the application we are going to describe does not rely on the use of multilingual dictionaries, from now on we will concentrate our attention uniquely on the former strategy.

2. IE and Natural Language Generation

As we mentioned, multilinguality becomes an issue as long as humans speaking two different languages are involved in the processing flow. Thus, two systems relying on the same template design do not, in principle, describe a multilingual system. However, as soon as the DB of templates is accessed by non-expert humans (i.e. persons unable to deal with the internal representations of the DB) the problem of displaying information in the language of the user raises its head again.

The obvious answer to such a problem is natural language generation. Templates and sets of templates represent by themselves an ideal input to natural language generation (compare for instance the description of the MUC templates with the input of the system described in [3]). In the simplest case they can feed a system of canned text generation. In more complex cases some variations on the syntactic structure might be required. Finally, the most interesting situation is raised by cases when the system has to deal with *sets* of templates that jointly provide an answer to the query posted by the user. Then there is room for a full-fledged system, exploiting sophisticate text planning strategies.

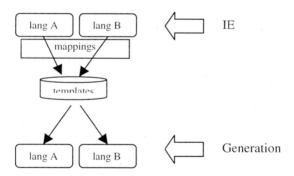

Figure 2: IE/NL-Generation

Notice that the exploitation of language independent template design and natural language generation is likely to solve some problems of multilingual DB maintenance, irrespective of an underlying IE system. Indeed, it is not uncommon to see Internet accessed DBs that are maintained in as many versions as the languages in which it is supposed to provide an output. In most cases, this effort of maintenance is due to the fact that the DB contains also small pieces of natural language information necessary to the visualization of the retrieved DB contents. By assuming a language independent template design coupled with natural language generation, such an effort of maintaining parallel DBs can be drastically reduced. Moreover, simple IE systems can provide a substantial help in bridging the language aware source DB into the new one, where natural language strings have been replace by more structured language independent information.

3.   MIETTA's Overview

The MIETTA[3] project (Multilingual Information Extraction for Tourism and Travel Assistance, LE4-8343) is aimed at building a prototype implementing the approach to multilingual IE described so far. The goal of the project is to provide Web users access to touristic information in their own languages. Four languages are considered, i.e. English, Finnish, German and Italian. Content providers such as municipalities and state/regional administrations provide the repository of documents.

As a final outcome, MIETTA will enable the end user to use a standard web browser (e.g. Netscape) as his/her interface to access "MIETTA enhanced tourist servers". From here, the user is able to express requests for documents by means of natural language queries. If the query expressed by the user is over- or underspecified from the system's point of view (or if it is expressed in such a way that the IE system for that language cannot get any useful information out of it), the system can ask the user for more precise information to constrain further retrieval. This strategy will be coupled with a search strategy based on advanced information retrieval techniques.

The set of documents retrieved for a query will be presented to the user in an informative way. The ranking of the documents will be determined either by the user profile (if any) or by standard ranking criteria. Moreover, for every document it will be possible to produce, through natural language generation, a short summary of the information it contains, presented in the user language.

3.1   THE ARCHITECTURE

The core of the MIETTA architecture is represented by four "aligned" systems for IE and NL-generation targeting English, Finnish, German and Italian. University of Helsinki will provide the modules for Finnish, DFKI the ones for German and CELI the ones for Italian. As for English, the MIETTA consortium will provide an adequate processing chain simply by putting together modules that are already available to the single participants.

From the point of view of input/output specification, the following functionalities will be implemented for every language:

- The *Template Extractor* will accept standard unprocessed (ASCII or HTML) texts as input. These texts come either from the user interface (query text) or they are delivered by the Document Manager (documents contained in the registered touristic sites). The template extractor returns either a single template or a set of templates, which are either stored in the template DB (if the source text belonged

---

[3] The definition of the objectives and the methodology of MIETTA was the result of a close cooperation among all partners. Thus what I expose in this section has to be considered a "shared" work. All errors are mine.

to some tourist site) or matched against the templates contained in the DB (if the source text comes from a user query);

- The *Template Matcher* takes as input a query template and, by using the global database of templates as a knowledge source, returns a list of pairs where each pair contains a pointer to some document in the relevant tourist site and a set of templates.
- The *Report Generator* takes input from the template matcher and returns a text to be passed to the user interface;
- The *Relevance Engine* applies advanced indexing methods (e.g. concept-based indexing) to store contents of all the relevant items in the document base, and co-operates with the template handler in order to provide better output to the users.

In fig. 3 the standard interaction with the user is described: the user addresses a Query which is immediately passed to the Query Manager. The Query Manager sends the query to the Template Extractor, which transforms it into a template, which is passed to the Template Matcher. The Template Matcher interrogates the Template DB to retrieve matching templates. Then it passes them back to the Query Manager. The Query Manager evaluates if the answer from the template system was satisfactory or not. If not, it addresses the Query to the low level Relevance Engine which performs a search in the index and send the retrieved documents back to the Query Manager. Once all this processing is done, it sends everything to the Report Generator, which, for every document, generates a short natural language description to be returned to the user through the User Interface.

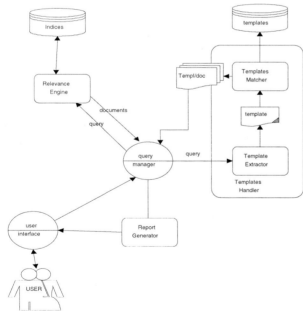

Figure 3: The process of querying a tourist server.

In fig. 4 we describe a typical process of updating the knowledge base of the server. The various regional DBs/web sites notify the Document Manager of the presence of new documents. These are retrieved by the Document Manager, which passes them to the Template Extractor. The templates that are obtained are stored into the template DB and enrich the knowledge base of the system. In parallel the documents are indexed through advanced technologies of information retrieval for later retrieval, in case the information extraction engine can not provide any satisfactory request to some user's query.

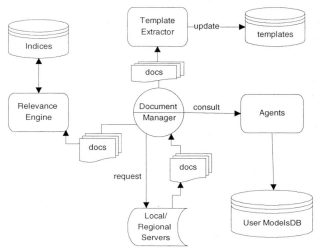

Figure 4: The process of updating the knowledge base of the tourist server

3.2 THE DESIGN OF TEMPLATES

It is evident that the templates used in the MIETTA prototype have to be language independent. This is a core assumption of the project, actually one which the whole system is based on. Its violation would cause both poorer retrieval performances (the template(s) extracted from the query would not match the ones contained in the DB of templates) and worse performance in the document presentation phase (not all templates would undergo generation in the target language as they could still contain information specific to the source language). Consequently, the design must conform a commonly agreed pattern. In order to make this possible, an abstract description language, such as the one of typed feature structure, will be used. This will enable the structuring of templates in a hierarchical way, in order to minimize the amount of information. Moreover it will allow the use of standard unification techniques (coupled with domain dependent inferences) in order to perform template matching for retrieving the templates satisfying the constraints expressed by the user query. Finally, such a structuring allows an easy porting of the set of templates into commercial databases (either relational or object-oriented) for more efficent retrieval in a real life setting. Figure 5 (mainly due to Paul Buitelaar, Klaus Netter, Feiyu Xu) shows an example of a hierarchical structuring of three templates, where principles of feature inheritance are applied in order to keep information compact.

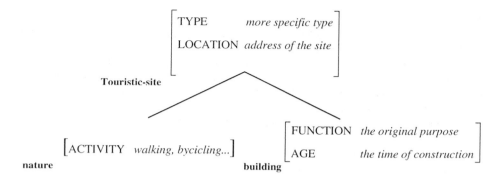

Figure 5: Some sample templates

In order to fulfill the goal of the MIETTA prototype, the consortium has rejected the use of deep analysis modules based on full parsing by competence grammars. Indeed, this kind of technology has been considered as not yet mature for the processing of large amounts of real-life texts. Shallow systems, on the contrary, have been proved to be accurate and efficient enough to be successfully applied in information extraction tasks. These methods usually combine well-mastered NLP technologies with statistical methods.

The generic category of shallow systems actually embraces two different kinds of analysis, which are in most cases integrated. On the one hand there is the shallow syntactic analysis, which mostly reduces to the identification of the main constituents of a sentence (VPs, NPs, PPs, etc.). On the other hand, shallow parsing techniques are almost invariably coupled with conceptual, or semantic, modules, which take advantage of the previous shallow representation in order to produce templates representing part of the information encoded in the input text. These modules are usually domain dependent and are implemented either as sets of rules which activate a certain semantic template when a certain syntactic pattern is retrieved, or as finite state automata which apply when a certain semantic trigger is encountered.

The three language technology partners of the MIETTA consortium have agreed to reuse for this particular task three information extraction engines (or parts thereof) which had been already developed for different purposes. In particular CELI will reuse IUTA (Italian Unrestricted Text Analyzer, cf. [2]), DFKI will reuse SMES (cf. [4]) and University of Helsinki will reuse a combination of already available NLP tools (cf. [5]).

SMES is an information extraction core system for real world German text processing. The basic design criterion of the system is of providing a set of basic, powerful, robust and efficient natural language components and generic linguistic knowledge sources, which can be easily customized for processing different tasks in a flexible manner. SMES consists of the morphological component MORPHIX, a declarative tool for expressing finite-state grammars, and an efficient and robust bidirectional lexically driven parser. The information extraction tool for Finnish is based on three existing modules: (1) FINTWOL: a morphological analyzer of Finnish, based on Kimmo Koskenniemi's language-independent two-level morphology; (2) FINCG: Fred Karlsson's Constraint Grammar parser for Finnish, which performs a morphological disambiguation of ambiguous word-form tokens and gives them a surface-syntactic analysis; and (3) a low-level syntactic Constraint Grammar tagger for Finnish which identifies the noun phrases.

As far as IUTA is concerned, the system will be described in section 4 of this chapter.

3.4   NATURAL LANGUAGE GENERATION

Hovy ([6]) distinguishes three types of generation task: text planning, sentence planning, and surface realization. "Text planners select what information to include in the output from a knowledge pool, and out of this create a text structure to ensure coherence. On a more local scale, sentence planners organize the content of each sentence, massaging and ordering its parts. Surface realisers convert sentence-sized chunks of representation into grammatically correct sentences." The processes used in NL generation are classified and ranked according to sophistication and expressive power, starting with inflexible canned methods and ending with maximally flexible feature combination methods.

In the context of the MIETTA system the template generator JTG/2 from CELI/DFKI will be adopted by the consortium. JTG/2 is a porting into Java™ of TG/2 (developed in LISP at DFKI [7]), a surface generator that can be easily hooked up to application systems or to language planning systems. In domains where no extensive utterance planning is needed, TG/2 can be parametrised according to preferred properties of utterances (style, wording, and contents). Parameters guide the selection of grammar rules, which are condition/action pairs allowing the grammar-writer to formulate canned text, templates, or context-free derivations, as well as any combination thereof.

The major challenge for the adopted NL generation engine will be represented by the language independent nature of the templates. Indeed, the set of templates retrieved after a query matching will be passed directly to the generator. Two main tasks can then be envisaged:

Presentation: the generator will have to arrange all the information in order to present the user with a reasonably structured page. In order to accomplish this goal, the generator will take advantage of the fact that the generated pages will be in HTML. In this way, some of the relations which would have to be expressed through natural language (e.g. enumeration, specification, topic proximity) can be encoded using typical hypertextual/graphic devices.

Sentence generation: since language independent templates are passed to the generation module, the task of the generator is particularly difficult, as it has to reconstruct every single piece of linguistic information. Moreover, given the "richness" of the input templates, it will have to keep a correspondence table mapping abstract concepts to language dependent lexemes. These lexemes will be used to perform lexical insertion after the phase of sentence plan.

4.   UTA and IUTA

UTA (Unrestricted Text Analyzer) is a general purpose Information Extraction engine developed at CELI. It is based on an object-oriented architecture in which language independent modules satisfying certain predefined interfaces can be plugged without any need to access their internal code (cf. [8]). In this sense it would be more reasonable to talk about an architectural model than a program in the proper sense of the word: indeed the design of every component and the interfaces they have to satisfy force a strictly modular view, which enables a fast prototyping of ready-to-deliver information extraction systems. [4]

UTA comes together with a set of *engines*, which includes a textual tokenizer, a class based morphology, a part of speech tagger, a shallow parser system (chunking), a system for extracting underspecified logical forms, and a module for coding simple domain dependent rules for template filling. In the design of the modules we stuck to the assumption that no language dependent information should be contained in the processing engines. So, they are in principle extensible to deal with any language for which there are enough linguistic resources.

UTA has been used to produce IUTA, an Italian information extraction system (cf. [2], [9]).[5] IUTA exploits all the modules currently available to UTA thus implementing a quite prototypical processing flow, as described in figure 6.

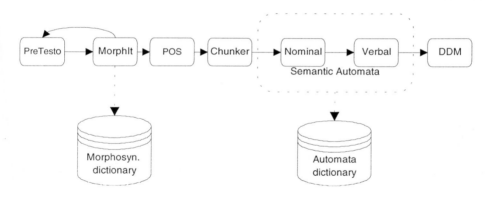

Figure 6: The architecture of IUTA.

---

[4] This result is achieved by designing every module as a JavaBean™ component and by establishing a common communication protocol. In the optimal case a whole system can be parametrized just with the help of a visual tool for JavaBean manipulation.

[5] Actually a version of IUTA was already existing when UTA was under development. The lingware of the previous version was just migrated in the new architectural model.

The first modules (Pretesto, Morphit, the POS tagger and the chunker) are quite standard, so they do not deserve a particular attention (cf. [10]). In the next section we will rather describe in some detail the core of the IE system, i.e. the set of semantic automata and the Domain Dependent Module).

4.1   THE IE CORE OF IUTA

The basic idea of IUTA is that effective and easily parametrizable IE engines should be based on two processing steps. During the first step a sequence of chunks is analyzed in order to produce a *Naif Logical Form* (*nlf*), i.e. an abstract representation level independent on the particular type of application for which the system has been designed. Information extraction in a proper sense is performed only during the second step of processing, where *nlf* are spanned by template driven rules which select the appropriate template and fill the relevant roles. The main advantage of this strategy over approaches which goes directly from syntactic structures to filled templates (cf. [11], [12], [13], [14]) is reusability: indeed part of the processing needed to bridge from syntax to templates is kept common to many applications, in such a way that linguistically motivated improvements are immediately reflected on all the system accessing the intermediate level of *nlfs*. The advantages from the point of view of software engineering are significative, as it is possible to maintain a certain degree of coherence among different application (cf. fig. 7)

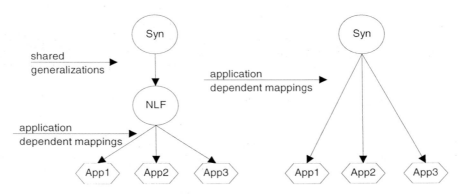

Figure 7: Advantages of NLF.

More technically, the module in charge of producing *nlfs* is implemented as a cascade of finite state automata (FSA) having as an alphabet indexed objects. The indices associated to every object can be referred to in a separate part of the FSA, called *action,* whose main task consists in building an appropriate output once the associated automaton reaches a final state. Consider for instance the following sample rule:

```
(PP$1|ADVP$1)* NP$2 VT$3 (PP$1|ADVP$1)*
Action = {TransNLF($1,$2,$3)}
```

The regular expression described in the first line is compiled into a deterministic FSA[6] to be applied to the input sequence, represented in this case by a sequence of objects of type `Chunk`. If the application has success, the objects referred to by the variables prefixed with "$" ($1 an array of chunks having category PP or ADVP, $2 a chunk with category NP and $3 a chunk with category VT)[7] are passed directly to the Java constructor `TransNLF` which takes care of producing the proper output.

In the actual setting we use only two levels of FSA, one for guessing the semantics of nominal expression and one for identifying the *nlf* to be associated with verbs. However more levels can be easily implemented, for instance for describing parenthetical expressions and subordinate clauses.

As for the modules for template filling, labeled *Domain Dependent Module* (DDM) in fig. 7, it is conceived as a set of template-driven structure matching rules, along the lines described in [15]. Basically, for each attribute of each template *type* there is a set of rules with the task of spanning the input structure (a *nlf*) in order to find a possible filler for that attribute. The rules are ordered in terms of preference and the first rule which fires determines the values that the attribute will have in the resulting template. For instance the following rule spans a *nlf*, looking for the subject of a verb of a certain kind and sets the role **company** accordingly.

```
[PRED ricercare,
ARG1
    [ PRED $1]
]
action = {theCurrentTemplate.setCompany($1)}
```

DDM is also the module where textual relations and anaphoric links are solved, depending on the requirements of the application. Rules for performing these tasks are encoded directly as Java methods taking as input the filled templates.

5. Conclusions

We have described a model architecture for a system of information extraction, showing how it is exploited in order to satisfy application needs emerging from a field such as tourism. The system is based on the interaction of a set of monolingual IE systems with a set of natural language generation engines. In order for the system to be truly multilingual we have shown that a big emphasis has to be put on the design of templates rather than on the processing chain itself. Indeed a project such as MIETTA proves that different IE system can cooperate to reach the goal of multilingual access to monolingual information just by virtue of an agreement on appropriate communication standards and on a strict design of templates.

---

[6] In order to increase the efficency of the system all the rules belonging to the same level are compiled together to form a unique deterministic FSA.

[7] Actually the expressive power of our rules goes beyond this, as it is possible to refer to any feature of the object, provided the set of the possible values of such a feature is finite.

190

References

[1] Salton, G. Automatic Processing of Foreign Language Documents. *Journal of the American Society for Information Sciences,* 21:187-194, 1970.

[2] Bolioli A., Dini, L., Di Tomaso, V., Goy A., Sestero D., "From IR to IE through GL", in *Proc. of Int. Workshop on Lexically-Driven IE,* Frascati, Italy, 1997.

[3] McKeown, K.R., *Text Generation: Using Discourse Strategies and Focus Constraints to Generate Natural Language Text.* Studies in Natural Language Processing. Cambridge University Press. 1985

[4] Neumann, G., Backofen, R., Baur, J., Becker, M., Braun, C., An Information Extraction Core System for Real World German Text Processing, in *Proceedings of ANLP 1997.*

[5] Koskenniemi, K. Representations and Finite State Components in Natural Language, In Roche, E. and Schabes, Y. (Eds) *Finite-State Language Processing,* pp 383-402., The MIT Press, 1997.

[6] Hovy E.,. *Natural Language Generation.* In: Varile G., Zampolli,A. (Managing Editors), *Survey of the State of the Art in Human Language Technology,* 1995.

[7] Busemann S., Best-first surface realization. In Donia Scott (Ed.), *Eighth International Language Generation Workshop. Proceedings.* 1996.

[8] Bolioli A., Dini L., Di Tomaso V. UTA: a light portable component architecture for IE, forthcoming.

[9] Bolioli A., Dini L., Di Tomaso V., Goy A. and Sestero D. MILK: a Hybrid System for Multilingual Indexing and Information Extraction. *Proccedings of Recent Advaces in Natural Language Processing (RANLP 1997),* Tzigov Chark, Bulgaria, 11-13 September 1997.

[10] Bolioli A., Dini L., Di Tomaso V., Goy A. and Mazzini M., IUTA: un sistema di estrazione dell'informazione per l'italiano. In *Apprendimento automatico e linguaggio naturale.* AI*IA. 1997.

[11] Hobbs J.R., Appelt D.E., Bear J., Tyson M., "Robust Processing of Real-World Natural-Language Texts", in *Proc.of 3rd Conference on Applied NLP,* Trento, Italy, March, 1992.

[12] Lehnert W., Cardie C., Fisher D., Riloff E., Williams R., "University of Massachussets: Description of the CIRCUS System as Used for MUC-3", *Proc. of 3rd MUC,* San Diego, CA, 1991.

[13] Appel D., Hobbs J., Bear J., Israel D., Tyson M., "FASTUS: A Finite State Processor for Information Extraction from Real World Text", in: *Proc. of IJCAI,* 1993.

[14] Hobbs J.R., Appelt D., Bear J., Israel D., Kameyama M., Stickel M., Tyson M., FASTUS: A Cascaded Finite-State Transducer for Extracting Information from Natural-Language Text. In Roche, E. and Schabes, Y. (Eds) *Finite-State Language Processing,* pp 383-402., The MIT Press, 1997.

[15] Dini L., CIRO User Manual, *DFKI Internal Report.* 1997

17

Using Information Extraction for Knowledge Entering

F. VICHOT, F. WOLINSKI, H.-C. FERRI, D. URBANI
Informatique-CDC / DTA
4, rue Berthollet, 94110 Arcueil, France
{fvichot, fwolinski, hcferri, durbani}@icdc.caissedesdepots.fr

1. Introduction

During the past decades, the field of applied Artificial Intelligence has experienced important growth. Numerous Decision Support Systems (DSS) have been created in order to improve experts' efficiency at work. DSS appeared to be particularly useful where a lot of data were needed in order to take the right decision. However, gathering all the data, often for economic reasons, is not always possible. In fact, the data may not be available in the right form. In that case, the high cost and the highly repetitive task of manually "feeding" a DSS have often been the bottleneck of a promising system.

At the same time, the wide amount of available electronic texts appeared to be a huge deposit of information. The Natural Language Processing (NLP) community has imagined several ways to benefit from this wealth. A major way is provided by Information Extraction (IE), whose aim is to convert the knowledge contained in texts into structured data. These data in turn, as they are in a more computable form, might be used in applications such as DSS.

This chapter describes and discusses an experiment in feeding a DSS with IE. This experiment has led to an operational system used at *Caisse des dépôts et consignations* (CDC). It concludes that IE is a powerful solution that allows real new possibilities for DSS. It also shows that the cooperation of the two technologies represents, in turn, a concrete challenge for IE, particularly in its ability as regards continuous adaptation.

Section 2 situates this experience in the fields of NLP and DSS. Section 3 gives an overview of SAPE, a Decision Support System used by fund managers in order to anticipate takeover bids on stock markets. Section 4 describes how IE is used to provide SAPE with data from the flow of economic dispatches from the Agence France-

Presse (AFP). It shows the improvements made to SAPE by such a cooperation for the building of two different kinds of knowledge bases. Section 5 describes and discusses how the natural evolution of the world on which the system operates constitutes a need for its perpetual adaptation. Section 6 concludes and presents our future work.

2. NLP and DSS

It is not a new idea that Artificial Intelligence would benefit from big knowledge bases formatted for computer. It was the goal of the Cyc project to build such a base [1]. At this time, much of the data was manually entered in the database by teams of Knowledge Enterers. The considerable amount of work needed to accomplish this task suggested the use of NLP. As early as 1992, Hayes [2] proposed to use IE to improve Expert Systems. Yet there are significant differences between the two approaches. With Cyc, the approach is to extract a little information about many different subjects. On the contrary, IE technologies allow the production of a great amount of information but on a small number of subjects.

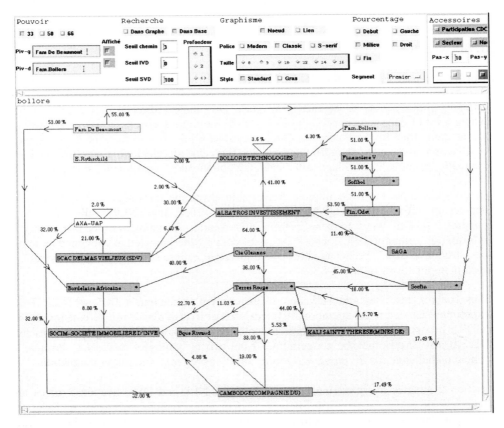

Figure 1 : The share-holding graph browser

Neither is it a new idea that information could be extracted from dispatches of press agencies. Many teams applied themselves to produce such extractions. The MUC contest [3] shows the high level of accuracy reached by current systems. Yet scientific research remains the main goal of these systems. As underlined by Costantino et al. [4], there are still few publications describing realizations that have reached the level of operational use, notably in the financial field.

3. SAPE : a Decision Support System dedicated to complex share-holding information

The growing complexity of financial technology, fiercer competition on financial markets, and the need for higher return on investment, has lead the equity fund managers of *Caisse des Dépôts et Consignations*, aware of technological progress in information management, to ask for new, more sophisticated, decision-making tools [5].

In this section we describe one of these software systems, the SAPE system (*Système d'Analyse des Participations des Entreprises* - a system to analyse the share-holdings between companies) [6]. This tool shows and analyses the information concerning the breakdown of companies' shareholders. So it helps the manager in his investment decision.

The connections between national and international companies are becoming more and more complex. The precise knowledge of these links is becoming a duty for fund managers. First for a better understanding of the firms' strategies, and secondly to anticipate a possible future move in shareholders' composition.

SAPE provides ways to manage such a complex information network and give the capability to simulate moves in the composition of a firm's shareholders. We describe the user interface and the associated financial module.

3.1. AN INTERFACE TO EXPLORE THE SHARE-HOLDING NETWORK

We visualize all the firms' stock connections as a huge network. We use a graph-oriented representation. The nodes of those graphs are the companies and the links between them represent the percentage of equity capital owned (see Figure 1).

SAPE also provides a way to generate and manipulate sub-networks associated with groups. The user can edit these graphs, change positions of the nodes, colors and fonts. He can save and print them. The tool uses a huge database concerning firms. However, the "links" information is gathered and fed by SAPE.

3.2. A FINANCIAL MODULE TO COMPUTE COMPANIES' POWER

SAPE is linked to a financial module. This module can compute the "potential power" of a company over others. The potential power of company A over a company B is viewed as a percentage of the stock of B which can be considered as belonging to A.

The resolution of the power of a company over a set of other companies linked together by direct or indirect links is solved as for a flow problem. So whatever the complexity of the graph (even with crosslinks) this module is able to tell you if a company is controlled or not. This control is either a veto control (33% of the stock), ordinary control (50 %), extraordinary control (66%).

This module is closely associated with SAPE and allows the construction of what we call controlling groups, or to compare the level of control of two companies over a third one.

Moreover, the tool is able to simulate changes in stock ownership, and then to recompute the power. These simulations can help the manager to appreciate the move of stocks and their consequences in terms of control.

4. Using IE to Feed SAPE

In order to anticipate future moves accurately, the system must cover a great part of the share-holding map. Though mainly concerned by France and Europe, the database already contains more than 7,000 companies. Collecting these data, keeping them up to date as and when new links are created and others are destroyed, is a considerable task. The first version of the system was fed by hand from companies' annual reports or from the daily press.

4.1 AUTOMATICALLY PROPOSING DATABASE UPDATES

The growing size and the continuous moves from the share-holding network, led us to search how to improve the database updates. Actually, important moves are mostly related in dispatches from press agencies. In our case, the AFP economic flow covers in depth the geographic zone traced by the system, i.e. France and Europe.

We have built a robot in charge of continuously reading this flow in order to detect dispatches on share-holdings. This robot extracts information about the event, that is : entities involved (people, families, companies, states) [7], the roles they play (parent company or target), the amount of the link (% or number of equities) and its kind (equity or voting rights) [8].

Then, another module decides whether the database manager should be informed about the dispatch [9]. The aim of the filtering is to reduce the number of alerts. Two criteria are used in the decision :

(a) The presence, among the entities involved in the share-holding, of at least one that is also in the database and thus subject to monitoring, so that the network can progress connectedly.

(b) The presence of a figure measuring the share-holding so that the network can progress numerically.

A final module allows the double display of the dispatch and the relative part of the database (see Figures 2 and 3). The informative text of the dispatch is high-lighted. The display of the share-holding network already inside the database is limited to the entities mentioned in the dispatch. The database manager may then compare the information brought by the news with that contained in the database.

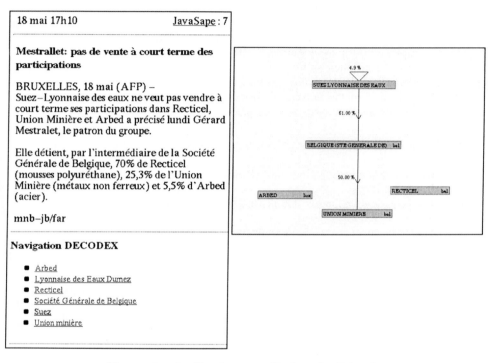

| 18 mai 17h10 | JavaSape : 7 |

Mestrallet: pas de vente à court terme des participations

BRUXELLES, 18 mai (AFP) –
Suez–Lyonnaise des eaux ne veut pas vendre à court terme ses participations dans Recticel, Union Minière et Arbed a précisé lundi Gérard Mestralet, le patron du groupe.

Elle détient, par l'intermédiaire de la Société Générale de Belgique, 70% de Recticel (mousses polyuréthane), 25,3% de l'Union Minière (métaux non ferreux) et 5,5% d'Arbed (acier).

mnb–jb/far

Navigation DECODEX

- Arbed
- Lyonnaise des Eaux Dumez
- Recticel
- Société Générale de Belgique
- Suez
- Union minière

Figures 2 et 3 : Simultaneous display of a dispatch
and of the relative part of the share-holding database

In our example, the fund manager is alerted that an important news has occurred. The text gives several numerical share-holdings involving 5 companies : *Suez-Lyonnaise, Société générale de Belgique, Arbed*, the *Union minière* and *Recticel*. SAPE user interface automatically searches and displays the relative sub-part of the database. The user notes that many links are out of date. The link between the *Société générale de Belgique* (SGB) and the *Union minière* should be decreased from 50% to 25.3%. Conversely, the one between SGB and *Recticel* should be 70%. Finally, the link between SGB and *Arbed* should be 5.5%. The user may then update the database with this information.

Would automatic updating be possible? The correctness of the data and consistency of the database are essential for the deductions of the financial module. Our choice is that the updates should be checked by the fund manager while the tool should facilitate his task as much as possible. Future improvements of technology will allow more profound controls within the database, but the ultimate decision will remain the user's responsibility.

4.2 INDEXING IE RESULTS TO EXTEND SAPE CAPABILITIES

Feeding a graph database is not the only use for IE results. The different parts of share-holding (parent company, target, measure, nature) are of great interest for indexing the documents that they come from. These by-products of IE have been used for the building of a large Information Retrieval (IR) system, dedicated to share-holdings. It allows to search the dispatch archive in order to obtain information about a company's past share-holdings. This is a useful complement to the graph browser because it adds a historical perspective to the current share-holding network. Being aware of the past strategy of a group makes it easier to anticipate its future acquisitions.

1 Restructuration financière en vue autour de la Générale de Belgique
  0,92, 07/04/98 17h49
2 Suez Lyonnaise croque la Générale de Belgique en ménageant l'orgueil belge pa
  0,87, 18/05/98 19h07
3 Suez–Lyonnaise des Eaux veut acquérir 100% de la SGB, selon Le Soir Revoici
  0,87, 07/04/98 11h26
4 Le rapprochement Fortis–Générale de Banque se concrétise par OPE de Fortis
  0,84, 18/05/98 09h16
5 Fred Chaffart réélu à son poste d'administrateur à Générale de Banque
  0,84, 23/04/98 13h37
6 Le rapprochement de Fortis et de la Générale de Banque toujours à l'étude par
  0,84, 22/04/98 15h51
7 Suez–Lyonnaise: la Belgique s'interroge sur le sort réservé à Tractebel REVOIC
  0,84, 18/04/98 09h42
8 Suez Lyonnaise confirme "éventualité" renforcement participation dans SGB
  0,84, 09/04/98 20h27
9 Suez Lyonnaise des Eaux: S and P confirme les notes A+ (long terme) et A–1
  0,80, 19/05/98 12h38
10 Deux symboles de l'économie belge disparaîtront de la cotation à Bruxelles par
  0,80, 18/05/98 18h31

TITRES SUIVANTS
▼

Figure 4 : Past share-holdings of *Générale de Belgique*

In practice, the user starts the search by clicking on the name of the company that he wants to study (see Figure 2, bottom).

The system selects and ranks the relevant dispatches containing information about its share-holdings (see Figure 4). In our example, the portfolio manager is interested in *Société générale de Belgique*. An additional panel allows the user to refine the search by specifying the company's role in the event (parent company or target).

A previous implementation of the system based on a relational database is described in Vichot et al. [10]. A new implementation has proved far more efficient. After a standardization process, the extracted information is reinserted in the text as HTML annotations. Then the documents are indexed with a classical search engine.

This approach, that relies on IE for improving IR, seems to us to be one of the more promising ways to improve this kind of DSS. Yet it should be noted that taking full benefit of this new abundance of data requires also the exploration of new ergonomic metaphors.

5. New boundaries for IE based DSS

Adapting an IE engine to a new application is well-known to be a difficult problem [11]. The most frequently mentioned cases are adaptation to a new scenario, to a new corpus or even to a new language. The daily use of IE in order to feed a DSS has underlined an adaptive problem of another kind. For us, it is a major issue, because the long-term life of the system itself is concerned. That is, *the IE engine's ability to follow the evolution of the environment in which it operates*.

We give below some examples of variations that the IE engine has to face. They are chosen from among the most frequent cases we have met. The variations come from three main sources: the continuous innovation of the language, the evolution of the underlying world and the focus of the user himself.

5.1. THE INNOVATION OF LANGUAGE

The profusion of language is what strikes you most when you tune an IE engine to a large corpus written by numerous authors. It is the case with the AFP economic flow, produced by professional journalists. One concept might be described in a seemingly unlimited number of forms. Consequently, the system can never be considered to be finished. Here are two particularly productive examples. One is about proper names. The second one concerns conceptual locutions.

(a) Thus, the well-known credit rating agency, *Standard and Poor's*, has been denoted in more than 10 ways (*Standard and Poors, S & Poors, S et P, SandP, SP...*). The phenomenon of intensive polymorphism, which exists as soon as a compound proper name is frequently used, is a challenge for classical name-finding techniques.

(b) A second case of intensive polymorphism occurs when a concept has not one single word to express it. Instead, a composite locution will be used. In practice, each element may be subject to variations of form as well as of position. Thus, in French, the concept of share-holding is frequently built with the verb *détenir* (*A détient x% du capital de B*). But this verb may be switched with many verbs containing the notion of possession (*avoir, posséder, acheter, vendre, prendre...*). If variations in position can, in principle, be left to the syntactic parser (with a lot of unsolved problems), variations in form remain a painstaking task.

5.2. THE REAL WORLD IS CHANGING

Slowly but surely, if we don't pay attention to it, the system loses its effectiveness. It is another striking point when supporting an IE-based DSS over a long period. Here again, sources of variation are multiple. Here are some examples chosen in the field of the life of companies showing that no week passes without an event impacting the system.

(a) Changing names: A short time ago, for instance, *Compagnie Générale des Eaux* (CGE) reported that it was changing its name for a more international one, *Vivendi*.

(b) The merger of two existing entities: Citicorp and Travelers have merged to form a new entity called Citigroup.

(c) The appearance of a new abbreviation: *Bayerische HypoBank* and *Vereinsbank* reported they were merging into a new entity, named *Bayerische HypoBank und Vereinsbank*, whose abbreviation, BHV, is already used by a French department store, *Bazar de l'Hôtel de Ville*.

In each case, the problem is (i) to detect a change that concerns one company among thousands (ii) to teach the system to recognize a new name and (iii) to prepare it to resolve possible ambiguities.

5.3. USER FOCUS EVOLVES

The DSS is a means of helping the user to discover new information. In return, as the user goes along, he wants more precise information or information about connected topics. Let's look at some examples :

(a) Extending the share-holding network: As the user explores the share-holding network, he discovers new connections and wants to add new companies to it. If an entity was unknown by the IE engine, once the system has learned to take it into account, months of news archives are to be processed again in order to extract all the company's previous share-holding.

(b) Covering connected subjects : Of course, the classical problem of adapting the system to a new scenario is also present. The study of the share-holding network leads to focus on other networks concerning the control of companies such as Boards of Directors or partnership agreements. Also managers' appointments, credit ratings, annual profits or market share are all useful information for the assessment of a company. Although all these data are available in AFP dispatches, and although extracting them depends on exactly the same techniques, discovering and coding again all the required patterns remains a considerable task.

Nevertheless, due to its importance, the adaptive problem has aroused a lot of research from the IE community. A well-shared idea is that the IE engine should be able to "reconfigure" itself, at least partially, with minimal effort from the human side. Many machine learning techniques tackling this issue from a concrete point of view are becoming available. For example, Paliouras et al. [12] describe a word disambiguator based on the C4.5 learning algorithm. As well, Cucchiarelli et al. [13] propose an incremental method for extending a dictionary-based name finder to a new corpus. A next step is to make these techniques available for a corpus that has a potentially infinite size, such as a continuously renewed flow of dispatches.

6. Conclusion and Future Work

In this chapter, we illustrate how we use IE for collecting the information needed by a DSS from a flow of dispatches. Resorting to IE has turned out to be a powerful solution, especially for this sort of application requiring, in a highly repetitive way, a well-specified unique kind of data. We also show that if IE may be used in order to feed a structured database, it may also improve a specialized IR system. This approach allows the exploitation of a dispatch archive as a vast knowledge base. Our project is to use this process extensively in order to build an economic encyclopedia dedicated to the fund managers of the CDC Group.

Nevertheless, a major success factor lies in the ability of the system to adapt to a continuously changing environment. We are analyzing the main causes of variations we have met. If the subject remains wide open, some machine learning techniques, used with success in the field of IE, should allow results in this direction to be obtained.

7. Acknowledgments

We would like to thank F. Pelletier, T. Pullman et F. Sillion, for the assistance they gave us in the course of this project.

8. References

[1] Lenat D., Guha R. (1990) Building Large Knowledge Based Systems, Addison Wesley.

[2] Hayes J. (1992) Intelligent High-Volume Text Processing Using Shallow, Domain-Specific Techniques, in Text-Based Intelligent System, Lawrence Erlbaum Associates.

[3] MUC-6 (1995) Proceedings of the 6th Message Understanding Conference, Morgan Kauffmann.

[4] Costantino M., Morgan R., Collingham R. (1996) Financial Information Extraction using pre-defined and User-definable Templates in the LOLITA system, International Meeting of Artificial Intelligence in Accounting, Finance and Tax, Huelva, Spain.

[5] Berthier F. (1992) SEVE: A Stock Portfolio Long Term Planning System, International Conference on Economics/Management and Information Technology, CEMIT'92/CECOIA3, Tokyo, Japan.

[6] Ferri H.-C., Berthier F. (1993) De nouveaux outils pour la planification du portefeuille action et l'évaluation du marché, Colloque Systèmes experts bancaires, Milan, Italy.

[7] Wolinski F., Vichot F., Dillet B. (1995) Automatic Processing of Proper Names in Texts, European Chapter of the Association for Computational Linguistics, EACL'95, Dublin, Ireland.

[8] Landau M.-C., Sillion F., Vichot F. (1993) Exoseme: a Document Filtering System Based on Conceptual Graphs, International Conference on Conceptual Structures, ICCS'93, Montreal, Canada.

[9] Landau M.-C., Sillion F., Vichot F. (1993) Exoseme: a Thematic Document Filtering System, Intelligence Artificielle'93, Avignon, France.

[10] Vichot F., Wolinski F., Tomeh J., Guennou S., Dillet B., Aidjan S. (1997) High Precision Hypertext Navigation Based on NLP Automatic Extractions, Hypertext Information Retrieval Multimedia, HIM'97, Dortmund, Germany.

[11] Grishman R. (1997) Information Extraction : Techniques and Challenges, in M.-T. Pazienza (Ed.) Information Extraction : a Multidisciplinary Approach to an Emerging Information Extraction Technology, Springer-Verlag.

[12] Paliouras G., Karkaletsis V., Spiropoulos C. (1998) Machine Learning for Domain-Adaptive Word Sense Disambiguation, Workshop on Adapting Lexical and Corpus Resources to Sublanguages and Applications, LREC, Granada, Spain

[13] Cucchiarelli A., Luzi D., Velardi P. (1998) Using Corpus Evidence for Automatic Gazetteer Extension, First International Conference on Language Resources and Evaluation, Granada, Spain

18

Eliciting Terminological Knowledge for Information Extraction Applications

Byron Georgantopoulos[1] [(*)(@)] and Stelios Piperidis[(*)(&)]

(*) Institute for Language and Speech Processing - ILSP
Artemidos & Epidavrou, Athens, GREECE
Tel: +30 1 6800959, fax: +30 1 6854270

(@) University of Athens

(&) National Technical University of Athens

Email: {byron, spip}@ilsp.gr

1   Introduction

In this chapter we present a method for automatically extracting terminological resources from text corpora. Automatic term extraction is of great interest nowadays, where huge volumes of texts are produced and published electronically, resulting in new requirements for their management and processing (automatic text classification, information retrieval, information extraction, etc.). The application of language technology systems in order to meet these requirements presupposes the customisation of the system to the domain of the processed text. A basic step towards this procedure is the improvement and enrichment of language resources by incorporating the appropriate domain terminology. The application of methods for automatic extraction of terminological resources provides a reliable, fast and low-cost solution to the customisation procedure.

There are many natural language processing applications that automatic term extraction supports:

- text indexing - the extracted terms are used to construct a typical back-of-the-book index
- text classification - texts with equal or similar term sets are classified in the same domain/text category
- information retrieval and information extraction - user queries are answered after comparing the terms of the query with the terms of the texts contained in a corpus of the same domain

[1] The author would like to thank the Greek Scholarships Foundation (IKY) for its financial support.

- text condensation and draft summarisation - sentences containing dominant terms can be considered as highly representative of the content of the original text
- bilingual text alignment - terms of one language usually are translated uniformly in another language, within the same domain.

2 Methodological approaches

A term is a linguistic realisation of a domain specific concept and usually is lexicalised in the form of a noun phrase. In bibliography, one can find two basic methods for extracting terms:

1. Using a term grammar (usually a context free grammar) which is applied to an appropriately annotated text and extracts all the phrases it recognises [1]
2. Using statistical tools similar to the ones developed in the field of information retrieval and text indexing. These tools include frequency counting, formulas from information theory, formulas which take into account the context of words, etc. [2],[9]

There are important differences between these two lines of action. A term grammar describes all the possible syntactic structures that a valid term should satisfy, but it is possible that phrases recognised by the grammar are not true terms. The weakness of a grammar is attributed to the fact that its rules, though a subset of NP rules, are general enough to generate a large number of potential terms[2]. Furthermore, a grammar cannot locate single word terms since such a term does not have any syntactic structure except part-of-speech information. In general, a term grammar can only produce a set of potential terms that remain to be validated by an expert or a module of different nature.

The statistical approach is based on the assumption that words and phrases indicative of the content and domain of a document tend to appear *frequently* (the same applies for phrases consisting of words that appear frequently *together*). The notion of frequency in this context can have two different interpretations: (1) a word/phrase is more frequent in the current text than in a representative collection of texts belonging to its domain and (2) a word/phrase is more frequent than others within the same text. Based on this "competitive" conception of frequency, each phrase[3] is assigned a score representing its significance in the current text[4]. Phrases at the top of this ranking will have the highest probability of being valid terms. This method can extract single-word terms as well as multi-word terms. On the other hand, it cannot locate terms which do not satisfy the statistical criteria, i.e. they are not frequent enough. This is partly due to the fact that it is difficult to draw the line

---

[2] For example, a typical rule for term extraction, Term ::- Adj+Noun, will definitely recognise many non-term phrases besides valid ones.
[3] Either extracted by a grammar, or a collocation of two or more words
[4] Functional words are not taken into account.

between middle frequency (non-termhood) and high frequency (termhood), not to mention the possibility that a true term may appear very rarely in a text and thus is out of the scope of frequency statistics. Finally, the selected statistical formula affects the performance of extraction in the same way that the selected rules of the grammar, i.e. its syntactical coverage, affect the performance of the grammatical method. Another factor that has an effect on the quality of results is the size and the representativity of the reference corpus. Corpora which are either small or not homogeneous may not provide enough statistical evidence for valid terms and "allow" semantically insignificant phrases to achieve better than expected rankings.

In between these methodologies stand other approaches which combine statistical processing with linguistic modelling [3], [4], [5], [6], [8], [9]. These hybrid systems initially construct a list of term candidates using a term grammar and then filter this list through statistical techniques in order to remove syntactically acceptable phrases that are not frequent enough to be assigned valid termhood. Combining the best of two worlds, such systems yield results closer to human intuition as:

- Linguistic processing can identify grammatically legal noun phrases, and
- Statistical processing accounts for heaviness (or sparsity) of use which is an inherent feature of human termhood assignment processes

3 System description

The developed approach is largely based on the linguistic analysis of the source text. The basic steps are:

1. morphosyntactic annotation of the domain corpus using (a) a below part-of-speech tagger based on a morphological lexicon and (b) a system for resolving morphosyntactic ambiguities
2. parsing based on a pattern grammar endowed with regular expressions and feature-structure unification[5]
3. lemmatisation using a morphological lexicon and the grammatical category obtained by the tagger

The following flowchart illustrates how the algorithm proceeds:

---

[5] Feature-structure unification is necessary in order to capture agreement between words (e.g. case agreement) in Greek. This is the reason why below part-of-speech tagging was performed.

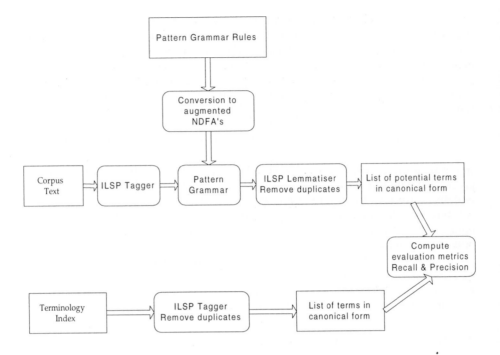

The **pattern grammar** used in the syntactic analysis is a subset of pattern rules presented in [10]. It is actually a grammar that utilises the feature structure unification formalism (typical in grammar theories like HPSG) and regular expression operators. For example, the pattern that describes terms of the form : NOUN (PRON) PREPOSITION (ART) NOUN has the following format:

term_ pattern :   (cat = No
        ^(cat = Pn
        type = Cl),
        [[(cat = Pp
          type = Sp);
          ^ (cat = At
            gender = G
            number = N
            case = C)] ;
        (cat = Pp
        type = Pa
        gender = G
        number = N
        case = C)],
        (cat = No
        gender = G
        number = N
        case = C)).

The '^' symbol denotes optionality (zero or one appearance), while the ';' symbol stands for the 'OR' operator and brackets are used for grouping elements together. The basic constraint posed by this rule is the number-case-gender agreement between the article (At), the preposition (Pp) and the noun (No). The agreement is expressed by using the same variable name for case, gender and number (namely variables C, G and N) throughout the rule, denoting that each of these variables should keep the same value for all the three constituents.

The term grammar consists of 18 rules recognising two to four-word terms[6]. Each rule is converted to a **non-deterministic finite state automaton** (NDFA), implemented in Prolog. NDFA's were used in preference to context-free grammar parsers (like Prolog Definite Clause Grammars) because (a) they are much faster, operating in linear time and (b) typical parsers do not support regular operators directly. Features used in unification include grammatical category as well as subcategorisation features like gender, case, tense, number, aspect, etc. Typical regular expression operators are optionality, star, disjunction, etc.

An example of such a non-deterministic automaton is illustrated below:

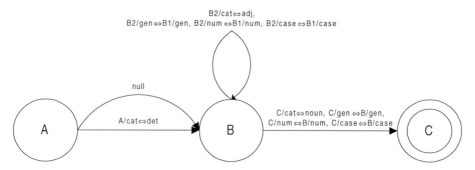

Grammar rule ::- Det[Gen,Num,Case]? (Adj[Gen,Num,Case])* Noun[Gen,Num,Case]

? = optional, * = zero or more times

In converting to Prolog, the main predicate is t(S1, A, B, S2), which expresses the requirements for transition from state S1 to state S2, reading symbol B from stack and having just read symbol A. The previous automaton is therefore coded as follows:

---

[6] By words, in this context, we refer to content words.

```
initial( A ).
t(A, null, B).
t(A, X, B):-     A/cat ⇔ Det.
t(B, X, Y, B):- Y/cat ⇔ Adj, Y/gen ⇔ X/gen, Y/num ⇔ X/num,
                   Y/case ⇔ X/case.
t(B, X, Y, C):- Y/cat ⇔ Noun, Y/gen ⇔ X/gen, Y/num ⇔ X/num,
                   Y/case ⇔ X/case.
final( C ).
```

The **corpus** on which the system was tested is a software manual of 90.000 words in size. Pre-processing of the corpus includes tokenisation, sentence boundaries identification, part-of-speech tagging and lemmatisation. The particular text was selected because it also contained a human-crafted **terminology index** against which the results of pattern matching would later be evaluated. During the evaluation procedure terms were reduced to their **canonical form** where each word was replaced by its lemma.

4 Results of term grammar parsing

In order to evaluate the efficiency of the system we calculated standard recall and precision figures. The set of terms extracted by the term grammar was compared against the target terminology index provided along with the manual. The target list was first cleared from single-word terms, resulting in 209 canonically-formed target terms in total. The grammar extracted 3596 candidate terms, 130 of them being correct, i.e. matching correctly one of the target 209 ones. Thus, recall and precision were calculated as follows:

Recall: 130/209   = 62,2%
Precision: 130/3596 = 3,6%

The recall figure is considered satisfactory. Study of the remaining 79 terms that were not identified by the grammar revealed that 17% of them were due to foreign words (no part-of-speech information assigned) or to tagger inaccuracies caused by unknown words or incorrect disambiguation. The rest of the unidentified terms followed syntactic structures not coded in the grammar rules, e.g. they contained verb phrases. On the other hand, the precision figure is considerably, but not unexpectedly, low, since grammars inherently produce numerous terms because of the generality of the rules. This problem initiated the consideration of further filtering the list of potential terms aiming at reducing its size without significant loss in recall.

5 Statistical evaluation

After the execution of the term grammar module, the extracted terms were statistically evaluated in order to remove items without adequate statistical evidence. This evidence comes from Daille's extensive work [6] on contingency tables and statistical measures used in valuing potential terms. These measures were applied on two-word terms since manipulation of 2×2 contingency tables is easier.

For a given pattern $w_i\ w_j$, (e.g. noun+noun) the contingency table is defined as:

	w_j	$w_{j'}, j\neq j'$
w_I	a	b
$w_{i'}, i\neq i'$	c	d

where:

a stands for the frequency of pairs involving both w_i and w_j,
   (the number of occurrences of a pair)
b stands for the frequency of pairs involving w_i and $w_{j'}$,
   (the number of occurrences of pairs where a given word appears as the first element of the pair)
c stands for the frequency of pairs involving $w_{i'}$ and w_j ,
   (the number of occurrences of pairs where a given word appears as the second element of the pair)
d stands for the frequency of pairs involving $w_{i'}$ and $w_{j'}$
   (the total number of occurrences of the pairs where none of the given words appears as an element)

Daille presents several statistical scoring formulas which test the strength of bond between the two variables w_i and w_j of a contingency table[7], taken from the fields of lexical statistics, information retrieval and other technical domains such as biology. All of these formulas were tested in a telecommunication corpus and the best scoring terms were examined against a terminological bank of the domain[8]. The best-performing formulas, i.e. the ones that assigned high values to valid terms, and low values to co-occurrences that are not valid, were found to be: (1) log-likelihood, (2) Fager and McGowan coefficient, (3) cubic association ratio, and (4) simple term frequency. It is worthy noticing that the first three scores fall under the "association criteria" category, while the fourth one is purely frequency-based.

We conducted the same experiment, using these scoring mechanisms (adding another one, NC-value [9], which takes into account the context of candidate terms), and found that the best results came from the following four statistical formulas:

---

[7] namely, Term Frequency, Simple Matching Coefficient, Kulczinsky Coefficient, Ochial Coefficient, Fager and McGowan Coefficient, Yule Coefficient, McConnough Coefficient,
[2] Coefficient, Association ratio, Cubic Association ratio and Log-likelihood Coefficient.
[8] Eurodicautom, terminology data bank of the EEC, telecommunication section.

1. Fager and McGowan Coefficient (FAG):

$$\frac{a}{\sqrt{(a+b)(a+c)}} - \frac{1}{2\sqrt{a+b}}$$

2. Cubic Association Ratio (IM3):

$$\log 2 \frac{a^3}{(a+b)(a+c)}$$

3. Log-Likelihood (LLH):

$$a \log a + b \log b + c \log c + d \log d - (a+b)\log(a+b) - (a+c)\log(a+c)$$
$$- (b+d)\log(b+d) - (c+d)\log(c+d) + (a+b+c+d)\log(a+b+c+d)$$

4. NC-value

These results are very close to the top-ranked formulas found in Daille's work. The following table summarises precision and recall figures for the top 200 terms produced by each one of the four different measures. Since the grammar located 77 out of 134 two-word terms in the terminology index we have calculated two recall figures, with two different denominators: the first referring to the full index (134), the second referring to the grammar results (77).

	FAG		IM3		LLH		NC	
precision	17%		14%		18%		20%	
recall	26%	45%	22%	38%	28%	48%	30%	52%

The improvement in precision is significant (from 3% to 14-20%, i.e. five to six times bigger) since the list of potential terms was dramatically reduced (from 3596 to 200 terms), while recall dropped by about 50%.

6   Conclusion and outlook to future work

We have presented a system for automatic term extraction using both linguistic and statistical modelling. Linguistic processing is performed through a unification-based term grammar, the results of which are statistically evaluated using frequency-based and associative statistical measures. The method was able to locate 62% of technical terminology in a software manual text.

The presented work is to be enriched with features, improving its efficiency by reducing the number of potential terms recognised by the grammar module as well as improving its coverage. These features include:

- Implementation of other statistical processing techniques (such as term weighting with TF·IDF scoring [14]) on the terms extracted by the grammar so that the frequency of the term in the domain is also taken into consideration[9].

- Utilisation of further syntactic information (e.g. NP head) in order to group together terms with the same semantic but slightly different syntactic structure.

- Extension of an already existing terminological base through linguistic operations such as overcomposition, modification, coordination, etc. [5], [11].

- Enrichment of the grammar with more pattern rules, so that terms containing verb phrases, etc can be recognised.

7 References

[1] Bourigault D. (1992), Surface Grammatical Analysis for the Extraction of Terminological Noun Phrases. Proceedings of the 14th International Conference on Computational Linguistics.

[2] Church K. W. and Hunks P. (1990), Word Association, Norms, Mutual Information, And Lexicography. Computational Linguistics, Vol 16, Number 1.

[3] Dagan I. and Church K. W. (1994), Termight: Identifying and Translating Techical Terminology. Proceedings of the EACL 1994.

[4] Daille B., Gaussier E., Lange J. M. (1994), Towards automatic extraction of monolingual and bilingual terminology, Proceedings of COLING 94, pp 515-521.

[5] Daille B. (1994), Study and implementation of combined techniques for automatic extraction of Terminology. in The Balancing Act: Combining Symbolic and Statistical Approaches to Languages, Workshop at the 32nd Annual Meeting of ACL, Las Cruces, Nouveau Mexique.

[6] Daille B. (1995), Combined approach for Terminology Extraction : Lexical statistics and linguistic filtering, TALANA, Université Paris 7

[7] Dunning T. (1993), Accurate methods for the statistics pf surprise and coincidence. Computational Linguistics, vol. 19, n° 1

[8] Frantzi K. and Ananiadou S. (1996), Extracting nested collocations, Proceedings of COLING 96, pp 41-46.

[9] Frantzi, K.T. and Ananiadou, S. (1997), Automatic term recognition using contextual clues, Proceedings of Mulsaic 97, IJCAI, Japan.

---

[9] Contingency tables utilise only the frequency of the word inside the processed text.

[10] Gavriilidou M, Lambropoulou P., Report on the Constituent Grammar, RENOS project, LREI- 62-048, Athens, 1994.

[11] Jacquemin C., A Symbolic and Surgical Acquisition of Terms through Variation, Proceedings of the 14th International Joint Conference on Artificial Intelligence, Montreal 1995.

[12] Hatcher A.J. (1960), An introduction to the analysis of English noun compounds. In Word, 16, 356-373.

[13] Smadja F. A. and McKeown K. R. (1990), Automatically Extracting and Representing Collocations For Language Generation, Proceedings of the 28th annual Meeting of the ACL.

[14] Salton, G. (1989), Automatic text processing : the transformation, analysis, and retrieval of information by computer, Reading : Addison-Wesley.

19

Natural Language Interface to an Agent

J. KONTOS, I. MALAGARDI AND D. TRIKKALIDIS
Department of Informatics
Athens University of Economics & Business
76 Patission St., 104 34 Athens, Greece
E-mail: jpk@aueb.gr

1. Introduction

This chapter presents the design and implementation of a motion command understanding system with a natural language interface for the communication with its user. The development of techniques like ours is necessary in order to allow untrained users to make efficient and safe use of a robot. Motion command understanding is a basic function required for mobile robots [1]. Designing interfaces for human-robot communication involves the following considerations:

1. First, the user must be provided with an interface that allows him to instruct the robot. Here, instruction involves translating the user's intention into correct and executable robot programs.
2. Second, feedback must be provided to the user so that he can immediately understand what is happening on the robot's side. This task requires translating internal, possibly low-level representations used by the robot into a representation that can be understood by the user.

The system described here accepts Greek and English as language of communication with an agent [2, 3, 4] which will be used as a testing device for human-robot communication. The system is applied to the communication with an artificial agent in a virtual environment and accepts commands and knowledge about the objects and about the actions that are possible in this environment. The development of the system is part of a research program initiated about 15 years ago [5] based on the results of our earlier work on procedural interpretation [6,7] of natural language sentences for human-computer communication. These publications preceded T. Winograd's [8] related publication which is often mistakenly mentioned as the first publication on the implementation of procedural interpretation of sentences. An extensive review of the subject is presented in [9].

The commands issued by a human to an agent in natural language may express three kinds of actions. The first kind of action concerns change in position e.g. movement of an object. The second kind concerns change in state e.g. opening or closing an object. The third kind concerns change of the relation between two objects e.g. putting an object on top or inside another object.

When the system is given a command like: "open the door", "open the bottle", "close the box", "put the book on the desk" etc., which specifies a task, the system has "to understand" the command before attempting to perform the task. Understanding such commands requires understanding the meaning of verbs such as "open", "close", "put" and the meaning of prepositional words such as "on". The meanings of the constituents of a sentence must be combined and the meaning of the sentence as a whole must be obtained, taking into consideration the knowledge of the environment or of the "microcosm" in which the commands must be performed.

Motion may be specified by a verb either directly or indirectly. The simplest way to specify the motion of an object is by using a verb that specifies motion directly. An example is the verb "open" as used in the sentence "open the box". This sentence implies that a motion must be executed by the system, the box being the affected object. Indirect specification of motion can be accomplished in two ways: either in terms of geometric goals, or in terms of force. Indirect specification of motion in terms of a goal involving physical relationship between objects is quite common. Consider the command "put the bottle on the table". This command requires that a physical object be moved e.g., "the bottle" with a goal to establish the physical relationship of "on" between it and another physical object e.g., "the table". Performance of such an instruction demonstrates that the goal of establishing a physical relationship drives the motion of the first object. For verbs such as "put" that specify motion in terms of a geometric goal, properties of the objects participating in the underlying action are of crucial importance. Indirect specification of motion in terms of a force uses verbs such as "push" and "pull". Objects affected by motion commands may be also specified either directly or indirectly. Direct specification is based on names of objects known to the system, such as box, table, etc. Indirect specifications can be accomplished by the use of complex noun phrases such as "the book on the desk".

The representation of the meaning of motion verbs using physical primitives is addressed in [10]. This is a component of a computer system that accepts natural language commands as input and produces graphical animation as output. A fixed lexicon written manually is used based on their representation method. The long-term goal stated in [9] is to investigate how semantic information for motion verbs can be automatically derived from machine-readable dictionaries. It is also stated, that at present, the system has no feedback from the graphical animation system to the semantic processor. Finally, their system has no learning capabilities. Our system satisfies all these three requirements.

Our system satisfies the above requirements using some novel features e.g. the automatic creation of its lexicon by use of a machine-readable dictionary, learning the correct interpretation of commands with more than one meaning which is accomplished using machine learning by supervision techniques based on visual feedback. One source of the multiplicity of meaning of a command is the multiplicity of the senses of a word as recorded in a machine-readable dictionary. Another source is the possibility of placing an object on the surface of another object in different ways.

The main contribution of the present chapter is based on the ability of the system implemented to learn from its user, to understand and to execute correctly motion commands that go beyond its initial capabilities. This learning takes place in cases when the system faces the problem of unknown words, unknown senses of words or of underspecified positions of objects as in the cases of on "on-top" placing.

The system was implemented with Turbo Prolog which has some simple facilities for computer graphics. Using these facilities, the system displays on the computer screen a room with a door, some furniture and some manipulatable objects. These are objects such as a table, a door, a desk, a bottle, a box and a book. The bottle, the box and the book are examples of manipulatable objects. It is supposed that, in the room, there is an invisible agent, who can move around and execute the user's motion commands. These commands may refer directly or indirectly to the movement of specific objects or to the change of their state. The agent knows the names of these objects displayed on the screen and their position in the room. The agent also knows how to execute some basic commands .

When the user submits a command, the agent, in order to satisfy the constraints of the verb's meaning, may ask for new information and knowledge about objects and verbs, which may be used in the future. A machine-readable dictionary with possibly ambiguous entries is used, which provides the analysis of complex verbs into basic ones obtained automatically as described in [11]. In particular, in the case of the Greek language about 600 motion verbs were analysed automatically in terms of about 50 basic verbs. Finally, every time a command is executed, which is amenable to more than one interpretation, the system allows the user to observe the graphical output and state its approval or disapproval, which helps the system learn by supervision. Verb classification is studied in [12]. Our class of motion verbs is a union of different motion-related verb classes presented in [12]. A number of motion related verbs the meaning of which cannot be visualised as agent actions have been ignored.

2. System Architecture and Operation

The system is composed of the modules:

- Machine readable dictionary processor
- Lexical processor
- Syntactic processor
- Semantic processor
- Basic motion command processor

- Graphics processor
- Learning module

The operation of these modules is supported by a number of databases and knowledge bases. These are:

- Machine readable dictionary
- Basic Lexicon
- Stems Base
- Objects Attributes Base
- Knowledge Base

The user enters his commands in natural language with the keyboard. The commands are imperative sentences with one motion verb which specifies the action and some other words like nouns, prepositions or adverbs. Prior to the syntactic and semantic analysis of the sentence, the system checks if each word of the sentence belongs to its lexicon. Stemming is used for morphological analysis because of the multiple forms of Greek words. When the command contains a word unknown to the system, then the system produces a message for the user and terminates the processing of the present command and waits for instructions.

After having recognised all the words in a command, the system performs the syntactic analysis of it. If the input sentence is syntactically correct, the system recognises the motion verb, the object or objects and the adverb or preposition. After this, the module for "processing of basic motion commands" tries to satisfy all the constraints and the conditions for the specified motion.This processing requires searching in the knowledge base from where the system retrieves information about the object's properties (e.g. position, weight, size, state, subparts etc.). At this point, when some information is unavailable or ambiguous, the system interacts with the user in order to acquire the missing knowledge. There are two different types of questions that the system asks: the first type concerns information such as object properties absent from the knowledge base. The second type refers to questions which demand a Yes or No answer. This happens when more than one interpretation of an input command is possible and the system cannot decide which is the correct one. In these cases, the system, using the machine learning mechanism, suggests each time one solution and requests an answer from the user. The Yes or No answers generate appropriate entries to the knowledge base and are used whenever a similar command is submitted again by the user.

Prolog predicates that retrieve each object's position from the object database have been implemented. Each predicate calculates the coordinates of the specified points that constitute the shape of the object. A more general predicate redesigns all the objects after the processing of the user's command. All the knowledge and the current state of each object can be saved in external files which are available for future use through the menu options of the interface.

3. Examples of Operation of the System

Suppose that the user enters the command "open the door". The system isolates the words of the command and recognises the verb "open" and the noun phrase "the door". The verb "open" appears in the lexicon with a number of different definitions. E.g. in the LDOCE Dictionary [13] we find, among others, the two senses of "open" a) to (cause to) become open, b) to make a passage by removing the things that are blocking it. The Greek dictionary we used contains similar sets of senses for this verb and the sense selection mechanism is practically the same for the two languages. The only difference is the wording of the sense selection rules for the two languages where the objects and their properties have different names. The system selects the sense "b" because it knows that a door blocks a passage. The next decision the system has to make concerns the way in which the opening action is executed.

The system finds in the knowledge base that there are two alternative ways of interpreting the verb "open", using either a "push" or a "pull" basic motion. Then, it selects the first motion and asks the user if this is the right one. If the answer is "No", the system selects the next available interpretation and prompts again the user to give an answer. When the answer is "Yes", a fact is recorded in the knowledge base denoting that for the verb "open" and the object "door" the appropriate motion is e.g. "pull" in case that the "Yes" answer was given for the "pull" interpretation.

The second example refers to the movement of a book that exists in the microcosm of the system. When the command "put the book on the desk" is given by the user, the system searches the knowledge base to find a side of the book that can be used as a base for it. The book has 6 sides and when the system selects one of them, it presents graphically an image with the book on the desk lying on this side. Then, it asks the user if the result of the motion is the correct one. When the user enters a "Yes" answer, this is recorded in the knowledge base and the process terminates. When the user enters a "No" answer, the process continues trying sequentially all the available sides of the book until a "Yes" answer is given by the user.

In particular let us consider the initial state of the microcosm as shown in Fig. 1. The user enters the command "put the book on the table". The agent executes initially this command in the wrong way and the new state of the microcosm resulting from this wrong execution is shown in Fig. 2.

The user observes that the book is placed on the table in an upright position, which was not what he meant and he inputs the word "no" as an answer to the agent which lets the agent know that a mistake has occurred during the execution of the command. As a result, the agent tries again and interprets the command differently so that the new state that results from the new interpretation where the book lies on a side with a large surface. The user now approves the agent's action by typing "yes".

From now on, the agent will interpret the command "put the book on the table" only in the way just approved of by the user. The graphical user interface implemented was very

216

helpful during the development of the system. It was much easier to see the result on the screen graphically rather than read lists of the knowledge to find the changes that were recorded during the program execution and the machine learning process.

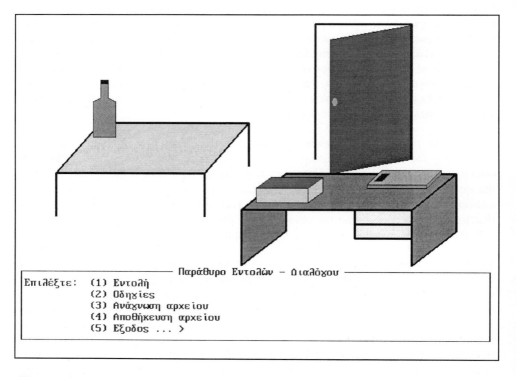

Figure 1

4. Conclusions

This chapter presents the design and implementation of a natural language interface for a motion command understanding system. The system accepts Greek and English as the user's natural language of communication with the system for the execution of motion commands. The system is applied to the communication between a user and an artificial agent moving in a virtual environment and accepting commands and knowledge about the objects and the actions possible in this environment. These commands may refer directly or indirectly to the movement of specific objects or to the change in their state. The agent knows the names of these objects and their position in the room displayed on the screen and knows how to execute some basic commands.

When the user submits a command, the agent, in order to satisfy the constraints of the verb's meaning, may ask for new information and knowledge about objects and verbs, which may be used for the execution of similar commands in the future. Indirect specification of motion is expressed in terms of a goal involving physical relationship among objects. Objects affected by motion commands may be specified either directly or indirectly. Direct specification is based on names of objects known to the system. Indirect specification may be accomplished by the use of complex noun phrases. The lexicon of the system is created automatically by the use of a machine readable dictionary while learning the correct interpretation of commands with more than one meaning is accomplished using machine learning by supervision based on visual feedback. The system has the ability of learning from its user to understand and execute correctly motion commands that go beyond its initial capabilities as shown by appropriate examples.

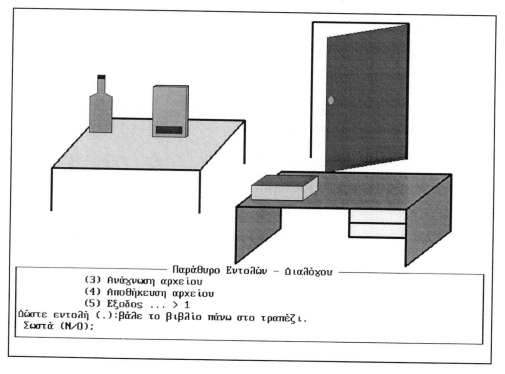

Figure 2

References

[1]. Klingspor, V., J. Demiris, and M. Kaiser. Human-Robot Communication and Machine Learning. *Applied Artificial Intelligence,* 11 pp. 719-746, 1997.

[2]. Vere, S.A., and T.W. Bickmore. A basic agent. *Computational Intelligence.* 6 pp. 41-60, 1990.

[3]. Webber, B.L., N.I. Badler, B. Di Eugenio, C. Geib, L. Levinson, and M. Moore. Instructions, intentions and expectations. *Artificial Intelligence,* 73, pp. 253-270, 1995.

[4]. Mueller, H. J. Towards agent systems engineering *Data and Knowledge Engineering,* No 23, pp. 217-245, 1997.

[5]. Kontos, J. Syntax-Directed Plan Recognition with a Microcomputer. *Microprocessing and Microprogramming.* 9, pp. 227-279, 1982.

[6]. Kontos, J., and G. Papakonstantinou. A Question-Answering System using Program Generation. *Proceedings of A.C.M. International Computing Symposium,* Bonn, Germany 1970.

[7]. Kontos, J., and A. Kossidas. On the Question-Answering System DELFI and its Application. *Proceedings of AGARD Symposium on Artificial Intelligence.* Rome, Italy 1971.

[8]. Winograd, T. *Natural Language Understanding.* Academic Press. New York. 1972.

[9]. Kontos, J. *Artificial Intelligence and Natural Language Processing,* (In Greek) E. Benou, Athens, Greece, 1996.

[10]. Kalita, J. K., and J.C. Lee. An Informal analysis of Motion Verbs based on Physical Primitives. *Computational Intelligence,* Vol. 13, N.1, pp. 87-125, 1997.

[11]. Kontos, J., I. Malagardi, and M. Pegou. Processing of Verb Dictionary Definitions by Computer. *3rd International Conference for the Greek Language. University of Athens,* 1997.

[12]. Levin, B. *English verb classes and alternations: a preliminary investigation.* The University of Chicago Press, Chicago, IL, 1993.

[13]. *LONGMAN DICTIONARY OF CONTEMPORARY ENGLISH. The up-to-date learning dictionary.* Editor-in-Chief Paul Procter. Longman group Ltd. UK, 1978.

20

Using Forward Temporal Planning for the Production of Interactive Tutoring Dialogues

T. Panayiotopoulos, N.Avradinis, C.C. Marinagi,
Department of Computer Science,
University of Piraeus,
80 Karaoli & Dimitriou Str, 18534 Piraeus, Greece
e-mail : themisp@unipi.gr, avrad@unipi.gr

1. Introduction

Dialogue is an effective means of communication between two or more participants. Its value has been acknowledged since ancient times, when it was used by Socrates as a method to discover truth and later evolved to an instruction tool in Plato's Academy. Its power lies into four attributes, inherent in its structure [1]:

a) It allows interlocutors to express different and often incompatible opinions, which allows the exchange of ideas between the participants and can unveil disagreements between them.
b) It is an incremental process of understanding, through series of questions and answers, which facilitates learning by gradually increasing and/or modifying the participants' knowledge.
c) It allows participants debate upon the issue in question by trying to verify a sequence of arguments, which helps creating a common knowledge base and eliminating disagreements.
d) It is a highly interactive process, providing the tutor with feedback from the student and allowing learning by forcing interlocutors to participate in the whole process rather than restrict themselves to a mere listener's role.

Dialogue can be defined as a process with dual form: a method for discovering disagreements as well as a tool for eliminating them. Although this may seem obvious, one can notice that settling disputes is not always feasible. It is rather difficult to achieve a common ground when dealing with issues that involve a great deal of vagueness, such as philosophical concepts. This task is even more difficult when issues of belief and not mere knowledge are dealt with. In such cases it is rather impossible for the participants to agree to a common basis due to disagreements on fundamental principles. Settling disagreements is, however, feasible when dealing with more formalized and better-defined domains, such as mathematics. This is possible because of the existence of a non-disputable theoretical background and a set of axioms and theorems, which can be used by the interlocutors in order to achieve mutual consent on the issue in question.

2. Dialogue as an instruction tool

The number of people participating in a discourse as well as the relationship between them affects the structure and the outcome of the dialogue process. It is far more difficult for a great number of people to be in accord than it is for two interlocutors or small groups, due to the plethora of conflicting opinions and alternative definitions of concepts.

Based on the above observation one can discern two general classifications of various forms of dialogue. The first classification is according to the number of participants in the dialogue and the second one is according to the underlying relationship between them. According to the number of participants three types of dialogue can be defined:

a) One-to-one. This is the simplest and most common form, where two people argue, having different opinions on the same subject.
b) One-to-many. In this form of dialogue, we assume that a group of people with common beliefs disagree with the opinion of another person.
c) Many-to-many. In a many-to-many dialogue, disagreement can occur among two or more groups of people, or even many people acting individually, each one expressing a different opinion.

According to the relationship between the participants one can discern two types:

a) Equal, where all of the participants are considered to have the same or comparable knowledge level. When disagreements occur, the opinion of none of the participants is considered correct or false and they have to agree to a common knowledge base upon which all verifications are to be made.
b) Expert/Non-expert, where one (or some) of the participants are assumed as experts on the domain in discussion and have, therefore an authoritative role in the dialogue. When disagreements occur, the opinion of experts is considered to carry greater weight than the opinion of non-experts.

One could also define a third form of dialogue, [2], that at first glance seems like an equal-type dialogue. In this special form, one of the interlocutors, although being an expert on the issue in discussion, pretends to be a non-expert and participates as equal to other interlocutors in a dialogue. His expertise is used in a subtle way; he does not directly present his knowledge but rather lets the other participants discover it by guiding them through a sequence of carefully selected questions. This feigned ignorance technique is well known since ancient times and was called "Socratic irony", after Socrates who was the first one to use it.

Dialogue has proved to be an invaluable tool in a specific form of interaction between two participants, where one of them is the advice giver (the tutor) and the other is the advice seeker (the student). This type of dialogue corresponds to an One-to-one, Expert/Non-expert dialogue, as defined previously. In this case dialogue can help understanding and learning through an evolving sequence of questions and answers, where the tutor tries to monitor the student's performance on a task and unveil his misbeliefs and misconceptions on certain concepts. This kind of dialogue is easier for implementation in a standalone intelligent tutoring system, compared to an one-to-many or many-to-many approach, which would require a distributed architecture over local or wide-area networks. In this chapter we describe how it is possible to exploit artificial intelligence techniques for the automatic generation of a tutoring dialogue aiming to monitor a student's performance on problem solving as well as guide him through the process in order to find out his misconceptions and provide him with advice when needed. Linear equation solving was used as a case study.

2.1. COMPUTER GENERATED DIALOGUES

Although dialogue may seem as a completely natural process to humans, things are quite different when it comes to computers. Natural dialogues have some inherent attributes that a knowledge engineer should have in mind when trying to create an intelligent tutoring system that uses computer-based dialogue as an instruction tool.

A basic fact one should always have in mind is that student errors are often caused due to misconceptions at an earlier point of the discussion. These misconceptions are not always obvious - they often remain latent and only emerge when the student makes a mistake at a later point. Trying to discover these misconceptions and providing advice is not always an easy task, as the tutor has to analyse not only the student's final answer, but also the course he followed in order to reach that answer. This requires a method to keep a history track of the student's solving attempts, in order to recognise the exact problem source.

The dialogue should also be versatile and adaptive enough allowing for many different solution versions of a given problem. The system should be able to change its course according to the student's answers, rather than constrain him within the bounds of a static, linear procedure where he is supposed to follow a predetermined path from problem definition to problem solution. Static processes may be adequate for simple procedural problems, such as the addition of two numbers, where the possible student errors are easy to model and recognise, by simply comparing the student's answer to the correct solution template, stored in the system's knowledge base. However, more complex problem domains such as high school algebra, which is our chosen domain, require an elaborate method to identify and eliminate misconceptions [3]. Consider the following hypothetical case:

System: Please solve 5x-2=3x+3
User (step 1): 5x=3x+5
System: Correct. Now go on.
User (step 2): 5x+3x=5
System: False! Please check again your answer for mistypes.
User (repeats error): 5x+3x=5
System: You seem not to have fully understood the process. When a term is being transferred from right to left you should change its sign.

The user seems to have made only one mistake-he did not change the sign of the term 3x when moving it from right to left. However, it is not clear whether the user's first answer (step 1) is correct. Considering that the user probably has misunderstood the process of separating variables from constants, it is possible that the user's answer occurred through the following sequence: a)5x-2=3x+3, b)5x=3x+3-2, c)5x=3x+5, where two consecutive mistakes lead to a correct answer. Although this example is extreme, one should not exclude the case that a user has incomplete knowledge even on simple arithmetic operations.

It is also essential for the knowledge engineer to have in mind three basic principles when designing a tutoring system. The system should fully cover the problem domain, it should be customized according to the needs of its target group and it should follow a specific educational philosophy.

The first of these principles means that the system should embody adequate and accurate knowledge. It should have a sound and complete problem solver that contains all the rules and knowledge required to solve every possible problem within the specified domain and also ensures that every solution produced is accurate [4].

The second principle implies that there should be a user model upon which system behavior is based. A classification by Self [5] defines six categories of user models, according to their functional properties:

a) Corrective, where the aim is to eliminate bugs in the student's knowledge.
b) Elaborative, where the aim is to fill in the gaps in the student's knowledge.
c) Strategic, where the aim is to initiate alternative tutorial strategies.
d) Diagnostic, aiming to diagnose bugs in the student's knowledge.
e) Predictive, where the aim is to help determine the user's most likely response to tutorial actions.
f) Evaluative, which aims to assess the student or the intelligent tutoring system itself.

The third principle corresponds to the general strategy the system will follow in order to achieve its educational goals as well as the level of control it allows to the user. Various classifications of intelligent tutoring systems have been made according to the degree of control, with a classic approach by Wenger [6] being very popular:

a) Monitoring systems, which monitor the student but do not allow him to take the initiative of the interaction
b) Mixed initiative systems, that also allow the student to pose questions and provide him with answers
c) Guided discovery systems, where full control is granted to the student and the system guides him by altering the environment parameters.

Our approach is towards a monitoring system, implementing the corrective, elaborative, diagnostic and evaluative user model functions. The system can monitor the student's actions while he tries to solve an equation, tries to diagnose his misconceptions based on his errors or the course he follows in order to solve the problem, provides him with help and advice when a weakness is diagnosed and also can evaluate the student's performance based on the feedback collected. The system's architecture is at present quite simple, implementing three of the four basic components of the classic ITS architecture [7]. In the current implementation we assume a static user model, focusing mainly on the advice generation and domain knowledge. A schematic representation of the system's architecture is shown in Figure 1.

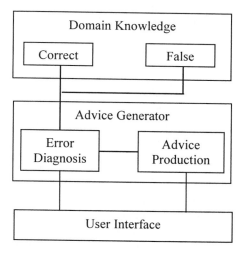

Figure 1. *TRL-Tutor Approach*

The main component is the Advice Generator. It performs monitoring, diagnosing and advising tasks, using a forward temporal planner as the core of the system. Considering monitoring and advice tasks as planning problems is rather suitable for our case, as equation solving closely resembles to planning. When a student tries to solve an equation, he has in mind an overall goal, which is getting the equation to a form of $x=c$. In order to achieve this goal, he tries to divide it to a sequence of smaller, easier to accomplish subgoals, [8], that correspond to the subsets of the equation solving algorithm.

3. Forward Temporal Planning in the TRL-Tutor

Traditional planning approaches use backward-chaining search strategies, starting from the goal and generating sequences of actions that lead to the initial world state. This approach, however, is not appropriate for a set of problems where the goal state is not precisely known and there are infinite initial world states leading to the goal state. The set of world states is also very difficult to be reduced by the use of heuristics. Consider the following problem, where we have to solve the equation $5x-2=3x+3$. A backward-chaining planner would start from the solution $x=2,5$ (which should be known in advance) and try to work its way up to the given equation by producing a world state logically preceding $x=2,5$. It is easily understood that there are infinite possible mathematical expressions which lead to the solution $x=2,5$. In TRL-Tutor we adopted a forward chaining search strategy, starting from the initial state and trying to work our way towards the goal. This approach has a number of clear advantages over backward strategies. The first one is that we do need to know the exact goal in advance-we just need to provide the planner with a generic, abstract description of the desired world state. In the case of equation solving it is adequate to define the goal as $x=c$. The planner will try to produce a mathematical expression of this form and will automatically assign the correct value to c.

The second advantage over backward strategies is that the set of world states following the current one is much easier to determine, provided that actions are properly defined. Equation solving is essentially a reduction process from complex to simpler mathematical expressions. The planning process is therefore less complex, as all possible solutions converge to the same point, in contrast to backward planning, where a continuously expanding graph is produced.

The third advantage of a forward-chaining strategy in the case of equation solving is that it accurately simulates the real-life process, following the same steps as a student in order to produce the final solution. This is essential for our case, where we want to monitor the student and provide advice, which would not be possible using alternative solving algorithms.

TRL-Tutor's planning mechanism is based on TRL-Planner [9], an earlier implementation of a backward-chaining temporal planner. A similar action representation is schema used, while the planning algorithm has been properly modified to a forward-chaining strategy. TRL-Tutor also uses the same reasoning mechanism as TRL-Planner, based on the TRLi temporal reasoning system [10,11]. The use of the TRLi temporal reasoning system allows for the use of elaborate diagnostic techniques by exploiting temporal information.

3.1. TEMPORAL PLANNING

The above representation of temporal information allows the definition of an action representation schema where actions, conditions and effects are temporally qualified. In TRL-Planner two types of preconditions are adopted, *assumptions* and *subgoals*, corresponding to "use-only-if" and "achieve" conditions, respectively. Effects are distinguished in *primary* and *general*. Primary are the effects that motivate (initiate) the execution of an action, while general are the ones that are produced as side effects when the action is executed.

However, in TRL-Tutor we have simplified the action description schema, maintaining only preconditions and effects. Actions, preconditions and effects are represented as TRLi extended atoms; they are predicates bearing temporal reference labels. Actions also contain duration information, determined by a temporal constant. The action representation schema is *action=<N,D,Cd,E,Cst>*, where N is the action's name, D is the duration, Cd is the preconditions list, E is the effects list, and Cst is the temporal constraints list.

3.2. FORWARD TEMPORAL PLANNING ALGORITHM

The proposed algorithm uses a forward planning technique. The algorithm is quite simple and is described as follows:

While the UG list is not empty:
> Select an action A; Extract from UG_List all the goals G that are satisfied due to the utilisation of A.

Select an action
> For each precondition PreC of the action call TRLi to verify that PreC is satisfied in WD (WD|$_{-TRLi}$PreC). Insert effects in the world description WD.

During the generation of the plan the following lists are used:

- *World Description* (WD list). This contains the initial world description as a set of properties holding over particular time intervals. The WD list is updated with the action's effects each time a new action is chosen. Entries in the WD list are TRLi clauses and are either temporal facts of the form $T_i:p_i$ or temporal rules of the form $T_0:p_0:-T_1:p_1,\ldots,T_n:p_n$, where T_i is a temporal reference and p_i is a classic predicate.
- *Current Plan list* (P list), which contains the actions that have been chosen up to the current point of the search space. Entries in the P list have the form T:action, where T is the temporal reference of the action.
- *Temporal Constraints list* (C list), which contains all constraints that are valid up to the current point. When an attempt to add a new constraint is made, consistency is first checked by the SCS symbolic constraint solver. Entries in the C list have the form T1 *rel* T2, where rel belongs to the set of relationships $\{<,>,=,\leq,\geq\}$.
- *Unsatisfied Goals list* (UG list). At the beginning of the planning process this contains the top goals that must be satisfied and is updated as planning proceeds. Satisfied goals are extracted from the list and new subgoals that come up when an action is selected are inserted.

These lists provide us with a status of the plan as it has been produced up to the current point. At each step of the planning algorithm two lists are maintained: Status_in=<WDin, Pin, Cin> and Status_out=<Wdout, Pout, Cout>. Changes to the status of the plan occur when an action is selected as well as when the WD list or the C list has been changed. Plan status does not only give a snapshot of the world at the current point, but also gives global information of the plan up to the current point.

4. Case study: linear equation solving

Our problem domain is linear algebra. We selected linear equations of the form $Ax+C=Bx+D$ as a toy example. Although this form of equations is rather simple, it allows us to experiment and draw some conclusions on the efficiency of the approach.

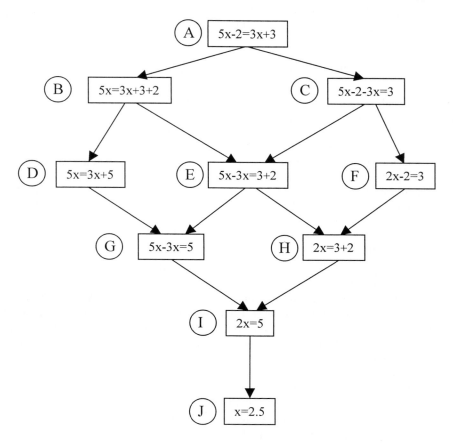

Figure 2. Graph depicting alternative solutions

A crucial step while trying to model the equation solving process was simulating it step-by-step, as a student would perform it. Three basic steps are recognized when trying to solve this kind of equations:

a) Separate constant from variable terms by moving all constants to the right and all variables to the left side of the equation.

b) Perform arithmetic operations among constants and variables
c) Divide all terms by the factor of the unknown variable when the equation has reached a form of Ax=G

Based on the above elementary steps (a,b,c) we tried to identify possible solutions of the equation 5x-2=3x+3 which we used as an example. Assuming that when separating constants from variables we move constants to the right side and variables to the left side of the equation as a heuristic rule to avoid excessive complexity, a graph depicting the possible steps and alternative solutions was created (Fig. 2). At first glance there seem to be six alternative solutions:

1) ABDGIJ   3) ABEHIJ   5) ACEHIJ
2) ABEGIJ   4) ACEGIJ   6) ACFHIJ

The number of possible solutions is, however, much larger if one includes solutions that skip steps. It is very possible that the student omits certain steps while solving the equation, as he may consider some of them too easy. For example, a possible solution is AEIJ, where the student omits a step by simultaneously moving the terms 3x and -2 to the left and to the right side of the equation, respectively. The same applies later, when the student simultaneously calculates 5x-3x=2x and 3+2=5. In cases like this, it is more difficult to monitor and advise the student. There are four possible correct solutions that derive from ADIJ.

4.1. PROBLEM DEFINITION

In order to create a model describing the above process we used the TRL-Tutor notation to define a proper world description to represent the initial equation, a set of final goals representing the solution and a set of actions that enable the production of a plan solving the equation. The world description representing the equation 5x-2=3x+3 is as follows:

Initial_world:{<0.0,T2>:left(5,1), <0.0,T3>:left(-2,0), <0.0,T4>:right(3,1), <0.0,T5>:right(3,0), <0.0,T6>:count(2,2)}.

The predicates left and right define the terms of the equation. For example, left (5,1) means that the term $5*x^1$ exists on the left side of the equation. The predicate count defines the number of terms on each side of the equation: count(2,2) means that we have two terms on the left and two terms on the right side of the equation. The same predicates are used for the definition of the set of final goals:

Final_Goals:{<T1,T2>:left(1,1), <T9,T5>:right(C,0), <T10,T6>:count(1,1)}.

Here we define that the planner has found a solution if it reaches a state where only two terms exist, $1*x^1$ on the left side and a constant C on the right side of the equation. This describes an equation form of x=c.

4.2. DEFINITION OF ACTIONS

The steps of the equation solving process are described in the TRL-Tutor as actions with specific assumptions, goals and constraints as follows:

```
action:{ <T1,T2>:movec(C,L,R,L1,R1), duration,
        [<T3,T1>:count(L,R), <T3,T1>:compute(C1,(-1)*C),
        <T3,T1>:compute(L1,L-1), <T3,T1>:compute(R1,R+1),
        <T4,T1>:left(C,0)], % preconditions
        [<T2,T5>:right(C1,0), <T2,T6>:count(L1,R1)], % effects
        [T2=T1+2]}. %constraints
```

In the above example, we define an action that moves constants from left to right. We define as primary effects that the selected term should be multiplied by (-1) in order to change its sign and appended to the right side of the equation after applying the action. This also affects the number of terms on both sides of the equation, so an additional effect is defined, stating that the number of terms on the left side should be reduced by one, while the number of terms on the right side should be increased by one.

4.3. SPECIAL CATEGORIES OF ACTIONS

In addition to actions representing the steps of the equation solving algorithm, we implemented a set of special actions which we call *Wrong* actions. These actions are used to represent possible user errors, in order to enable diagnosis and advice provision. Wrong actions are selected after all possible standard actions fail to produce a successful plan. An example of a wrong action definition is shown below, where the user forgets to change the sign of a term when moving to the other side of the equation:

```
action:{ <T1,T2>:w1movec(C,L,R,L1,R1),
        duration,
        [<T3,T1>:count(L,R), <T3,T1>:compute(C1,C), /*this is the error cause*/
        <T3,T1>:compute(L1,L-1), <T3,T1>:compute(R1,R+1),
        <T4,T1>:left(C,0)], % preconditions
        [<T2,T5>:right(C1,0), <T2,T6>:count(L1,R1)], % effects
        [T2=T1+2]}. %constraints
```

wrong(w1movec(_,_,_,_,_), 'You seem to have made a mistake here. When we move a term to the other side of the equation we have to change its sign').

4.4. CONSULTING THE PLANNER

In our current implementation we use the following method to monitor student actions. After presenting the user with the equation, the system awaits his first step. After confirming that the user's answer is correct, the system returns a message and waits for the next step.

The validation of the user's answer is achieved by calling the planner many times. Each step the user enters is treated as the planner's next goal, while the previous state is treated as the initial world. This process continues until either a solution is found, which means the user's answer was recognized as correct or false, or the planner fails, which means that user input is not possible to produce using TRL-Tutor's knowledge base. Calling the planner many times consecutively allows step-by-step monitoring of the student's efforts in solving the equation. Using the user's answer as a goal and the previous state as the initial world ensures consistency between the steps of the solution. It also reduces execution time, compared to using the given equation as the initial world description.

4.5. TRACING A WRONG USER RESPONSE AND PROVIDING ADVICE

Tracing user errors is done using wrong actions. In case the planner fails to find a solution with the default set of actions, it retries using wrong actions. If the goals (which correspond to the user's input) are satisfied with a wrong action then the system decides that the user has made an error.

In our current implementation, when an error is detected, the system simply provides the user with a message stating that his answer is wrong, informs him about the error type and requires him to re-enter the answer. The produced plan can, however, be used for a more elaborate diagnostic process. The diagnostic module of the Advice Generator can be expanded in order to discover previous misconceptions by backtracking the plan and trying to discover the student's rationale.

TRL-Tutor's capability of holding temporal information can also be exploited in various ways. It can be used to temporally identify the various steps of the solution. Temporal points and intervals can also be related to real time, allowing the system to judge the student's efficiency based on the time he needs to solve the equation. Conclusions can also be drawn on whether the student encounters difficulty on a certain step of the solving process, if unusual delay is noticed.

5. Conclusions

In the current chapter we described how forward temporal planning techniques can be exploited in order to monitor and advise users in a dialogue process, using linear equation solving as a case study. Classifications of dialogue according to the number of participants and the relationship among them were made.

Various categories of interactive tutoring dialogue systems were also presented. TRL-Tutor, a prototype interactive tutoring system on equation solving falls into the corrective, diagnostic, elaborative and evaluative categories and follows an architecture slightly different from the classic ITS architecture. A forward temporal planning algorithm has been described, as well as the way it is used to reproduce the user's solution. Planning results are used to provide advice to the user, although in the current implementation this is done in a simple way.

In future implementations we intend to extend TRL-Tutor's diagnosis module to discover previous misconceptions. Additional work has to be done on the user module component, in order to allow for a more adaptive system. We also intend to work on the many-to-many dialogue notion, towards the direction of a virtual collaborative educational environment aiming to enhance interaction and provide a visual user interface to distributed users on remote locations.

Acknowledgements

This work was partially funded by the Greek General Secretariat of Research and Technology under the project "TimeLogic" of ΠΕΝΕΔ '95, contract no. 1134.

References

[1]. J.D. Moore, 'Participating in Explanatory Dialogues', MIT Press, 1995.

[2]. A. Panayotopoulos, T. Panayiotopoulos, N. Avradinis, 'LOGOS: An intelligent multimedia tutoring dialogue system for the teaching of Ancient Greek Philosophy', Proceedings of the 16th IASTED Int. Conf. on Applied Informatics (AI'98), M.H. Hamza (Ed.), pp.237-239, IASTED/ACTA Press, 1998.

[3]. M.R. Genesereth, 'The role of plans in intelligent teaching systems', in 'Intelligent Tutoring Systems', D. Sleeman, J.S.Brown (Eds.), Academic Press Ltd, pp.137-155 1982.

[4]. F. Bacchus, F. Kabanza, 'Planning for Temporally Extended Goals', Proceedings of the 13th Nat. Conf. on Artificial Intelligence-AAAI '96, AAAI Press/MIT Press, pp.1215-1222, 1996.

[5]. J.A. Self, 'Student models: What use are they?', in 'Artificial Intelligence tools in education', P. Ercoli, R. Lewis (Eds.), North Holland, pp.73-86, 1988.

[6]. E. Wenger, 'Artificial Intelligence and Tutoring Systems', Morgan Kaufmann, 1987.

[7]. M. Virvou, 'Models and Techniques of Intelligent Tutoring and Help', Proceedings of the National Symposium of Artificial Intelligence, SETN'96, Hellenic Association of Artificial Intelligence (HAIS), Piraeus, Greece, December 1996.

[8]. J. Bonar, B. Liffick, 'Communicating with high-level plans', in 'Intelligent User Interfaces', J.W. Sullivan, S.W. Tyler (Eds.), ACM Press , pp.129-156, 1991.

[9]. C. C. Marinagi, T. Panayiotopoulos, C.D. Spyropoulos, 'Planning through the TRLi temporal reasoning system', appears in 'Applications of Artificial Intelligence on engineering X', G. Rzevski, R.A. Adey, C.Tasso (Eds.), Computational Mechanics Publications, pp. 19-27, 1995.

[10]. T. Panayiotopoulos, M. Gergatsoulis, 'A Prolog like temporal reasoning system', Proceedings of the 13th IASTED Int. Conf. on Applied Informatics (AI'95), M.H. Hamza (Ed.), pp.123-126, IASTED/ACTA Press, 1995.

[11]. "Intelligent Information Processing using TRLi", T.Panayiotopoulos, M.Gergatsoulis, DEXA'95, London, U.K., September 4-8, 1995, N.Revell, A. M. Tjoa (Eds.), pp.494-501, 1995.

21

Content-Based Audio Retrieval Using a Generalized Algorithm

PUNPITI PIAMSA-NGA
S. R. SUBRAMANYA
NIKITAS A. ALEXANDRIDIS
SANAN SRAKAEW
GEORGE BLANKENSHIP
Department of Electrical Engineering and Computer Science
George Washington University
Washington D.C. 20052 USA
e-mail: {punpiti,subra,alexan,srakaew,blankeng} @seas.gwu.edu

G. PAPAKONSTANTINOU
P. TSANAKAS
S. TZAFESTAS
Department of Electrical and Computer Engineering
National Technology University of Athens
Athens Greece

1. Introduction

Content-based indexing of audio (and multimedia) data has become more important since conventional databases cannot provide the necessary efficiency and performance [1,2]. However, there are three main difficult problems. First, the content of audio data is subjective information; it is hard to give the descriptions in words. The recognition of data content requires prior knowledge and special techniques in Signal Processing and Pattern Recognition, which usually require long computing time. Second, since several audio features can be used as indices [3] (such as pitch, amplitude, and frequency), a method or processing technique designed and developed for one feature may not be appropriate for another. Third, the extremely large data size and the use of a similarity search require extensive computation. Similarity matching is based upon the computation of the distance between a query and each record in the database; the best match is in the data set with the smallest distances. To solve these three problems, we

231

use a histogram-based feature model to represent subjective features[4], a unified model [5] to represent the data structures of the multimedia data, and a fast, generalized comparison algorithm to reduce the retrieval time.

To reduce the retrieval time, we proposed two techniques in our previous reports, one is a weighted cascade (pipeline) for multiple-feature querying, and the other is the use of a parallel search algorithm [3,4]. The algorithm yielded lower retrieval time and acceptable results. However, if there is only one feature and parallel processing is not available, the retrieval time will be very long. To reduce retrieval time when there is only one feature or parallel processing is not possible, a new fast comparison algorithm is required

In this chapter, we propose a new generalized algorithm for audio (and multimedia) data retrieval using an unrestricted-format query. The algorithm is called the "Virtual-Node (VN)" algorithm. The VN algorithm is adapted from our "Partial-Matching (PM)" algorithm [6]. We have developed and tested the PM algorithm successfully on image retrieval. Here, we extend our previous work by applying the same algorithm on audio databases and then improve the PM algorithm to the new VN algorithm. The results of the VN algorithm are perceptibly more acceptable, and have a smaller retrieval time, than the previous PM algorithm.

The rest of this chapter is organized as follows. Section 2 gives an overview of similarity search types. Section 3 describes the histogram-based binary tree for audio features. Section 4 presents an algorithm for restricted-format queries. Section 5 presents the details of the PM and VN algorithms: Experimental results are given in Section 6, followed by a summary and future directions.

2. Retrieval algorithm

In this Section, the general algorithm for audio (and multimedia) retrieval by content is introduced. Section 2.1 reviews similarity searching. A classification of queries is given in Section 2.2.

2.1 SIMILARITY SEARCH

Exact keyword matching database systems are inefficient and inappropriate for audio (and multimedia) data since the search is based upon the use of a static subjective summary of the data rather than the data itself [1]. Similarity searching is a more appropriate approach. Just as in the exact match approach, the crucial issues are the building of an index table and the retrieval algorithm. The basis of the algorithm is a function that measures the distance between the query and each of the multimedia objects and reports the best matching data, which are those that have the smallest distances to the query.

Two main concerns in the field of similarity searching are the enormous computation requirement and efficiency of the index. Because the similarity search is based on finding the minimal distances between the query and each of the records, the computation time grows with the size of the database. Retrieval from a multimedia database is based upon similarity searches of the index space. A multimedia index entry should contain the salient features, which have been extracted from the raw data, and encompass the spatio-temporal relationships. The usefulness of an index entry is a function of the computation time used to extract the features that define the entry, the space required to store the entry, and the ease with which the entity identifies the data.

2.2 CLASSIFICATION OF SEARCH ALGORITHMS

There are two types of multimedia data queries: query-by-example (QBE) and query-by-content (QBC). The QBE approach searches for data that are similar to the query. An example of QBE query is "Which records in an audio database are similar to an audio query clip?" The QBC takes a qualitative description of data as query. The example of QBC query is "Which audio clips in an audio database contain laughter?" In this chapter, all query types are QBE. QBE can be divided into two subtypes by application approaches: search-by-restricted-format (SBRF) and search-by-unrestricted-format (SBUF). A search-by-restricted query looks for data that is similar to the query globally, without regard to the scale of the query. In other words, all records in a database must be resampled into a uniform sampling rate. An example of query using SBRF is "Which are the data items that are similar to the given query, as a whole?" In contrast, a query using SBUF is searches for data with similarities to the query. An example of SBUF query is "List all items that have 'portions' that are similar to the given query."

3. An audio feature

In this section, a histogram-based k-tree feature is used as an index for audio data. A summary description of the k-tree model is given in Section 3.1 and the construction of the binary tree of a histogram-based feature is explained in Section 3.2.

3.1 THEK-TREE MODEL

A k-tree model [5] is used to unify the data characteristics of multimedia data. A k-tree is a directed graph; each node has 2^k incoming edges and one outgoing edge with a balanced structure. A *binary* k-tree is used for 1-dimentional data; a *quadtree* for 2-dimension data; and an *octtree* for 3-dimensional data. There are three main benefits for exploiting a k-tree. First, the spatio-temporal information of the data is embedded into the tree structure. It reduces the time to compute distances between two nodes when spatio-temporal relationships are required in a query. Second, a k-tree can exploit multiresolution processing by computing small, global information first and then large, local information when more accurate resolution is required. Third, the complexity of the data structure affects only the degree of the tree. Consequently, an algorithm for a particular type of feature can be reused for a feature of other media types. The comparison of two features is based upon the distance between the histograms that define the features.

Since a histogram-based feature is used as index, we have investigated the effects of using different existing histogram distance functions. Three distance functions were selected to determine a distance between two histograms: Euclidean (d_E), Histogram Intersection (d_I) [6], and Histogram Quadratic (d_Q) [8] distance functions, which are defined below:

$$d_E(h,g) = \sum_{m=0}^{M-1} (h[m] - g[m])^2 \tag{1}$$

$$d_I(h,g) = 1 - \frac{\sum_{m=0}^{M-1} \min(h[m], g[m])}{\min(\sum_{m_1=0}^{M-1} h[m_0], \sum_{m_1=0}^{M-1} g[m_1])} \tag{2}$$

$$d_Q(h,g) = \sum_{m_0=0}^{M-1}\sum_{m_1=0}^{M-1} (h[m_0] - g[m_0]) \cdot a_{m_0,m_1} \cdot (h[m_1] - g[m_1]) \tag{3}$$

The h and g are M-dimensional histograms. The $h[m]$ and $g[m]$ are the frequencies of an element in bin m of histogram h and g, respectively. The *Euclidean distance* function (Eq. 1) only compares the identical bins in respective histograms. All bins contribute equally to the distance and the differences of content between two bins of histogram are not ignored. The *histogram intersection distance* function (Eq. 2) compares only the elements, which exist in the query. Note that the result from the intersection is not symmetric. The *histogram quadratic distance* function (Eq. 3) uses matrix $\{a_{m_0,m_1}\}$, where $\{a_{m_0,m_1}\} = |m_0 - m_1|$. This gives a weight to the histogram distance; bins that are closer are given less weight in the product term than those that are farther apart. In the algorithms, the histogram intersection is used. Intuitively this gives the number of samples common to both histograms.

3.2   A HISTOGRAM-BASED BINARY TREE OF AUDIO FEATURE

In this chapter, we use audio amplitude as the feature of interest and a key index for an audio database. Generally, a normalization technique is used for generating the index. The domain of a feature is reduced to a set of selected values from a universe of potential values of the feature. We assign an identification number of each element in the reduced set. When data is inserted into the system, it is converted or transformed into the selected domain then a histogram is generated. Thus, the feature is represented by a histogram of data in the selected domain. For an audio feature, the amplitude values are picked from the whole infinite universe and then mapped into a set of 256 pulse-code modulation (PCM) values.

The amplitude feature of an audio clip is organized as a binary tree ($k=2$) of histograms. A histogram is constructed from the counts of PCM values that occur in an audio clip. Since there are 256 existing PCM values, a histogram feature of an audio clip is a 256-element vector; each element i contains the number of samples with a PCM value of i. To construct the binary tree, the audio clip is divided into m segments, where $m=2^l$ and l is an integer. The histograms for each segment form the nodes of the tree. Pairs of histograms from consecutive segments are summarized into a new histogram forming a parent node. The process is repeated until a global histogram (at the root of the tree) is generated.

4. Algorithms to search by restricted-format query (SBRF)

In the search-by-restricted-format approach, the size of the data in the multimedia database and the query must be normalized to a uniform format. Normalized audio data has a similar sampling rate and a similar number of samples. Any audio data must be quantized and resampled to the normalized configuration. This is useful for applications that require the search by restricted-format query or the query, which disregards the scale of the data. The pseudocode of SBRF is given below:

Algorithm 1: Search by Restricted-format (SBRF)

```
SearchByRestrictedFormat (In Query, In FeatureOfData [n], In K-TreeLevel,
                          Out Record (Distance [n], Data [n]))
1)   Begin
2)   NormalizedQuery = NormalizeSize (Query)
3)   FeaturesOfQuery = FeatureExtraction(NormalizedQuery)
4)   For i=0 to n do
5)       Record.Distance[i] = DistanceSBRF (FearureOfQuery,
                                            FeatureOfData[i],
                                            K-TreeLevel).
6)       Record.Data[i] = Data[i]
7)   End for
8)   Sort(Record(Distance[n],Data[n]))
9)   End
```

Algorithm 2: DistanceSBRF

```
Out DistanceSBRF (In FeatureA, In FeatureB, In K-TreeLevel)
1)   Begin
2)   Distance=0;
3)   For i=0 to NumberOfNodesAt (K-TreeLevel) do
4)       X = HistogramDistance(FeatureAtNode ( i, FeatureA),
                               FeatureAtNode ( i, FeatureB))
5)       Distance = Distance + (X * X )
6)   End for
7)   Return sqrt(Distance)
8)   End
```

5. Algorithms of search by unrestricted-formatted query (SBUF)

In the search-by-unrestricted-format (SBUF) approach, the queries are assumed to be in a continuous form. The algorithm determines the best match for the query in any portion of each of the audio clips and reports a number of close matches. The simple algorithm for searching by unrestricted-formatted query is shown in Algorithm 3 and Algorithm 4. Note that Algorithm 4 uses brute-force searching method, which has $O(n)$

time complexity. Faster algorithms have been developed to replace the Algorithm 4 and are given in Section 5.1 and 5.2.

Algorithm 3: Search by unrestricted-formatted query (SBUF)

SearchByUnrestrictedFormatted*(In Query, In K-TreeLevelQuery,*	
In FeatureOfData[n], In K-TreeLevelData[n],	
Out Record (Distance[n], Data[n]))	
1)	**Begin**
2)	FeatureOfQuery = FeatureExtraction(Query)
3)	**For** i=0 to n **do**
4)	Record.Distance[i] = DistanceSBUF(FeatureOfQuery,
	K-TreeLevelQuery,
	FeatureOfData[i],
	K-TreeLevelData[i])
5)	Record.Data[i] = Data[i]
6)	**End for**
7)	Sort(Record(Distance[n],Data[n]))
8)	**End**

Algorithm 4: The Simple DistanceSBUF

Out DistanceSBUF *(In FeatureA, In K-TreeLevelA,*	
In FeatureB, In K-TreeLevelB)	
1)	**Begin**
2)	NumberOfNodesLevelA = NumberOfNodesAtLevel (KTreeLevelA)
3)	NumberOfNodesLevelB = NumberOfNodesAtLevel (KTreeLevelB)
4)	NumberOfComparisons = NumberOfNodesLevelB-
	NumberOfNodesLevelA+1
5)	Distance=0
6)	**For** i to NumberOfComparisons **do**
7)	**For** j to NumberOfNodesLevelA **do**
8)	Distance = min(Distance , HistogramDistance(
	FeatureAtNode(j, FeatureA),
	FeatureAtNode(i+j, FeatureB))
9)	**End For**
10)	**End For**
11)	**Return** Distance
12)	**End**

Two algorithms for searching using and unrestricted-formatted query have been developed, tested, and presented using the k-tree model. The algorithms are called "partial-matching" [6] and "virtual-node." The mathematical analysis and experimental results on both algorithms show a much faster response time than using the brute-force search technique of Algorithm 4. Both methods have a common prologue derived from Algorithm 3 and Algorithm 4.

5.1   THE PARTIAL-MATCHING (PM) ALGORITHM

The Partial-Matching (PM) algorithm [6] is a generalized algorithm for searching a multimedia database by content using queries with unrestricted formats. This algorithm allows us to search for data in a multimedia database and find those entries that contain the same "object" as the query; since no scaling is performed on the query, it represents the *"unrestricted query format"* approach. This algorithm has been developed and tested for retrieval using image database. The results are very satisfactory, when measured by both retrieval time and perceptual quality. We have adapted this algorithm for an audio database. Algorithm 5 below shows the details of the partial-matching algorithm. An illustration of searching using the partial matching algorithm is depicted in Figure 1.

Algorithm 5: A partial-matching algorithm for DistanceSBUF

PartialMatchingComparison(***In*** *FeatureOfQuery*, ***In*** *FeatureOfData*, ***Out*** *Distance*)

Step 1: *Find candidates*
1)   Calculate the "feature distances" between the <u>root of the query tree</u> and the <u>all nodes in data trees, which have the same height as the query tree</u> in the database, (we assume that data size are larger than the query size.)
2)   Select those data types for which the distances are below a threshold value. These candidates will be searched at a higher-resolution level in Step 2: below.

Step 2: *Determine the position*
1)   On the candidate nodes, generate virtual nodes (nodes that locate across the boundary of real nodes)
2)   Calculate the distances between root of the query tree and all virtual nodes around the candidate nodes.
3)   Sort the distances from 2).
4)   Return the minimum distances with node positions.

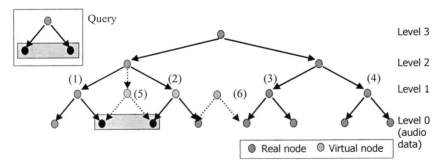

Figure 1: Illustration of searching using the partial-matching algorithm

The process begins at Step 1: of Algorithm 5 which finds the candidate matches – illustrated at node (1), (2), (3), and (4) in Figure 1. Figure 1 demonstrates the comparison between the root of the query and the nodes that have the same height as the query. Suppose we found that node (2) is a candidate. A culling of candidates that do not match (Step 2:) creates virtual nodes at node (5) and node (6). Finally, similar data to the query is found at the position of node (5).

5.2   THE VIRTUAL-NODE (VN) ALGORITHM

The Virtual-node algorithm is a newly proposed generalized algorithm for searching data using an unrestricted-format query. This algorithm is an extension of the partial-matching algorithm. It exploits the multiresolution capabilities of the hierarchical structure to eliminate some unnecessary computations. The detail is shown in the pseudo-code of Algorithm 6.

Algorithm 6: A virtual-node algorithm for DistanceSBUF

---

*VirtualNodeComparison(**In** FeatureOfQuery, **In** FeatureOfData, **Out** Distance)*

Step 1: *Find candidates*
1)   Calculate the "feature distances" between the root of the query tree and the roots of all data trees in the database, (assume that data tree height is larger than the query height.) In our experiments, the "histogram intersection function" has been used as the "Feature distance" to determine whether the query histogram is contained within the searched multimedia database and the "histogram quadratic function" to determine whether the query histogram is similar to the data.
2)   Select those data types for which the distances are below a threshold value. These candidates will be searched at a higher-resolution level in Step 2: below.

Step 2: *Determine the position*

Case a) If query's tree aligns within the k-tree structure of data:
1)   Find the feature distances between feature in root of query tree and nodes of data at level L_{i-1} – nodes with solid-line link in Figure 2. If the distance between a child node is equal to the distance between the query and its parent (L_i), the query may be found within that child node.
2)   Repeat Case a) recursively on the found child node. If there is no distance at level L_{i-1} close to the distance of the parent, the query is "not aligned". Follow Case b) below.

Case b) If the query data falls in between two or more nodes:
1)   If no "real node" in tree (nodes connected by solid lines in Figure 2 can be a candidate, "virtual nodes" (node connected by dot lines) between two nodes have to be generated from the parts of their child nodes.

2)   Repeat the algorithm using a new tree of virtual nodes i.e. use the algorithm within the dashed box in the Figure 2.

Case c) If height of query is equal to a node height:

1)   Use Algorithm 1: Search by Restricted-format (SBRF) with histogram quadratic distance function to calculate the distance and then:

2)   Return the distance.

An example of a search using the virtual node algorithm is illustrated in Figure 2. The root node of the query is compared with the root node of the data to determine if the query is part of the data in the tree under node (1) at level 3. If the result shows that the target is in the tree under node (1), the comparison between root of the query and the lower-level nodes continues using the nodes in level 2. If the result at level 2 shows that the query is under node (2) and not under node (3), then, the subtree under node (3) is ignored and the process continues with the node at level 1, node (4) and node (5). If results from both node (4) and (5) do not indicate that at match is below either, node (6) is generated dynamically and then compared with the query.

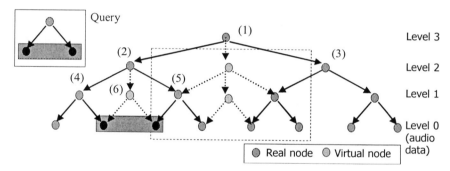

Figure 2 Illustration of example for searching by virtual-node algorithm

5.3   MATHEMATICAL ANALYSIS

Let k be a number of spatio-temporal dimensions of the data; S_d be the average size of all records in the database; and S_q be the size of query. (The size of the data is the number of leaf nodes in a query or data record tree.) The total time complexity to compare two records is given in the following discussion.

For the linear simple DistanceSBUF, the time complexity is $(\sqrt[k]{S_d} - \sqrt[k]{S_q} + 1)^k$. For the partial-matching algorithm, the time complexity is $(S_d / S_q) + 3^k$. For the virtual-node algorithm, the time complexity is $3^k \cdot \log_{2^k}(S_d / S_q)$ plus time of generating the virtual node, which is $(3^k - 2^k) \cdot \log_{2^k}(S_d / S_q)$. The DistanceSBUF algorithm has linear time complexity; it grows with S_d. The new algorithm has time complexity $O\left(\log\left(\dfrac{S_d}{S_q}\right)\right)$. However, a k grows, the constant term may have may have

the dominant effect on the processing time.

6. Experiments

6.1 COMPUTATIONAL PLATFORM

The audio database used in this experiment consists of over hundred files downloaded from Sunsite at University of North Carolina [9]. The database represents 3.5 hours of audio clips and requires about 100MB of mass storage. The average length of a clip is 10 seconds. Each clip has the same sampling rate of 8KHz. A Linux/Pentium-II machine was the computation platform. Several audio clips in the database were selected as a set of queries for whole clip searches (the *whole-clip* set). A one-second portion from each clip in the *whole-clip* set was randomly selected to be used for partial clip searches (the *part-of-clip* set). During the search, the query is compared only with the data that is equal to or longer than the query.

In this experiment, the results from the *whole-clip* set was retrieved using a search-by-restricted-format query, while the *part-of-clip set* results was retrieved using an unrestricted-format query.

6.2 EXPERIMENTAL RESULTS

We have performed the experiments for searching audio data using the proposed "Virtual-node" algorithm and compared the results to those using the "Partial-matching" algorithm. The experiments focused on two primary issues: the quality of search results and the retrieval time.

The results from each query are a list of audio clip identified with distances less than a given threshold. The retrieval time, the number of candidate results, and the average position of the expected results in the candidate lists were measured and are shown in Table 1.

Table 1 shows that the retrieval time using "part of clip" is slightly more than the retrieval time using whole clips. We conclude that a search-by-unrestricted-format query (search by "part of clip") does not require a significantly greater processing time than a search-by-restricted-format query (search by "whole clip".)

Table 1 also shows better metrics from the virtual node: the average retrieval time, the number of candidate results, and the position of expected results in the list of candidates.

Table 1   Comparisons between the "partial matching" and the "virtual node" algorithms

Set of queries	PARTIAL MATCHING		VIRTUAL NODE	
	Part of clip	Whole clip	Part of clip	Whole clip
Average query length (sec.)	1	10	1	10
Average retrieval time (sec.)	38	33	35	32
Number of candidates	218	34	22	1
Average position of target clip in candidate list	109	1	8	1

7.   Summary and future directions

We have presented a generalized algorithm for content-based audio retrieval using the concept of a "virtual-node." The algorithm is an improvement over our previous "partial-matching" algorithm and also exploits the multiresolution data structure of the unified k-tree model. The experimental results of retrieval from an audio database using the virtual-node algorithm show that the retrieval time is less than, and the perceptive accuracy is better than, the results from the partial-matching algorithm. The results of a restricted-format query using this generalized virtual-node algorithm do not require a significantly longer time than the conventional restricted-format algorithm.

The unified k-tree model can be used for various types of multimedia data; we intend to extend the virtual-node algorithm for other multimedia data types, such as images and video, in the near future. Preliminary experimental results of retrieval from an image database show that the virtual-node approach improves retrieval performance.

Moreover, we are also extending the k-tree model and developing a new search algorithm to allow the searching for data that have a different resolution than that of the query, such as a different sampling rate (audio clip) and different scale (image). Currently, the preliminary results in audio retrieval exhibit a requirement for intensive computation in order to retrieve acceptable results.

8.   References

[1] V. Gudivada and V. Raghavan, *"Special issue on content-based image retrieval systems,"* IEEE Computers, Vol. 28, No. 9, September 1995.
[2] Z. Kemp, *"Multimedia and spatial information systems,"* IEEE Multimedia, Vol. 2, No. 4, 1995.

[3] E. Wold et al., "Content-based classification, search and retrieval of audio data," IEEE Multimedia, 1996.

[4] P. Piamsa-nga, N. A. Alexandridis, S. Srakaew, G. Blankenship, G. Papakonstantinou, P. Tsanakas, and S. Tzafestas, "Multi-feature content based image retrieval," in *International Conference on Computer Graphics and imaging*, 1998

[5] P. Piamsa-nga, N. A. Alexandridis, G. Blankenship, G. Papakonstantinou, P. Tsanakas, and S. Tzafestas, "A unified k-tree model for multimedia retrieval," in *International Conference on Computers and their applications*, Hawaii, March 1998.

[6] S. R. Subramanya, P. Piamsa-nga, N. A. Alexandridis, and A. Youssef, *"A Scheme for Content-Based Image Retrievals for Unrestricted Query Formats,"* International Conference on Imaging Science, Systems and Technology (CISST'98), Las Vegas, July 1998

[7] M. J. Swain and D. H. Ballard, *"Color Indexing"* International Journal of Computer Vision, 7:1, 1991.

[8] J. R. Smith, *"Integrated spatial and feature image systems: retrieval, analysis, and compression,"* Ph.D. Thesis, Columbia University, 1997

[9] Sunsite at University of North Carolina, *"FTP archive,"* available URL: http://sunsite.unc.edu/pub/multimedia

22

Intelligent Retrieval of Temporal and Periodic Data

L. Baxevanaki*, E. Ioannidis*, T. Panayiotopoulos **
*Software and Knowledge Engineering Laboratory
Institute of Informatics and Telecommunications,
N.C.S.R. "Democritos",
e-mail : stathis@iit.nrcps.gr
**Department of Informatics,
University of Piraeus, e-mail:themisp@unipi.gr

1. Introduction

During the last decade, the need of special manipulation of temporal data has become obvious. Most of the real world trivial applications consider storage and manipulation of temporal data as essential. Examples are applications having to do with scheduling, banking, sales, reservations, transportation, patient medical files etc. However, efficient storage and manipulation of temporal data has always been a great challenge and a problem too.

Time management systems fall into two major categories : temporal databases and temporal logics. Temporal databases are an extension of conventional databases which attempt to grasp the notion of change of data in relation with time. Representatives of temporal databases are the HDRM [1], the temporal extensions in the relational model proposed in [2], Gadia's Homogeneous Relational Model [3], the Historical Relational Algebra [4], Tansel's model [5] and the conceptual model "BCMD" proposed in [6]. However, temporal databases do not support any kind of reasoning. On the other hand, temporal logics are defined in an attempt to add the notion of time in the knowledge dimension and the intelligent reasoning in the temporal dimension. Temporal logics are very powerful tools but their application is limited due to their low efficiency in bulky domains. Representatives of such systems are Allen's temporal Logic [7], McDermott's temporal logic [8] and Kowalski and Sergot's temporal logic [9].

Periodicity is a property of many real world phenomena. Periodic phenomena are of particular interest since they occur in scheduling, planning, medical records, calendars, scientific databases, multimedia databases etc. . An adequate representation, storage and reasoning for the periodicity of certain phenomena is desirable in terms of space economy and information modelling. Although the need of supporting periodic phenomena in temporal databases and temporal query languages is recognised, limited approaches can be met in the literature [10], [11].

The system described in this chapter was designed having as requirements the efficient storage and retrieval of temporal and periodic data.

2. Background Work

2.1 PERIODIC DATA

When we use the term 'periodic phenomenon' we have in mind something that is iterated over a specific period in certain intervals. The time intervals, over which the instances of the phenomenon happen, recur every n units of time, starting from a specific time point. The arithmetic value of n is the "period" T of the periodic phenomenon. We call these iterating time intervals, "valid time intervals", V, of the instances of the periodic phenomenon. The greater time interval over which the phenomenon keeps iterating is the one that the periodic phenomenon lasts. This greater interval is the "cover interval" C [12]. The valid time interval of any specific instance of the phenomenon defines its duration, D. This is the same for all the instances of the phenomenon inside C.

A very common everyday life periodic phenomenon is the switching of the red traffic light. A red traffic light is on for a specific time period and then is off for another specific time period. Say for example that, a red traffic light was on at time point 0 and stays on for five more time points. During the next five time points it is off. The light turns on every 10 time points (units) and so on infinitely. Therefore, the period of this periodic phenomenon is 10 time units, i.e. T=10, the valid time interval of its first instance [0, 4], its duration D = 5 units and the cover interval that the phenomenon lasts from time point 0 to infinity, i.e. C = [0, ∞).

There are cases of periodic phenomena much more complex than the above example. Let' s consider a periodic phenomenon, the cover interval of which is itself another periodic phenomenon. The times that a bus starts its route can be an example of a complex periodic phenomenon. A bus sets off every two hours from Monday to Friday for the whole year of 1997. The first bus route, i.e. the first instance, happens at [6/1/1997 00:00, 6/1/1997 01:59].

In this case, we distinguish three correlated periodic phenomena that constitute the complex periodic phenomenon of the bus routes. The first one expresses the iteration of 5-days periods (Monday-Friday) every 7 days. This periodic phenomenon has T=7 days, D=5 days, C = all the year 1997 and first instance at [6/1/1997, 10/1/1997]. We say that this periodic phenomenon falls in the level of periodicity of weeks (7-days). The second periodic phenomenon expresses the iteration of a day over a Monday to Friday period. This periodic phenomenon has T=1 day, D=1 day, C= 5 days and first instance at [6/1/1997, 6/1/1997]. We say that this periodic phenomenon falls in the level of periodicity of 1-day. There are many such identical phenomena iterating through the year 1997, taking place within the instances of the previous periodic phenomenon in the level of 7-days. The third periodic phenomenon expresses the iteration of the bus routes every two hours within one day. This periodic phenomenon has T=2 hours, D=2 hours, C=1 day and first instance at [6/1/1997 00:00, 6/1/1997 1:59]. We say that this periodic phenomenon falls in the level of periodicity of hours. There are many such identical phenomena iterating every day, from Monday to Friday, for all the year 1997 taking place within the instances of the previous periodic phenomenon in the level of 1-day.

It has been shown in [13], that what we want to know about a periodic phenomenon are the following characteristics : the cover interval C that the phenomenon lasts, one specific instance of the phenomenon along with its duration D, and its period T for each of its levels of periodicity.

2.2 TEMPORAL REFERENCE LANGUAGE INTERPRETER (TRLI)

TRLi [14] is a temporal reasoning system with a syntax and a deduction procedure similar to Prolog. This system takes as input the temporal data and a query and gives as output the answers to the query. The meta-interpreter implements the temporal reasoning rules with the help of a Symbolic Constraint Solver.

TRLi considers time to be discrete. Time points are totally ordered. An interval is considered to be a convex set of time points. Time is unbounded in both directions (in the past and in the future). Time can be determinate or indeterminate. In TRLi it is assumed that assertions hold at specific time points or during a temporal interval or that they may hold at some temporal instances.

Temporal information is expressed through temporal references. The most general temporal reference is the indeterminate interval : $<[T_1,T_2], [T_3,T_4]>$. This interval expresses a temporal interval $[t_1,t_2]$ such that $t_1 \in [T_1,T_2]$ and $t_2 \in [T_3,T_4]$. T_i can be a time point, a variable or a compound term. The indeterminate intervals $<[T_1, T_2], T_3>$ and $<T_1, [T_2, T_3]>$ can also be used as temporal references.

Events and properties defined by Allen are considered as elements in the temporal logic. Their basic difference is their behaviour as far as it concerns their validity over a temporal interval.

2.3 TEMPORAL REFERENCES RELATIONAL MODEL (TRRM)

The TRRM [15] is a temporal extension of the classical relational schema. TRRM considers time to be discrete. It supports both valid and transaction time. The data type that is used to present valid time and transaction time is the temporal interval. TRRM uses tuple stamping for both transaction and valid time. Valid time can be determinate or indeterminate.

The tables of this temporal relational schema fall into two categories : *Data Tables* and *Temporal References Tables (TRTs)*. Data tables are tables containing data. This category is further subcategorised in three kinds of tables : *Static*, *Temporal* and *Periodic Tables*. Static tables contain time invariable data. Temporal tables contain non-periodic time variable data. Periodic tables contain periodically time variable data. The TRTs contain the temporal information of the temporal and periodic tables.

All the data tables consist of the normal user-defined attributes, as when they are represented in a snapshot relation, plus some additional attributes. One additional attribute is **"RecordCode"**, the unique code defining a record in a table. This attribute has the same semantics in all the tables in which it appears.

A TRT refers to one specific temporal table. It consists of the following attributes : **"RecordCode"**, **"ReferenceCode"** and **"VtStart"**, **"VtEnd"**. Each record of a TRT contains a temporal reference of one record in the specific temporal table. The record code of the referred record is stored in **"ReferenceCode"**. **[VtStart, VtEnd]** defines the valid time interval of the record defined in **"ReferenceCode"**.

Periodic tables contain data that are temporal periodic. There are one or more TRTs that correspond to a periodic table. Each TRT containing temporal references of a periodic table consists of the following attributes : **"RecordCode"** **"ReferenceCode"** **"TS"**, **"TE"**, **"Begin"**, **"Finish"**, **"T"**. Such TRTs may refer to the specific periodic table or to a TRT that expresses a higher level of periodicity of the same periodic phenomenon. The two attributes **"TS"**, **"TE"**, define the valid time interval [TS,TE] of the first instance of the periodic data. The attributes **"Begin"** and **"Finish"** define the cover interval [Begin, Finish] of the periodic phenomenon. The attribute **"T"** contains the period of the periodic phenomenon.

All the above tables have some additional attributes that help in storing the history of the tuples and the objects from the database point of view. Their functionality concerns neither the retrieval on valid time nor the representation or handling of periodic data.

3. Specification and System Design

3.1 PROBLEM DEFINITION AND REQUIREMENTS

Our target was the development and implementation of a database that could efficiently store and retrieve temporal and periodic data. The requirements of the system were set as it follows:
- a database that supports valid time and transaction time, represents periodic data and valid time indeterminacy and that is based on the relational model
- a temporal reasoning system that can process and retrieve temporal information and indeterminate temporal information.
- a language for the insertion of both non-temporal and temporal queries.

3.2 SPECIFICATIONS

TRLi is used for temporal reasoning. Temporal reasoning is needed in two cases. The first case occurs when data are inserted, deleted or updated. In this case the TRLi has to perform several tests to assure the consistency of the database. The second case occurs at the retrieval of data that possibly involves reasoning on the valid time of data. The language used for the insertion of queries from the user is TRL [16].

The database that is implemented is based on the TRRM. The data stored in the database are data of the application domain together with their valid time information and their type (events or properties). This information is essential to TRLi in order to provide output at retrieval queries that involve reasoning on valid time.

It was decided that the implementation of a TRLi-TRRM meta-interpreter was essential. This meta-interpreter provides the connection between the database and TRLi and the user.

A TRRM supervisor is needed in the case of insertion, deletion or updates of data. This supervisor should produce queries for the checking of the consistency of the data.

The system architecture is depicted on diagram 1. We have focused on the database design and the design of the TRLi-TRRM, i.e. on the part that concerns the intelligent retrieval from a database of type TRRM with the help of TRLi.

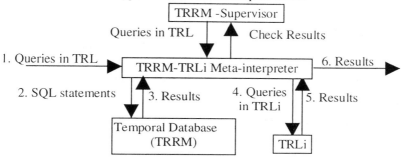

Diagram 1: System Architecture

3.3 TRRM-TRLI META-INTERPRETER

A key component of the system is TRRM-TRLi meta-interpreter. This tool provides a graphical interface to the user for inserting queries in TRL and getting the corresponding results. However, its core functionality is the interpretation of the queries in language that TRLi can understand, the retrieval of the referring to the query data from the database and the interpretation of the retrieved data to language that TRLi can understand. The meta-interpreter inputs the data and the query to TRLi.

A sample query "When did the salary of the employee e1 increase" in TRL has the following form :

$<X_1,Y_1>$:salary(e1,S_1), $<X_2,Y_2>$:salary(e1,S_2), $<X_2,Y_2>$after$<X_1,Y_1>$, $\{S_1>S_2\}$

with return parameter X_2. The "$<X_1,Y_1>$:salary(e1,S_1)" clause states that e1 took salary S_1 at the temporal interval from time point X_1 to time point Y_1. X_1, Y_1, X_2, Y_2, S_1, S_2 are variables. "after" is one of Allen's well known relations. The expression inside curly brackets, { }, is a constraint. On receiving this query, the meta-interpreter would retrieve from the tables "salary" and its corresponding temporal references table, the records concerning only employee e1. The retrieved records would be translated into the right prolog statements forming the knowledge of TRLi. The meta-interpreter has also to translate the query in TRLi statements. The data and the query are inserted to TRLi. In that way TRLi has only to cope with information concerning just employee e1. TRLi will return the values of the requested parameters and the meta-interpreter will show them on the screen.

In the case of retrieval from periodic tables, the meta-interpreter has to retrieve the lines from the database describing the periodic phenomenon. Therefore, the meta-interpreter has to retrieve the values of T, C and the valid time interval of the first instance for each level of periodicity of the periodic phenomenon. Using these values, it calculates all the instances of the periodic phenomenon. This information will be transformed in TRLi statements.

4. Implementation Issues

4.1 IMPLEMENTATION TOOLS

TRLi had already been implemented in Sicstus Prolog 3.0. TRLi is called by a program written in C language. This program requests from TRLi to read a text file and answer the query contained in that file based on the data contained in the same file. The C program, takes one by one the results from TRLi and puts them in another text file (output.txt). We used MS Visual C++ 4.00 to implement the C program. The Meta-interpreter is implemented in Powerbuilder 4.0. The temporal database was stored in Sybase System11 for Solaris.

4.2 APPLICATION DOMAINS

The system was tested in two application domains.

The first implementation was applied to the domain of transportation. The information kept in this database concerns the employees of a transport's company and the vehicles of the company. This information is considered as static. The database contains information about what is the salary of an employee at a certain time and who is the manager of each department of the company each time. This information is considered to be temporal. The database also contains information about who is driving which vehicle and when. The bus routes are stored as periodic data.

The second implementation was applied to the domain of an equivalent database described in [17]. This database contains information about employees working in an organisation. This information is considered as static data. It also contains information about who is the manager of a department at a certain time and which employee works at which department at a certain time. It also contains information about what skills a certain employee has at certain time. The results in [17] are compared with those extracted from our system.

4.3 SAMPLE DATABASE TABLES

In the first set of sample tables, we show the temporal table Salary and its corresponding TRT, TRT1. Salary table contains the information about the salaries certain employees earn. TRT1 contains the valid time information of the records in Salary table. For instance, both lines in TRT1 refer to the record of Salary table with RecordCode = 1, i.e. the value stated in the of the ReferenceCode attribute. By reading the two tables, we extract the following information : The salary of employee e1 was 20 from 1/2/82 to 31/5/82 and from 1/9/82 to 31/1/85.

RecordCode	EmpCode	Salary
1	e1	20

Temporal Table : Salary

RecordCode	ReferenceCode	VtStart	VtEnd
1	1	1/2/82	31/5/82
2	1	1/9/82	31/1/85

Temporal References Table for Salary : TRT1

The second set of sample tables are the BusRoute periodic table and its corresponding TRTs : TRT2, TRT3 and TRT4. The following tables express the complex periodic phenomenon of the routes of a bus that sets off every two hours from Monday to Friday with first instance at [6/1/1997 00:00, 6/1/1997 01:59].

As it was stated in 2.1, this periodic phenomenon has three levels of periodicity. One TRT is needed per level of periodicity. In the following example, TRT2 represents the level of periodicity of 7-days, TRT3 the level of periodicity of 1-day and TRT4 the level of periodicity of hours. The TRT of the higher level of periodicity, TRT2, refers to records of the periodic table. TRTs of lower levels of periodicity refer to records of a TRT of a higher level of periodicity. In our case, TRT3 records refer to TRT2 records and TRT4 records refer to TRT3 records.

The line in TRT2 refers to line with RecordCode=1 in the BusRoute table. It contains the characteristics of the periodic phenomenon in its first level of periodicity, i.e. T = 7 days, C = [1/1/1997, 31/12/1997] and first instance at [6/1/1997, 10/1/1997]. The line in TRT3 refers to line with RecordCode =1 in TRT2. It contains the characteristics of the periodic phenomenon in the second level of periodicity, i.e. T = 1 day, C = [6/1/1997, 10/1/1997] and first instance at [6/1/1997, 6/1/1997]. The line in TRT4 refers to line with RecordCode =1 in TRT3. It contains the characteristics of the periodic phenomenon in the third level of periodicity, i.e. T = 2 hours, C = [6/1/1997 00:00, 6/1/1997 23:59] and first instance at [6/1/1997 00:00, 6/1/1997 1:59].

The instances of this periodic phenomenon are approximately 2880. If we recorded every single instance without taking into consideration the periodicity, we would require approximately 2880 number of lines in a table in order to store it!

RecordCode	Route	Bus
1	r1	b1

Periodic Table : Bus-Route

Record Code	Reference Code	TS	TE	T	Begin	Finish
1	1	6/1/1997	10/1/1997	7	1/1/1997	31/12/1997

Temporal References Table, Level of Periodicity 7-Days :TRT2

Record Code	Reference Code	TS	TE	T	Begin	Finish
1	1	6/1/1997	6/1/1997	1	6/1/1997	10/1/1997

Temporal References Table, Level of Periodicity 1 Day : TRT4

Record Code	Reference Code	TS	TE	T	Begin	Finish
1	1	6/1/1997 00:00	6/1/1997 01:59	2	6/1/1997 00:00	10/1/1997 23:59

Temporal References Table, Level of Periodicity Hours : TRT5

5. Performance Evaluation

In [17] there is a number of queries and their expected results from a database with a given content. The very same data content of that database was inserted in our equivalent database. All the queries that could be expressed in TRL were entered in the system and the output was compared to the corresponding one in [17]. From the 146 queries entered in the system 109 were answered.

The system can answer queries that involve valid time selection and non-temporal selection. The temporal reasoning is based on the indeterminate interval. Projection on valid time, as well as on the rest of the attributes, given their valid time is possible. The supported time granularity is "seconds". The system can give results to queries involving comparison of the duration of the valid time but only when there is the constraint of continuous duration. This is due to the fact that the reasoning is based on temporal intervals and not on temporal elements. The system provides support for Allen's 13 relationships and also for some of their extensions that implicate the notion of indeterminacy [18].

The wrongly answered queries revealed the following inadequacies in the system. User defined time is not supported. As a consequence of this, a field containing for instance the birthday of an employee has no extra meaning to the system other than a plain string. The reasoning is based on the temporal interval and not on the temporal element. Queries that contain phrases like : longest continuous time, shortest continuous time, longest total time, shortest total time, cannot be answered. No counting of events is supported.

6. Conclusions

The system described in this chapter combines the advantages of both a temporal database and a temporal logic. Data are stored in a database and not all of the contents of the database need to be processed by the TRLi when a query is inserted. Only the relevant data to the query are going to be given as input to TRLi. Given the fact that in a temporal database data is always being accumulated since nothing is literally deleted, the amount of data processed by TRLi is significantly reduced.

TRL was selected as the query language due to its rich expressiveness. The expressiveness of the system is also enriched with the capability given to the user to define dynamically which relations (tables) are to be taken as properties or events during reasoning. Moreover, the user can define reasoning rules when inserting a query.

The insertion of queries and the manipulation of the reasoning rules, as well as the presentation of results is done through a graphical user interface (GUI). During the insertion of the queries, the tables' format is given to the user. In that way mistakes can be avoided since the user does not have to remember the attributes of the tables.

The implementation of a component named TRRM-Supervisor is pending. This component is responsible for the maintenance of the data integrity in the database during insertions, deletions and updates. Depending on the operation requested, the supervisor has to forward queries to the TRRM-TRLi interpreter. According to the results of these queries the corresponding operations are triggered in the database.

The retrieval of periodic data is done only at the lower level of granularity. The retrieval can also be expanded in further levels other than the lower level. An expressive representation of exceptions in periodic data at various levels of periodicity, as well as a way of coping with it, is within our scopes of future work. In order to cope with multiple time granularities, TRLi needs to be extended to support multiple time units.

Another important issue which implicitly emerges from the representation of periodic data, is the issue of valid time. In the representation of periodic data, we are defining valid time intervals and cover intervals, which can be both considered as valid times with different semantics. The notion of periodic valid time is also implicitly introduced. Issues concerning multiple valid times have also been discussed elsewhere [19], but through the current approach the notion of valid time is given additional semantics.

Acknowledgement

We would like to thank Dr. D. Spyropoulos, Prof. N. Lorentzos, Dr S. Kokkotos for their valuable remarks and guidelines. The system described in this chapter was part of the PENED'95 561 project with title "TEDRAS" and was funded by the European Community and the GSRT of the Greek Ministry of Development.

References

[1] J. Clifford, A. Crocker, "The Historical Relational Data Model (HRDM) Revised", appears in [20].

[2] S. Navathe, R. Ahmed, "Temporal Extensions to the Relational Model and SQL", Appears in [20].

[3] S. K. Gadia, "A Homogeneous Relational Model and Query Languages for Temporal Databases", ACM Transactions on Database Systems, Vol. 13, No. 4, December 1988, pp. 418-448.

[4] N. Sarda, "HSQL: A Historical Query Language", Appears in [20].

[5] A. Tansel, "A Generalised Relational Framework for Modelling Temporal Data" appears in [20].

[6] R. Snodgrass, I. Ahn, G. Ariav. D. Baton, J. Clifford, C. Dyreson, R. Elmarsi, F. Grandi, C. Jensen, W. Kafer, N. Kline, K. Kulkarni, T. Cliff Leung, N. Lorentzos, J. Roddick, A. Segev, A. Soo, S. Spirada, "A TSQL2 Tutorial" SIGMOD RECORD, Vol. 23, No 3, September1994.

[7] J. F. Allen "Maintaining knowledge about temporal intervals", Comm. ACM, 26, pp 832-843.

[8] D. McDermott, "A Temporal Logic for Reasoning About Precesses and Plans", Cognitive Science 6, pp 101-155.

[9] F. Sadri, "Kowalski and Sergot's Calculus of Events", in "Temporal Logics and their applications", Antony Galton, Chapter 4, pp 122-139, 1987.

[10] N. Lorentzos, "The Interval Extended Relational Model and Its Applications to Valid-Time Databases", appears in [20].

[11], A. Tuzhilin, J. Clifford, "On Periodicity in temporal databases", Information Systems, Vol. 20, No. 8, pp. 619-639, 1995.

[12] T. L. Anderson, "Modelling time at the conceptual level", Proceedings of the International Conference on Databases : Improving Usability and Responsiveness, P. Sheuerman (ed.), Academic Press, pp. 273-297, Jeurasalem , Israel, (1982).

[13] T. Panayiotopoulos, L. Baxevanaki, M. Maniathakis, "Representing, Storing and Retrieving Periodic Phenomena with Exceptions", working paper.

[14] T. Panayiotopoulos, M. Gergatsoulis, "Intelligent Information Processing using TRLi", Database and Expert Systems Applications, 6th International Conference London, United Kingdom, September 1995.

[15] L. C. Baxevanaki, "Temporal Data Management Systems and the Temporal References Relational Model", Internal Report, Department of Informatics, University of Piraeus, October 1996 (In Greek).

[16] T. Panayiotopoulos, C.D. Spyropoulos "TRL : A formal language for Temporal References", in "Temporal Logic" Proceedings of the ICLT Workshop, H. J. Ohlbach (Ed), MPI-I-94-230, pp. 99-109, 1994.

[17] C. S. Jensen et al., "A Consensus Test Suite of Temporal Database Queries", Institute for Electronic Systems, Department of Mathematics and Computer Science, AALBORG University, November 1993.

[18] K. T. Frantzi, T. Panayiotopoulos, C. D. Spyropoulos, "Extending Allen's relations for uncertain time points and uncertain intervals", AICS, Dublin, September 1994.

[19] S.Kokkotos, E.Ioannidis, T.Panayiotopoulos, C.D.Spyropoulos, "On the Issue of Valid Time(s) in Temporal Databases", SIGMOD RECORD, Vol.24, No 3, pp. 40-43, September 1995.

[20] Tansel, Clifford, Gadia, Jajodia, Segev, Snodgrass, "Temporal Databases : Theory, Design and Implementation", Benjamin/Cummings Publ. 1992.

PART III

IMAGE PROCESSING AND VIDEO-BASED SYSTEMS

23

A Java-Based Image Processing System

P. Androutsos†, D. Androutsos†, K. N. Plataniotis‡,

A. N. Venetsanopoulos†

†*Digital Signal & Image Processing Lab,*
Department of Electrical and Computer Engineering,
University of Toronto,
10 King's College Rd.,
Toronto, Ontario, M5S 3G4, Canada.

‡*School of Computer Science,*
Ryerson Polytechnic University,
350 Victoria Street,
Toronto, Ontario, M5B 2K3, Canada.

1. Introduction

The recent Internet phenomenon has impacted all walks of life. E-Commerce, Networking, and the World-Wide-Web have become household words. Programming, which is the heart of all computing, has not been invulnerable to the Internet boom, and has changed for the better. *Java* [1] as an entity has led to the coining of the name *Applet*, and more importantly spearheaded the architecture-independent programming revolution. The whole idea of having a single version of machine code for all computing platforms is extremely important because it saves both time during the development stage, as well as disk space further down the road. This is because binary executable files that are exclusive to individual machine architectures are eliminated, and the end result is a network of computers all utilizing a shared binary. In addition to all the the practical advantages involved with its use, *Java* is an object-oriented, multi-threaded, and above all, easy to learn programming language. Many routines that are either difficult or laborious to do in other

---

[1] *Java* is a registered trademark of *Sun Microsystems*

255

programming languages, have been included in Java libraries. These libraries facilitate file access, networking, multimedia, and most importantly, user-interface development. Finally, Java also permits programming for the World-Wide-Web. This means that programs are no longer need to be limited to local networks, and can be included in web pages so that their functionality can be accessed from a browser [2].

2. Image Processing with Java

2.1 *IMAGEnius*

The aforementioned characteristics make programming in *Java* extremely conducive to learning, and it is for this reason that it is important for the existence of software packages that can encourage the understanding, implementation and development of computational algorithms in all fields. *IMAGEnius* is just such a software package that is aimed at providing a hands-on approach to programming for image processing. Written exclusively in *Java* at the Digital Signal and Image Processing Lab at the University of Toronto, *IMAGEnius* is an architecture-independent program, which is unique due to the fact that it has been left completely open for anyone wishing to incorporate changes to, or create new and improved algorithms to existing program code[3]. This software package has been slated for incorporation into the graduate course in Digital Image Processing at the University of Toronto as a learning aid for students. Since *IMAGEnius* was developed to act as a testbed for new image processing algorithms, it has the ability to allow new users to explore and experiment with algorithm and filter implementation, thus providing immediate hands-on experience to theory. As an ever-growing program, *IMAGEnius* is a highly flexible and simple, yet powerful tool for the development and implementation of image processing routines. It is important to stress here that *IMAGEnius* is by no means complete. Since the program code has been left open, *IMAGEnius* grows as eager users incorporate their own code into its functionality.

2.2 OVERVIEW

The remainder of this document will deal with an in depth-look at *IMAGEnius*. Chapter 3 will overview the functions that are included in the current version of *IMAGEnius*, and Chapter 4 will cover aspects of structure and implementation. In Chapter 5, a discussion regarding the extension of existing classes and code to incorporate new routines is given. Finally Chapter 6 will close with Future Considerations and Conclusions.

---

[2]Current browsers that support Java Applets are *Netscape Navigator 4.0* and *Microsoft Internet Explorer 4.0*

[3]Complete *IMAGEnius* source code can be obtained upon request from the authors.

3. *IMAGEnius* Structure and Implementation

3.1 MAIN COMPONENTS

IMAGEnius is organized in a layered manner. Although an object-oriented approach allows for all objects to communicate to each other, this can get somewhat disorganized, and cluttered. The layered approach only allows objects in a particular layer to communicate with objects in the layers immediately above or immediately below them. Such a formal scheme maintains a strict level of order as far as inter-object commination is concerned. A single class makes up the topmost layer of the program structure. This is the *IMAGEnius* Class. The second layer is comprised of the *ImageCanvas Class, MouseHandler Class*, and *MenuHandler Class*. Layer three is made up of a collection of classes for the creation of Dialog, Popup, and other windows, as well as three different *abstract* classes containing the actual methods required to perform manipulation of the image pixels. These abstract classes are the *Analysis Class, Filter Class*, and the *Image Manipulation Class*. Finally, in the bottom layer, are the *Supplemental Classes* which include commonly used miscellaneous objects and methods.

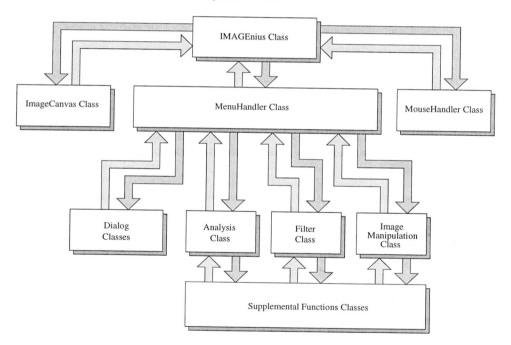

Figure 1: *IMAGEnius* Organization

3.2 LAYER ONE

The uppermost layer consists of only the *IMAGEnius Class*. This class is the foundation for the entire software package. It is responsible for the initialization of the program, for the generation of the user-interface, and creation of the event-handlers, In addition, this class contains within it variables that are used to keep track of window dimensions, image characteristics, and pixel data.

3.3 LAYER TWO

IMAGECANVAS CLASS - The *ImageCanvas Class* is a derivative of the *Canvas* class. It is responsible for updating the screen, painting new pixels, drawing the selection box, and maintaining the correct application dimensions. The *ImageCanvas Class* communicates only with the *IMAGEnius Class*, from which it accepts requests for screen updates and to which it sends new values for window and image-related variables.

MOUSEHANDLER CLASS - All mouse movements and clicks are monitored by the *MouseHandler Class*. The mouse handler has been set up such that it is able to detect right and left clicking, as well as clicks that are modified using the *CTRL*, *ALT*, and *META* keys.

MENUHANDLER CLASS - The menu that is set up by the *IMAGEnius* class during program initialization is monitored by the *MenuHandler Class*. In the most basic terms, this class is simply a very long switch that keeps track of what menu event has been generated by the user, and performs the appropriate section of code. The menu handler is perhaps the most important class because of the fact it initiates the methods found in the lower layers to perform the actual computations for processing.

3.4 LAYER THREE

DIALOG CLASSES - The *Dialog Classes* are a number of different classes encompassing popup windows, dialog boxes, and interactive windows.

ANALYSIS CLASS - The *Analysis Class* is responsible for operations that result in the extraction of features or information contained within the image. Examples are *Channel Separation*, *Edge Detection*, *Segmentation*, and *Fourier Analysis*.

FILTER CLASS - The abstract *Filter Class* contains methods which perform some sort of image filtering. Some of the filters that *IMAGEnius* currently supports are *Morphological Filters*, *Low-Pass* and *High-Pass Filters*, *Fuzzy Filters*, and *Vector-Based Filters*.

IMAGE MANIPULATION CLASS - The final class found in the third layer is the *Image Manipulation Class*, which is responsible for routines that change the way the image at hand looks. Examples include the generation of noise, zoom factor, and image size.

3.5 LAYER FOUR

The bottommost layer in the *IMAGEnius* structure is made up of the *Supplemental Functions Class*. These classes are simply a collection of methods that are frequently used by the algorithms found in the Analysis, Image Manipulation, and Filter classes.

4. *IMAGEnius* **Command Functions**

At the moment, *IMAGEnius* supports a variety of different image formats. On top of the functionality provided by *Java* to support GIF and JPG images, *IMAGEnius* has been programmed with additional classes that allow it to display and process images in MVP, RAW, and PPM formats.

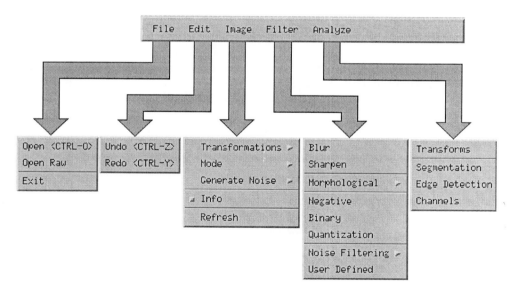

Figure 2: *IMAGEnius* Command Menu

Image files are obtained through the use of the *getDefaultToolkit().getImage()* method in Java's Toolkit object. Along with the use of a *MediaTracker* object, the MenuHandler returns an *Image* object that can subsequently be processed by the various functions to provide pixel data. This is done through the use of the *PixelGrabber.grabPixels()* method. The pixel data is then put into an integer

260

array and sent to the appropriate abstract class for either filtering, analysis, or processing.

As shown in Figure 3, the Image Menu contains commands for the generation of noise, the changing of color modes and also provides an *Information Popup*. This popup gives information such as spatial coordinates and pixel color. Figure 4 depicts this information window.

NEGATIVE - Image negation is performed through a simple subtraction of the current pixel's intensity from the maximum intensity. For a color image, each separate channel is treated individually, and then the components are merged at the end.

BINARY - Only grayscale (single channel) images are allowed to be made binary. If the current image is a color one, then the user is prompted as to whether he/she would like to convert the image to grayscale. Once a grayscale image is obtained, the user is prompted for a threshold value used to determining which pixels should be set high and which should be set low.

QUANTIZATION - In a similar manner to the function that creates a binary image, the Quantization function requires a grayscale image in order to function properly. Input from the user is required for information on the number of quantization bins desired.

NOISE - Gaussian and Impulsive noise is generated according to the algorithms provided by Myler and Weeks in [1].

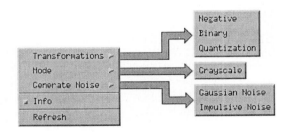

Figure 3: *IMAGEnius* Image Menu

Most of the functionality of this software package is controlled from the *Filter Menu*. Referring to Figure 5, we see that some of the preprogrammed filters include Low-Pass, Morphological, and Fuzzy Filters.

LOW-PASS FILTERS - The most basic of low-pass filters has been implemented in *IMAGEnius*, namely, the *Averaged Blur*.

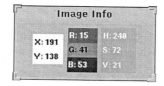

Figure 4: *IMAGEnius* Information Popup

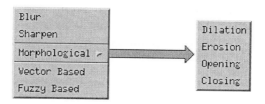

Figure 5: *IMAGEnius* Filter Menu

HIGH-PASS FILTERS - Two sharpening filters are included included here. They are the traditional high-pass filter and the *High-Pass Boost* (or *Unsharp Mask*). The high-pass filter uses a convolution kernel where all coefficients equal to -1, except for the center pixel whose value is set such that the sum of all the coefficients is equal to zero. The high-pass boost filter has a similar convolution mask, but the center coefficient is multiplied by a constant.

MORPHOLOGICAL FILTERS - Four basic morphological filters are supported in this software package. These are *Dilation, Erosion, Opening*, and *Closing*. The Dilation and Erosion operators each have their own dialog windows for parameter selection. Such parameters include Morphological mask selection, and size. Furthermore, custom masks are also possible using the *CustomDialog* class. Figure 6 depicts the Dilation, Erosion dialog boxes, and an example 5x5 Custom Mask dialog.

VECTOR-BASED FILTERS - Since color pixels are very commonly represented as vectors, Vector-Based Filters become absolutely necessary in many situations for color image processing. *IMAGEnius* has been programmed with a Vector Median Filter that calculates the median vector from a pixel neighborhood via a sum-of-angles measure. The Vector Median Filter has shown to be extremely effective in the removal of impulsive noise [2]. This ability is illustrated in Figure 9.

262

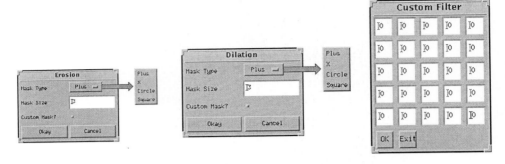

Figure 6: Dilation, Erosion Dialogs and an example Custom Filter Dialogs

FUZZY FILTERS - Fuzzy Filters utilize the angular distance x between color vectors in a given image neighborhood as input into a fuzzy membership function. Within a neighborhood, the output from the membership function determines each vector's contribution to the final filter output [3] . A Fuzzy filter using the membership function shown in Equation 1 has been incorporated into existing program code.

$$MembershipFunction = \frac{1}{1 + e^x} \tag{1}$$

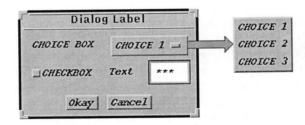

Figure 7: Sample Generic Dialog

5. Hands-on Learning

IMAGEnius has been written so that users can always re-use and recycle existing code. Difficult tasks such the opening, and displaying of images have already been taken care of through the use of built-in Java functions and additional code which supports other image formats. Also, interactivity has been incorporated into the program through the menus, the information popup, and a selection box. Perhaps

Figure 8: Screenshot of *IMAGEnius* performing a selective Morphological Dilation operation

Figure 9: Screenshot of Vector Median filtering of Impulsive Noise

the biggest advantage to using *IMAGEnius* as a means to provide experience in image processing implementation, is that it contains many methods that simplify the process of acquiring image pixels and drawing to the screen. Since the source code is freely available, people who are interested in implementing new routines are able to use method templates for filter kernel convolution, transformations, etc. In addition, *IMAGEnius* allows for the user to do away with much of the difficulty involved in setting up interfaces. This is done through the use of a *GenericDialog* class. This multi-purpose dialog requires only to be given an ordered set of parameters defining the dialog's elements. As shown in Figure 7 such elements include text fields, check boxes, and labels.

6. Conclusions and Future Work

IMAGEnius is still in its infancy, and has yet to be tested in a learning environment. Hopefully, with the future incorporation of this software package into the Digital Image Processing course at the University of Toronto, more feedback on this program's effectiveness will be able to be obtained from actual external sources. Another obvious future consideration for this program lies in the incorporation of a wider variety of image processing algorithms, tools, and functions. Of greater importance, however, is the ability of this package to be easily altered to be made into an *Applet*. Minor changes could be made to the current program code such that it could be run via a web browser similar to work done in [4] for DSP education.

References

[1] H.R. Myler, A.R. Weeks, *The Pocket Handbook of Image Processing Algorithms in C*, Englewood Cliffs, New Jersey, Prentice-Hall, 1993.

[2] I. Pitas, A.N. Venetsanopoulos, *Nonlinear Digital Filters: Principles and Applications*, Kluwer Academic Publishers, Norwell, Massachusetts, 1990.

[3] K.N. Plataniotis, D. Androutsos, S. Vinayagamoorthy, and A.N. Venetsanopoulos, *Color Image Processing Using Adaptive Multichannel Filters*, IEEE Transactions on Image Processing, Volume 6, Number 7, pp 933-949, IEEE, July 1997.

[4] A. Clausen, A. Spanias, A. Xavier, M. Tampi, *A Java Signal Analysis Tool for Signal Processing Experiments*, Proceedings of the IEEE International Conference on Acoustics, Speech, and Signal Processing (ICASSP) 1998, pp 1849-1852. Seattle, Washington, May 12-15, 1998.

[5] J. Zukowski, *JAVA AWT Reference*, O'Reilly & Associates, Inc., Sebastopol, California, 1997

[6] JavaSoft, *Java Development Kit 1.1.4 Documentation*, Sun Microsystems, Mountain View California, 1997.

[7] G. Cornell, C.S. Horstmann, *Core Java, 2nd Edition*, Sunsoft Press, Mountain View California, 1997.

24

Wavelet Image Compression : A Comparative Study

K. Friesen, N.D. Panagiotacopulos, S. Lertsuntivit, J.S. Lee
Department of Electrical Engineering, College of Engineering
California State University, Long Beach CA 90840, USA

1. Introduction

We are making a comparative study of image compression methods using 5 transform methods; 3 standardized and 2 wavelet based transforms. After the image is transformed with each of the 5 transforms, the transformed image is quantized to zero-out all transform values that correspond to the high-frequency components of the image. Quantization is followed by image encoding that minimizes the data storage requirements for the quantized residuals. Each transform method is compared using the criteria: (1) compression ratio, (2) signal to noise ratio (SNR), and (3) mean squared error (MSE) of the reconstructed image compared to the original. The objective of this comparative study is to determine if either or both of the wavelet transforms out-perform the three standardized transforms.

2. Background

The three standardized transforms are (1) Discrete Cosine transform (DCT), (2) the Hadamard transform, and (3) the Walsh transform. Each of these transforms have gained wide usage in image compression because they are efficient in decorrelating highly correlated images and compacting the image energy in much smaller scope than the original. Of the three, the DCT actually decorrelates the original image by transforming it to a series of independent basis images, and it is the transform currently used by JPEG (Joint Programmers Expert Group) in data compression. The DCT is easy to implement on a computer and has fast execution time. The Hadamard transform and the Walsh transform have an advantage over the DCT in ease of computer implementation and execution time. The wavelet transforms (WT) used are (1) the 2D Haar WT (2D Haar WT) and the 2D Daubechies WT of 4th order (2D D4 WT). The 2D Haar WT and the 2D D4 WT both act as multiple branch filter banks upon the image. In the case of the 2D Haar WT the number of branches for the 2D filter bank equals $\log_2 N$, where N equals the image dimension, and the number of branches for the 2D D4 WT equals $\log_2 N - 1$. (It is necessary for N to equal 2^n for some n. If $N > 2^n$ for some n then the image must be "zero-padded" to dimensions M x M where $M = 2^{n+1}$). Each branch of these filter banks decomposes the image into 2 disjoint subimages, one

265

containing high-frequency components the other containing low-frequency components. The dimension of each subimage is one half the dimension of the parent image. The high-frequency subimages are orthogonal and hence independent to the low-frequency subimages and consequently are decomposed no further. In the next branch of the filter bank the low-frequency subimage is decomposed into its high-frequency and low-frequency components. This successive decomposition continues until the filtering operation can continue no further, which occurs when the dimensions of the resultant subimage are equal to one half the order of the WT used. In the next section we present a detailed description of the filter bank operation for any 2D wavelet that generates what is known as a Multi-Resolution Analysis (MRA) decomposition of the Hilbert space of finite energy 2D signals.

3. Filter Banks [1, 2]

Both the 2D Haar WT and the 2D D4 WT act as multiple branch filter banks upon the original image transforming it to a sequence of independent bases images preserving the decorrelation property of the DCT. For a 256 x 256 image the 2D Haar WT is a 8 branch filter bank and the 2D D4 WT is a 7 branch filter bank. (The 2D D4 WT has 4 tap weights while the 2D Haar WT has 2 tap weights). Each branch of the filter bank *down samples* the input image by a factor of 2 which results in the output subimage having its dimensions reduced by a factor of 2. The filter bank stops when the dimensions of the output subimage equal one half the tap weights of the filter.

The original image contains superimposed signals in all frequency bands (correlation). The first branch of the filter bank separates the original image into high-frequency and low-frequency subimages which are orthogonal hence independent. The low-frequency subimage still contains superimposed signals in multi frequency bands. The second branch of the filter bank operation separates the low-frequency components and high-frequency components of the correlated subimage output from the first branch into orthogonal low-frequency and high-frequency subimages. This successive decomposition of the correlated low-frequency output subimages from each branch of the filter bank continues through the last branch of the filter bank. The result is a sequence of independent subimages each containing 2D signals in a distinct frequency range.

3.1. Two-Dimesional MRA Decomposition

Both the 2D Haar WT and the 2D D4 WT generate a Multi Resolution Analysis Decomposition of the Hilbert space $L^2(\Re^2)$, the space of finite energy 2D signals. A signal f(x, y) is contained in $L^2(\Re^2)$ if the following condition holds:

$$\int_{-\infty}^{\infty} \int_{-\infty}^{\infty} f(x,y)^2 \, dxdy \; < \infty$$

Any finite image is contained in $L^2(\Re^2)$ since the above integral always exists for functions defined on a 2D finite interval $\{[a, b] \times [c, d]\}$. Any 2D wavelet constitutes a MRA decomposition of $L^2(\Re^2)$ if it consists of a pair of basis functions for $L^2(\Re^2)$ $\{\varphi(x, y), \psi(x, y)\}$ that have special algebraic properties and as a result impose a special algebraic structure upon the space $L^2(\mathfrak{R}^2)$. The basis function $\varphi(x, y)$ is called the *scaling* function and the basis function $\psi(x, y)$ is called the *wavelet* function. Before presenting a detailed description of the 2D MRA generated by the 2D Haar WT and the 2D D4 WT, we note that both $\varphi(x, y)$ and $\psi(x, y)$ are separable functions in the variables x, y. That is, $\varphi(x, y) \equiv \varphi_1(x)\varphi_2(y)$, where $\varphi_1(x)$ and $\varphi_2(y)$ are 1D scaling functions which generate 1D MRA decompositions of the space of 1D finite energy signals: $L^2(\mathfrak{R})$; and $\psi(x, y) \equiv \psi_1(x)\psi_2(y)$, where $\psi_1(x)$ and $\psi_2(y)$ are the 1D wavelet functions corresponding to the scaling functions $\varphi_1(x)$ and $\varphi_2(y)$ respectively. The properties of the 2D MRA decomposition for the space $L^2(\mathfrak{R}^2)$ will be defined as the Cartesian product of 2 independent 1D MRA decompositions one acting on the x-axis the other on the y-axis.

The key function generating a MRA decomposition is the scaling function $\varphi_1(x)$. A scaling function is the solution to the following *dilation equation*:

$$\varphi_1(x) = \sum_k a_k \varphi_1(2x - k)$$

where the coefficients a_k have the property $<a_k, a_{k-2l}> = 2\delta(k-l)$ which makes the set of integer translates of $\varphi_1(x)$: $\{\varphi_1 (x-k)\}$ orthogonal. This property is expressed:

$$<\varphi_1(x-k), \varphi_1(x-l)> = \delta(k-l)$$

As such the integer translates of the scaling function form a basis for a closed subspace V_0. If $\varphi_1(x)$ is scaled by a factor of 2, the integer translates of $\varphi_1(2x)$, $\{\varphi_1(2x-k)\}$, also form a basis for another subspace of $L^2(\mathfrak{R})$ V_1 which contains V_0 :

$$V_0 \subset V_1$$

If $\varphi_1(x)$ is scaled by the factor 2^2 then the integer translates of the function $\varphi_1(2^2x)$, $\{\varphi_1(2^2x-k)\}$, form a basis for another closed subspace of $L^2(\mathfrak{R})$ V_2 which properly contains both the subspaces V_0 and V_1:

$$V_0 \subset V_1 \subset V_2$$

The set of integer translates of $\varphi_1(x)$ scaled by a positive integer power of 2 (2^j) forms the basis for a closed subspace of $L^2(\mathfrak{R})$ V_j which properly contain all the subspaces 'V' of a lower order:

$$V_0 \subset V_1 \subset \ldots \subset V_j \subset \ldots$$

If the scaling of the function $\varphi_1(x)$ is continually scaled by increasing powers of 2 a "ladder" of subspaces is generated whose infinite union comprises all of the space $L^2(\mathfrak{R})$:

$$L^2(\mathfrak{R}) = clos \bigcup_{j \in Z} V_j$$

Hence any finite energy 1D signal f(x) can be expressed as a series expansion of the basis functions of these component spaces:

$$f(x) = \sum_j \sum_k c_{jk} \varphi_{jk}(x)$$

where the coefficient c_{jk} is given by:

$$c_{jk} = \int_{-\infty}^{\infty} f(x)\varphi_{jk}(x)dx \; ; \quad \varphi_{j,k}(x) = \varphi_1(2^j x - k)$$

Since V_0 is a closed subspace in a Hilbert space and it is also contained in V_1, another closed subspace in a Hilbert space, there must exist another closed subspace denoted W_0 which is the orthogonal complement to subspace V_0 such that the following relationship between the subspaces V_0, W_0, and V_1 hold:

$$V_1 = V_0 \oplus W_0$$

That is, the subspace V_1 has as a basis the union of the bases for the 2 subspaces on the right hand side and the bases for the respective subspaces are orthogonal. This relationship is known as the "decomposition" of subspace V_1. The subspace W_0 is generated by integer translates of the wavelet function $\psi_1(x)$, $\{\psi_1(x-k)\}$, corresponding to the scaling function $\varphi_1(x)$. Since W_0 is orthogonal to V_0 for all $j, k \in Z$,

$$<\varphi_1(x-j),\psi_1(x-k)> = 0$$

Since the integer translates $\{\varphi_1(2x-j)\}$ are also a basis for the closed subspace V_1 which is contained in a closed subspace there must exist the closed subspace that is the orthogonal complement of V_1, W_1, and the closed subspace V_2 is the direct sum of the closed subspace V_1 and its orthogonal complement subspace W_1:

$$V_2 = V_1 \oplus W_1$$

If one substitutes the decomposition of subspace V_1 in the above expression one obtains:

$$V_2 = V_0 \oplus W_0 \oplus W_1$$

Since closed subspace V_2 is contained in closed subspace V_3, there exists the orthogonal complement subspace to V_2, W_2, such that $V_3 = V_2 \oplus W_2$. Then,

$$V_3 = V_0 \oplus W_0 \oplus W_1 \oplus W_2$$

The decomposition relation

$$V_j = V_{j-1} \oplus W_{j-1}$$

can be extended to all order of the V subspaces. If one substitutes the direct sum decomposition for all V_j in

$$L^2(\Re) = clos \bigcup_{j\in Z} V_j$$

then $L^2(\Re)$ has an alternate subspace composition:

$$L^2(\Re) = clos \{ \bigoplus_{j\in Z} W_j \}$$

this is known as the "direct sum" composition for space $L^2(\Re)$. This expression is more refined than the composition of V spaces since each of the generating subspaces is orthogonal. This is the special algebraic structure for the Hilbert space $L^2(\Re)$ created by a MRA basis: it produces 2 sets of generating subspaces as defined in the two previous relations. The latter being composed of mutually independent (orthogonal) generating subspaces. The WT in the space $L^2(\mathfrak{R})$ is the transformation from the generating expression (1) onto the generating expression (2); that is the WT of any 1D finite energy signal $f(x)$ is the conversion of its series expansion relative to the subspaces $\cup_{\{j \in z\}} V_j$:

$$f(x) = \sum_j \sum_k c_{jk} \varphi_{jk}(x)$$

to a series expansion relative to the subspaces $W_j \oplus, j \in Z$:

$$f(x) = \sum_j \sum_k d_{jk} \psi_{jk}(x)$$

We now extrapolate the definition of a 1D MRA decomposition to the 2D Hilbert space $L^2(\mathfrak{R}^2)$ generated by separable wavelet functions. The scaling function $\varphi(x, y)$ is defined $\varphi(x)_1 \varphi_2(y)$, where $\varphi_1(x)$ and $\varphi_2(y)$ are 1D scaling functions in $L^2(\mathfrak{R})$. The 2D wavelet function $\psi(x, y)$ is defined as $\psi_1(x) \psi_2(y)$, where $\psi_1(x)$ and $\psi_2(y)$ are the 1D wavelet functions corresponding to the scaling functions $\varphi_1(x)$ and $\varphi_2(y)$ respectively. The subspace $V_0^{(2)}$ is defined as the Cartesian product space $V_0^{(x)}$ x $V_0^{(y)}$ of the two 1D subspaces which for the sake of notational convenience will be referred to as $V_0^{(x)} V_0^{(y)}$. The subspace $V_0^{(x)}$ is generated by the basis $\{\varphi_1(x-k)\}$ and $V_0^{(y)}$ is generated by the basis $\{\varphi_2(y-k)\}$. The space $W_0^{(2)}$ is the Cartesian product $W_0^{(x)} W_0^{(y)}$ where $W_0^{(x)}$ is generated by the basis $\{\psi_1(x-k)\}$ and $W_0^{(y)}$ is generated by the basis $\{\psi_2(y-k)\}$. The decomposition of the space $V_1^{(x)} V_1^{(y)}$ is given by:

$$V_1^{(x)} V_1^{(y)} = (V_0^{(x)} \oplus W_0^{(x)})(V_0^{(y)} \oplus W_0^{(y)}) = \{V_0^{(x)} V_0^{(y)} \oplus V_0^{(x)} W_0^{(y)} \oplus W_0^{(x)} V_0^{(y)} \oplus W_0^{(x)} W_0^{(y)}\}$$

Each of these 2D subspaces is orthogonal. The subspace $V_0^{(x)} V_0^{(y)}$ contains only low-frequency components in both variables x and y, the subspaces $V_0^{(x)} W_0^{(y)}$ contains low-frequency components in the variable x and high-frequency components in the variable y, the subspace $W_0^{(x)} V_0^{(y)}$ contains high-frequency components in the variable x and low-frequency components in the variable y, and the subspace $W_0^{(x)} W_0^{(y)}$ contains high-frequency components in both variables x and y.

3.1.1. Decomposition Procedure

In our case, we are wavelet transforming a 256 x 256 image. By its dimension the original image is contained in the $L^2(\mathfrak{R}^2)$ subspace $V_8^{(x)} V_8^{(y)}$. The WT of this image is the transformation of subspace $V_8^{(x)} V_8^{(y)}$ onto the most refined "direct sum" subspace possible for the WT used. In the case of the 2D Haar WT this transformation is:

$$V_8^{(x)} V_8^{(y)} \rightarrow \{V_0^{(x)} V_0^{(y)} \oplus V_0^{(x)} W_0^{(y)} \oplus W_0^{(x)} V_0^{(y)} \oplus \bigcup_{0 \le j,k \le 8} W_j^{(x)} W_k^{(y)}\}$$

since 0 is the stopping order for the 2D Haar WT. For the 2D D4 WT, the transformation is:

$$V_8^{(x)} V_8^{(y)} \rightarrow \{V_1^{(x)} V_1^{(y)} \oplus V_1^{(x)} W_1^{(y)} \oplus W_1^{(x)} V_1^{(y)} \oplus \bigcup_{1 \le j,k \le 8} W_j^{(x)} W_k^{(y)}\}$$

since 1 is the stopping order for the 2D D4 WT. As described above the target subspace for either transformation is composed of mutually orthogonal subspaces. Each of these respective decomposition procedures takes place in a series of levels; 8 levels for the 2D Haar WT and 7 levels for the 2D D4 WT. Since both wavelet operators are separable, each level in turn is composed of 2 distinct steps: (1) the decomposition of subspace $V_k^{(x)}$ onto the orthogonal subspace $V_{k-1}^{(x)} \oplus W_{k-1}^{(x)}$ and (2) the decomposition of the subspace $V_k^{(y)}$ onto the orthogonal subspace $V_{k-1}^{(y)} \oplus W_{k-1}^{(y)}$ for $1 \leq k \leq 8$ (2D Haar WT) and $0 \leq k \leq 8$ (2D D4 WT).

The first step in each transformation level is accomplished by applying the 1D wavelet operator to each row of a subimage representing a subspace $V_k^{(x)}$. The second step in each transformation level is accomplished by applying the 1D wavelet operator to each column of the subimage that is the output of the first step in the level decomposition. This second step decomposes the subspace $V_k^{(y)}$. We illustrate the first two levels of this decomposition:

Level 1: The first step of the first level operates on the subspace $V_8^{(x)}$ in the product subspace $V_8^{(x)}V_8^{(y)}$ and creating an intermediary subspace:

$$V_8^{(x)}V_8^{(y)} \rightarrow (V_7^{(x)} \oplus W_7^{(x)}) V_8^{(y)} = V_7^{(x)} V_8^{(y)} \oplus W_7^{(x)} V_8^{(y)}$$

This decomposition produces two orthogonal subspaces labeled L: $V_7^{(x)} V_8^{(y)}$ and H: $W_7^{(x)} V_8^{(y)}$ (Labels L and H represents low and high frequencies, respectively. (Fig. 1))

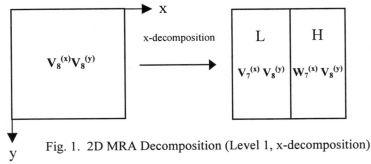

Fig. 1. 2D MRA Decomposition (Level 1, x-decomposition)

The second step of level 1 decomposition operates on subspace $V_8^{(y)}$ contained in the product subspace $(V_7^{(x)} \oplus W_7^{(x)}) V_8^{(y)}$ creating the orthogonal subspace:

$$(V_7^{(x)} \oplus W_7^{(x)})V_8^{(y)} \rightarrow (V_7^{(x)} \oplus W_7^{(x)}) (V_7^{(y)} \oplus W_7^{(y)}) =$$

$$V_7^{(x)}V_7^{(y)} \oplus W_7^{(x)}V_7^{(y)} \oplus V_7^{(x)}W_7^{(y)} \oplus W_7^{(x)}W_7^{(y)}$$

This resultant subspace is composed of mutually orthogonal subspaces containing either low-frequency components in x and y, high-frequency components in x and y, or mixtures of low-frequency components in x and high-frequency components in y or high-frequency components in x and low-frequency components in y. This decomposition produces four subspaces with the following labeling: (Fig. 2)

$$LL : V_7^{(x)}V_7^{(y)}, \ HH: W_7^{(x)}W_7^{(y)}, \ HL : W_7^{(x)}V_7^{(y)}, \ LH : V_7^{(x)}W_7^{(y)}$$

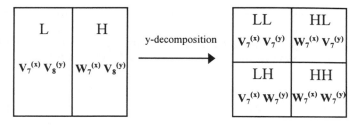

Fig. 2. 2D MRA Decomposition (Level 1, y-decomposition)

Level 2: The first step in the second level decomposition decomposes the subspaces $V_7^{(x)}$ onto $V_6^{(x)} \oplus W_6^{(x)}$, and the second step decomposes the subspaces $V_7^{(y)}$ onto $V_6^{(y)} \oplus W_6^{(y)}$. The affected subspaces in step 1 are: $(V_7^{(x)} V_7^{(y)})$ and $(V_7^{(x)} W_7^{(y)})$, and in the second step are: $(V_7^{(x)} V_7^{(y)})$ and $(W_7^{(x)} V_7^{(y)})$. The subspace $W_7^{(x)}W_7^{(y)}$ contains only high-frequency components and, therefore, is not decomposed further. Each step is illustrated: Step one is accomplished by application of the 1D WT to the rows of the subimage 128 x 256 of the output subimage in level 1. This x-decomposition produces: (Fig. 3)

$$V_7^{(x)} V_7^{(y)} \rightarrow (V_6^{(y)} \oplus W_6^{(y)}) V_7^{(y)} = \{V_6^{(x)}V_7^{(y)} \oplus W_6^{(x)} V_7^{(y)}\}$$

$$V_7^{(x)} W_7^{(y)} \rightarrow (V_6^{(y)} \oplus W_6^{(y)}) W_7^{(y)} = \{V_6^{(y)} W_7^{(y)} \oplus W_6^{(y)} W_7^{(y)}\}$$

Fig. 3. 2D MRA Decomposition (Level 2, x decomposition)

Step 2 is accomplished by application of the 1D WT to the columns of the subimage 256 x 128 of the output subimage of step 1 in level 2. This y-decomposition produces:

$$V_6^{(x)}V_7^{(y)} \rightarrow V_6^{(x)} V_6^{(y)} \oplus V_6^{(x)}W_6^{(y)}, \quad W_6^{(x)} V_7^{(y)} \rightarrow W_6^{(x)}V_6^{(y)} \oplus W_6^{(x)} W_6^{(y)}$$

These 2 decompositions are graphically illustrated:

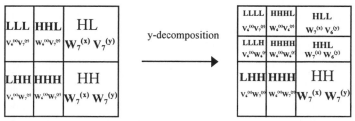

Fig. 4. 2D MRA Decomposition (Level 2, y decomposition)

3.2. Two Dimensional MRA Reconstruction

Image reconstruction is the inverse of image decomposition. The reconstruction process is the inverse WT of the subspace $\{V_\rho^{(x)}V_\rho^{(y)} \oplus V_\rho^{(x)}W_\rho^{(y)} \oplus W_\rho^{(y)} \oplus \bigcup_{\{\rho \leq j, k \leq 7\}} W_j^{(x)}W_k^{(y)}\}$ onto the subspace $V_8^{(x)}V_8^{(y)}$, ($\rho = 0$ for the 2D Haar WT, and $\rho = 1$ for the 2D D4 WT), where all the component 2D signals are superimposed over all frequency ranges. Since image reconstruction is the inverse of image decomposition, all the steps of the decomposition process are inverted. The step involving x-decomposition: $V_k^{(x)} \rightarrow (V_{k-1}^{(x)} \oplus W_{k-1}^{(x)})$ becomes $(V_{k-1}^{(x)} \oplus W_{k-1}^{(x)}) \rightarrow V_k^{(x)}$, and the step involving y-decomposition $V_k^{(y)} \rightarrow (V_{k-1}^{(y)} \oplus W_{k-1}^{(y)})$ becomes $(V_{k-1}^{(y)} \oplus W_{k-1}^{(y)}) \rightarrow V_k^{(y)}$. The two step procedure in each level of decomposition is inverted: (1) y-reconstruction and then (2) x-reconstruction.

4. Image Quantization [3, 4]

Image quantization is the conversion of the elements of the transformed image into a data representation that requires less data storage than the original image. Quantization is achieved by the following two step process: (1) normalization and (2) integer conversion. Since all transform methods concentrate most of the image's energy in the low-frequency regions (Fig. 4), most of the data in the high-frequency regions have values that are negligible with respect to the values of the low-frequency regions. These negligible values represent details of the original image. They are of such fine resolution that the human visual system cannot distinguish (the difference) when one compares the perfectly reconstructed image and the reconstructed image in which only transformed values of significant magnitude were retained.

Normalization converts all negligible values to '0' and simultaneously reduces the magnitude of the next level of lower high-frequencies while preserving the magnitude of the lowest frequency components (see LLLL in Fig. 4). Thus, normalization is achieved by dividing each element in the transformed image by a value known as a "quantizer". It is important to select a quantizer that will reduce all negligible values to '0'. The second step in quantization is the conversion of these normalized values to integer values. This process is achieved by the following operation

$$f_q(x, y) = \text{int} \, [f(x, y)/q(x, y)]$$

where $f(x, y)$ is the transformed coefficient, $q(x, y)$ is the quantizer, and $f_q(x, y)$ is the normalized value converted to an integer. Integer conversion changes the normalized data to a set of integer values that can be encoded in such a way as to minimize their storage requirements. The quantizer must be adjusted to different frequency regions of the transformed image. In the low-frequency region, the quantizer must have relatively low value and, in the high-frequency region, the quantizer must have a relatively high value. This quantizer adaptation is accomplished by the use of "non-uniform" quantization. Non-uniform quantization utilizes two parameters (1) the shifting size, and (2) the step size. The first parameter is used to adjust the value of the quantizer for a given step size to different regions of the transformed image. The second parameter is used to determine the range of quantization. A minimum shifting size of 1 corresponding to a step size of 1 allows the quantizer to be adjusted for each element in the transformed image. In this research, the effect of the shifting and the step size changes on the storage requirements was examined. Highly localized quantization (small shifting size) mainly concentrates the non-zero elements in low-frequency region (upper left hand corner), but also increases the detail loss of quantization. Less localized quantization (larger shifting size) concentrates the non-zero elements less, but it reduces its detail loss of quantization. In this study, the step size is fixed at 1 and shifting sizes as large as 128 and as small as 4 are used.

5. Image Encoding [5]

Image encoding is the final step in the image compression process where integer values of the quantized image are converted to binary symbols that require on the average less storage than the integer representation. Zero and positive integer less than 255 require only 1 byte of storage, while all other integer values require at least 2 bytes of storage. Binary symbols are composed of sequences of '0' and '1' each requiring only a single bit of storage. Thus, the bit requirements for the storage of a binary symbol equal the number of 0's and 1's that compose it. The object of encoding is to assign short binary symbols to integer values with relatively high probability (frequency of occurrence) and longer binary symbols to integer values with relatively low probability.

There are a number of different encoding schemes available. This study deals with the variable length coding (VLC) and the run length coding (RLC) methods. The VLC method assigns binary symbols to the integer values the length of which is a function of its probability value. The RLC methods condenses streams of '0' values to special symbols called "tokens" whose binary representation require less average storage than the assignment of a single bit to each of the consecutive 0's. The VLC methods is applied to the non-zero values while the RLC method is applied to the '0' values. In this study, a combination of these two methods was used; taking advantage of the quantization effect of concentrating non-zero integer values in a small region in the upper left hand corner of the image.

In the case of the VLC method, a histogram analysis was first performed on the quantized image in order to determine the probability of all integer values. For the elements with the highest probability, the binary symbol '0' was used while for the elements with the second highest probability, the binary symbol '1' was used. The elements with the third, fourth, fifth, and sixth highest probabilities were assigned the binary symbols '00', '01', '10', and '11', etc. For those values of lowest probabilities, longer binary symbols were assigned. However, the average number of bits per binary symbol was always equal to the total entropy of the encoded image. Thus, the long binary symbols assigned to the integers with low probability did not significantly increase the average storage requirements. Therefore, the RLC method reduces the number of 0's. If, for example, the following pattern of integers occurs; "1324000000000345", then the token (0:9) is assigned to the 9 consecutive 0's. The resulting representation is 1324(0:9)345. The token (0:9) will be converted to a binary symbol using the VLC method. When quantized integers and tokens for the sequences of 0's have been encoded, the data can be stored in a binary stream in which each binary symbol is distinguished from its adjacent symbols by a special prefix code. The quantized image can be recovered without loss by reversing this coding operation.

6. Results

Table 1 shows the performance of five transform coding methods for six different quantizers. The comparison was made on the basis of image size, compression ratio, signal to noise ratio (SNR) and mean square error (MSE).

Table 1. Comparison Table

Comparison of Compressed Size						
Shifting size	128	64	32	16	8	4
DCT	54332	45631	37620	28412	18849	11989
Walsh	58148	48980	40463	31883	21965	13784
Hadamard	58621	49228	40288	32529	23386	14297
Haar	30128	23101	15951	9610	5660	3966
Daubechie 4	47384	39782	31585	23564	16273	10918
Comparison of Compression Ratio						
DCT	1.20:1	1.43:1	1.74:1	2.30:1	3.49:1	5.46:1
Walsh	1.12:1	1.33:1	1.62:1	2.05:1	2.98:1	4.75:1
Hadamard	1.11:1	1.33:1	1.62:1	2.01:1	2.80:1	4.58:1
Haar	2.17:1	2.83:1	4.11:1	6.82:1	11.58:1	16:53
Daubechie 4	1.38:1	1.64:1	2.07:1	2.78:1	4.02:1	6.00:1
Comparison of Signal to Noise Ratio						
DCT	22980.51	8304.59	2437.27	698.75	266.15	138.63
Walsh	23079.83	8303.37	2439.89	662.4	219.03	100.25
Hadamard	22887.14	8293.45	2432.08	665.41	193.79	76.03
Haar	1558.61	620.81	224.88	90.17	42.76	24.26
Daubechie 4	28316.56	9668.74	2968.09	957.77	351.14	160.97
Comparison of Mean Square Error						
DCT	0.75	2.08	7.09	24.72	64.91	124.62
Walsh	0.75	2.08	7.08	26.08	78.87	172.23
Hadamard	0.75	2.08	7.1	26.36	89.15	227.23
Haar	11.08	27.83	76.82	191.85	404.06	712.25
Daubechie 4	0.61	1.79	5.82	18.04	44.2	107.32

In Fig. 5, we present an example of an image subjected to five transform methods and a non-uniform quantizer corresponding to shifting size = 4.

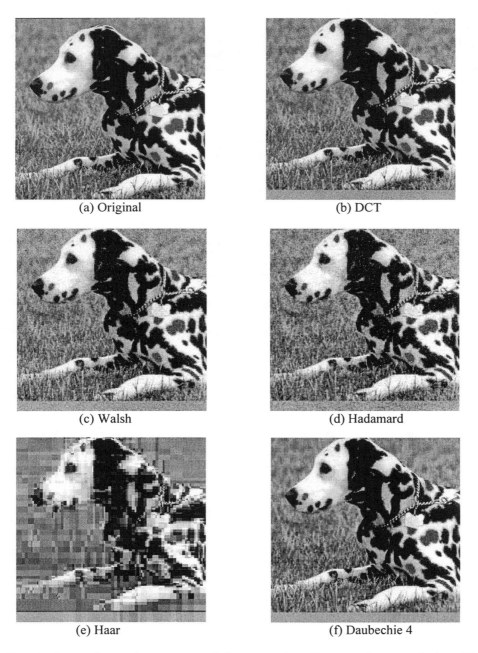

(a) Original

(b) DCT

(c) Walsh

(d) Hadamard

(e) Haar

(f) Daubechie 4

Fig. 5. Comparison of reconstructed images using five transform methods with quantization parameters shifting size = 4 and step size = 1. The image dimensions are 256 x 256 (65536 bytes) with 256 gray levels.

7. Discussion and Conclusion

As the table in result section shows, the 2D D4 WT outperformed all other transform methods in signal to noise ratio (SNR) and mean square error (MSE) and was second to the 2D Haar in data compression. The three standardized transforms had nearly equal performance rating, and the 2D Haar WT performed the best in the area of data compression and worst in the other areas.

Acknowledgment

Fig. 5 (a) is the courtesy of Eastman Kodak Co.

References

1. Vetterli M and Kovacevic J, *Wavelet and subband coding*. Englewood Cliffs: Prentice Hall PTR, 1995
2. Walter GG, *Wavelets and other orthogonal system with applications*. Boca Raton: CRC Press, 1994
3. Weeks AR, *Fundamentals of electronic image processing*. Bellingham: SPIE Press, 1996
4. Nelson M and Gailly JL, *The data compression book,* New York: *M&T Books*, 1996
5. Jain AK, Fundamentals of digital image processing, Englewood Cliffs: Prentice Hall, 1989

25

Effective Image Expansion Using Subband Filterbanks

Vassilis Alexopoulos, Anastasios Delopoulos and Stefanos Kollias

Computer Science Division,

Department of Electrical Engineering,

National Technical University of Athens,

Athens GR-15773, GREECE

1   Introduction

In a wide range of image processing applications digital images available in a certain resolution level have to be expanded to larger dimensions. This expansion is required to obey certain constraints related to the smoothness of the produced image, the similarity of the latter to an original continuous space image and the complexity of the employed expansion algorithm.

Typical areas of applications that utilize image expansion techiques, include but are not limitted to:

- Processing of sattelite or aerial images of earth landscape. Creation, for example, of high detail maps requires magnification (i.e., expansion) of the entire image or of particular regions of interest. In this case improved resolution is necessary for the better understanding of the map, particularly for automatic or semi-automatic region characterization in digital chartography

277

applications.

- Analysis and processing of medical images. Enlargement of x-ray, CT and ultrasonic images is necessary for easier visual interpretation and diagnosis.

- Computer graphics, texture mapping in 3-D graphics and digital typography [4].

In its general formulation the problem of image expansion is ill-posed. Certainly, starting from an $M \times N$ image, i.e., MN samples it is impossible to reconstruct a $qM \times qN$ ($q > 1$) image of $q^2 MN > MN$ samples in total, unless some a priori knowledge is available regarding the nature of the specific images.

In fact, the most celebrated a priori assumption usually made is that of Nyquist Criterion (see e.g. [2]). Under this assumption the original $N \times N$ image is considered as a uniformly sampled version of a band-limited continuous space image with sampling frequency at least twice the maximum frequency of the latter. If this condition is fulfilled the expansion procedure can be performed by virtually re-synthesizing the original continuous space image and then re-sampling it with a sampling rate q times the original sampling rate.

Although this approach is mathematically concrete its practical application is limited for two main reasons: (i) Nyquist Constraint is not always met in practice and particularly it is not met localy, (ii) the computational load needed for the interpolation via the inverse formula imposed by Nyquist theory is extremely high.

In practice, zero-order hold, linear and cubic interpolation are usually employed corresponding to fitting a polynomial of zero-th, first and third degree respectively to the available sample values [5], [1], [2], [3], [14]. This family of interpolation methods is of relatively low complexity but does not take into account the statistical properties of the particular image.

Recently a new interpolation technique was developed by Schultz and Stevenson [6] that assuming a Gibbs distribution of the pixel data performs optimal interpolation in the sense of maximizing a likelihood function of the reconstructed high resolution image.

Present work can be considered as an extension and enhancement of [6] in the sense that (i) we introduce a more a accurate model for the relation between the original low- and resulting high-resolution images, (ii) offer a computationally attractive algorithm for the implementation of the interpolation procedure. Both (i) and (ii) are succeeded via subband filter representations and operations respectively.

2 Problem Formulation - Image Modeling

Let $x(m, n)$ be an $M \times N$ discrete space gray-scale image with $(m, n) \in [0 \cdots M - 1] \times [0 \cdots N-1]$. For an integer $q > 1$ we seek a higher-resolution version of $x(m, n)$, denoted as $y(m, n)$ such that $(m, n) \in [0 \cdots qM-1] \times [0 \cdots qN-1]$. Equivalently, for a given 1-D column vector $\mathbf{x} \triangleq vec\{x(m, n)\}$ of length MN we seek $\mathbf{y} \triangleq vec\{y(m, n)\}$ of length $q^2 MN$. The operator $vec\{a(m, n)\}$ denotes stacking of the columns of matrix $a(m, n)$.

As already mentioned in the introduction, $y(m, n)$ cannot be uniquely determined unless we pose certain constraints. In the sequel we introduce the following requirements regarding $y(m, n)$ in order to obtain a unique solution:

[A1] The original low-resolution image $x(m, n)$ is related to $y(m, n)$ through a decimating operation consisting of an ideal low-pass filtering followed by $q : 1$ subsampling, i.e.,

$$x(m, n) = \sum_{k,l} h(k, l) y(qm - k, qn - l) \tag{1}$$

Equation (1) corresponds to subband filtering of $y(m, n)$ (see e.g. [13]). We further assume that the decimation filter is seperable, i.e.,

$$h(k, l) = h_c(k) h_r(l) , \tag{2}$$

and each of the 1-D column- and row-filters is ideal low pass with normalized cut-off frequencies $f_{max} = \frac{\pi}{q}$.

Using vector notation, eq. (1) can be re-written, ignoring border inaccuracies, in the form,

$$\mathbf{x} = \mathbf{H}\mathbf{y} \ , \tag{3}$$

where the $MN \times q^2 MN$ decimation matrix $\mathbf{H}$ can be expressed as the Kronecker product of two matrices representing column and row-wise decimation:

$$
\begin{aligned}
\mathbf{H} &= \mathbf{H}_c \otimes \mathbf{H}_r \\
\mathbf{H}_c(i,j) &= h_c(qi - j) \quad i, j = 0, \cdots N - 1 \\
\mathbf{H}_r(i,j) &= h_r(qi - j) \quad i, j = 0, \cdots M - 1
\end{aligned}
$$
$$\tag{4}$$

[A2] High resolution image $y(m, n)$ maximizes the log-likelihood function,

$$
\begin{aligned}
L(\mathbf{y}|\mathbf{x}) &= log(\ Prob(\mathbf{y}|\mathbf{x})) \\
&= log(\frac{Prob(\mathbf{x}|\mathbf{y})Prob(\mathbf{y})}{Prob(\mathbf{x})}) \\
&= log(Prob(\mathbf{x}|\mathbf{y})) + log(Prob(\mathbf{y})) - log(Prob(\mathbf{x}))
\end{aligned}
\tag{5}
$$

Under assumption [A1], $Prob(\mathbf{x}|\mathbf{y}) = 1$ provided that $\mathbf{x} = \mathbf{H}\mathbf{y}$ and 0 else. Hence, within the class of $\mathbf{y}$ that satisfy eq. (3), the maximizer of the RHS of eq. (5) is,

$$\mathbf{y}_0 = argmax_{\mathbf{y}} log(Prob(\mathbf{y})) \tag{6}$$

Maximization of (6) requires an expression of $Prob(\mathbf{y})$ w.r.t $\mathbf{y}$. To this end we adopt Gibbs probability distribution function [7],

$$Prob(\mathbf{y}) = \frac{1}{Z} exp\left(-\frac{1}{\lambda} \sum_{c \in C} \sum_{m=0}^{qM-1} \sum_{n=0}^{qN-1} \rho(\mathbf{d}_c[\mathbf{y}](m, n)) \right) \tag{7}$$

In the above expression Z is a normalization factor, λ is the temperature coefficient of Gibbs distribution, $\mathbf{d}_c[.]$, $c \in C$ is a set of linear operators that applied on $\mathbf{y}$ produce output vectors corresponding to directional derivatives of $y(m, n)$, C is the set of directions used (also known as clices) and $\rho(.)$ is a point-wise function of its vector argument. The effect of each $\mathbf{d}_c[.]$ on $\mathbf{y}$ is equivalent to,

$$\mathbf{d}_c[\mathbf{y}](m, n) \overset{\triangle}{=} [D_c * y](m, n) \tag{8}$$

where $*$ denotes 2-D convolution of $y(m, n)$ by the matrix mask D_c. Typically, $c \in C = \{0, 1, 2, 3\}$ and

$$
D_0 = \begin{bmatrix} 0 & 0 & 0 \\ 1 & -2 & 1 \\ 0 & 0 & 0 \end{bmatrix} \qquad D_1 = \begin{bmatrix} 0 & 0 & 1/2 \\ 0 & -1 & 0 \\ 1/2 & 0 & 0 \end{bmatrix}
$$

$$
D_2 = \begin{bmatrix} 1/2 & 0 & 0 \\ 0 & -1 & 0 \\ 0 & 0 & 1/2 \end{bmatrix} \qquad D_3 = \begin{bmatrix} 0 & 1 & 0 \\ 0 & -2 & 0 \\ 0 & 1 & 0 \end{bmatrix}
$$

Regarding $\rho(.)$ we adopt hereafter the Huber - Markov function, [12], [10], defined as

$$
\rho(x) = \left\{ \begin{array}{ll} x^2 & |x| \leq T, \\ T^2 + 2T(|x| - T) & |x| > T. \end{array} \right\} , \tag{9}
$$

where T is a scalar threshold determining the imposed smoothness of the reconstructed image $y(m, n)$, [11], [8], [9].

In view of (6) and (7) the optimally expanded image $\mathbf{y}$ is the minimizer of,

$$
G_y = \sum_{c \in C} \sum_{m=0}^{qM-1} \sum_{n=0}^{qN-1} \rho(\mathbf{d}_c[\mathbf{y}](m, n)), \tag{10}
$$

under the constraint (3).

3   Expansion Iteration

Since $\rho(.)$ is convex the global minimum of G_y can be obtained via constrained gradient search using the recursion,

$$
y_{n+1} = y_n - \mu P g_n, \tag{11}
$$

where $g_n \triangleq \left. \frac{\partial G_y}{\partial y} \right|_{\mathbf{y} = \mathbf{y}_n}$ is the gradient of the cost function G_y at y_n and the projection matrix $P = I - H^T (H H^T)^{-1} H$ guarantees that in each step the resulting y_n

fulfils the constraint (3). Scalar μ determines the speed of steepest descent and the convergence properties of the recursion.

Even for relatively small sized images x and y computations required in each step of (11) involve extremely large matrices P and $\mathbf{H}$. In the sequel we derive computationally atractive equivalents of (11) by converting matrix multiplications into linear filtering and up/sub-sampling operations.

3.1 EFFICIENT COMPUTATION OF g_n

Computation of g_n in each iteration $n = 0 \cdots$ turns out to be simple. In fact, we show in the sequel that it corresponds to two filtering operations and a hard-limiter.

Combining eqs. (10) and (8) we obtain the following expression for the gradient g_n:

$$
\begin{aligned}
g_n(i,j) &= \left. \frac{\partial}{\partial y(i,j)} \sum_{c \in C} \sum_{k=0}^{qM-1} \sum_{l=0}^{qN-1} \rho(D_c * y)[k,l] \right|_{y=y_n} \\
&= \sum_{c \in C} \sum_{k=0}^{qM-1} \sum_{l=0}^{qN-1} \left. \frac{\partial \rho(s)}{\partial s} \right|_{s=(D_c*y_n)[k,l]} \left. \frac{\partial (D_c * y)[k,l]}{\partial y(i,j)} \right|_{y=y_n} \\
&= \sum_{c \in C} \sum_{k=0}^{qM-1} \sum_{l=0}^{qN-1} \dot{\rho}((D_c * y_n)[k,l]) D_c(k-i, l-j) \\
&= \sum_{c \in C} \tilde{D}_c * \dot{\rho}(D_c * y_n)[i,j] \,,
\end{aligned}
\tag{12}
$$

where $\tilde{D}_c(m,n) = D_c(-m,-n)$ and

$$
\dot{\rho}(x) = \frac{\partial \rho(x)}{\partial x} = \left\{ \begin{array}{ll} -2T & x < -T \\ 2x & -T \le x \le T \\ 2T & x > T \end{array} \right\}
\tag{13}
$$

Consequently g_n can be computed in each iteration by successively filtering $y_n(i,j)$ by each mask D_c, then applying the hard-limiter $\dot{\rho}(.)$ and finally filtering by the mask $\tilde{D}_c$.

3.2 EFFICIENT COMPUTATION OF Pg_n

As indicated by (11) the next step is to project g_n on the subspace of P. Under the assumption that $h(m, n)$ is a separable ideal low-pass filter it can be proved that the general form of P reduces to,

$$P = I - \mathbf{H}^T \mathbf{H} . \qquad (14)$$

In view of (14) the projection Pg_n can be computed as,

$$Pg_n = g_n - \mathbf{H}^T (\mathbf{H}g_n) . \qquad (15)$$

The second term of the above expression can be computed by successively applying to g_n a decimation and an interpolation mechanism since it obviously resembles the structure of eq. (3). More precisely, $\mathbf{H}g_n$ is computed by applying the ideal low-pass filter $h(m, n)$ on g_n and the subsampling by q. As a next step, $\mathbf{H}^T(\mathbf{H}g_n)$ can be computed by first up-sampling the output of the previous operation (inserting $q - 1$ zero-rows and columns) and then applying the same low pass filter $h(m, n)$. Both decimation and interpolation can be implemented very efficiently in terms of computation speed using polyphase implementation.

4   Simulation Results

In order to demonstrate the performance of the proposed method we applied the expansion procedure to the 128×128 image of Figure 1.a. This image was created by synthetically re-sizing the 256×256 image of Figure 1.b. We next applied the proposed recursion (with $q = 2$) in order to reconstruct the high resolution image based on its low-resolution version. The resulting image is illustrated in Figure 1.c. For comparison purposes Figure 1.d illustrates the output of a cubic interpolation algorithm implemented by using a typical commercial software package. The image expanded by the proposed scheme is closer to the original high-resolution image in terms of Mean Squared Error and free of pixel artifacts introduced by the cubic interpolation.

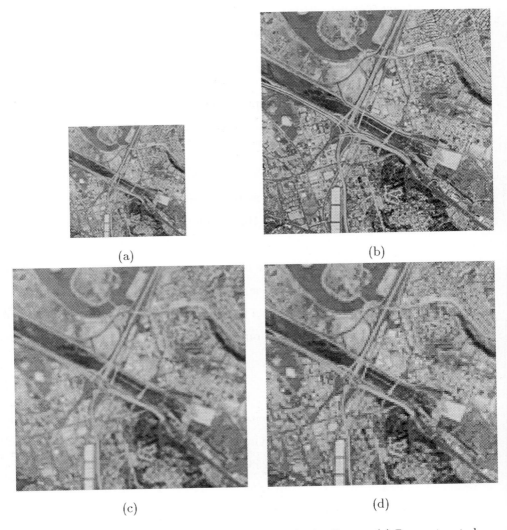

(a)

(b)

(c)

(d)

Figure 1: (a) Low Resolution Image, (b) High Resolution Image, (c) Reconstructed High Resolution Image, (d) Conventionally Reconstructed High Resolution Image

5 Conclusions - Further Research

We introduced a new maximum likelihood method for increasing the resolution of digital images. The maximization procedure is carried out in a recursive manner by a constrained gradient search algorithm. In each iteration both the gradient and its projection on the subspace imposed by the constrain are computed very efficiently via linear filtering, and down/up-sampling.

Extensions of the proposed scheme include treatment of noisy low-resolution images and also the case that more than one instances of a low-resolution image are available as in the case of video sequences.

References

[1] A.J. Parker, R.V. Kenyon and D.E. Troxel, "Comparison of interpolating methods for image resampling," IEEE Trans Med. Imaging, vol. MI-2, no. 1, pp. 31-39, 1983.

[2] D.E. Dudgeon and R.M. Mersereau, Multidimensional digital signal processing, Prentice hall, 1984.

[3] A.K. Jain, Fundamentals of Digital Image Processing, Prentice Hall, 1989.

[4] J.D. Foley, A. van Dam and J.F. Hughes, Computer graphics: principles and practice, Addison-Wesley, 1990.

[5] H.H. Hou and H.C. Andrews, "Cubic splines for image interpolation and digital filtering," IEEE Trans. ASSP, vol. ASSP-26, no. 6, pp. 508-517, 1978.

[6] R.R. Schultz and R.L. Stevenson, "A Bayesian Approach to Image Expansion for Improved Definition," IEEE Trans. Image Proc., vol. 3, no.3, pp. 233-241, May 1994.

[7] S. Geman and D. Geman, "Stochastic relaxation, Gibbs distributions and the Bayesian restoration of images," IEEE Trans. PAMI, vol. 6, no. 6, pp. 721-741, 1984.

[8] C. Bouman and K. Sauer, A generalized Gaussian image model for edge-preserving MAP estimation," IEEE Trans. Image Processing, vol. 2, no. 3, July 1993.

[9] R.L. Stevenson, B.E. Schmitz and E.J. Delp, "Discontinuity preserving regularization of inverse visual problems," IEEE Trans Syst. Man Cybern., vol. 24, no. 3, Mar. 1994.

[10] P.J. Huber, Robust Statistics, New York: Wiley, 1981.

[11] R.R. Schultz and R.L. Stevenson, "Improved definition image expansion," Proc. ICASSP-92, vol. III, pp. 173-176, 1992.

[12] A.N. Tikhonov and V.Y. Arsenin, Solutions of ill-posed problems, Washington DC: V.H. Winston, 1977.

[13] M. Vetterli and H. Kovacevic, Wavelets and Subband Coding, Prentice Hall, New Jersey, 1995.

[14] N.B. Karayiannis and A.N. Venetsanopoulos, "Image interpolation based on varational principles," Signal Processing, vol. 25, no. 3, pp. 259-288, 1991.

26

Color Image Segmentation for Multimedia Applications

N. Ikonomakis†, K. N. Plataniotis‡, A. N. Venetsanopoulos†

†*Digital Signal & Image Processing Lab,*
Department of Electrical and Computer Engineering,
University of Toronto,
10 King's College Rd.,
Toronto, Ontario, M5S 3G4, Canada.

‡*School of Computer Science,*
Ryerson Polytechnic University,
350 Victoria Street,
Toronto, Ontario, M5B 2K3, Canada.

1. Introduction

Image segmentation refers to partitioning an image into different regions that are homogeneous or "similar" in some image characteristics. It is usually the first task of any image analysis process module, and thus, subsequent tasks rely heavily on the quality of segmentation. The quality of segmentation determines the eventual success or failure of the analysis. For this reason, considerable care is taken to improve the probability of a successful segmentation.

From the applications point of view, digital photography, electronic imaging, digital television, image libraries, and multimedia have given image segmentation 'a boost', so to speak, so that the field has become a principal area of research not only in electrical engineering but also in other academic disciplines such as computer science, geography, medical imaging, criminal justice, and remote sensing. Image segmentation has taken a central place in numerous applications, including, but not limited to, multimedia databases, color image and video transmission over the Internet, digital broadcasting, interactive TV, video-on-demand, computer-based training, distance education, video-conferencing and tele-medicine, and with the development of the hardware and communications infrastructure to support visual applications.

Most attention on image segmentation has been focused on grey-scale (or monochrome) image segmentation. A common problem in segmentation of a grey-scale image occurs when an image has a background of varying grey level, such as gradually changing shades, or when regions assume some broad range of grey levels. This problem is inherent since intensity is the only available information from monochrome images. It is known that the human eye can detect only in the neighborhood of one or two dozen intensity levels at any point in a complex image due to brightness adaptation, but can differentiate thousands of color shades and intensities [1].

There are currently a large number of color image segmentation techniques that are available. These can be categorized into three general groups: pixel-based techniques, edge-based techniques, region-based techniques, and model-based techniques. These techniques are either based on concepts of similarity (edge-based) or on discontinuity (pixel-based and region-based) of pixel values.

Region-based segmentation algorithms include region growing, merging and splitting/merging techniques. The focus of this chapter will be on region-based segmentation. In particular, we will examine a region growing and region merging technique. These region-based segmentation techniques will be examined on multimedia-based color images. In particular, results on video-conferencing type color images will be discussed. It will be shown that these interactive techniques have the advantage of controlling the degree of segmentation which is related to the number of regions (i.e. the number of colors) that are in an image. This is a useful property in every image process module.

2. Techniques of Color Image Segmentation

Because of the uncertainty problems encountered while trying to model the human visual system, there are currently a large number of image segmentation techniques that are available. However, no general methods have been found that perform adequately across a varied set of images. The early attempts at grey-scale image segmentation are based on three techniques: pixel-based, region-based, and edge-based techniques. Even though these techniques were introduced three decades ago, they still find great attention in color image segmentation research today. Model-based segmentation techniques have also become popular in the last decade. The following sections will survey the various techniques of color image segmentation starting with pixel-based, edge-based, region-based and finally model-based techniques.

2.1. PIXEL-BASED TECHNIQUES

Pixel-based techniques do not consider the spatial context but only decide solely on the basis of the color features at individual pixels [1]-[6]. This attribute has its advantages and disadvantages. Simplicity of the algorithms are an advantage to pixel-based techniques while lack of spatial constraints make them susceptible to noise in the images. Model-based techniques which utilize spatial interaction models to model images are used to further improve pixel-based techniques. The simplest technique of pixel-based segmentation is histogram thresholding. It is one of the oldest and most popular techniques for image segmentation. If an image is composed of distinct regions, the color histogram of the image usually shows different peaks, each corresponding to one region and adjacent peaks are likely to be separated by a valley. For example, if the image has a distinct object on a background, the color histogram is likely to be bimodal with a deep valley. In this case, the bottom of the valley is taken as the threshold so that pixels that belong above and below this value on the histogram are grouped into different regions. This is called bilevel thresholding [2]. For multithresholding the image is composed of a set of distinct regions. In this case, the histogram has one or more deep valleys and the selection of the thresholds becomes easy because it becomes a problem of detecting valleys. However, detection of the valleys is not a trivial job.

Clustering is another pixel-based technique that is extensively used for image segmentation [7]-[12]. The rationale of the clustering technique is that, typically, the colors in an image tend to form clusters in the histogram, one for each object in the image. In the clustering-based technique, a histogram is first obtained by the color values at all pixels and the shape of each cluster is found. Then, each pixel in the image is assigned to the cluster that is closest to the pixel color. Many different clustering algorithms are in existence today [13][14]. Among these, the K-means and the fuzzy K-means algorithms have received extensive attention [9]-[12].

The pixel-based segmentation techniques surveyed above do not consider spatial constraints which make them susceptible to noise in the images. The resulting segmentation often contains isolated, small regions that are not present in noise-free images. In the past decade, many researchers have included spatial constraints in their pixel-based segmentation techniques using statistical models (Section 2.4).

2.2. EDGE-BASED TECHNIQUES

Edge-based techniques [1][15]-[23] focus on the discontinuity of a region in the image. Most edge detection techniques are based on finding maxima in the first derivative of the image function or zero-crossings in the second derivative of the image function. Edge-based segmentation techniques are very sensitive to texture variations and impulsive noise.

In a monochrome image, an edge is defined as an intensity discontinuity. Early approaches to color edge detection comprise of extensions from monochrome edge detection techniques which utilize the Sobel operator for finding the first derivative or the Laplacian or Mexican Hat operators to find the second derivative. These techniques were applied to the each of the color components independently and then the results were combined using certain logical operations [15].

One common problem with the approaches mentioned previously is that they fail to take into account the correlation among the color channels, and as a result, they are not able to extract certain crucial information revealed by color. For example, they tend to miss edges that have the same strength but in opposite direction in two of the three color components. Consequently, the approach to treat the color image as vector space has been proposed [17]-[23]. A color image can be viewed as a two-dimensional three-channel vector field which can be characterized by a discrete integer function $\vec{f}(x, y)$. The value of this function at each pixel is defined by a three dimensional vector in a given color space.

The vector gradient operator employs the concept of a gradient operator on a three channel color vector space. Di Zenzo [17] proposed a combination of three chromatic gradients for getting a global gradient. He implemented this operator in the RGB color space.

2.3. REGION-BASED TECHNIQUES

Region-based techniques [1][24]-[27] focus on the continuity of a region in the image. Segmenting an image into regions is directly accomplished through region-based segmentation which makes it one of the most popular techniques used today. Unlike the pixel-based techniques region-based techniques consider both color distribution in color space and spatial constraints. Standard techniques include region growing and split and merge techniques. Region growing is the process of grouping neighboring pixels or a collection of pixels of similar properties into larger regions. The split and merge technique constitutes iteratively splitting the image into smaller and smaller regions and testing to see if adjacent regions need to be merged into one. The process of merging pixels or regions to produce larger regions is usually governed by a homogeneity criterion, such as a distance measure linked to color similarity.

Region growing is the process of grouping neighboring pixels or a collection of pixels of similar properties into larger regions. Testing for similarity is usually achieved through a homogeneity criterion. Quite often after an image is segmented into regions using a region growing algorithm regions are furthered merged for improved results. A region growing algorithm typically starts with a number of *seed* pixels in an image and from these grows regions by iteratively adding unassigned neighboring pixels that satisfy some homogeneity criterion with the existing region of the seed pixel. That is, an unassigned pixel neighboring a region, that started from a seed pixel, may be assigned to that region if it satisfies some homogeneity criterion. If the pixel is assigned to the region, the pixel set of the region is updated to include this pixel. Region growing techniques differ in choice of homogeneity criterion and choice of seed pixels. Several homogeneity criteria linked to color similarity or spatial similarity can be used to analyze if a pixel belongs to a region. These criteria can be defined from local, regional, or global considerations. The choice of seed pixel can be supervised or unsupervised. In a supervised method the user chooses the seed pixels while in an unsupervised method the choice is made by the algorithm.

As opposed to the region growing technique of segmentation, where a region is grown from a seed pixel, the split and merge technique subdivides an image initially into a set of arbitrary, disjointed regions and then merge and/or split the regions in an attempt to satisfy a color homogeneity criterion between the regions. Gonzales and Wood [1] describe a split and merge algorithm that iteratively works toward satisfying these constraints. They describe the split and merge algorithm for grey-scale images proposed by Horowitz and Pavlidis [24]. It will be described here for color images. The image is subdivided into smaller and smaller quadrant regions so that for each region a color homogeneity criterion holds. That is, if for region R_i the homogeneity criterion does not hold, divide the region into four subquadrant regions, and so on. This splitting technique may be represented in the form of a so-called *quadtree* (that is, a tree in which each node has exactly four descendants). The quadtree data structure is the most common used data structure in split and merge algorithms because of its simplicity and computational efficiency [28]. Note that the root of the tree corresponds to the entire image. Merging of adjacent subquadrant regions is allowed if they satisfy a homogeneity criterion. The procedure may be summarized as:

1. Split into four disjointed quadrants any region where a homogeneity criterion does not hold.

2. Merge any adjacent regions that satisfy a homogeneity criterion.

3. Stop when no further merging or splitting is possible.

Most split and merge approaches to image segmentation follow this simple procedure with varying approaches coming from different color homogeneity criteria.

2.4. MODEL-BASED TECHNIQUES

Recently, much work has been directed toward stochastic model-based techniques [29]-[38]. In such techniques, the image regions are modeled as random fields and the segmentation problem is posed as a statistical optimization problem. Compared to previous techniques, the stochastic model-based techniques often provide more precise characterization of the image regions. In fact, various stochastic models can be used to synthesize color textures that closely resemble natural color textures in real-world natural images [29]. Most of the techniques introduced use spatial interaction models like Markov Random Field (MRF) or Gibbs Random Field (GRF) to model digital images. The reports by Cross and Jain [29], Geman and Geman [30], Cohen and Cooper [31], Derin and Elliott [32], Lakshmanan and Derin [33], Panjwani and Healey [34], Liu and Yang [35], Pappas [36], and Chang *et al.* [37] all make use of the Gibbs distributions for characterizing MRF. Model-based techniques tend to be computationally intensive and thus are only used in such cases where complexity is not a problem.

3. Region Growing and Merging

Region growing and merging algorithms, which we have developed in the past [27], are employed to segment color images into disjoint regions by segmenting the chromatic pixels separately from the achromatic ones. The HSI (hue, saturation, intensity) color space is used in the algorithms. This color space is closely related to the way in which people describe the perception of color and is thus ideal for image segmentation [1].

The region growing approach starts with a set of seed pixels and from these grows regions by appending to each seed pixel those neighboring pixels that satisfy a certain homogeneity criterion which will be described later. The algorithm is summarized in the following steps:

1. The color values of the pixel are converted from the standard RGB (red, green, blue) color values to the HSI color values using well known transformation formulas.

2. Assign the first pixel of the image under consideration as the first seed pixel.

3. The seed pixel is compared to its 8-connected neighbors: eight neighbors of the seed pixel. Any of the neighboring pixels that satisfy a homogeneity criterion are assigned to its region and its pixel value would change to the seed pixel value.

4. Repeat the neighbor comparison step for every new pixel assigned to the first region until the region is completely bounded by the edge of the image or by pixels that do not satisfy the criterion.

5. The next seed pixel for the next region is determined by choosing the first unassigned (to the previously grown region) pixel while moving through the image in a right-to-left and top-to-bottom fashion. If there are any unassigned pixels left then go to step 3.

Steps 3 through 5 are repeated until every pixel in the image belongs to a region. The homogeneity criterion used is that if the value of the Square Euclidean Distance metric between the unassigned pixel and the seed pixel is less than a threshold value than the pixel is assigned to the region. That is, if:

$$\sum \left((H_s - H)^2 + (S_s - S)^2 + (I_s - I)^2\right) \tag{1}$$

is less than a certain threshold, where H, S, I and H_s, S_s, I_s are the HSI values of the unassigned pixel and the seed pixel, respectively, than the criterion is met and the pixel is assigned to the region.

The proposed color segmentation scheme is also applicable in all other color spaces, such as the RGB, Lab, and Luv. However, best results are given when performed in the HSI color space.

4. Results

A series of video-phone type color images were tested with the segmentation algorithm. Because of space limitations we present only the results for the 'claire' image. Figure 1 is the original 'claire' color image. This image is representative of the kinds of images encountered in important applications, such as video conferencing. The segmentation of such images is important because it reduces the transmission overhead by reducing the number of regions in the image. Experimental analysis, of various images, showed good results as shown in the Figure 2 and Figure 3. The threshold used to obtain the segmented image in Figure 3 is lower than the one used to obtain Figure 2. The segmented image can be improved further by changing the powers of all three HSI values along with the threshold value. A different form of (1) was once again attempted. This time the powers of the hue, saturation, and intensity differences in (1) were all variables. It was found that if more emphasis is placed on the saturation values of the pixels in the homogeneity criterion better results are obtained. Figure 4 shows a segmented claire image with the powers of the hue, saturation, and intensity differences changed to 1, 3, and 2 respectively. Arguably, this image is one of the best segmented images of 'claire'.

5. Conclusion

The proposed method of image segmentation via region growing and region merging was shown to be very effective when an image needed to be partitioned into different homogeneous regions. Usually, this is the first task necessary in an image analysis process module and therefore needs to be effective since it determines the eventual success or failure of the analysis.

The segmentation method proposed is interactive. The threshold value and the three variable HSI power values can all be adjusted to control the number of regions that a segmented image is to contain. This allows for control of the degree of segmentation.

294

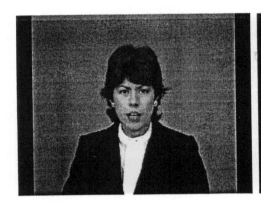

Figure 1. *Original "claire" image*

Figure 2. *Segmented image*
(threshold=0.1)

Figure 3. *Segmented image*
(threshold=0.05)

Figure 4. *Segmented image for variable*
HSI powers (H power of 1, S
power of 3, and I power of 2)
(threshold=0.1)

References

[1] Gonzales, R.C., Wood, R. E., Digital Image Processing. Addison-Wesley, Massachusetts, 1992.

[2] Pal, N., Pal, S.K., "A Review on Image Segmentation Techniques", *Pattern Recognition*, vol. 26, no. 9, 1993, pp. 1277-94.

[3] Ohlander, R., Price, K., Reddy, D.R., "Picture Segmentation Using A Recursive Splitting Method", *Computer Graphics and Image Processing*, vol. 8, 1978, pp. 313-33.

[4] Ohta, Y., Kanade, T., Sakai, T., "Color Information for Region Segmentation", *Computer Graphics and Image Processing*, vol. 13, 1980, pp. 222-41.

[5] Holla, K., "Opponent Colors as a 2-Dimensional Feature within a Model of the First Stages of the Human Visual System", *Proc. of the 6th Int. Conf. on Pattern Recognition*, Munich, Germany, Oct 19-22, 1982, pp. 161-3.

[6] Tominaga S., "Color Image Segmentation Using Three Perceptual Attributes", *Proc. CVPR'86*, Miami Beach, Florida, USA, June 22-26, 1986, pp. 628-30.

[7] Tominaga, S., "A Color Classification Method for Color Images Using a Uniform Color Space", *Proc. 10th Int. Conf. on Pattern Recognition*, vol. 1, 1990, pp. 803-7.

[8] Celenk, M., "A Color Clustering Technique for Image Segmentation", *Computer Vision, Graphics, and Image Processing*, vol. 52, 1990, pp. 145-70.

[9] Weeks, A.R., Hague, G.E., "Color Segmentation in the HSI Color Space Using the K-means Algorithm" *Proceedings of the SPIE*, vol. 3026, 1997, pp.143-54.

[10] Huntsberger, T.L., Jacobs, C.L., Cannon, R.L., "Iterative Fuzzy Image Segmentation", *Pattern Recognition*, vol. 18, no. 2, 1985, pp. 131-8.

[11] Trivedi, M., Bezdek, J.C., "Low-level segmentation of aerial images with fuzzy clustering", *IEEE Transactions on Systems, Man, and Cybernetics*, vol. 16, no. 4, 1986, pp. 589-98.

[12] Lim, Y.W., Lee, S.U., "On the Color Image Segmentation Algorithm Based on the Thresholding and the Fuzzy c-Means Techniques", *Pattern Recognition*, vol. 23, no. 9, 1990, pp. 1235-52.

[13] Hartigan, J.A., Clustering Algorithms. John Wiley and Sons, USA, 1975.

[14] Tou, J., Gonzalez, R.C., Pattern Recognition Principles. Addison-Wesley Publishing, Massachusetts, USA, 1974.

[15] Koschan, A., "A Comparitive Study on Color Edge Detection", *Proc 2nd Asian Conference on Computer Vision, ACCV'95*, Singapore, 5-8 December, Vol. III, 1995, pp. 574-8.

[16] Marr, D., Hildreth, E., "Theory of Edge Detection", *Proc. of the Royal Society of London*, B207, 1980, pp. 187-217.

[17] Di Zenzo, S., "A Note on the Gradient of a Multi-image", *Computer Vision Graphics, Image Processing*, CVGIP, vol. 33, 1986, pp. 116-26.

[18] Chapron, M., "A New Chromatic Edge Detector used for Color Image Segmentation", *Proc. 11th Int. Conf. on Pattern Recognition*, ICPR, Vol III: Conf C, 1992, pp. 311-4.

[19] Trahanias, P.E., Venetsanopoulos, A.N., "Vector Order Statistics Operators as Color Edge Detectors", *IEEE Transactions on System, man and Cybernetic*, Part-B, vol. 26, no. 1, Feb. 1996, pp. 135-43.

[20] Scharcanski, J., Venetsanopoulos, A.N., Edge Detection of Color Images Using Directional Operators", *IEEE Transactions on Circuits and Systems for Video Technology*, vol. 7, no. 2, April 1997, pp. 397-401.

[21] Shiozaki, A., "Edge Extraction Using Entropy Operator", *Computer Vision Graphics, Image Processing*, CVGIP, vol. 33, 1986, pp. 116-26.

[22] Cumani, A., "Edge Detection in Multispectral Images", *CVGIP: Graphical Models and Image Processing*, vol. 53, 1991, pp. 40-51.

[23] Alshatti, W., Lambert, P., "Using Eigenvecors of a Vector Field for Deriving a Second Direcional Derivative Operator for Color Images", *Proc. of the 5th Int. Conf. on Computer Analysis of Images and Patterns*, CAIP'93, Budapest, Hungary, Sept. 1993, pp. 149-56.

[24] Horowitz, S.L., pavlidis, T., "Picture Segmentation by a Directed Split-and-Merge Procedure", *Proc. 2nd International Joint Conf. on Pattern Recognition*, Copenhagen, 1974, pp. 424-33.

[25] Gauch, J., Hsia, C., "A Comparison of Three Color Image Segmentation Algorithms in Four Color Spaces", *Proceedings of the SPIE: Visual Communications and Image Processing*, vol. 1818, 1992, pp. 1168-81.

[26] Tremeau, A., Borel, N., "A Region Growing and Merging Algorithm to Color Segmentation", *Pattern Recognition*, vol. 30, no. 7, 1997, pp. 1191-203.

[27] Ikonomakis, N., Plataniotis, K.N., Venetsanopoulos, A.N., "Grey-Scale and Colour Image Segmentation via Region Growing and Region Merging", *Canadian Journal of Electrical and Computer Engineering*, vol. 23, no. 1-2, 1998, pp. 43-7.

[28] Samet, H., "The Quadtree and Related Hierarchical Data Structures", *Computer Surveys*, vol. 16, no. 2, 1984, pp. 187-230.

[29] Cross, G.R., Jain, A.K., "Markov Random Field Texture Models", *IEEE Transactions on Pattern Analysis and Machine Intelligence*, vol. PAMI-5, Jan. 1983, pp. 25-39.

[30] Geman, S., Geman, D., "Stochastic Relaxation, Gibbs Distributions, and the Bayesian Restoration of Images", *IEEE Transactions on Pattern Analysis and Machine Intelligence*, vol. PAMI-6, Nov. 1984, pp. 721-41.

[31] Cohen, F.S., Cooper, D.B., "Simple, Parallel, Hierarchical, and Relaxation Algorithms for Segmenting Noncausal Markovian Random Field Models", *IEEE Transactions on Pattern Analysis and Machine Intelligence*, vol. PAMI-9, no. 2, Mar. 1987, pp. 195-219.

[32] Derin, H., Elliott, H., "Modeling and Segmentation of Noisy and Textured Images Using Gibbs Random Fields", *IEEE Transactions on Pattern Analysis and Machine Intelligence*, vol. PAMI-9, no. 1, Jan. 1987, pp. 39-55.

[33] Lakshmanan, S., Derin, H., "Simultaneous Parameter Estimation and Segmentation of Gibbs Random Field Using Simulated Annealing", *IEEE Transactions on Pattern Analysis and Machine Intelligence*, vol. PAMI-11, no. 8, Aug. 1989, pp. 799-813.

[34] Panjwani, D.K., Healey, G., "Markov Random Field Models for Unsupervised Segmentation of Textured Colour Images" *IEEE Transactions on Pattern Analysis and Machine Intelligence*, vol. 17, no. 10, Oct. 1995, pp. 939-54.

[35] Liu, J., Yang, Y.-H., "Multiresolution Color Image Segmentation", *IEEE Transactions on Pattern Analysis and Machine Intelligence*, vol. 16, no. 7, Jul. 1994, pp. 689-700.

[36] Pappas, T.N., "An Adaptive Clustering Algorithm for Image Segmentation", *IEEE Transactions on Signal Processing*, vol. 40, no. 4, Apr. 1992, pp. 901-14.

[37] Chang, M.M., Sezan, M.I., Tekalp A.M., "Adaptive Bayesian Segmentation of Colour Images", *Journal of Electronic Imaging*, vol. 3, no. 4, Oct. 1994, pp. 404-14.

[38] Baraldi, A., Blonda, P., Parmiggiani, F., Satalino, G., "Contextual Clustering for Image Segmentation", International Computer Science Institute, Berkeley, California, TR-98-009, Mar. 1998.

27

Transformations Between Pictures from 2D to 3D

Milan SIGMUND, Pavel NOVOTNY
Institute of Radioelectronics
Purkynova 118, 612 00 Brno
Czech Republic
E-mail: sigmund@urel.fee.vutbr.cz

1. Introduction

With a development of three-dimensional (3D) engraving machines controlled by personal computer started up a claim on a simply program for a fast data preparation for these machines. There are already programs for engraving on a market, but they allow only 2D engraving (applicable for texts or patterns) or they are large products those are user complicated and expensive. These absences a next program attempts.

2. Three-Dimensional Engraving

2.1. ORDER REQUIREMENTS

It was required to use a picture in grey scale for a three-dimensional engraving. A pixel brightness determines an engraving depth. A source picture was required in format BMP (*Windows Bitmap*). A program must convert the picture from 2D to 3D and the output model must convert to controlling commands for an engraving machine. The program will mainly use with machines MIMAKI (they use HP-GL language), a generating the G-code is also requirement.

2.2. ENGRAVING MACHINE MIMAKI ME-500

The plotter MIMAKI ME-500 is for the three-dimensional engraving. The plotter is a portal construction, an engraved material is chuck to a plane that is moving in y axis direction. The head with a spindle is moving in x axis direction. An engraving tool is chucked in the spindle, the tool is moving in z axis direction. The engraving is in a raster, default settings is 40 steps per mm in x-y axes and 100 steps per mm in z axis direction. The tool moving speed is possible to set independent in x-y axes and z axis direction.

The plotter is controlled by the HP-GL language version 2.0 [1] that is extended about 3D commands. The plotter is connected to a computer by a serial port.

Fig. 1: Engraving machine Mimaki ME-500.

2.3. LANGUAGE HP-GL (HEWLETT-PACKARD GRAPHICS LANGUAGE)

The HP-GL language is used for a vector controlling of 2D output devices. The language has a text syntax. To a device are sent only commands for draw (point, line, circle...) and these commands are processed by a processor in the device. It was important to add new commands for controlling the third axis for the machine MIMAKI, as follows:

!PZ   Command for setting upper and lower z co-ordinate.
!MC   Controlling the spindle motor.
!VZ Z axis speed setting.
ZM   3D tool move.

An example of commands in the HP-GL language is bellow:
IN;!MC1;VS40;!VZ20;ZM0,0,-
100;VS5;!VZ2;ZM0,0,0;ZM400,0,200;ZM400,400,0;ZM200,200,200;
VS40;!VZ20;ZM200,200,-100;ZM0,400,-100;!MC0;

2.4. G-CODE

This code is for a controlling CNC machines. It has a text-syntax, too. For controlling are used commands for 3D move only, the co-ordinates are in inches.

The same example of commands generated in the G-code:

```
G0 G90
F40.0000
G0 X0.0000 Y0.0000 Z0.0394
F5.0000
G1 X0.0000 Y0.0000 Z-0.0000
G1 X0.3937 Y0.0000 Z-0.0787
G1 X0.3937 Y0.3937 Z-0.0000
G1 X0.1969 Y0.1969 Z-0.0787
40.0000
G0 X0.1969 Y0.1969 Z0.0394
G0 X0.0000 Y0.3937 Z0.0394
```

3. Program *Photo Engrave 3D*

3.1. DESCRIPTION

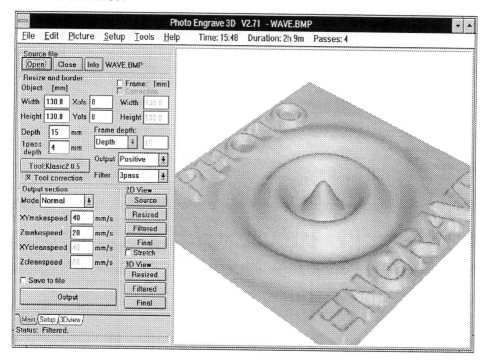

Fig. 2: The *Photo Engrave 3D* desktop.

The program loads a picture in grey scale in BMP format. The z co-ordinate is assigned to every pixel. A white pixel responds to a product surface and a black pixel is in an entered depth. Other pixels in grey scale are between these levels. It is possible to change an orientation of outer points, it is important to make a negative form. A frame round the product for a product clipping also can be added. The program engraves the 3D surface in

302

the raster, the engraving step is determines from the diameter of the tool peak and this value is scaled down about a lap, adjacent tracks will be overlapped. It is important to select the required tool from the tool library. The program also makes possible to select an engraving mode.

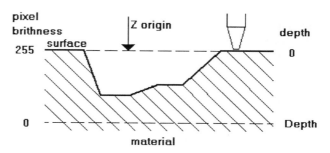

Fig. 3: Relation between pixel brightness and *z* co-ordinate.

The program was made in Delphi 1 for Windows 3.11 or later (16-bit). It is characterise by small hardware requirements and quick data preparation. The basic requirements are: processor 486 with 8MB RAM and display 1024x768x256 colours. This program makes possible to full change of language during execution. By default you can select English or Czech, but all texts in this program are in a text file and no problem is to translate the file to any other language [2].

3.2. PROGRAM BLOCK STRUCTURE

The whole conversion algorithm was divided into next blocks:

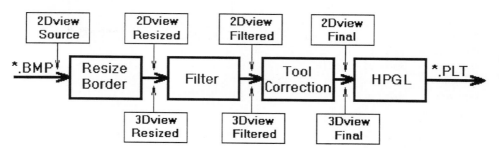

Fig. 4: Block diagram of the program *Photo Engrave 3D*.

There is a picture in the BMP format on the input of the whole algorithm. The first data processing is in the *Resize&Border* block in which is the source picture resized to the required size and step. In the next *Filter* block is the picture filtered for a cleaning the product surface. The *Tool Correction* block makes a data correction through a geometric tool dimensions. The last *HPGL* block converts prepared data to controlling commands of the machine. The data can be checked in output of every block and displayed as a picture or as a three-dimensional view, respectively.

3.3. BLOCKS DESCRIPTION

Block *Resize&Border*

This block is for a resizing of picture. The ratio of the required product size and the engraving step determines the picture size. The block makes possible to increase or decrease the picture size, a linear interpolation is used. If the engraving mode *Rough* is selected, it is not possible to use the interpolation. The program must search points with a minimum depth, it is important to cut out the material for a final cleaning. This block can invert the picture for the engrave a negative form and also can be added a frame round the product with an arbitrary depth.

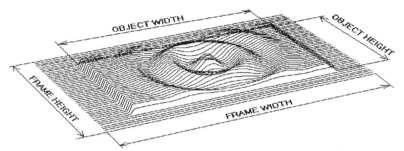

Fig. 5: The dimensions of the product and the frame.

Block *Tool Correction*

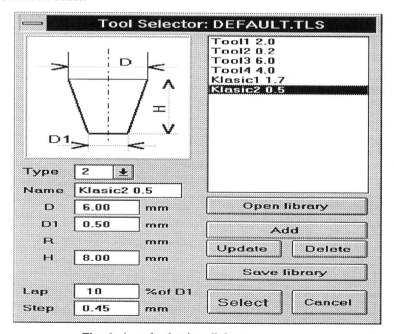

Fig. 6: A tool selection dialogue.

The *Tool Correction* block makes data correction through a geometric tool dimension. 3D tool model is made in a tool selection dialogue. This dialogue is for a selection of required tool type and its dimensions. We can also work with a tool library: add new tool, modify its parameters, delete tools, etc.

During the correction, it is tested how the tool can be let down to the material without a collision with the material round this point. If it is not able to let down the tool to full depth, the algorithm fills the rest under the tool.

The program makes possible to use four types of engraving tools. After the selection of the required tool is made its 3D model.

Tools are defined by the next equations (1) to (6):

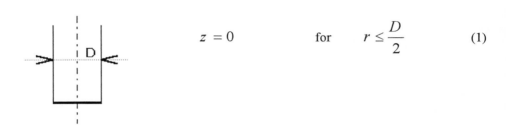

$$z = 0 \qquad \text{for} \qquad r \le \frac{D}{2} \qquad (1)$$

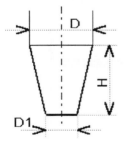

$$z = 0 \qquad \text{for} \qquad r \le \frac{D_1}{2} \qquad (2)$$

$$z = \frac{2*H*(R - \frac{D_1}{2})}{D - D_1} \qquad \text{for} \quad r \le \frac{D}{2} \qquad (3)$$

$$z = R - \sqrt{R^2 - r^2} \qquad \text{for} \qquad r \le R \qquad (4)$$

$$z = R - \sqrt{R^2 - r^2} \qquad \text{for } r \le R \qquad (5)$$

$$z = \frac{(H-R)*(r-R)}{\dfrac{D}{2} - R} + R \quad \text{for } r \le \frac{D}{2} \qquad (6)$$

Block *HPGL*

The main bitmap picture conversion to a vector language is made in the HPGL block. All calculations are in millimetres, only output values are printed in selected units. The engraving is in layers, it is important for hard materials. The thickness of the layer must be entered. This block also enables various engraving modes. It is possible to edit all parameters for this block, the output data format can be adapt to arbitrary machines. The Fig. 7 shows the dialogue for the machine and the language setup.

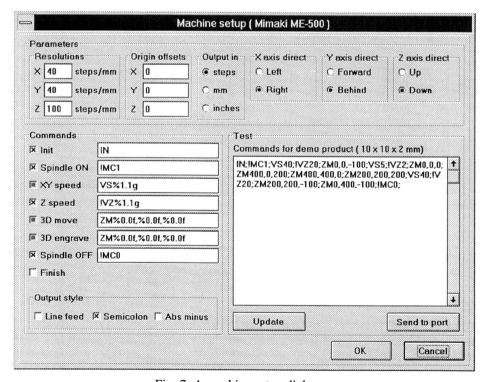

Fig. 7: A machine setup dialogue.

Block *Filter*

The resized picture is filtered in the next *Filter* block. The filtering using a 3x3 average filter is applicable for the surface cleaning. If needed, this filter can be used more times. For a soft filtering the 2x2 filter is more suitable.

3.4. ENGRAVING MODES

The engraving is possible in more different modes. Some of the modes are for abbreviating of engraving time and some of them are for surface cleaning. There is one mode for engraving a printed circuit directly from a picture.

Mode *Normal*

This mode is specified for making the final product. Engraving is in one direction only and in more passes. Maximum depth of one pass is determined by 1pass depth constant. The number of passes is displayed in a status line. *XYmakespeed* and *Zmakespeed* set engraving speeds.

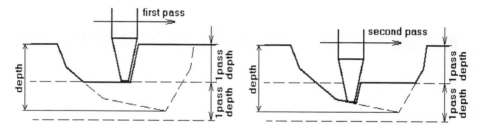

Fig. 8: The passes of the engraving.

Mode *Cleaning*

This mode is specified for cleaning surface of the product in *Y*-axis direction in one pass only. This mode is used after engraving in the *Normal* mode. These *XYcleanspeed* and *Zcleanspeed* constants determine cleaning speeds.

Mode *Normal+Cleaning*

This mode comprises the combination of the *Normal* and *Cleaning* modes. The product is made in the *Normal* mode and then in the *Cleaning* mode.

Mode *Rough*

This mode is used together with the next two modes in case of acceleration of engraving time. A rough tool must be selected. Engraving is making in more passes like in the *Normal* mode but the product is not engraved to full depth and a thin layer is kept for cleaning (*clearlayer*) in the *Final* mode. These *XYmakespeed* and *Zmakespeed* constants determine engraving speeds.

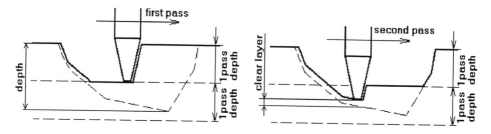

Fig. 9: The rough mode engraving passes.

Mode *Final*

Use the *Final* mode for finishing your product after engraving in the *Rough* mode. The thin layer will be cleaned. The engraving is in one pass with *XYmakespeed* and *Zmakespeed* speeds only.
After finishing the *Rough* mode chuck a thin tool, select a responded model and set *Z*-origin again. It is possible to enter greater engraving speeds.

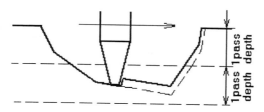

Fig. 10: The finishing of the product.

Mode *Final+Cleaning*

This mode is the same as the previous mode and in addition will make a cleaning in the *Cleaning* mode.

Mode *Printed circuit*
This mode is for the engraving of a printed circuit. The source picture must be drawn in two colours - black colour for wires and white colour for engraved regions. In this mode is guaranteed that the wires will not be engraved.

4. Program *Photo Layer 3D*

Program *Photo Engrave 3D* is prefered for a fast and simply 3D engraving but the data preparation as a grey-scale picture is not so easy for complicated 3D surfaces. For the data preparation is suitable the program *Photo Layer 3D*.

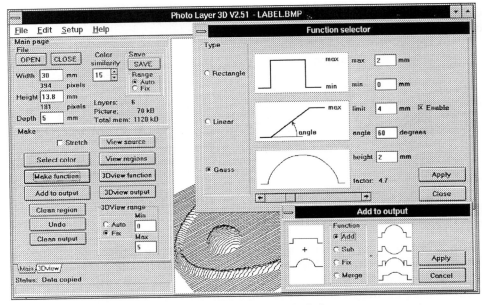

Fig. 11: The *Photo Layer 3D* desktop.

This program loads pictures in BMP or WMF *(Windows MetaFile)*. These pictures can contain maximal 16 different colours. Separate regions are selected according to their colours and than they are projected to the third dimension. The output model is saved as a picture in grey scale. The block diagram is shown in Fig. 12.

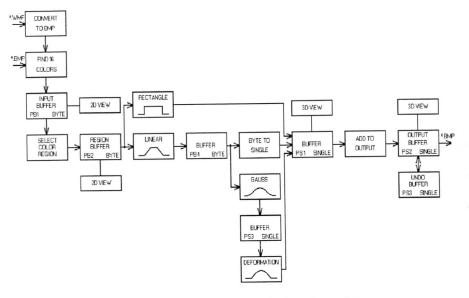

Fig. 12: The block diagram of *Photo Layer 3D*.

4.1. 3D MODEL CREATION

After loading a source picture, the program will check the number of colours. The *Select colour* button executes a dialogue for the colour region selection. It is possible to select more colours at the same time but the output region will be processed at one pass. Any selected 3D function will project the region to the third dimension. The dialogue for the function selection is executed by the *Select function* button. After the calculation, the region will be a part of a 3D model. This part must be added to the output model. The dialogue for this process is executed by the *Add to output* button. It is possible to select one of four types: *Add* - add the region to the output model; *Sub* - sub the region from the output model; *Fix* - replace the region in the output model and *Merge* - merge the region with the output model. The output model can be shown as a three-dimensional view., These actions are repeated after a selection a new colour region. The finish output model is saved as a grey scale picture, this picture can be used in the program *Photo Engrave 3D*.

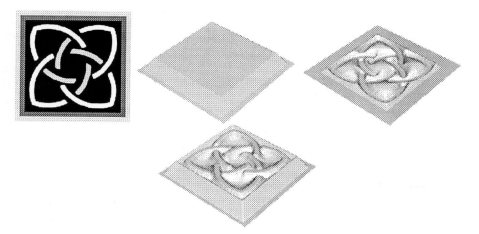

Fig. 13: An evaluation of a 3D model from a bitmap picture.

5. Program Protection

Both programs use hardware key *Sentinel Super Pro* for protection. The key must be connect to a printer port. The program tests the validity of the key, if the key is not valid then the program stops itself. The key has time and execute limits, it allows to run a maximum of 30 hours or 30 times. Then the program will not generate data to a port or to a file.

For presentations and tests are free versions available. A demo version does not allowed to generate data to output. These programs can be downloaded from the address *www.electron.cz*.

6. Conclusion

Both programs allow fast data preparation for a three-dimensional engraving. The program *Photo Engrave 3D* is for the engraving of 3D products from grey scale picture. The program *Photo Layer 3D* is for a preparation of this picture. Both programs work with a raster graphics and the engraving is also in the raster. But with the raster engraving is a problem with labels, it is important to select small engraving step and a production time is too long. A dynamic step control would be good for speed up of the production.

These programs were tested by users in Czech Republic and in France. Some users sent any useful reminds. The final version of programs is finished at present times.

This work was supported by the Ministry of Education of Czech Republic - Project No.: VS97060.

References

[1] Mimaki Command Language, Tokio 1995.
[2] Novotny, P.: Converting a Scan Picture from 2D to 3D. Project FEI VUT Brno, 1998 (in Czech).

28

A Multicamera Active 3D Reconstruction Approach

Theodore Lilas, Stefanos Kollias
Dept. of Electrical and Computer Engineering
National Technical University of Athens
Zografou, Athens, Greece
thodoris@theseas.ntua.gr

1. Introduction

Manual object digitizing is a tedious task and for small objects can be replaced by 3D scanners or CNC machines with a suitable probe. On the other hand large areas can be modeled automatically using photogrammetry. However, objects, which range from a few meters to several meters, are not modeled automatically. In this chapter we present a methodology for three-dimensional modeling, which can be applied to a wide range of objects. The approach is a synthesis of techniques and algorithms used in stereo vision together with methods used in active laser range scanners. Finally an artificial neural network enhances the created model.

In stereoscopy two different views of the object are processed and features extracted from the images are matched against each other. Based on the disparity of the corresponding features depth is estimated. Several problems occur during feature extraction and matching resulting in many cases in an inaccurate and ill-defined object model. On the other hand in active laser scanning systems depth is computed by triangulation, that is by measuring the disparity of the trace of the laser beam on the object. Laser triangulation requires very precise machining of the mechanical structure which oscillates and moves around the object. An additional problem is that shiny parts of the object reflect the beam and then the measured position is incorrect.

In the presented approach several cameras survey the object in order that all sides of it are visible by at least two cameras. Then a laser beam scans the object and measurements are taken by processing the images of the suitable cameras. We do not require any knowledge of the position of the laser beam. All calculations are made by processing the images after having calibrated the cameras and compensated any errors and distortions of the lens and the sensor. Processing involves tracing of the laser beam and performing stereoscopic matching based on the detected beam trace. Object reconstruction follows the processing step. During reconstruction the object is also processed by an artificial neural network, which reduces noise and determines areas that require higher spatial sampling frequency.

2. Stereo Vision

Perspective projection maps three-dimensional world to a two-dimensional image. It maps an infinite line onto a point in the image. Therefore it is not directly invertible. However, multiple images taken from different viewports can be combined to derive depth information, by using the inverse perspective transformation. The inverse perspective transformation determines the equation of a line of sight in three-dimensional space given the image coordinates and the camera model.

One approach is to identify features in two images, which correspond to the same point. Then derive the two lines of sight for each feature points using inverse perspective. Finally intersect the two lines to derive the 3D coordinates. The most difficult part is to identify the corresponding parts in the two images. One way is to perform block matching, however the matching areas are not identical and in some cases a feature of one image is occluded on the other. Another approach is to work on the edges. More recent techniques address the problem of 3D-structure and motion estimation simultaneously [1,8]. The mutual relationship between stereo disparity and motion estimation is utilized, improving the accuracy of both estimations.

3. Laser Triangulation

Laser triangulation is based on the following principle. The laser probe projects a dot of light on the object, which is to be measured. The scattered light from the dot is focused on a light sensitive device. Depending on the distance to the object, the dot is focused at different positions on the light sensitive device. The triangle formed by the laser source, the illuminated point on the object and the trace on the light sensitive device is used to estimate the distance to the object. Then the beam oscillates in order to scan the object.

Structured light is used in many cases to derive depth information. The principle is to illuminate the scene using a geometric pattern, which will help extract the geometric information of the object [7]. Structured light allows deriving depth and surface orientation information even from dull areas where we cannot extract any features.

Laser light striping is a case of structured light illumination. Laser light has the advantage that is coherent and when focused properly can be modeled more accurately than any other light source as a line or a plane. It provides also good contrast and can be detected even if the scene is also illuminated with ambient light. Moreover, if a camera is fitted with the proper filter, a laser-based system can operate near strong light sources like the arc of arc welding. Deriving three-dimensional coordinates from laser striping is done in the following way. A plane of light is projected on the scene, which causes a stripe of light to appear on the scene. The plane of light has a known equation in 3D space and every image point defines a line in 3D space, therefore their intersection gives the world coordinates of each point on the stripe. The above is always true since the camera's focal point is not in the light plane. Therefore any point that can be seen by both the camera and the laser stripe can be measured. However, concavities are not always visible and therefore cause troubles. Another problem is that stripes are not evenly placed on the surfaces, but the density depends on the surface orientation relative to the light plane. This situation can be improved by performing an additional set of measurements by stripping at a different angle.

4. 3D Reconstruction

4.1 CALIBRATION

Depth perception can be derived by processing multiple images taken from different viewports. In order to extract depth it is necessary to know the relative position of the cameras. Precise camera position and orientation is not easily available from outside measurements, because the optical center we are looking for is inside the camera. Moreover, if we are using a zoom lens the optical center changes according to the zoom factor and does not correspond to a fixed point.

Of-the-self cameras compared to metric cameras require additional information in order to model and compensate for distortions. The above problem is solved using camera calibration techniques. The algorithm used estimates the extrinsic parameters of the camera, describing its position and orientation, and the intrinsic parameters describing its internal viewing characteristics. Calibration is based on viewing points on a calibration device whose 3D coordinates are known with great accuracy. Although, it is possible to calibrate the camera using self-calibration methods, a special calibration pattern is necessary in order to achieve high accuracy. In order to describe the position and orientation of each camera we need to estimate the transformation matrix, which specify the relation between the world coordinate system and the camera coordinate system.

The camera coordinate system C is a right handed reference system defined as follows: The origin f of the camera reference system is positioned at the center of projection of the lens. The z-axis points away from the camera and corresponds to the optical axis. The x-axis is parallel to the horizontal lines of the camera image from left to right. The y-axis is parallel to the vertical axis of the camera image from top to bottom.

A simple camera model is the pin hole model, where the camera transformation from the 3D world coordinates to the 2D image coordinates is considered a perspective projection. The center of projection is the camera coordinate system origin. The center of the image is defined as the intersection between the optical axis of the lens and the image sensor.

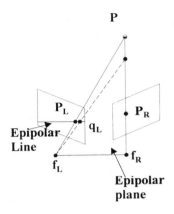

Let $p_L = (x_{1L}, x_{2L})$, $p_R = (x_{1R}, x_{2R})$ be the image coordinates of the point $P = (X_1, X_2, X_3)$.

The coordinates of the point P, in the right camera and left camera coordinate systems, are given by:

$$P_R = R_P P + T_R$$

$$P_L = R_L P + T_L$$

where R and T represent the rotation and translation transformation from the world to camera coordinate system.

Combining the equations we have:

$$P_L = R_L R_R^{-1} P_R - R_L R_R^{-1} T_R + T_L$$
$$= M P_R + B$$

where M and B are known as relative configuration parameters.

Assuming equal focal length on both cameras:

$$x_{1L} = f \cdot \frac{X_{1L}}{X_{3L}}$$

$$x_{1R} = f \cdot \frac{X_{1R}}{X_{3R}}$$

$$x_{2R} = f \cdot \frac{X_{2R}}{X_{3R}}$$

$$x_{2L} = f \cdot \frac{X_{2L}}{X_{3L}}$$

Combining the equations we have:

$$\frac{X_{3L}}{f}\begin{bmatrix} x_{1L} \\ x_{2L} \\ f \end{bmatrix} = \frac{X_{3R}}{f} M \begin{bmatrix} x_{1R} \\ x_{2R} \\ f \end{bmatrix} + B$$

The real image is quite different from the image derived by the ideal model. In order to model distortions we add non-linear terms to the perspective projection of a 3D world coordinate point. In the complete model in order to describe the 3D to 2D transformation we have to combine the camera translation and rotation, the perspective projection, the distortion and the sampling [2,9].

The radial distortion produces the most important effect and is modeled as

$$x_{1d} = x_1 + ax_1(x_1^2 + x_2^2), \; x_{2d} = x_2 + ax_2(x_1^2 + x_2^2)$$

and the decentering of the lens

$$x_{1d} = x_1 + ax_1(3x_1^2 + x_2^2) + 2bx_1x_2, \; x_{2d} = x_2 + ax_2(3x_1^2 + x_2^2) + 2ax_1x_2$$

Based on the calibration setup accurate estimation was achieved by modeling and compensating radial and decentering of the lens distortions. The intrinsic camera parameters are the scale factor (s), the effective focal length (f), the principal point (Cx, Cy) which is the intersection between camera's coordinate frame z-axis with the sensor plate, and the radial distortion coefficients.

The calibration device is a cube with a chessboard pattern on each side. The image is processed and the features extracted are the corners of the squares forming the

316

pattern. The features are detected with 1/10th of a pixel accuracy [5,6], which leads to an estimation accuracy of a few millimeters of the camera position with respect to the calibration target.

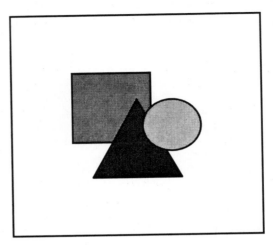

Calibration Cube

4.2   3D MEASUREMENTS

Once the cameras have been calibrated the algorithm is applied again whenever the position of a camera changes. In this case we either use five points with known world coordinates in order to compute only the external parameters of the camera or use the intrinsic parameters from the first calibration as a starting point.

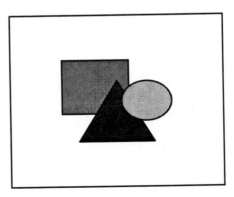

Camera Setup

The next step is to take 3D measurements of the object. Stereoscopic analysis has sometimes given poor results depending on the light conditions and workpiece texture. We were able to overcome these problems by scanning the workpiece with a laser beam. Two approaches were examined, one utilized a laser stripe and the other a laser spot. Laser striping is advantageous over spot triangulation, because it provides information about the continuity of the surface along the stripe. On the other hand laser spot is more easily detectable especially at distant objects and provides better accuracy on the measurement of the spot. The laser beam is traced with accuracy by taking into account the luminance gradient of the spot in the image. The accuracy is much better than the size of the laser spot and the camera pixel.

The scene is uniformly sampled using vertical laser beams. Using inverse perspective as described in the previous section the image coordinates of the center of the laser trace are projected in the 3D world along the line of sight. The lines of sight from the different cameras, which see the trace, are intersected and the intersection coordinates gives and object point coordinates. However, due to noise and physical constraints, lines do not intersect in the geometrical sense so we compute the point, which is at minimum distance from the lines of sight. The sampling frequency along the object surface depends on the orientation of the surface relative to the laser beam. Therefore areas with low sampling frequency along their surface are detected and are sampled again using beams perpendicular to the surface.

4.3 NOISE FILTERING

Visual reconstruction from noise corrupted data is a fundamental problem in computer vision. Most approaches to surface reconstruction are based on generalized spline models on a bounded domain in the (x, y) image plane. The continuous spline surface is represented as a single-valued function z(x, y) and the computational grid is uniform. The nodes of the surface can move only in the vertical or z direction during the iterative reconstruction procedure. More advanced techniques use nonuniform sampling and reconstruct input data using adaptive meshes [3,10].

A neural network is used in order to filter out noise. The parametric function represented by a network with one hidden layer has the form $\hat{z} = s(ax + by + c)$ where $s(\ )$ is a non-linear activation function. Networks with more than one hidden layer can model complex surfaces using a variety of basis functions. The advantage is that the geometric characteristics of the basis functions are learned from data instead of being specified by the user.

The training algorithm is based on Back Propagation algorithm modified according to Chen [4] so that it can handle errors in the training data set. Instead of minimizing the sum of squared errors, it minimizes a new function, which is adjusted with the progressively refined knowledge of noise in the data. The algorithm minimizes $\sum_{p=1}^{P} \phi_t(z_p - \hat{z}_p)$ where the shape of ϕ_t resembles the tanh estimator [4] whose derivative is given as

$$\psi_t(r) = \begin{cases} r & |r| \le a_t \\ a\tanh(\beta(b_t - |r|))\operatorname{sgn}(r) & a_t < |r| \le b_t \\ 0 & |r| > b_t \end{cases}$$

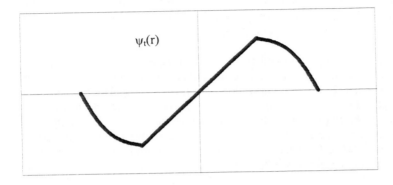

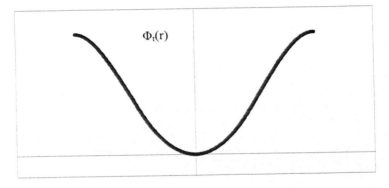

When minor noise exists in the training data, the algorithm is similar to back propagation algorithm. However, when noise level increases the influence of noise on the learning process is significantly reduced, because the amount of weight adjustment at the output layer during leaning is proportional to $\psi_t(r)$ instead of the residual r. We need then to specify q, which is the percentage of bad data to be tolerated by the algorithm. Using the bootstrap method the confidence interval of the residuals is computed that separates the good from the bad data.

5. Experiments

Experiments have been conducted in a laboratory and in an industrial environment. In the laboratory small objects were scanned. During calibration the target was illuminated with ambient light in order to increase the image contrast. The calibration pattern was placed one meter away from each camera. This was done so that the calibration data was distributed across the field of view and the range of the object model in order to accurately estimate the radial lens distortion and image center parameters.

Line Scan   **Spot Scan**

In the industrial environment 3D measurements were taken in order to verify the accuracy of the system on large steel structures. It also demonstrated the robustness of the system under poor lightning conditions on mat objects and shiny ones too. Stereoscopy is not accurate on dull areas, because features can be detected. Laser scanning is not accurate on shiny surfaces because the system may detect a secondary reflection of the beam and erroneously assume that this point lies on the laser plane.

Steel Structures

In order to scan objects at different angles, we have utilized a robotic system, which is capable of moving the laser in all directions in space. This has been also a way of examining the accuracy of the measurements which were accurate to 1mm in the range of a 500 mm object. The images were acquired from three CCD cameras, using a frame grabber which supports four input channels for cameras. Depending in the complexity of the object more cameras can be used in order to completely cover the object directly [11].

6. Conclusions

The proposed methodology has the following advantages: -can be applied to large objects -provides high accuracy by fusing several sensor data and by super sampling, - does not require position measurements of moving parts, - any defects and distortions are compensated accurately by software calibration, - proper placement of the cameras provides full coverage of the object.

References

[1] A. Murat Tekalp, "Digital Video Processing", *Prentice Hall*, 1995.

[2] R. Y. Tsai, "A versatile Camera Calibration Technique for High-Accuracy 3D Machine Vision Metrology Using Off-the-Shelf TV Cameras and Lenses", *IEEE Journal of Robotics and Automation*, Vol. RA-3, No. 4, August 1987, pp 323-344.

[3] D.Terzopoulos and M Vasilescu, " Sampling and Reconstruction with adaptive Meshes",*CVPR'91 IEEE Computer Vision and Pattern Recognition 1991*, pp 70- 75.

[4] D. Chen, R. Jain and B. Schunk, "Surface Reconstruction Using Neural Networks", *CVPR'92 IEEE Computer Vision and Pattern Recognition 1992*, pp 815-817.

[5] Robert J. Valkenburg, A. M. McIvor, and P. Wayne Power,"An evaluation of subpixel feature localisation methods for precision measurement", *SPIE Vol. 2350 Videometrics III* (1994) pp. 229-238.

[6] M. R. Shortis, T. A. Clarke, and T. Short, "A comparison of some techniques for the subpixel location of discrete target images", *SPIE Vol. 2350 Videometrics III* (1994) pp. 239-250.

[7] Yang, Z.M., Wang, Y.F., "Error Analysis of 3D Shape Construction from Structured Lighting", *PR(29), No. 2, 1996*, pp. 189-206.

[8] A. Delopoulos and Y. Xirouhakis, "Robust Estimation of Motion and Shape based on Orthographic Projections of Rigid Objects", *Tenth IMDSP Workshop 98*, July 1998.

[9] J. Heikkila, O. Silven, "Calibration procedure for short focal length off-the-shelf CCD cameras", Proc. of The 13th International Conference on Pattern Recognition. (1996) Vienna, Austria. pp. 166-170.

[10]M. Maed, K.Kumarumaru, H.Zha, K. Inoue and S. Sawai, "3D Surface Recovery from Range Images by Using Multiresolution Wavelet Transform", *IEEE Conference on Systems, Man, and Cybernetics,1997, pp.3654-3659.*

[11] Ch. Schutz, T. Jost, H. Higli, "Semi-Automatic 3D Object Digitizing System Using Range Images", *ACCV'98*, Hong Kong, 1998.

29

Video Object Segmentation Using the EM Algorithm

N.D. Doulamis, A.D. Doulamis and S.D. Kollias
Electrical and Computer Engineering Department
National Technical University of Athens
Email: adoulam@image.ntua.gr

1. Introduction

Video object segmentation is an important task towards the framework of MPEG-4 and MPEG-7 standardization phase. Especially, the MPEG group has adopted the concept of Video Objects (VOs) and Video Object Planes (VOPs) for improving the coding efficiency and providing multimedia functionalities to the future encoders [1][2]. While frame-based coding and description schemes provide limited capabilities in terms of access, identification and manipulation of individual objects, object-based coding and description offers a new range of capabilities, where objects are separately described giving a new dimension to playing with, creating or accessing video content. Video objects correspond to meaningful entities of arbitrary shape in a digital video stream, such as a human, a chair, while VOPs are the projection of VOs into a plane. Such an extraction plays an important role in many other image analysis problems: content-based indexing and retrieval [3][4], reconstruction of a 3D human model from several 2D images [5], human face detection and recognition [6][7], video surveillance of specific areas and identification of objects by industrial robots [8].

However, although humans effortlessly "perform" object segmentation, it remains one of the most difficult problems for computer systems, apart perhaps from the case of video games or graphics applications, where object segmentation is a-priori available. Current approaches to unsupervised image segmentation are based either on spatial or motion homogeneity criteria or a combination between them. Unfortunately, none of these techniques can provide satisfactory results towards the goal of video object segmentation since a video object generally contains regions of different colors and motions [9].

Recently, some other approaches have been proposed for achieving a more efficient video content segmentation. In particular, in [10] morphological connected operators are presented for performing video object classification while in [6] human face extraction is achieved based on a probabilistic model at multiple scales. Color characteristics are considered sufficient in [11] for human face detection, even in very complicated video sequences. An algorithm which fuses motion and color information and then peforms object extraction using a ruled-based decision module has been proposed in [12]. Furthermore, a neural network scheme is used in [13] for extracting humans in videoconference or videophone applications.

In this chapter, unsupervised video object segmentation is performed based on an iterative maximum-likelihood (ML) scheme. In particular, the proposed method is applied for extracting humans even from complex background, mainly in videophone applications. We consider that the probability density function (pdf) of the image to be segmented can be represented as a mixture of pdfs of individual objects (in our case human and background) [14]. Moreover, two approaches are examined. The first assumes that the image pixels and the respective classes are independent. The second considers that the segments are modeled as Markov Random Fields (MRFs), following a Gibbs distribution. In this case, improvement of object segmentation can be accomplished since the local correlation of color information, which encountered in an image, is better tackled. Then, the Expectation-Minimization (EM) algorithm [15] is applied to carry out the estimation of model parameters. However in the second approach, it is more difficult to obtain simple expressions for the marginal expectation, involved in the EM algorithm. Reduction of complexity can be achieved if approximation techniques are used for the marginal expectation, such as pseudo-likelihood approximation, which it seems to work well in practical applications [16].

The initial conditions of the EM algorithm are provided through an estimation module, which approximately indicates the human object location. In particular, the human face is modeled as a unimodal Gaussian density where its mean value and variance are both estimated using a known training set. Then, based on previous probabilities, the face region is detected by taking into account all subimages (blocks) whose the probability exceeds a certain threshold and then the face area and center are calculated. For the human body, a simplified model is considered as a product of two independent one-dimensional Gaussian pdfs, the parameters of which are estimated based on the face detection module. The performance of the proposed scheme is evaluated using the MPEG-4 test video sequences.

This chapter is organized as follows: Section 2 presents the system overview, while in section 3 the initial estimates for the EM algorithm are described. The algorithm for maximum likelihood estimate of model parameters is proposed in section 4. Experimental results are provided in section 5 and conclusions are given in section 6.

2. System Overview

Let $y_{i,j}, i = 0,...,M_1 - 1, j = 0,...,M_2 - 1$ denote the pixel intensity of an image of size $M_1 \times M_2$ at i, j location. However in the following, for notional convenience, we denote the image as 1-D pixel array $\mathbf{y} = \{y_1, y_2, \cdots, y_N\}$ where N is the total number of pixels in the image (i.e., $N = M_1 \times M_2$). In video object segmentation problems, the target is to categorize each image pixel to one of K available classes, say, $\omega_j, j = 1,2,\cdots, K$ each of them corresponds to an individual video object. A map $\mathbf{z} = \{\mathbf{z}_1, \mathbf{z}_2, \cdots, \mathbf{z}_N\}$ is then formed, the elements of which indicate the class that the respective ith pixel belongs to. Particularly, $\mathbf{z}_i$ is a vector, which indicates the class status of y_i. For a K-class segmentation problem, $\mathbf{z}_i$ has all its elements equal to zero, apart from a single component which is equal to one. The index of unity component corresponds to the respective class. Based on the array $\mathbf{y}$ and map $\mathbf{z}$, the set $\mathbf{x} = (\mathbf{y}, \mathbf{z})$ is formed defining the image and the segmentation results respectively. In our case, we are interested in a two-class classification problem, where classes ω_1, ω_2 refer to foreground (human) and background object in an image.

In the following, the probability density function (pdf) of the image to be segmented is considered as a mixture of multi-modal pdfs, each of them corresponds to an individual video object, i.e., background/foreground in our case. Given a set of model parameters, each image pixel is assigned to the object with the highest associated probability. The EM algorithm is used to jointly estimate the model parameters, using a number of proper observations. Considering a video object (e.g., a human) as a union of several regions, which are characterized by different color properties, the three-color components of an image (Y, U, V) are used to form the observation vector.

Let us first define as $f(\mathbf{x} \mid \boldsymbol{\theta})$ the probability density function of the set $\mathbf{x}$, where $\boldsymbol{\theta}$ indicates the model parameters that characterize the mixture model. The Expectation-Minimization (EM) algorithm has been adopted for estimating the parameters $\boldsymbol{\theta}$ through an iterative procedure including two main steps. In the first, which is called E-step, a function, say, Q is computed using the parameters obtained from the previous iteration, while in the second (M-step) the new parameters are estimated through minimization of Q. Particularly, the algorithm performs as follows [15]

- E-Step: Calculate the function

$$Q(\boldsymbol{\theta} \mid \hat{\boldsymbol{\theta}}^{(n)}) = E[\log f(\mathbf{x} \mid \boldsymbol{\theta}) \mid \mathbf{y}, \hat{\boldsymbol{\theta}}^{(n)}] \tag{1}$$

- M-Step: Find

$$\hat{\boldsymbol{\theta}}^{(n+1)} = \arg \max_{\boldsymbol{\theta}} Q(\boldsymbol{\theta} \mid \hat{\boldsymbol{\theta}}^{(n)}) \tag{2}$$

where in the previous equations the superscript n indicates the nth iteration of the algorithm.

Initial estimates of parameters $\boldsymbol{\theta}$, say, $\boldsymbol{\theta}^{(0)}$ are required to start the algorithm. Then, the function $Q(\cdot)$, given in Eq. (1), can be computed and model parameters are updated based on Eq. (2). The process is iterated until convergence is achieved. In the approach presented in this chapter, convergence is obtained by calculating the segmentation mask of the image, after each iteration, and then counting the pixels which have been reassigned from the one iteration to another. When the number of pixels falls below a pre-determined threshold the iteration stops and the algorithm is assumed to have converged.

In supervised video object segmentation the initial estimates for the EM algorithm are given through the user's interaction, by selecting, for example, proper areas of foreground/ background object. However, most widely used video applications require unsupervised object segmentation. In this case, the initial parameter estimates should be automatically provided. Therefore, two additional modules have been embedded in the proposed architecture, suitable for giving an initial approximation of

324

foreground/background object, in case of videoconferencing applications; the human face and body detection modules.

Human Face Detection Module: This module performs a human face detection task, using a unimodal Gaussian density function for modeling human face regions.

Human Body Detection Module: The human body is approximately localized using a probability distribution model whose the parameters are estimated based on the output of the human face detection module.

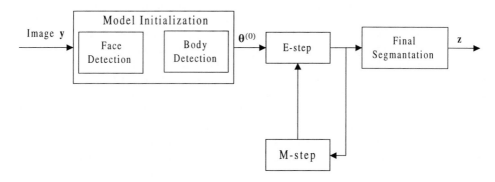

Figure 1: The proposed scheme.

Figure 1 illustrates a block diagram of the proposed architecture. Firstly, the initial parameters $\theta^{(0)}$ of the image to be segmented are estimated using human face and body models. Then, the EM algorithm is applied for maximum likelihood estimate of the vector parameters θ. Finally, the model-based segmentation is performed, assigning each image pixel to one of the K-available objects using the results obtained by the EM algorithm.

3. Initialization of Model Parameters

In this section the procedure, providing an initial estimate of the model parameters, is described. In particular in subsection 3.1 the human face detection task is presented, while in 3.2 the human body one.

3.1. HUMAN FACE DETECTION

A unimodal Gaussian density function is used to approximately model the face region. In particular, this density is provided using the following equation

$$P(\mathbf{u}_i \mid \Omega_f) = \frac{\exp(-\frac{1}{2}(\mathbf{u}_i - \boldsymbol{\mu})^T \mathbf{S}^{-1}(\mathbf{u}_i - \boldsymbol{\mu}))}{(2\pi)^{N/2}|\mathbf{S}|^{1/2}} \tag{3}$$

where the vector $\boldsymbol{\mu}$ corresponds to the mean value vector and $\mathbf{S}$ to the correlation matrix. In the previous equation, we denote as Ω_f the face class and $\mathbf{u}_i$ a subimage of the image $\mathbf{y}$. A common choice for the size of the subimage $\mathbf{u}_i$ is a block consisting of 8x8 pixels. Better face detection can be achieved if smaller resolution for the subimage

$\mathbf{u}_i$ is used. However, in this case the computational load increases more rapidly than the improvement of the face localization. Thus, in this chapter an 8x8 pixel block has been adopted for the size of $\mathbf{u}_i$.

The vector $\boldsymbol{\mu}$ and matrix $\mathbf{S}$ are estimated based on a training set consisting of several data of different human face images. In Eq. (3), the Mahalanobis distance, $(\mathbf{u}_i - \boldsymbol{\mu})^T \mathbf{S}^{-1}(\mathbf{u}_i - \boldsymbol{\mu})$ is involved. However, direct calculation of the matrix $\mathbf{S}^{-1}$ is very computational. Thus, a principal component analysis is used for providing an estimation of the matrix $\mathbf{S}^{-1}$ [6]. In particular, using eigenvector and eigenvalue decomposition, an approximation of the Mahalanobis distance can be obtained by retaining the M principal coefficients.

Thus, for every subimage $\mathbf{u}_i$, the probability $P(\mathbf{u}_i | \Omega_f)$, defined in (3), is used for detecting face regions. In particular, all subimages $\mathbf{u}_i$, whose the probability $P(\mathbf{u}_i | \Omega_f)$ exceeds a certain threshold, are considered that belong to face areas with a high degree of confidence. Then, a postprocessing technique is applied to the above selected regions similar to that described in [11] for eliminating the "noisy" regions. This is due to the fact that human face is connected and elliptic object and thus isolated subimages or regions with shape quite different from that of human face should be discarded.

3.2. HUMAN BODY DETECTION

Human body detection is performed by exploiting information provided by the human face detection module. In particular, the area and the center of gravity of the face region are first calculated. Then, a simplified model is considered, for the human body, as a product of two independent 1-D Gaussian probability density functions whose the parameters depend on the face location.

$$P(y_{i,j} | \Omega_b) = \frac{\exp(-\frac{1}{2\sigma_i^2}(i-\mu_i)^2)\exp(-\frac{1}{2\sigma_j^2}(j-\mu_j)^2)}{(2\pi)\sigma_i\sigma_j} \tag{4}$$

where we recall that $y_{i,j}$ is the image pixel at i, j location. The Ω_b corresponds to the human body class, while the mean values μ_i, μ_j and the variances σ_i, σ_j are computed according to the center of gravity and area of face regions. Equation (4) indicates that the probability of pixel $y_{i,j}$ to belong to the class Ω_b depends on the distance between this pixel and the estimated center of human body. Combing the two previously described tasks about the human face and body location, we form an approximate object segmentation mask, the elements of which indicate the probability for each pixel belonging to the human class.

3.3. INITIAL MODEL ESTIMATES

The previously described modules provide an approximation of foreground/ background object in video conferencing applications. In particular, based on the approximate segmentation mask, a training data set, for each video object, is formed. This training set is used to obtain an initial estimate of the parameters for each video object pdf. To achieve this, it is assumed that the parameters of $f(\mathbf{x}|\boldsymbol{\theta})$ depend only on $\mathbf{y}$, i.e., the intensity of the image pixels. This means that there is a pre-defined relationship among the pixel class status $\mathbf{z}_i$ the parameters of which are a priori known and cannot be estimated by the EM algorithm, as it is discussed in subsection 4.2.

Modeling the probability density function of image to be segmented as a sum of Gaussian distributions, the elements of vector $\boldsymbol{\theta}$ will correspond to the mean and variance of each Gaussian mode. An initial estimate of these parameters can be obtained by segmenting the approximate foreground/background regions, provided by the aforementioned modules, according to color homogeneity criteria. This is accomplished using for example the watershed segmentation algorithm. Since the number of Gaussian modes are assumed to be fixed, the number of the segments should be equal to the fixed number of modes. Thus, starting with a fine watershed segmentation, color segments are merged based on their mean value, until the number of image regions remaining equals the required number of modes. The resultant mean observation vectors and the associated variance are taken as the initial estimates of the parameters of the objects' pdf.

4. Model Parameters Estimation

Given a fixed number of model classes, say, K and an initial parameter estimate, say, $\boldsymbol{\theta}^{(0)}$ the EM procedure is activated, yielding a maximum likelihood estimate of the model parameters for each video object, $\hat{\boldsymbol{\theta}}_{ML}$. Then, the model-based segmentation is applied using the $\hat{\boldsymbol{\theta}}_{ML}$ estimates. In our approach, we assume that the number of model classes is a priori available and are equal to 2, that is the foreground/background object. In case that the number of classes are unknown, a validation scheme based on the Akaike information criterion (AIC), described in [17], can be used for their estimation.

In the following, we consider two approaches as far as the condition of pixels y_i and their respective class status $\mathbf{z}_i$ is concerned. The first, which is the simplest case, assumes that y_i and $\mathbf{z}_i$ are independent and it is described in subsection 4.1, while the second models $\mathbf{z}_i$ as a Markov Random Field (MRF), following a Gibbs distribution and is discussed in subsection 4.2.

4.1. THE INDEPENDENT MODEL

In this case, y_i's and $\mathbf{z}_i$'s are independent for all i. Then, the log-likelihood probability density function for each video object can be written as

$$\log f(\mathbf{x}|\boldsymbol{\theta}) = \log f(\mathbf{y},\mathbf{z}|\boldsymbol{\theta}) = \sum_{i=1}^{N}\log f(y_i|\mathbf{z}_i,\boldsymbol{\theta}) + \sum_{i=1}^{N}\log f(\mathbf{z}_i|\boldsymbol{\theta}) \qquad (5)$$

where Eq. (5) have been obtained using the independence assumption and the Bayes' conditional probability formula. In this case, it is possible to derive analytical expression for the function $Q(\boldsymbol{\theta}|\hat{\boldsymbol{\theta}}^{(n)})$. In particular, taking into consideration the properties of the class status $\mathbf{z}_i$ we can conclude to the following expression for the function $Q(\boldsymbol{\theta}|\hat{\boldsymbol{\theta}}^{(n)})$

$$Q(\boldsymbol{\theta} \mid \hat{\boldsymbol{\theta}}^{(n)}) = \sum_{i=1}^{N} E[\mathbf{z}_i^T \mid \mathbf{y}, \hat{\boldsymbol{\theta}}^{(n)}] \cdot \mathbf{a}(y_i \mid \boldsymbol{\theta}) + \sum_{i=1}^{N} E[\mathbf{z}_i^T \mid \mathbf{y}, \hat{\boldsymbol{\theta}}^{(n)}] \cdot \mathbf{b}(\boldsymbol{\theta}) \tag{6}$$

where the $\mathbf{a}(y_i \mid \boldsymbol{\theta})$ and $\mathbf{b}(\boldsymbol{\theta})$ are computed using the following equation

$$\mathbf{a}(y_i \mid \boldsymbol{\theta}) = (\log f(y_i \mid \mathbf{e}_1, \boldsymbol{\theta}), \log f(y_i \mid \mathbf{e}_2, \boldsymbol{\theta})) \tag{7a}$$

and

$$\mathbf{b}(\boldsymbol{\theta}) = (\log f(\mathbf{e}_1 \mid \boldsymbol{\theta}), \log f(\mathbf{e}_1 \mid \boldsymbol{\theta})\} \tag{7b}$$

where $\mathbf{e}_i$, $i=1,2$ a unity vector whose the ith component is 1 and all the other ones are 0.

The conditional expectation $E[\mathbf{z}_i^T \mid \mathbf{y}, \hat{\boldsymbol{\theta}}^{(n)}]$ in (6) is a vector the elements of which are expressed through the following equation

$$E[z_{ik} \mid \mathbf{y}] = f(\mathbf{z}_i = \mathbf{e}_k \mid \mathbf{y}) = \frac{f(\mathbf{y} \mid \mathbf{z}_i = \mathbf{e}_i) f(\mathbf{z}_i = \mathbf{e}_k)}{\sum\limits_{j=1}^{K} f(\mathbf{y} \mid \mathbf{e}_j) f(\mathbf{e}_j)} \tag{8}$$

where z_{ik} is the kth component of $\mathbf{z}_i$. In Eq. (9) we have dropped the dependence upon the vector $\hat{\boldsymbol{\theta}}^{(n)}$ for simplicity purposes. If the $\mathbf{z}_i$'s are independent of y_i for all $j \neq i$, the previous expectation can be simplified to

$$E[z_{ik} \mid \mathbf{y}] = \frac{f(y_i \mid \mathbf{z}_i = \mathbf{e}_k) f(\mathbf{z}_i = \mathbf{e}_k)}{\sum\limits_{j=1}^{K} f(y_i \mid \mathbf{z}_i = \mathbf{e}_j) f(\mathbf{z}_i = \mathbf{e}_j)} \tag{9}$$

Knowing the probability $f(\mathbf{z}_i = \mathbf{e}_k)$ of the kth class in the image, the function $Q(\boldsymbol{\theta} \mid \hat{\boldsymbol{\theta}}^{(n)})$ is calculated. The next step is to update the model parameters by minimizing $Q(\boldsymbol{\theta} \mid \hat{\boldsymbol{\theta}}^{(n)})$ with respect to vector $\boldsymbol{\theta}$ (M-step). This is achieved by setting the derivatives to be equal to zero. Thus, the following conditions should be satisfied:

$$\sum_{i=1}^{N} \left[\nabla_{\boldsymbol{\theta}} \cdot \mathbf{a}^T(y_i \mid \boldsymbol{\theta}) \cdot E[\mathbf{z}_i \mid \mathbf{y}, \hat{\boldsymbol{\theta}}^{(n)}] \right] = 0 \tag{10}$$

where $\nabla_{\boldsymbol{\theta}}$ expresses the gradient of the quantity $\mathbf{a}(y_i \mid \boldsymbol{\theta})$ with respect to $\boldsymbol{\theta}$.

Assuming that the probability $f(y_i \mid \mathbf{z}_i = \mathbf{e}_k, \boldsymbol{\theta})$ is modeled as a Gaussian distribution with mean value m_k and variance σ_k^2 with $k=1,2,\cdots,K$, it can be shown [16] that the necessary conditions described in (10) take the form of

$$\hat{m}_k^{(n+1)} = \frac{1}{\hat{N}_k^{(n)}} \sum_{i=1}^{N} z_{ik}^{(n)} y_i \tag{11a}$$

and

$$\hat{\sigma}_k^{(n+1)} = \frac{1}{\hat{N}_k^{(n)}} \sum_{i=1}^{N} \hat{z}_{ik}^{(n)} (y_i - \hat{m}_k^{(n+1)})^2 \tag{11b}$$

where

$$\hat{z}_{ik}^{(n+1)} = E[z_{ik} \mid \mathbf{y}, \hat{\boldsymbol{\theta}}^{(n)}] = \frac{\hat{\pi}_k^{(n)} f(y_i \mid \mathbf{e}_k, \hat{\boldsymbol{\theta}}^{(n)})}{\sum \hat{\pi}_j^{(n)} f(y_i \mid \mathbf{e}_j, \hat{\boldsymbol{\theta}}^{(n)})} \tag{12a}$$

and

$$\hat{N}_k^{(n)} = \sum_{i=1}^{N} \hat{z}_{ik}^{(n)}, \quad \hat{\pi}_k^{(n)} = \hat{N}_k^{(n)} / N \tag{12b}$$

Equations (11a), (11b) is used for estimating the model parameters for each video object. In particular, starting with an initial estimate provided by the algorithm described in section 3, new parameters are computed based on Eq. (11). Then, the model-based segmentation is performed, classifying each image pixel to one of the K available classes, according to $\hat{\boldsymbol{\theta}}_{ML}$ estimates.

However, video object segmentation obtained by the independent model approach tends to produce noisy segmentation results, since the initial estimates of the model parameter may be erroneous. For this reason, a Markov Random Field (MRF) model is adopted, in the following, for the pixel class status $\mathbf{z}_i$, providing a spatial continuity constraint in an effort to obtain smooth segmentations. This however makes the evaluation of the conditional expectation $E[\mathbf{z}_i^T \mid \mathbf{y}, \hat{\boldsymbol{\theta}}^{(n)}]$ and the subsequent maximization more difficult to be computed as described in the following.

4.2. THE MARKOV RANDOM FIELD MODEL

In this approach we assume that image pixels are independent but class status $\mathbf{z}_i$ follows a Markov Random Field model with a Gibbs distribution. In particular, video objects are composed of compact regions, due to the connectivity property, and thus, the elements of the map $\mathbf{z}$, i.e., $\mathbf{z}_i$, are strongly correlated to their spatial neighbors. Thus, for two image pixels their respective class status will satisfy the following property [18].

$$f(\mathbf{z}_i \mid \mathbf{z}_j, j \neq i) = f(\mathbf{z}_i \mid \mathbf{z}_j, j \in \partial_i), \ \forall i \tag{13}$$

where ∂_i denotes a neighborhood of the ith image pixel. Such a property characterizes a Markov Random Field (MRF) and hence a Gibbs distribution can be used for modeling $f(\mathbf{z})$ [19].

$$f(\mathbf{z}) = \Gamma \exp\left(-\frac{\sum_{c \in C} V_c(\mathbf{z})}{\gamma}\right) \tag{14}$$

where Γ is a normalizing constant, γ is the "temperature" parameter of the density, $V_c(\cdot)$ is any function of a local group of points c called clique and C is the set of all such local groups.

As already mentioned, images are locally smooth. Consequently, it is not probable that an image pixel belongs to a specific class, if all its neighbors belong to another one.

Thus, function $V_c(\cdot)$ should award image pixels that satisfy the smoothness property and discourage the rest. A common choice for the $V_c(\cdot)$ is to be a squared form function concluding to

$$\sum_{c \in C} V_c(\mathbf{z}) = \sum_{i=1}^{N} \sum_{l \in \partial_i} \rho(\mathbf{z}_i - \mathbf{z}_l) \tag{15}$$

with

$$\rho(\mathbf{z}_i - \mathbf{z}_l) = (\mathbf{z}_i - \mathbf{z}_i^{(0)})^2 + \tau(\mathbf{z}_i - \mathbf{z}_l)^2 \tag{16}$$

where parameter τ controls the contribution of each of the two terms in the right hand side of (16) to the minimization of the cost function. A typical value of τ is equal to 0.3. The $\mathbf{z}_i^{(0)}$ represents the class the respective ith pixel belongs to, at the first iteration of the algorithm. Based on (14), $\log f(\mathbf{x} \mid \boldsymbol{\theta})$ can take the following form

$$\log f(\mathbf{x} \mid \boldsymbol{\theta}) = \sum_{i=1}^{N} \mathbf{z}_i^T \cdot \mathbf{a}(\mathbf{y}_i \mid \boldsymbol{\theta}) - \frac{1}{\gamma} \sum_{c \in C} V_c(\mathbf{z}) - \log \frac{1}{\Gamma} \tag{17}$$

and thus

$$Q(\boldsymbol{\theta} \mid \hat{\boldsymbol{\theta}}^{(n)}) = \sum_{i=1}^{N} E[\mathbf{z}_i^T \mid \mathbf{y}, \hat{\boldsymbol{\theta}}^{(n)}] \cdot \mathbf{a}(y_i \mid \boldsymbol{\theta}) - E[\frac{1}{\gamma} \sum_{c \in C} V_c(\mathbf{z}) + \log \frac{1}{\Gamma} \mid \mathbf{y}, \hat{\boldsymbol{\theta}}^{(n)}] \tag{18}$$

It is observed from (18) that model parameters are estimated similarly to the independent model (Eq. (6)). However, in this case it is difficult to obtain expression for the marginal distribution since $\mathbf{z}_i$ are not independent. Reduction of complexity can be achieved if approximation techniques are used for the marginal expectation, such as pseudo-likelihood approximation, which it seems to work well in practical applications. In particular the model of class status $\mathbf{z}_i$ is written as follows [20]

$$f(\mathbf{z}) \approx \prod_{j=1}^{N} f(\mathbf{z}_j \mid \mathbf{z}_l, l \in \partial_i) \tag{19}$$

where we recall that ∂_j is the local neighborhood of element j. Then, it follows that

$$f(\mathbf{z}_i) = \sum_{\mathbf{z}_j, j \neq i} f(\mathbf{z}) = \sum_{\mathbf{z}_j, j \neq i} f(\mathbf{z}) \prod_{j=1}^{N} f(\mathbf{z}_j \mid \mathbf{z}_l, l \in \partial_i) \tag{20}$$

Suppose that using the previous defined parameters a model-based segmentation is obtained, assigning for each image pixel y_i one of the K available classes. Assuming that $\hat{\mathbf{z}}_l$ is the previously obtained estimate for the pixel class status $\mathbf{z}_l$ we can say that,

$$f(\mathbf{z}_i) = \sum_{\mathbf{z}_j, j \neq i} \coprod_{j=1}^{N} f(\mathbf{z}_j \mid \hat{\mathbf{z}}_l, l \in \partial_j) \tag{21}$$

In this case, Eq. (9) is modified as follows

$$E[z_{ik} \mid \mathbf{y}] = \frac{f(y_i \mid \mathbf{z}_i = \mathbf{e}_k) f(\mathbf{z}_i = \mathbf{e}_k \mid \hat{\mathbf{z}}_l^{(n-1)}, l \in \partial_i)}{\sum_{j=1}^{K} f(y_i \mid \mathbf{z}_i = \mathbf{e}_j) f(\mathbf{z}_i = \mathbf{e}_j \mid \hat{\mathbf{z}}_l^{(n-1)}, l \in \partial_i)} \tag{22}$$

330

where $\hat{\mathbf{z}}_l^{(n-1)}$ is the state estimate obtained in the preceding iteration for the l pixel. This leads to a recursive scheme for obtaining the required conditional expectation $E[\mathbf{z}_i^T \mid \mathbf{y}, \hat{\boldsymbol{\theta}}^{(n)}]$. It also provides an approximate method for estimating the new parameter updates. In particular, the recursion is similar to Besag's iterative conditional mode technique (ICM).

5. Experimental Results

In the following, the performance of the proposed scheme was examined for detection and extraction of humans, and particularly of the upper part of human body including the head, shoulders and arms areas, in video sequences. Such an extraction plays an important role in many image analysis problems, such as object-based video coding, content-based indexing and retrieval and face recognition systems.

The Akiyo videoconference sequences were used to evaluate the proposed system. This sequence was at QCIF format (144x176 pixels) while the ratio of luminance to chrominance components was 4:2:2.

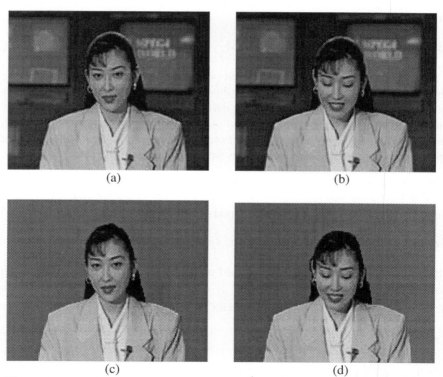

(a)　　　　　　　　　　　　　　　　(b)

(c)　　　　　　　　　　　　　　　　(d)

Figure 2: The segmentation results (a) the 10th original frame of Akiyo (b) the 100th original frame of Akiyo (c) the segmentation output for the 10th frame (d) the segmentation output for the 100th frame.

Figure 2 illustrates the segmentation results for two frames of the Akiyo. Figures 2a, 2b present the original frames (10 and 100 respectively) while Figures 2c, 2d the final classification provided by the proposed algorithm. For clarity of presentation, when a pixel belongs to the foreground area (human) it is left as it is in the figures. Instead, the background pixels are presented with a gay color. The MRF was used for modeling the class status z . Thus, better segmentation results are generated since the erroneous noisy areas existing in the initial estimates are eliminated.

It is observed that, although the background area consists of a large amount of color information, the segmentation obtained is very satisfactory. This is due to the fact that the proposed scheme "unifies" different color regions where there is an indication that they belong to the human object. This indication stems from the initial approximation of the human face and body location. In cases that the initial estimates generates large erroneous regions, the improvement providing by the EM algorithm may not be so satisfactory.

6. Conclusions

Video object segmentation was examined in this chapter based on a probabilistic framework. The object-based segmentation is very useful for a variety of applications, including efficient video coding, multimedia capabilities, indexing and retrieval from large image/video databases, construction of 3D models by several 2D ones or even face recognition systems.

The EM algorithm was proposed for performing the video object segmentation. The pixels of the image were assumed to be independent while the array that represents the class status (segments) was considered to follow either the independent model or a Markov Random Field (MRF). The second approach eliminates probable noisy pixels located in the foreground as well as in the background areas and as a result it better behaves than the Gaussian model.

The initial estimates of the EM algorithm were provided by using two additional modules. The first is responsible for approximately estimating the location of the human face. The second module indicate the initial estimates for the human body.

References

[1] MPEG Video Group, "MPEG-4 Video Verification Model-Version 2.1," ISO/IEC JTCI/SC29/WG11, May 1996.
[2] T. Sikora, "The MPEG-4 Video Standard Verification Model," *IEEE Trans. on Circuits and Systems for Video Techn.* vol. 7, no. 1, pp. 19-31, Feb. 1997.
[3] Special issue on visual information management, *Communications of ACM,* Dec. 1997. Guest Editor: Ramesh Jain.
[4] N. Doulamis, A. Doulamis, Y. Avrithis and S. Kollias, "Video Content Representation Using Optimal Extraction of Frames and Scenes" *Proc. of IEEE Inter. Confer. on Image Processing (ICIP)*, Chicago, USA, Oct. 1998.
[5] J. D. Foley, A. van Dam, S. K. Feiner and J. F. Hughes. *Computer Graphics: Principles and Practice.* Addison-Wesley Publishing Company, 1995.

332

[6] B. Moghaddam, A. Pentland, "Probabilistic Visual Learning for Object Representation," *IEEE Trans. Patt. Anal. Machine Intell.*, vol. 19, pp. 696-710.

[7] N. Tsapatsoulis, N. Doulamis, A. Doulamis and S. Kollias, "Face Extraction from Non-uniform Background and Recognition in Compressed Domain," *Proc. of IEEE Inter. Confer. on Acoustics Speech and Signal Processing, (ICASSP)*, vol. 5, pp. 2701-2704, Seattle, USA, May 1998.

[8] D. H. Ballard and C. M. Brown. *Computer Vision*. Prentice Hall, Englewood Cliffs, New Jersey.

[9] A. Doulamis, N. Doulamis and S. Kollias, "Retrainable Neural Networks for Image Analysis and Classification," *Proc. of IEEE Inter. Confer. on Systems, Man & Cybern. (SMC)*, pp. 3558-3562, Orlando Oct. 1997.

[10] P. Salembier and H. Sanson, "Robust Motion Estimation Using Connected Operators," *Proc. of IEEE Inter. Confer. on Image Processing, ICIP*, pp. 77-80, vol. 1, Santa Barbara, USA, October 1997.

[11] H. Wang and Shih-Fu Chang, "A Highly Efficient System for Automatic Face Region Detection in MPEG Video Sequences," *IEEE Trans. on Circuits & Systems for Video Techn.*, special issue on Multimedia Systems and Technologies, 1997.

[12] A. Aydin Alatan, E. Tuncel and L. Onural, "Object Segmentation via Ruled-based Data Fusion," *Proc. of Work. On Image Analysis for Mult. Inter. Services, (WIAMIS)*, pp. 51-55, Louvain-la-Neuve Belgium, June 1997.

[13] A. Doulamis, N. Doulamis and S. Kollias, "A Neural Network Based Scheme for Unsupervised Video Object Segmentation," *Proc. of IEEE Inter. Confer. on Image Processing (ICIP)*, Chicago, USA, Oct. 1998.

[14] J. Zhang, J. Modestino and D. Langan, "Maximum-Likelihood Parameter Estimation for Unsupervised Stochastic Model-Based Image Segmentation," *IEEE Trans. on Image Processing*, vol. 3, nol. 4, pp. 404-429, July 1994.

[15] A. Dempster, N. Laird and D. Rubin, "Maxium Likelihood from Incomplete Data via the EM Algorithm," *J. Roy. Soc. Statist.*, Series B., no. 1, pp. 1-38, 1977.

[16] D. Titterington, A. Smith and U. Makov. *Statistical Analysis of Finite Mixture Distributions*. New York, Wiley, 1985.

[17] J. Zhang and J. Modestino, "A Statistical Model-fitting Approach to Cluster Validation with Applications to Image Processing," *IEEE Tran. Patt. Anal Machine. Intell.*, vol. 12, pp. 1009-1017, Oct. 1990.

[18] R. Kinderman and J. L. Snell. *Markov Random Fields and their Applications*. Providence. America Mathematical Society 1980.

[19] S. Geman and D. Geman, "Stochastic Relaxation, Gibbs Distributions, and the .Bayesian Restoration of Images," *IEEE Trans. on Patt. Anal. Machine Intell.*, vol. PAMI-6, pp. 721-741, 1984.

[20] J. Besag, "On the Statistical Analysis of Dirty Images," *J. Roy. Soc. Statist.*, Series B. vol. 48, pp. 259-302, 1986.

30

Estimation of 3D Motion and Structure of Human Faces

Y. Xirouhakis, G. Votsis and A. Delopoulos
Computer Science Division,
Department of Electrical Engineering,
National Technical University of Athens,
Athens GR-15773, GREECE

1 Introduction

The extraction of motion and shape information of three dimensional objects from video sequences emerges in various applications especially within the framework of the MPEG-4 and MPEG-7 standards. Particular attention has been given to this problem within the scope of model-based coding and knowledge-based $3D$ modeling. In this chapter, a novel algorithm is proposed for the $3D$ reconstruction of a human face from $2D$ projections. The obtained results can contribute to several fields with an emphasis on $3D$ modeling and characterization of human faces.

The problem of modeling human faces from their projections has been tackled before by several authors such as [1, 2]. However, the accuracy of the obtained results is still far from the desirable. In [1] an algorithm is provided for the estimation of $3D$ motion and structure from two orthographic projections (two frames) having available a relatively accurate initial guess of the object structure, for example a generic wireframe model for the case of human faces. In [2] a modification to the above algorithm is proposed in order to increase its tolerance to errors in the initial guess.

In the aforementioned approaches it is assumed that the human face is a generally rigid object. Local deformations of the face, such as mouth and eye movement, are generally tackled separately as local motions. This constraint is also set in this work, i.e. the human head is assumed to be globally rigid. Similarly to other approaches, it is assumed that there is no camera movement and that the video is obtained under orthographic projection. The latter, as it can be seen in [6], yields

333

334

satisfactory results and significantly reduces computational cost.

In the present work, the algorithm presented in [5] is utilized in order to obtain an accurate solution for the object's $3D$ motion and structure (without any initial guess for the object's $3D$ structure) employing three orthographic projections. The algorithm exploits the information provided by the fact that the object is a human head, in the sense that keypoints and keylines are extracted increasing the accuracy of the $2D$ motion estimates. As it will be quoted in the sequel, accuracy in the $2D$ motion estimates plays a significant role to the accuracy of depth estimates.

2 Overview

The first step of the presented algorithm is to obtain $2D$ motion estimates of the most characteristic face elements (iris and edges of the eyes, nasal points, lips edges, etc.). Reliable estimation of these feature correspondences forms a crucial presumption for $3D$ information extraction. In this work, reliable estimation of corresponding points is obtained through the use of adjustable template matching involving correlation coefficient calculations for $2D$ feature templates (see Section 3).

The second step uses these $2D$ point correspondences in order to yield estimates of the motion parameters corresponding to the actual $3D$ movement of the human face. These include the rotation angles and axes as well as the translation vectors that describe the movement of the object in $3D$ space (see Section 4).

The third step computes $3D$ shape information (depth and relative position) of the initial characteristic points. Computation is based on the $2D$ point correspondences obtained in the first step and the rotation/translation estimates of the second step. A variant of block-based $2D$ motion estimation algorithms that exploit the previously estimated motion parameters of the face are utilized for all points on the image (see Section 5). The estimates that these algorithm yield for the remaining - non characteristic - points of the face are next used for the accurate extraction of $3D$ position information of the human face.

Simulated experiments illustrating the performance of the proposed algorithms have been included.

3 Facial Feature Extraction

Research on face recognition has shown that some specific features of the face can be easily detected. Among these, we choose to extract those that can be considered as rigid points, throughout an arbitrary video sequence including the face. The relative solidity of the feature points is a crucial issue, since it affects the correct calculation of the rotation matrices. Such features can, for example, be the eyes and nasal points. As it will be seen, the accuracy of computation of motion parameters depends upon the reliability of the extraction of a sufficient number of characteristic points (at least four) on the complex surface of the face.

Additionally, the above set is enriched by features that may deform along time, so as to achieve $3D$ shape recovery by attaining a dense wireframe mask of the human face. Thus, supplementary points may be placed, e.g. on the chin and cheek contours and the lips outline.

A variety of techniques have been proposed for extracting characteristic points in images containing faces. These techniques comprise simple or more elaborate tools such as curvature extrema, Fourier descriptors, generalized symmetry operators, the KL transform, the Hough transform, adjustable template matching, Gabor wavelet decomposition and knowledge-based vision systems [9]. In our approach, a few variations of template matching techniques are adopted.

3.1 MATCHING 2D FEATURES

Firstly, some characteristic 3D regions of a generic head model are isolated and projected onto the $2D$ space, in order to derive generic templates containing the features to be tracked. In fact, $3D$ generic templates are utilized in order to obtain a set of views (affine transformations) for each characteristic region of the face. This task aims at finding the region in the 1st frame that matches best each generic template. Since in the first frame we may not have a frontal view of the human face, one element from each set is appropriate. In this way, each characteristic region is derived in the first frame. In the remaining frames the characteristic regions can be obtained similarly. Equivalently, the generic templates can be subsituted by the ones matched in the 1st frame, after they are subjected to a number of affine transformations forming new sets of characteristic templates.

Template matching is being performed with the use of correlation coefficient. Supposing that images are viewed as intensity matrices, the correlation coefficient between the prototype (matrix **A**) and the repeatedly audited blocks of an appropriately selected area of the face (matrix **B**) is computed as follows:

$$r = \frac{\sum_{n_1} \sum_{n_2} \mathbf{A}(n_1, n_2) \, \mathbf{B}(n_1, n_2)}{\sqrt{\sum_{n_1} \sum_{n_2} \mathbf{A}^2(n_1, n_2) \sum_{n_1} \sum_{n_2} \mathbf{B}^2(n_1, n_2)}} \tag{1}$$

The template of each group that achieves maximum correlation coefficient with its corresponding block of the image of interest, is the one used. If, however, the maximum correlation coefficient for a specific group doesn't exceed a confidence threshold, the specific feature is considered as not tracked. The ease for each feature's detection has been determined through years of research by computer engineers and neuro-physiologists [9]. The most reliable features are extracted first. The knowledge of their position restricts the search area for the rest of the features, making their tracking faster and more confident.

Subsequently, hierarchical filtering through 'steerable filters' is used to detect facial keypoints. The term 'steerable filters' is used to describe the reconstruction of deformed filter kernels that are created by superimposing a small number of orthogonal basis functions, as described by the following formula:

$$\mathbf{F}_\alpha(\mathbf{x}) = \sum_{k=1}^{N} b_k(\alpha)\, \mathbf{A}_k(\mathbf{x}) \qquad (2)$$

where $\mathbf{F}$ denotes the kernel, $\mathbf{A}_k$ denote the basis functions, b_k denote their weights and α is a general multi-deformation [10]. In this way, more complex filters are composed by simple ones, while filtering with more sophisticated filters is reduced to convolving the image with a small number of basis functions. Additionally, any number of basis functions reconstructs all deformed kernels, affecting only the reconstruction quality. In our case, low quality approximations are adequate, a fact that results in a small number of basis functions. For the detection of the iris, a model driven approach is used as in [10].

Alternatively, easier processing tools are being practiced, such as the integral projections towards vertical and horizontal orientation, as shown by the following relations respectively:

$$\mathbf{v}(x) = \sum_{y=y_1}^{y_2} \mathbf{I}(x,y) \qquad (3)$$

$$\mathbf{h}(y) = \sum_{x=x_1}^{x_2} \mathbf{I}(x,y) \qquad (4)$$

Simple image processing tools proved adequate for plain detection, while the more elaborate tool of hierarchical filtering is used for complex point extraction. Thus, characteristic face points are detected accurately within the tracked templates.

3.2 LINE MATCHING

Adjustable templates consisting of parabolas [11] are used for representing features, such as the chin contour. In this chapter, such features will be utilized in order to restrict the search area during the motion estimation procedure for all (non-characteristic) points on the face.

The area of the jaw that is visible in each frame can be automatically restricted, since the blocks already detected are enough to indicate its approximate position. By performing edge detection to this restricted area we find a set of points that are candidates for the description of the jaw line. Given these points, the estimation of the jaw line is provided by its approximation by a parabola or a second order line. In the case of the second order line approximation, a line described by $y = \alpha x^2 + \beta x + \gamma$ is derived by computing the unknowns α, β, γ. Thus the candidate points located by edge detection techniques are approximated with the appropriate second order line. This is performed with a Least Squared distance approximation of that set. By defining the distance of each point from the curve as $d_i = y_i - \alpha x_i^2 - \beta x_i - \gamma$, and by minimizing the squared sum for all the candidate points, we get:

$$\theta_{LS} = (\mathbf{X}^T \mathbf{X})^{-1} \mathbf{X}^T \mathbf{y} \tag{5}$$

where $\mathbf{X}$ is the $N \times 3$ table containing in line i the vector $[x_i^2\ x_i\ 1]$, $\mathbf{y}$ is the $N \times 1$ vector containing in line i the element y_i, while θ_{LS} is the vector containing the factors α, β, γ.

The above procedure is performed two times. After the first iteration and the determination of the unknown parameters, the points whose distance from the curve exceeds a threshold are discarded. In the second iteration, the calculations are performed using only the remaining points. Thus, the curve is now approximated precisely.

4 Computation of Motion Parameters

Earlier approaches in the problem of estimating 3D motion include [7, 8] in the case of perspective projections, and [4, 1, 2, 5] in the orthographic case. In this paper, the algorithm proposed in [5] is employed, since it provides an exact, computationally attractive solution even in the presence of noise. It should be mentioned that orthography poses no considerable limitation in the application of the proposed algorithm, as it is a reasonable approximation to perspective projection, when the object is relatively far away from the camera.

4.1 DEFINITIONS

The movement of a rigid object in $3D$ space can be seen as superposition of a $3D$ rotation and a $3D$ translation. Concequently, when a point (x, y, z) on the object moves to (x', y', z'), it holds:

$$\begin{bmatrix} x' \\ y' \\ z' \end{bmatrix} = \mathbf{R} \begin{bmatrix} x \\ y \\ z \end{bmatrix} + \mathbf{T} \tag{6}$$

where $\mathbf{R}, \mathbf{T}$ are the rotation and translation matrices respectively. Generally:

$$\mathbf{R} = \begin{bmatrix} r_{11} & r_{12} & r_{13} \\ r_{21} & r_{22} & r_{23} \\ r_{31} & r_{32} & r_{33} \end{bmatrix}, \quad \mathbf{T} = \begin{bmatrix} T_1 & T_2 & T_3 \end{bmatrix}^T \tag{7}$$

Assuming a rotation about an arbitrary axis, matrix $\mathbf{R}$ can be written in cartesian coordinates in terms of the rotation axis $\mathbf{n} = \begin{bmatrix} n_1 & n_2 & n_3 \end{bmatrix}^T$ and rotation angle a, as:

$$\mathbf{R} = \begin{bmatrix} n_1^2 + (1 - n_1^2)\cos a & n_1 n_2(1-\cos a) - n_3 \sin a & n_1 n_3(1-\cos a) + n_2 \sin a \\ n_1 n_2(1-\cos a) + n_3 \sin a & n_2^2 + (1 - n_2^2)\cos a & n_2 n_3(1-\cos a) - n_1 \sin a \\ n_1 n_3(1-\cos a) - n_2 \sin a & n_2 n_3(1-\cos a) + n_1 \sin a & n_3^2 + (1-n_3^2)\cos a \end{bmatrix} \quad (8)$$

The computation of motion parameters involves the estimation of matrices $\mathbf{R}$, $\mathbf{T}$. In fact, as far as the rotation matrix $\mathbf{R}$ is concerned, the estimation of the rotation axis $\mathbf{n}$ and rotation angle a is necessary. It must be noticed that under orthography, the component T_3 of the translation matrix, which is perpendicular to the projection plane, cannot be estimated by any scheme. The computation of the structure parameters involves estimation of the z, z' coordinates for each point in (6), given the point correspondences (given in fact coordinates x', x, y', y) and having computed the motion parameters. However, absolute depth cannot be estimated by any scheme under orthographic projections, as in this case absolute depth information is lost due to the type of projection. In fact, relative point depth is estimated, providing $3D$ shape information.

4.2 REVIEW OF THE ALGORITHM

In this subsection, we will follow the notations used in [5] for compatibility matters.

Ullman in his classical work [3] proved that four point correspondences over three frames are sufficient to yield a unique solution to motion and structure up to a reflection. The latter is verified in all proposed algorithms [4, 5]. In this sense, a small set of at least four relatively exact point correspondences, as described in the previous section, are needed for computing the rotation and translation matrices.

According to Stokes assertion, any movement such as (6) can be described by a translation $\mathbf{T}$ equal to the point correspondence of an arbitrarily chosen point P on the object, and a rotation $\mathbf{R}$ about an appropriate axis and considering point P as the rotation origin. Concequently, by picking a point P and assigning P to the world origin $(0,0,0)$, translation $\mathbf{T}$ is computed as the point correspondence of P. In this sense, the computation of $\mathbf{T}$ is straightforward.

The estimation of matrix $\mathbf{R}$ is achieved through the use of the so-called 2x2 matrices $\tilde{\mathbf{K}}$ which can be expressed in terms of the differential reference and motion vectors for every three points as,

$$\Delta \mathbf{v} = \tilde{\mathbf{K}} \, \Delta \mathbf{r} \quad (9)$$

where $\Delta \mathbf{r} = [\mathbf{r_3} - \mathbf{r_1} \ \ \mathbf{r_2} - \mathbf{r_1}]$, and $\Delta \mathbf{v} = [\mathbf{v_3} - \mathbf{v_1} \ \ \mathbf{v_2} - \mathbf{v_1}]$. Vectors $\mathbf{v_i}$, $\mathbf{r_i}$ are the motion and the reference vectors respectively for a point on 1st frame.

Similarly, matrix $\mathbf{K}$ can be used, defined as:

$$\Delta \mathbf{r}' = \mathbf{K} \, \Delta \mathbf{r} \tag{10}$$

where $\Delta \mathbf{r}' = [\mathbf{r}'_3 - \mathbf{r}'_1 \ \ \mathbf{r}'_2 - \mathbf{r}'_1]$. Vector $\mathbf{r}'_i$ is the reference vector of the corresponding point on the 2nd frame. It can be seen that matrices $\mathbf{K}$, $\tilde{\mathbf{K}}$ are the same for any three points on the same $3D$ planar surface. The rotation matrices can then be obtained using some manipulations, in terms of eigenvalues and eigenvectors, as it will be quoted in the sequel. In order to simplify the following equations, let matrix $\mathbf{L}$ be the adjoint matrix:

$$\mathbf{L} = \mathrm{adj}(\tilde{\mathbf{K}} + \mathbf{I}) = \mathrm{adj}\mathbf{K} \tag{11}$$

where $\mathbf{I}$ denotes the 2x2 identity matrix. Since $\mathbf{L}$ also characterizes the projection of a $3D$ planar surface to the projection plane, let $\mathbf{L}_{Rj}$ be the $\mathbf{L}$-matrix that corresponds to rotation $\mathbf{R}$ of the plane j. In accordance to Ullman's assertion, four $\mathbf{L}$-matrices contain sufficient information, in order to estimate the motion and structure parameters. In fact, employing four point correspondences over 3 frames, two $\mathbf{L}$-matrices are derived corresponding to movement frame 1-2 and another two corresponding to movement from frame 1-3. Without loss of generality, let for simplicity matrix $\mathbf{R}$ correspond to movement frame 1-2 and matrix $\mathbf{S}$ correspond to movement frame 1-3. Using more than four $\mathbf{L}$-matrices, a Weihgted Least Squares scheme is provided in [5] to improve the estimation of the motion parameters.

Let the rotation matrix $\mathbf{R}$ be written as:

$$\mathbf{R} = \begin{bmatrix} \mathbf{r}_0 & \mathbf{r}_1 \\ \mathbf{r}_2{}^T & r_{33} \end{bmatrix} \tag{12}$$

where $\mathbf{r}_0 = \begin{bmatrix} r_{11} & r_{12} \\ r_{21} & r_{22} \end{bmatrix}$, $\mathbf{r}_1 = \begin{bmatrix} r_{13} & r_{23} \end{bmatrix}^T$ and $\mathbf{r}_2 = \begin{bmatrix} r_{31} & r_{32} \end{bmatrix}^T$.

Let matrix $\Delta \mathbf{L}_{Rk}$ be the difference between two arbitrary $\mathbf{L}$-matrices corresponding to the same rotation.

$$\Delta \mathbf{L}_{Rk} = \mathbf{L}_{Ri} - \mathbf{L}_{Rj} \tag{13}$$

Then, the unknown vector $\mathbf{r}_1$ is computed within a scalar ambiguity ρ as the unit-norm eigenvector of the summation of matrices:

$$\mathbf{Y}_R = \frac{1}{M} \sum_{k=1}^{M} \Delta \mathbf{L}_{Rk}{}^T \, \Delta \mathbf{L}_{Rk} \tag{14}$$

corresponding to its smallest eigenvalue. In this sense, let $\mathbf{r}_1 = \rho \, \mathbf{c}_1$. Subsequently, $\mathbf{r}_2$ is estimated within the same scalar ambiguity ρ as

$$\mathbf{r}_2 = -\frac{1}{N} \sum_{j=1}^{N} \mathbf{L}_{Rj} \, \mathbf{r}_1 = -\frac{1}{N} \sum_{j=1}^{N} \rho \, \mathbf{L}_{Rj} \, \mathbf{c}_1 = \rho \, \mathbf{c}_2 \tag{15}$$

In the above equations notice that N, M are the numbers of $\mathbf{L}, \mathbf{\Delta L}$ matrices involved in the computations respectively. Generally $M \cong \frac{N}{2}$. Let also λ_R denote the larger eigenvalue of matrix $\mathbf{Y}_R$. Notice as well that $r_{33}^2 = 1 - \rho^2$, since norm($\mathbf{c}_1$) = norm($\mathbf{c}_2$) = 1. Similar equations hold for rotation matrix $\mathbf{S}$. Let similarly λ_S denote the respective eigenvalue and $\mathbf{s}_1 = \sigma \, \mathbf{d}_1$, $\mathbf{s}_2 = \sigma \, \mathbf{d}_2$. It can be proved that the ratio $\frac{\sigma}{\rho}$ can be estimated up to a sign ambiguity as

$$w^2 = (\frac{\sigma}{\rho})^2 = \frac{\lambda_S}{\lambda_R} \tag{16}$$

where one of the solutions can be easily rejected. Finally, r_{33} and s_{33} are computed from:

$$\begin{bmatrix} w\mathbf{J}\mathbf{c}_2 & -\mathbf{J}\mathbf{d}_2 \end{bmatrix} \begin{bmatrix} r_{33} \\ s_{33} \end{bmatrix} = -\frac{1}{N} \sum_{j=1}^{N} \begin{bmatrix} \mathbf{L}_{Sj} & \mathbf{L}_{Rj} \end{bmatrix} \begin{bmatrix} \mathbf{J}\mathbf{d}_1 \\ -w\mathbf{J}\mathbf{c}_1 \end{bmatrix}. \tag{17}$$

where $\mathbf{J} = \begin{bmatrix} 0 & -1 \\ 1 & 0 \end{bmatrix}$. The rotation matrices $\mathbf{R}$, $\mathbf{S}$ are estimated on the basis of $\mathbf{r}_1, \mathbf{r}_2, r_{33}, \mathbf{s}_1, \mathbf{s}_2, s_{33}$. Thus all the motion parameters are obtained.

4.3 ESTIMATION IN THE PRESENCE OF NOISE

In order to tackle the problem of noisy motion vector estimates, the proposed algorithm is extended in [5] to employ a large number of projected points. However, in this work only a small number of facial points is used. Moreover, these point correspondences are given great confidence, so the employment of the solution in noisy environments is not necessary. In fact, in order to take into consideration the respective confidence with which each facial characteristic is estimated, confidence weights are incorporated in the sums of (14) and (15).

5 Computation of Structure Parameters

Given the rotation matrices, the relative depth of the points on each of the three frames, can be estimated employing Equation (6). The z, z' coordinates are subsequently computed. In fact, it is possible to recover depth only for points whose correspondences are known, while accuracy in depth estimation, in general, requires accuracy in the estimation of all respective point correspondences. Thus, in order to recover depth for all points on the human face, a block-based motion estimation scheme is performed for all points on the images. To achieve faster and more accurate motion estimation, we have to exploit all information gained in the first step, which involves relatively precise detection of certain facial keypoints and keylines. For example, to confine the area of the cheeks -which is generally smooth, and is therefore very sensitive as far as motion estimation is concerned-we used the limits of the neighbouring blocks tracked during the first step of the algorithm. At the same time, the rotation matrices obtained in the second step yield an a-priori estimation of the dense motion field and impose a constraint to the orientation of motion vector estimates. The acquired motion field, although not perfectly accurate, is appropriate for 3D reconstruction of the human face. In order to improve the reconstructed mask, the latter can then be compared to a generic wireframe model.

6   Simulations

The model depicted in the figures is the model 'Conor' found in Headus (http: //www.headus.com.au/ vrml/conor10.wrl). It is a scanned face of a child mapped with its texture. The model itself was not available, so it was rotated on-line in the vrml environment and grabbed. For this reason, the following results are not arithmetically verified. However, the efficiency of the algorithm is visually apparent.

In Figures $1(a) - (c)$, the three input frames are depicted with the tracked templates indicated. As it was mentioned in Section 3, $3D$ templates are employed for the production of sets of affine-transformed $2D$ templates. Such a generic template (eye template) is depicted in Figure $2(d)$. Localization of specific points on the human face with the use of hierarchical filtering and simple tools is performed. The derived point correspondences between frames 1 and 2 are depicted in Figure $1(d)$.

Using the aforementioned methodology for the computation of rotation parameters, we obtain angles $a = 37.8^o$, $b = 39.7^o$ and rotation axes $\mathbf{n}_a = [\ 0.54\ 0.84\ 0.07\ ]^T$ and $\mathbf{n}_b = [\ 0.27\ 0.96\ 0.03\ ]^T$ respectively. The obtained motion parameters were experimentally verified subjecting a generic wireframe model to the same rotations.

In the sequel a block matching technique is employed in order to obtain the final structure parameters. Significant help for this scheme is provided by the previously described line-matching scheme as well as by the tracked templates and the $3D$ motion parameters. Results of the line-matching scheme for the jaw line are depicted in Figure $2(a)$. The restricted search area for the cheek points is depicted in Figure $2(b)$. The motion field derived through the motion estimation procedure between frames 1 and 2 is illustrated in Figure $2(c)$, where only an indicative set of motion vectors is depicted. The reconstructed human mask is depicted in Figure $2(e)$. Since the motion parameters were estimated in accurate fashion, the reconstructed model resembles well the human face apart from an amount of noise. This noise in depth coordinates is due to the noise in the motion field estimates. Using some heuristical smoothing techniques, the model is slightly improved in Figure $2(f)$. Study and development of more systematic $3D$ smoothing techniques is currently in progress.

7 Conclusions - Future Work

In the current work, the problem of $3D$ motion and structure extraction is treated in the case of human faces. This is achieved using a generalized algorihtm for the estimation of $3D$ motion and structure parameters of rigid objects under orthography. In the particular case of human faces, the algorithm performance is aided by facial feature extraction schemes that employ adjustable template matching techniques. $2D$ point correspondences of reliably tracked facial feature points is applied to determine the rotation matrices, whereas the obtained $2D$ structure information of the human face is used for the extraction of $3D$ structure through a variant block-based motion estimation technique. Relevant experimental results were presented using views of a head model.

Study and development of systematic $3D$ smoothing and texture-mapping techniques is currently in progress in order to obtain realistic presentations of the obtained results.

References

[1] K.Aizawa, H.Harasima, and T.Saito, "Model-based analysis-synthesis image coding (MBASIC) system for a person's face," *Signal Proc.: Image Comm.,* no.1, pp.139-152, 1989.

[2] G.Bozdagi, A.M.Tekalp, and L.Onural, "An Improvement to MBASIC Algorithm for 3-D Motion and Depth Estimation," *IEEE Trans. Im. Proc.,* vol.3, no.5, pp.711-716, Sept. 1994.

[3] S.Ullman, "The Interpretation of Visual Motion," Cambridge, MA, MIT Press, 1979.

[4] T.S.Huang and C.H.Lee, "Motion and structure from orthographic projections," *IEEE Trans. PAMI,* vol.11, pp.536-540, May 1989.

[5] A.Delopoulos and Y.Xirouhakis, "Robust Estimation of Motion and Shape based on Orthographic Projections of Rigid Objects," *accepted in Tenth IMDSP Workshop (IEEE),* Alpbach Austria, July 1989.

[6] A.M.Tekalp, "Digital Video Processing," *Prentice Hall,* 1995.

[7] R.Y.Tsai and T.S.Huang, "Uniqueness and Estimation of Three-Dimensional Motion Parameters of Rigid Objects with Curved Surfaces," *IEEE Trans. PAMI,* vol.6, pp. 13-27, Jan. 1984.

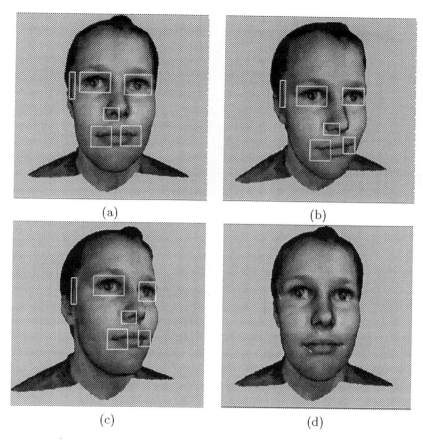

Figure 1: (a)Matched templates (frame 1), (b)Matched templates (frame 2), (c)Matched templates (frame 3), (d)Characteristic point correspondences

[8] J.Weng, N.Ahuja and T.S.Huang, "Optimal Motion and Structure Estimation," *IEEE Trans. PAMI,* vol.15, no9, pp. 864-884, September 1993.

[9] R.Chellappa, C.L.Wilson, and S.Sirohey, "Human and Machine Recognition of Faces: A Survey," *Proc. of IEEE,* vol.83, no.5, pp.705-740, May 1995.

[10] M.Michaelis, R.Herpers, L.Witta, and G.Sommer, "Hierarchical Filtering Scheme for the Detection of Facial Keypoints," *Proc. of ICASSP,* 1997.

[11] M.Kampmann, "Estimation of the Chin and Cheek Contours for Precise Face Model Adaptation," *Proc. of ICIP,* 1997.

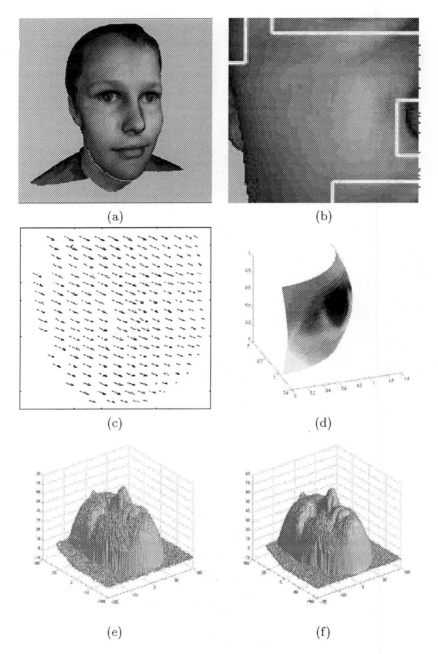

(a)

(b)

(c)

(d)

(e)

(f)

Figure 2: (a) Detected left chin line (frame 2), (b) Detected left cheek region (frame 2), (c) Indicative Motion Estimates, (d) 3D eye template, (e) 3D Reconstructed Human Face, (f)Reconstructed after 3D smoothing

31

Face Recognition Based on Multiple Representations : Splitting the Error Space

Miltiades Leonidou, Nicolas Tsapatsoulis, and Stefanos Kollias
Department of Electrical and Computer Engineering
National Technical University of Athens
e-mail:ntsap@image.ntua.gr

1. Introduction

In recent years numerous algorithms have been proposed for face recognition [1] and much progress has been made toward this direction. Most of these algorithms achieve high rates of correct recognition only under very small variations in illumination, scale, facial expressions and perspective angle or pose transformations [2, 3]. The inefficiency of the presented algorithms under extreme variations of the above factors is quite reasonable. Recent investigations have shown that the proposed face representation schemes derive greater variability in a given face under changes in scale, illumination, perspective angle and expression, than different faces when these three factors are held constant. In other words intra-class variance is larger than the inter-class one [2].

A generic block diagram of a face recognition scheme is illustrated in Figure 1 [4]. The main processing stage consists of the feature extraction and the recognition part. Feature extraction can overcome the problem of dimensionality, improve the generalization ability of classifiers and reduce the computational requirements of face classification [5, 6, 7]. The adopted features should describe each face in a unique way suitable for recognition in a less dimensional space. The main difficulty is the selection of a suitable and unique set of features describing each face in a way that will help the computerized system reach the right conclusions.

When choosing appropriate features, extensive processing of these features takes place, in order to eliminate the impact of all factors that increase intra-class variance. The feature extraction problem leads to a real puzzle when problems caused by position,

345

orientation, luminance, scale and texture variances are to be taken into account. In other words, the unique feature set has to be extensively processed so as to acquire invariance to all these distortions but yet, retain all necessary descriptive properties leading to successful identifications.

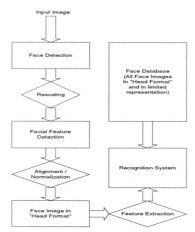

Figure 1: Architecture of a generic face recognition system.

The proposed scheme aims at recognizing faces irrespectively of position, orientation, scale, luminance, and texture variances in a different approach. We claim that the selection of multiple and mutually exclusive representations can overcome the difficulty of finding a global set of features. Multiple and complementary representations are adopted so as to split the error domain (the union of orientation, scale, luminance and texture variances) leading to successful parallel partial solutions for the different disciplines of face recognition problems. This is accomplished by embedding each representation with a suitable identification algorithm. In a subsequent stage a rule-based decision mechanism selects the most reliable solution, taking into account the partial identification results and an estimation of the variance that possibly caused the difference between the test face and its stored version.

2. Face Location and Segmentation

In the preprocessing stage the aim is to detect and isolate the face in the input image. This is not a trivial problem but we do not actually emphasize on it. Assuming that the face is the largest compact foreground object (e.g. mug-shot photos), the following procedure is adopted:

- The contrast of the image is enhanced and a threshold, obtained from the local image statistics, is applied in order to convert to a binary image. A further step utilizing morphological processing, including the application of the *closing* operator [8], is required to fill small holes in the foreground objects. By the end of this step foreground objects are appropriately masked.

- The *Distance Transform* of the masked area is computed using *chamfer metrics* [8]. Under the assumption of face being the largest compact foreground object the point with maximum distance from the background should be within the face area. Let us call this point 'marker'.

- Starting from the 'marker' and using an octagon-like structuring element we apply successive dilations until the critical size area is reached. The octagon-like structuring element is an eight-side polygon whose opposite vertical and horizontal sides are equal and the ratio between them is 5:7. This structuring element was selected so as to approach as close as possible the face anatomy. Critical size is defined as the maximum size of the structuring element that fits into the face area, while critical size area is the area occupied by this size of structuring element.

At the end of the procedure described above the face is efficiently isolated and the recognition task that follows can be applied independently of the face position in the input image (see Figure 2).

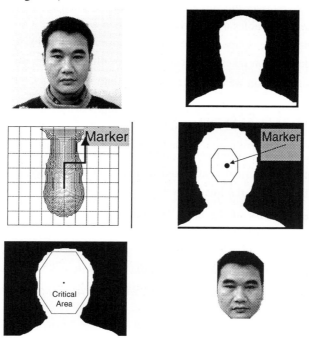

Figure 2: Face location segmentation

3. Feature Extraction

The face database, which is used in the recognition procedure, includes three feature sets extracted from each face. The first set is based on geometrical moments and the other two are derived from the Singular Value Decomposition of the image.

3.1 Scale Invariant Representation

Internal information of objects can be derived from geometrical moments [9]. Once the masked face is extracted ("head format"), the calculation of a set of geometrical moments, taking into account pixel values and their position in the "head format" area leads to a space of reduced dimensionality that describes facial characteristics. In the proposed scheme the calculation of moments up to the third order create a 7-D space and each face is represented in this space by a 7-D vector. For a gray-scale digital image space with intensity function $g(x,y)$, geometrical moments up to the third order are expressed according to the following equation:

$$m_{uv} = \sum_x \sum_y g(x, y) x^u y^v \qquad\qquad u, v=0,1,2,3 \qquad\qquad (1)$$

The sum is calculated over the sub-area specified by the facial mask. The definition of geometrical moments enables us to derive information about the correlation of areas and their texture within the shape. These moments are invariant to scale and rotation. However, this description takes into account the absolute position of the face. In order to eliminate this restriction we define geometrical central moments, μ_{uv} :

$$\mu_{uv} = \sum_x \sum_y g(x, y)(x-\bar{x})^u (y-\bar{y})^v \quad \text{where } \bar{x} = \frac{m_{10}}{m_{00}} \text{ and } \bar{y} = \frac{m_{01}}{m_{00}} \qquad (2)$$

These moments take as their reference point the center of mass of the face. The above definition enables geometrical centralized moments to be independent of the absolute position of the face within the image. In the case of digital, binary images ($\bar{x}, \bar{y}$) is the geometric center of mass of the face. The geometrical centralized moment vector is analytically calculated as follows:

$$\mu_{00} = m_{00}, \qquad \mu_{10} = \mu_{01} = 0, \qquad \mu_{20} = m_{20} - \bar{x}m_{10}, \qquad \mu_{02} = m_{02} - \bar{y}m_{01},$$

$$\mu_{11} = m_{11} - \bar{y}m_{10}, \qquad \mu_{30} = m_{30} - 3\bar{x}m_{20} + 2\bar{x}m_{10}, \qquad \mu_{03} = m_{03} - 3\bar{y}m_{02} + 2\bar{y}m_{01},$$

$$\mu_{12} = m_{12} - 2\bar{y}m_{11} - \bar{x}m_{02} + 2(\bar{y})^2 m_{10}, \qquad \mu_{21} = m_{21} - 2\bar{x}m_{11} - \bar{y}m_{20} + 2(\bar{x})^2 m_{01}.$$

The geometrical central moments can be normalized as follows:

$$n_{ij} = \frac{\mu_{ij}}{(\mu_{00})^k}, \qquad k = \frac{i+j}{2} + 1 \tag{3}$$

The creation of a normalized 7-D space is being generated from the above equation. In this space each face is represented by a vector, thus the face identification problem changed into a matching decision problem in a small dimensional space. Normalized geometrical centralized moments can give information about the face texture independently of variances to the absolute position, scale and size of the face within the image. The vector describing each face in the specified space (Scale Invariant Representation-SIR) is $SIR = [n_{00}, n_{20}, n_{02}, n_{30}, n_{03}, n_{12}, n_{21}]^T$

It has being suggested that clustering of the feature vectors provide more reliable results in the problems of face identification [5]. In this scheme clustering is achieved by using Self-Organized Maps (SOM). SOM, introduced by Kohonen [10], is an unsupervised learning process, which learns the distribution of a set of patterns without any class information. A pattern is projected from an input space to a position in the map; information is coded as the location of an activated node. SOM was selected, among other classification or clustering techniques, because it provides a topological ordering of the classes. Similarity of input patterns is preserved in the output of the process making SOM especially useful in the classification of data, which includes a large number of classes.

The training set for the SOM consists of the SIR vectors of the faces. Clustering is provided by the recognition of correlation and hidden similarities among the input vectors. In this way each face possesses one position in a specific cluster (Figure 3). SOM is trained according to the Kohonen rule and thus preserves the topology of the input space in a way similar to the function of the human mind in similar cases.

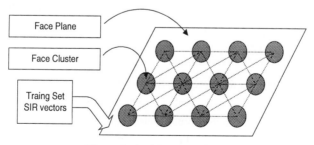

Figure 3: Architecture of SOM

3.2 'Content' Invariant Representation

Every image can be regarded as a two dimensional rectangular matrix. It is also known that any real symmetric matrix can be transformed into a diagonal matrix by means of orthogonal transformation and similarly for any general rectangular matrix $\mathbf{A}$ with dimensions $\mathbf{mxn}$ by means of so-called Singular Value Decomposition (SVD) [7]. SVD theory is very useful in cases where it is required to generate a less dimensional space, which preserves most of the energy in the original image. We briefly discuss some of basic theorems of SVD and properties that enable Singular Values to be a suitable representation of the face texture.

Let $\mathbf{A}$ be a real rectangular matrix with dimensions $\mathbf{mxn}$ (suppose m>n, without loss of generality) and rank $(\mathbf{A})=k$. Then there exist two orthonormal matrices $\mathbf{U}_{mxm}$, $\mathbf{V}_{nxn}$ and a diagonal matrix $\mathbf{\Sigma}_{mxn}$ satisfying the equation $\mathbf{A} = \mathbf{U\Sigma V}^{\mathrm{T}}$, where $\mathbf{\Sigma} = diag(\lambda_1, \lambda_2,..., \lambda_k, 0,..0)$ and $(\lambda_1 > \lambda_2 > .. > \lambda_\kappa)$.

Each λ_i^2, is one of the k eigenvalues of $\mathbf{A}^{\mathrm{T}}\mathbf{A}$ as well as $\mathbf{A}\ \mathbf{A}^{\mathrm{T}}$, λ_i is called singular value of matrix $\mathbf{A}$; $\mathbf{U} = (u_1, u_2,.., u_k, u_{k+1},..., u_m)$ and $\mathbf{V} = (v_1, v_2,..., v_k, v_{k+1},..., v_n)$ where u_i, v_i, (i =1,..,k), are column eigenvectors of $\mathbf{A}^{\mathrm{T}}\mathbf{A}$, $\mathbf{A}\ \mathbf{A}^{\mathrm{T}}$ corresponding to eigenvalue λ_i^2, respectively.

We may construct the following column vector consisting of the principal diagonal entries of matrix $\mathbf{\Sigma}_{mxn}$ i.e., $x_{nx1} = \mathbf{\Sigma} \cdot e = [\lambda_1 \lambda_2 .. \lambda_\kappa\ 0..0]$ where $e = (1,1,..,1)_{nx1}^T$.

We call x_{nx1} Singular Value (SV) feature vector of image $\mathbf{A}$. For any real rectangular matrix $\mathbf{A}$, under the constraint of $\lambda_1 > \lambda_2 > .. > \lambda_\kappa$, singular value vector $\mathbf{x}_{nx1}$ is unique. The uniqueness and other important properties [7], such as translation and rotation invariance of vector $\mathbf{x}_{nx1}$ have made it a suitable choice for the 'Content' Invariant Representation (CIR), i.e., $CIR = [\lambda_1 \lambda_2,..., \lambda_\kappa]$.

3.3 Luminance Invariant Representation

This representation attempts to overcome illumination variances between an input face to be identified and its stored version. The method relies on singular value decomposition of both input and stored images in order to isolate their texture and content information respectively. In [11] we have demonstrated that content and high frequency information of the stored faces is extracted from their singular vectors. Therefore singular vectors can successfully represent facial content information. Storing of the three most significant singular vectors from matrices $\mathbf{U_S}, \mathbf{V_S}$ used for the reconstruction of a "face version" leads to a luminance invariant representation matrix (LIR), i.e., $LIR = [u_1, u_2, u_3, v_1, v_2, v_3]$.

4. Identification Algorithm

The architecture of the identification algorithm is illustrated in the block diagram of Figure 4. The newly presented face is located, segmented and its scale, luminance, texture invariant representations are calculated. Each representation follows a different identification procedure and the obtained results are combined in the rule-based machine in order to reach the final identification.

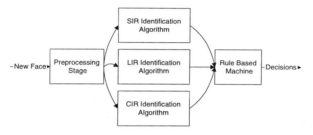

Figure 4: Architecture of Identification Algorithm

4.1 Scale invariant identification procedure

SIR is the descriptive vector, compound with a clustering of stored vectors provided by SOM, used for the scale invariant identification procedure. In order to reach the first identification target the SIR of the input face is used for the simulation of the SOM. Through the networks matrices the nearest cluster of faces is obtained as well as the class members which represent the nearest faces to the newly presented face according to the SI representation (see Figure 5).

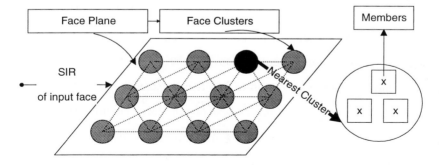

Figure 5: Simulation of SOM

Members of this cluster should be separated and the nearest one chosen. This procedure is accomplished by an appropriate Learning Vector Quantization (LVQ) network that is automatically initialized, trained, and simulated according to the respective cluster. The reason for using both SOM and LVQ networks was derived through experimental results, as shown in Table 1. The competitive measures were the l_1-norm and the correlation coefficient.

Data	Corr. Coef. (%)	l_1-norm(%)	LVQ(%)
Not Clustered	67	79	85
Clustered (SOM)	70	83	96

Table 1. Recognition Rates

4.2 Luminance invariant identification procedure

It has been illustrated in [11] that singular vectors provide the content information within an image. This identification procedure focuses on finding the best matching face in the database according to the content information (invariance to luminance conditions). The method is based on the successful adaptation of the luminance information of all stored faces according to the input image [11].

Let $\mathbf{I} = \mathbf{U}_\mathbf{I} \mathbf{\Sigma}_\mathbf{I} \mathbf{V}_\mathbf{I}^\mathbf{T} = \sum_{j=1}^{k} \lambda_{Ij} \mathbf{u}_{Ij} \mathbf{v}_{Ij}^\mathbf{T}$ be the input image to be identified. In order to

compare $\mathbf{I}$, with stored faces given by $\mathbf{S} = \mathbf{U}_\mathbf{S} \mathbf{\Sigma}_\mathbf{S} \mathbf{V}_\mathbf{S}^\mathbf{T} = \sum_{j=1}^{k} \lambda_{Sj} \mathbf{u}_{Sj} \mathbf{v}_{Sj}^\mathbf{T}$, independently of

illumination variations, instead of applying a similarity measure comparison between $\mathbf{I}$ and $\mathbf{S}$ we reconstruct an approximation matrix $\hat{\mathbf{S}}$, called "face version" of $\mathbf{S}$ according to the equation:

$$\hat{\mathbf{S}} = \mathbf{U}_\mathbf{S} \mathbf{\Sigma}_\mathbf{I} \mathbf{V}_\mathbf{S}^\mathbf{T} = \sum_{j=1}^{k} \lambda_{Ij} \mathbf{u}_{Sj} \mathbf{v}_{Sj}^\mathbf{T} \qquad (4)$$

As it can be seen from the above equation $\hat{\mathbf{S}}$ uses $\mathbf{U}_\mathbf{S}, \mathbf{V}_\mathbf{S}$ as reconstruction singular vector matrices, and $\mathbf{\Sigma}_\mathbf{I}$ as reconstruction singular value matrix. The successful adaptation of the illumination conditions of $\mathbf{I}$ in $\hat{\mathbf{S}}$ is depicted in Figure 6.

The best matching face $\mathbf{S}_\mathbf{m}$ is obtained by selecting the face that minimizes the l_1-norm of the difference between matrices $\hat{\mathbf{S}}_\mathbf{k}$ and $\mathbf{I}$, i.e.,

$$\mathbf{S}_\mathbf{m} = \arg\min_{k} \left| \hat{\mathbf{S}}_\mathbf{k} - I \right| = \arg\min_{k} \left| \mathbf{U}_{\mathbf{S}_\mathbf{k}} \mathbf{\Sigma}_\mathbf{I} \mathbf{V}_{\mathbf{S}_\mathbf{k}}^\mathbf{T} - \mathbf{U}_\mathbf{I} \mathbf{\Sigma}_\mathbf{I} \mathbf{V}_\mathbf{I}^\mathbf{T} \right| \qquad (5)$$

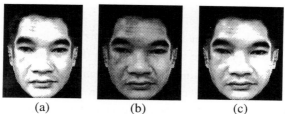

| (a) | (b) | (c) |

Figure 6: (a) Stored face (b) Input face (c) "Face version"

4.3 Recognition based on texture information

It has been illustrated in [11] that the content information within an image is provided by its singular vectors while texture information is retained by preserving its singular values. This part of the algorithm focuses on finding the best matching face in the database according to the texture information and independently of differences in content information (invariance to face translation, rotation and content variance).

Let $\mathbf{I} = \mathbf{U_I \Sigma_I V_I}^T$ be the face image to be identified. In order to concentrate only to the texture information we reconstruct all images $\mathbf{Y}_i$ stored in the database using their singular values and the singular vectors of image $\mathbf{I}$, i.e., $\hat{\mathbf{Y}}_i = \mathbf{U_I \Sigma_{y_i} V_I}^T$.

The best matching face $\mathbf{Y}_m$ is obtained by selecting the face that minimizes the Frobenius norm of the difference between matrices $\hat{\mathbf{Y}}_i$ and $\mathbf{I}$, i.e.,

$$\mathbf{Y}_m = \arg \min_i \left\| \hat{\mathbf{Y}}_i - \mathbf{I} \right\| = \left\| \mathbf{U_I \Sigma_{y_i} V_I}^T - \mathbf{U_I \Sigma_I V_I}^T \right\| = \left\| \mathbf{U_I} (\mathbf{\Sigma_{y_i}} - \mathbf{\Sigma_I}) \mathbf{V_I}^T \right\| \qquad (6)$$

Since $\left\| \hat{\mathbf{Y}}_i \right\| = \left\| \mathbf{Y}_i \right\| = \left\| \mathbf{\Sigma}_{y_i} \right\|$, using equation (6) we obtain:

$$\mathbf{Y}_m = \arg \min_i \left\| \mathbf{\Sigma}_{y_i} - \mathbf{\Sigma_I} \right\| = \arg \min_i \left\{ \sqrt{\sum_j (\lambda_{yj} - \lambda_{Ij})^2} \right\} \qquad (7)$$

where λ_{yj} is the *j-th* singular value of image $\mathbf{Y}_i$.

By using equation (7), reconstruction of images in the database is avoided since the Frobenius norm of the difference between matrices $\hat{\mathbf{Y}}_i$ and $\mathbf{I}$ is computed directly from their singular values.

5. Rule Based Machine

Since the algorithm includes three different recognizing schemes it is necessary to incorporate a rule based final stage to combine them. Rules were created according to the assumptions made, while creating the whole scheme. The success in experimental identification proved the suitability of the whole approach in the face recognition task. The rule-based machine takes as inputs the results of the partial identifications and selects, if any, the most reliable one. In case of a new face it reorganizes the whole database by adding the new face's representative vectors and reorganizing the SOM. This decision-making system is compound by a set of rules IF-THEN-ELSE applied in the sets generated by experimental results.

Let I_S, I_L and I_C represent the identification results of SI, LI and CI representations respectively. We also define the following sets:

S : Stored faces belonging to the same class with the newly presented one, according to the SI representation, obtained from the SOM classification.

L : Set of the m most probable faces according to LIR (m =5% of database size).

C : Set of the m most probable faces according to CIR.

$\mathbf{C}_S$: Subset of **C** containing the k ($k<<m$) most probable faces according to SIR.

$\mathbf{L}_S$: Subset of **L** containing the k most probable faces according to SIR.

$\mathbf{S}_L$: Subset of **S** containing the k most probable faces according to LIR.

$\mathbf{C}_L$: Subset of **C** containing the k most probable faces according to LIR.

$\mathbf{S}_C$: Subset of **S** containing the k most probable faces according to CIR.

$\mathbf{L}_C$: Subset of **L** containing the k most probable faces according to CIR.

Rule's structure is hierarchical one and the whole procedure stops when a value is given to the output X.

Rule1:   IF ($I_S = I_L$ OR $I_S = I_C$) THEN X = I_S ELSEIF $I_L = I_C$ THEN X = I_L

Rule2:   IF ($I_S \in \mathbf{L}_S$ OR $I_S \in \mathbf{C}_S$) THEN X = I_S

Rule3:   IF ($I_L \in \mathbf{S}_L$ OR $I_L \in \mathbf{C}_L$) THEN X = I_L

Rule4:   IF ($I_C \in \mathbf{S}_C$ OR $I_C \in \mathbf{L}_C$) THEN X = I_C

Rule5:   IF X = $\varnothing$ THEN *reorganize database including the newly presented face*

In the above rules an attempt has been made in order to combine the theory and the assumptions made throughout the designing of the scheme, along with the identification techniques and the obtained results.

The first rule confirms identifications in cases where no significant variances exist among the newly presented face and its already stored version and also cases where the only significant difference is at the positioning of the two faces. The rule is based on the fact that in such cases, at least two out of the three parallel identification algorithms will provide a successful identification since the positioning problem is solved during the preprocessing stage. The robustness of the rule is based on the assumption that, since the identification algorithms are designed to give correct results in different conditions, the possibility to give the same false result in different conditions is negligible.

Rule two checks the identification result based on the SIR representation dealing with problems caused by variances in scale and orientation. In such case the subset $\mathbf{L}_S$ or $\mathbf{C}_S$ should include I_S provided that $\mathbf{L}$ or $\mathbf{C}$ include it. Rules three and four rely on the same principle and check the identification results based on the LIR and CIR representations respectively. Rule five refers to the case that the newly presented face has no stored version and therefore the face database should be reorganized to encompass it.

6. Experimental Results

Real images, taken out of the face database created at the University of Bern, are used to illustrate the performance of the proposed method. Facial images to be identified were tested independently through the three identification algorithms and through the overall scheme. A total of 100 faces were used, divided according to the variation applied to them. The experimental results, shown in Table 2, depict the effectiveness of using an exclusive representation for each variation. Through the rule-based machine, an overall recognition rate of 93% was achieved.

Variation	Scaling (%)	'Content' (%)	Luminance (%)	Total Rate (%)
SIR	96	60	50	69
CIR	63	95	55	71
LIR	66	52	99	72
Overall scheme (Rule-based machine)				93

Table 2: Experimental Results

7. Conclusions

An innovate scheme creating a dynamic face storage database coupled with an identification algorithm producing a computerized system suitable for the solution of the face recognition problem in real time is proposed in this chapter. Multiple and complementary representations are adopted so as to split the error domain (the union of orientation, scale, luminance and texture variances) leading to successful parallel partial solutions for the different disciplines of face recognition problems. This is accomplished by embedding each representation with a suitable identification algorithm. Furthermore, parallel processing allows identifications to be performed in real time.

References

[1]. P. Chellapa, C. Wilson and S. Sirohey, "Human and Machine Recognition of Faces: A Survey," *Proc. IEEE,* vol. 83, no. 5, pp. 705-740, 1995.

[2]. J. Daugman, "Face and Gesture Recognition: Overview," *IEEE Trans. on PAMI*, vol. 19, no 7, pp 675-676, 1997.

[3]. B. Moghaddam and A. Pentland, "Probabilistic Visual Learning for Object Representation," *IEEE Trans.on PAMI*, vol. 19, pp. 696-710, 1997.

[4]. N. Tsapatsoulis, N. Doulamis, A. Doulamis, and S. Kollias, "Face Extraction from Non-uniform Background and Recognition in Compressed Domain," *Proc. of ICASSP'98*, Seattle WA, May 1998.

[5]. G. Cottrell and M. Flemming, "Face recognition using unsupervised feature extraction," *Proc. Int. Conf. Neural Network*, pp 322-325, Paris 1990.

[6]. S. Lawrence, et. al,. "Face Recognition: A Convolutional Neural Network Approach," *IEEE Trans. on Neural Networks*, Vol. 8, No. 1, pp. 98-113, 1997.

[7]. Z. Hong, "Algebraic feature extraction of image for recognition," *Pattern Recognition*, vol.24, pp. 211-219, 1991.

[8]. P. Maragos, "Morphological Signal and Image Processing," *Digital Signal Processing Handbook*, V. Madisetti and D. Williams, eds., IEEE Press.

[9]. David Vernon, *Machine Vision*, Prentice Hall, 1991.

[10]. T. Kohonen, "Self-Organisation and Associative Memory," Berlin: Springer, 1988.

[11]. M. Leonidou, N. Tsapatsoulis and S. Kollias, "An Illumination Invariant Face Recognition Algorithm," accepted for presentation to *JCIS '98*, North Carolina, USA, October 1998.

32

An Efficient Algorithm for Rendering Parametric Curves

S.G. Tzafestas and J. Pantazopoulos
Intelligent Robotics and Automation Laboratory
Department of Electrical and Computer Engineering
National Technical University of Athens
9, Iroon Polytechniou
Zografou 15773, Athens, Greece.

1. Introduction

Catmull's method [1] is known to be the first succesfull method for rendering parametric curves. Catmull's idea was that of subdividing the space to be rendered until it is reduced to the size of a pixel. Generally, the idea of subdividing the space to be rendered was used extensively since then [2]. Later, Lane, Carpenter, Whitted and Blinn [3] presented some scan-line methods for rendering surfaces. The basis of these algorithms was that of finding the intersection of the surface with the plane of horizontal lines, which were processed in a sequential way. Although the convergence to the solution of the intersection is fast enough, one has to deal with stability matters, leading to complicated, in implementation algorithms.

In the following sections, a simplified algorithm for rendering parametric curves will be presented with extension to rendering (wireframe) parametric surfaces. The basic idea is that of subdividing the space with a stopping criterion similar to that used by Catmull. The added features are the avoidance of stack use in the recursive rendering. Also, the disadvantages of other algorithms (like the crack in scan-line techniques) are eliminated.

2. Sequential versus recursive algorithms for rendering parametric curves

Consider a parametric curve described by the following equation:

$$p(u) = [x(u), y(u), z(u)], \quad u \in [0,1] \tag{1}$$

Two different algorithms have been proposed for rendering the parametric curve given in (1). In the first, one can (pre)-calculate a list of points of (1) with size n, that have a constant distance between them. So, we can approximate the curve (1) by a set of linear segments that connect successive points of the set:

$$S(n) = \{ p(i/n) \mid 0 \le i \le n \}$$

Algorithm 1:

```
render(int n)
{
    for (i=0; i<n; i++)
        draw_3d_linear_segment(p(i/n),p((i+1)/n));
}
```

In the second algorithm, the initial space [0,1] is divided into two (or more) subspaces of equal size. Now, the rendering of the entire space [0,1] is reduced to that of rendering the two subspaces, [0,0.5] and [0.5,1]. The subdivision is continued until a limit is reached and the rendering of a final elementary subspace takes place. The stopping criterion is a measure of the error that is produced when the subspace is linearly well approximated.

Algorithm 2:

```
render(lower_limit,upper_limit)
{
    if (stopping_criterion)
    {
        draw_3d_linear_segment(p(lower_limit),p(upper_limit));
    }
    else
    {
        mid_limit=(lower_limit+upper_limit)/2;
        render(lower_limit,mid_limit);
        render(mid_limit,upper_limit);
    }
}
```

The advantage of the second algorithm is obvious. It is more adaptive to the shape of the curve with no evaluation time wasting when the subspace image of the curve is nearly linear, and presenting a better rendering when the linearization of the subspace image produces a big error. However, two drawbacks arise: the *time-cost* of the *stopping criterion* along with the stack use that may lead to memory problems and speed drop.

3. Fixed arithmetic

As we know, real numbers can be stored in a computer, either in floating point or fixed point form. The processing of fixed point numbers is very fast because it is essentially integer arithmetic and is implemented directly in the CPU. Of course we have major drawbacks ; the quantization error during the convertion follows the same distribution (which means that the accuracy is the same in the space), overflow errors easily occur, and the accumulated error after many processes may become very big. Sometimes though, when the space of the numbers that are to be processed is *a priori* known and the rounding and cutting-off process yields no significant error, we can take advantage of the speed of such arithmetic.

Our case, where we want to subdivide the space [0,1] into half spaces, belongs to one of these classes. The type of subdivision we use is binary, and it follows that no rounding error occurs.

In a machine with integer description, with n bits, we store the initial limits of the space as shown below:

$u = 0$: $0\ 0\ ......0$

$u = 1$: $\underbrace{1\ 0\ ......0}_{n\ bits}$

Generally, if we have a subspace [a,b] with size 2^{-k} (i.e. $0 \leq a < b \leq 1$, $b-a=2^{-k}$), then the subdivision yields two halved sub-spaces:

$$[a,a+2^{-(k+1)}], \text{ with size } 2^{-(k+1)}$$
$$[a+2^{-(k+1)},b], \text{ with size } 2^{-(k+1)}$$

If the numbers a and b are described by the same scheme:

a,b: $\underbrace{x\,x......x\,0\,0......0}_{(k+1)\,bits}$

then the half point will be described by the scheme:

$a+2^{-(k+1)}$: $\underbrace{x\,x......x\,1\,0......0}_{(k+1)\,bits}$

In order to avoid a cutting-off error it should be:

$$k + 2 \leq n \implies k \leq n - 2$$

And the smallest size of a subspace during the above process will be $2^{-(n-2)}$. In a PC-386 with 32-bit integer arithmetic the smallest size is $2^{-(32-2)}$ ($\approx 0,000000000931$), which is a sufficiently small size to guarantee that the rendering process will finish without any error.

Consider now a subspace [a,b] described by two fixed point real numbers u, d where u is the upper limit (u=b) and d is the size of the subspace (d=b-a). In a subdivision we first work on the left subspace (left child) and then on the right subspace (right child) (depth first algorithm). So, in a subdivision we set a new value for the midpoint u of [a,b], u=(a+b)/2, and set d half of the previous value. If no longer a subdivision is needed, we have to decide whether we are at a left child (which means that we have to visit a right child), or at a right child which means that we have to ascend in depth. The process to identify is simple ; if in a sub-space [u-d,d] the (k+1)-th bit from left is true then the subspace is a left child, otherwise it is a right child.

The total scheme of the algorithm is as follows:

The Curve Rendering Algorithm:

```
if (stopping_criterion)
{

    while (! (u & d))
            /*test the (k+1) bit from left is equal to «and» with d=2-k */
        {                       /* We are at a right child so            */
            depth--;            /* ascend in depth                       */
            d <<= 1;            /* the next sub-space to test is [u-2*d,u] */
        }
        u += d;                 /* We are at a left child (at the end) so */
                                /* the next sub-space is [(u+d)-d,(u+d)]  */

}
else
{
    depth++;
            /*The stopping_criterion is off so we continue sub-division */
    d >>= 1;                /* the next sub-space is [(u-d/2)-d/2,(u-d/2)] */
    u -= d;
}
```

As one can see from the above listing, fixed arithmetic gives a very efficient way to build the binary tree of subdivision because only additions and shifts of integers are needed, operations that are elementary and therefore very fast.

4. The stopping criterion

As already pointed out, every recursive procedure needs a criterion to tell when to stop. Most of the recursive algorithms that are used for parametric curve rendering stop at a subspace where the curve can be well aproximated by a linear segment. However, in this work, use is made of a stopping criterion that works in the image space rather than the parametric curve space. So the subdivision will stop when the limits of a subspace correspond to neighbour pixels (along with 8-connectivity neighbourness). In this way, it is guaranteed that the image of the curve will be continuous on screen, without the need to calculate the slope of the curve or other relevant parameters. This leads to simple algorithms with the best possible output results, but with the drawback of more parametric function evaluations (for every image pixel).

But this criterion doesn't hold by itself. The reason is that in a subspace, its limits can correspond to neighbour pixels, but the curve segment needs more subdivisions to appear correctly on screen. An example is provided in Fig. 1.

33

Gesture Recognition : The Gesture Segmentation Problem

M. K. Viblis, K. J. Kyriakopoulos
Control Systems Laboratory, Mechanical Engineering Department,
National Technical University of Athens,
Greece

1. Introduction

Automated visual hand-gesture understanding could serve as a very practical tool in cases such as interfacing with intelligent machines (e.g. robots) ([1], [2]) or serving social purposes in the case of sign-language understanding [3].

The most common methodologies in human-computer interaction are based on simple devices such as keyboards and mice. Although these are carefully designed for simple and easy interaction with the user, they suffer from inherent difficulties in massive data inputs at reasonable speed. This limitation has become even more apparent with the dramatic evolution of computers in the fields of storing capacity and processing speed. For those reasons, we need new input methodologies assimilating at a greater degree the way people communicate i.e. speech and gestures. Hand-gestures could nicely serve as an additional means of providing instructions to a computer, robot, etc.

On the other hand, the sign-language understanding problem, besides being very challenging, is of great social interest since it will help deaf-mute people to get direct, discrete and effective service in general purpose public services. We envision a system of visual interpretation of hand gestures, where a computer will automatically translate the sign-language of a customer to a sign-language ignorant employee.

The gesture recognition problem deals with the detection, analysis and recognition of gestures from sequences of real images. The required hardware set-up is simple and inexpensive with the advent of embedded systems technology : 2-3 cameras, a frame-grabber and a computer. A single camera may not be enough because there are gestures that use all degrees of freedom of the human hand and every camera will provide an additional view.

Gesture recognition is a very complex problem, that only recently (last four years) has attracted some attention. Initial efforts considered setups appropriate only for laboratory environments i.e. a person wearing a wired glove so that finger movements are converted into voltage signal sent to the computer ([4], [5]). Obviously, such a setup is not appropriate for generalized use because of the limitations imposed by the length of the wires and the awkward form of the device. Later, the glove was replaced with a camera and a frame grabber ([6], [7]). Current research concentrates at making systems of visual gesture recognition as efficient and robust as possible.

In this paper we propose a method for gesture segmentation as the first step towards solving the whole problem. Due to the fact that gestures of sign-languages have a dynamic character i.e they incorporate hand motion, our gesture segmentation scheme is composed of two steps :

- accurate gesture contour tracking in space domain, and

- continuous tracking in time domain

In section 2, we present the major difficulties concerning the design of a Gesture Recognition System. In Section 3, our approach in gesture segmentation is described. The next two sections contain experimental results and implementations issues. The paper concludes with suggestions for future work.

2. The Difficulty of the Gesture Recognition Problem

The problems encountered towards on building a robust gesture recognition system are numerous and complicated. Those pertain :

- Image segmentation: The goal is to retain on the foreground only the gesture contour. The representation has to be accurate, which means that we have to deal successfully with problems like 3-D hand movement and occlusions. If only a small number of cameras are used then hidden spots could be created leading to performance degradation because the gesture vocabulary is very wide and there are minor variances between certain gestures. ([9], [12])

- Object analysis: One way to represent a gesture is via feature vectors i.e. a collection of well defined primitives (geometrical, etc.). The feature vector should serve as an accurate and compact representation. Thus, feature selection is application-dependent and the designer must spend a lot of time studying the performance of every possible feature vector. The final selection can be based on statistical and mechanical performance measures : minimizing the within-class pattern variability (compact classes), maximizing the between-classes pattern variability (separable classes) and finally, achieving robust classification of unknown patterns. Gesture is a complicated pattern so the classifier needs detailed representation (high dimensional feature vector) in order to proceed to a correct decision. For example, two gestures can have similar morphology with sole difference the hand twist angle. Also in sign languages, where the vocabulary is broader, there are gestures who demand the use of both hands and in some cases hands' movement. ([10], [11])

- Pattern recognition: We present three problems : (i) The discrimination gap between two successive gestures is not easy to define. In Optical Character Recognition (OCR), a word is at least one blank character away from the previous one. In technical terms, "one blank character" means a block of pixels with background intensity value. A similar definition for the separating gap between two successive gestures is not so easily defined, especially for image sequences in time-domain. (ii) Every incoming pattern (gesture) has to be tested for similarity, with a large number of classes. In the case of a zip-code recognition system, the patterns are tested for matching against ten classes (0÷9), if the zip-code contains only digits. In the case of our gesture recognition system, which is intended for sign-language understanding, every gesture of the vocabulary defines a new class. Considering the wide range of concepts described in a natural language, the real dimension of the problem becomes evident. (iii) Gesture recognition procedures have a strong probabilistic nature. No incoming pattern will be identical to the stored classes. Just like in writing, people develop personal ways of reproducing the signs. So thorough statistical considerations (a priori and a posteriori probabilities) have to be taken into account before the final decision. [11]

3. Gesture Localization and Contour Tracking

We have divided the procedure for gesture recognition into four phases :

- Gesture localization
- Gesture tracking
- Gesture analysis, and
- Gesture recognition

In this paper we have only treated the first two phases :

Phase 1: Gesture localization

The part of the input image containing the gesture has to be found. The initial objective is not to produce an accurate representation of the hand contour, but to allocate a fraction of the image with some rough characteristics e.g. two or three fully stretched fingers, a part of an arm and a palm etc. We propose a gesture localization scheme where a deformable template-matching technique is combined with a training algorithm so that the system begins searching for the most possible templates.

Gesture Localization - Step 1 : Model Building

The used human hand-model is a simplified version of a typical anatomical prototype because it represents the fingers with oblong rectangulars and the palm with a square (Fig. 1). This simplification of this model assumes the absence of knuckles (i.e. the fingers behave like rigid bodies). The hand model can be either a free-hand sketch or a bitmap image. We have selected the first form based on the following thinking : The control points of our hand model are the edges of the corresponding geometric primitives. The major attribute of this model is the fact that all transformations of the hand are fully described by the transformations applied to the control points. In computer graphics a large number of different human hand-models exist (Fig. 2). Since

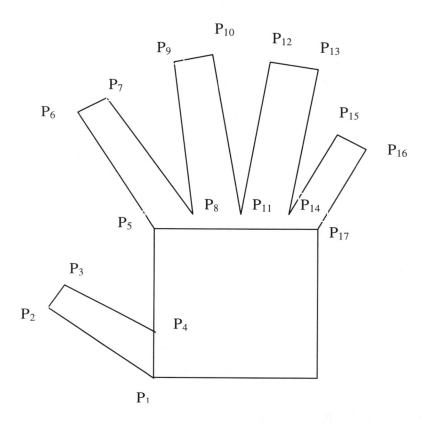

Fig. 1: Simplified Human-Hand Model

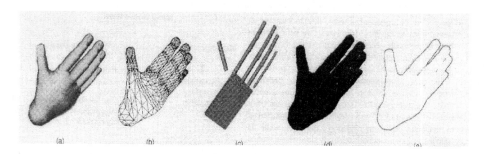

Fig. 2 : Various Human-Hand Models ([14])

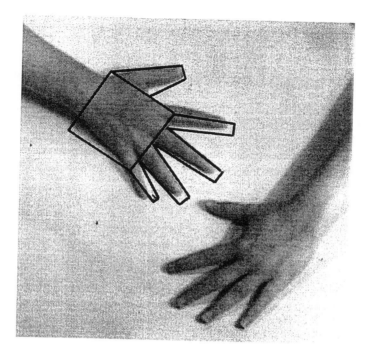

Fig. 3a

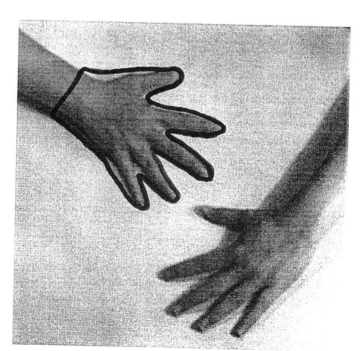

Fig. 3b

those are 3-D models, they are more accurate but at the same time more complicated inducing except from transformations also deformations.

Gesture Localization - Step 2 : Model Matching

The image has to repeatedly be scanned until a match is made with one of the possible models. In the first scan the human hand-model has undeformable shape, which means that no translation, rotation or scaling has been done. In most cases, matching is established after a long series of scanning (Fig. 3a, Fig. 5a). At the beginning of each scanning, the model is deformed by a transformation matrix, comprising of terms representing the three basic rigid transformations. Specifically, the new form of the human hand-model is the product of the initial form and the transformation matrix.

Gesture Localization : The computational - time problem

Since a practical gesture recognition scheme should be implemented in real-time, the approach of sequentially testing all the values of a variable is not satisfactory. Actually, the problem is more complicated because, during the transformation of the initial model, there are three vector variables (translation, rotation and scaling) which change value independently. In order to minimize the computational time, we decided to replace the trial-and-error logic with an intelligent scheme. The idea is simple: Among all possible deformations of the initial human hand-model, there are some of higher probability to appear. Prior to the gesture recognition system testing, we feed the learning system with a large number of gestures so that, after some processing, it can understand which are more probable. In this way, the matching process can begin from the group of transformations with the highest probability of appearance. For this scope we have selected a three-layer fully-connected feed forward neural network which uses backpropagation algorithm to tune its weights. We will not proceed further in this subject because the experimentation phase is still in progress.

Phase 2: Gesture Contour Tracking

Active contours, also known as snakes [8], are used for dynamic contour tracking.

A series of experiments involving gestures, lead us to the conclusion that active contours are among the most reliable and persistent edge detectors. If the pre-processing is correct and the "snake" encloses the right object, contour tracking will be very accurate. Active contours can follow lines, circles and arcs, as well as corners and shadows. Since the human hand has a complex geometrical form, especially at points where the fingers meet the palm, we decided on including active contours in a short list of candidates for gesture segmentation (Fig. 4a & b, Fig. 3b, Fig. 5b).

What counted in our final decision, in favor of active contours, is their ability to detect motion and remember their previous position. The intensity level features are not adequate for segmenting the object of interest in a real image. Other features, natural or user-imposed, should be considered. For example, a natural feature is the movement of the hand during the sign implementation, while a user-imposed feature can be the hand

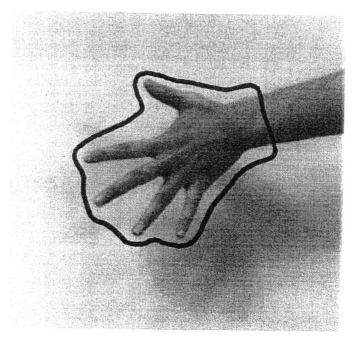

Fig. 4a

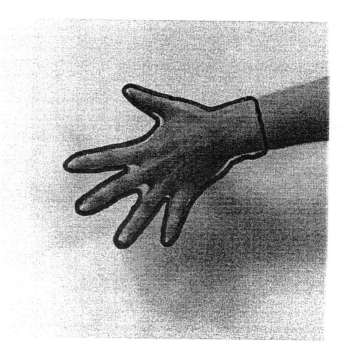

Fig.4b

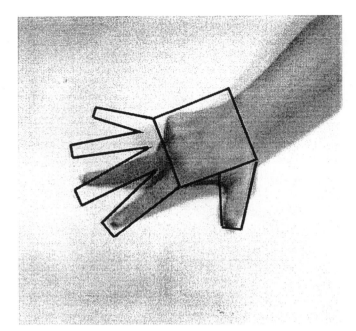

Fig. 5a

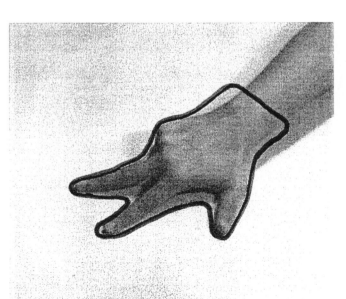

Fig, 5b

wearing a glove so that it is different from the other parts of the image due to the texture. Excessive experimentation lead us to the fact that a combination of intensity values and motion detection is very promising. For a number of segmentation tools, to achieve that, we must follow a three-steps procedure: (i) select image boundaries with an edge-detector, (ii) through a motion detection technique, determine all moving parts of the image and, finally, (iii) extract the moving boundaries by combining the two kinds of information. On the other hand, active contours have a built-in motion detection capability because one term of their partial differential equation model incorporates the velocity vector making possible the extraction of only the moving boundaries.

There is one last reason favoring active contours for gesture contour tracking: Active contours have memory. This means that the next time they will try to locate the gesture in an image sequence, they will remember the previous frame and instead of searching the whole new image, they will be confined in an area around the previous contour tracking. This is a direct consequence of the partial differential equation type of the mathematical model of active contours. The numerical solution of partial differential equations requires two kinds of conditions: initial and boundary. The computations move with small steps from the boundary surface and initial time, to the interior of the space domain and later time instances, thus giving hints for the next movement based on previous experience.

On the other hand, active Contours exhibit pour performance when the object of interest is located far away from the initial position of the "snake". For example, when the initial position of the "snake" encloses only part of the gesture, or when it is located somewhere in the interior of the gesture (Fig. 6a & 6b), it is possible to miss the required contour, because of its tendency rather to shrink than to expand. This is the reason we incorporated in our gesture recognition scheme the first phase, gesture localization to ensure that the initial snake curve encloses the object of interest from a small distance.

4. Discussion and future work

In this paper, we presented the first steps towards gesture recognition. There is still a lot of work to be done for the gesture localization phase. We currently treat this problem using Neural Networks, a very popular and successfully tested tool in the areas of optimization, pattern recognition, and machine learning. A more classical approach for data discrimination problem is this of clustering techniques. Very complex applications of machine learning handle efficiently the training and testing data using nearest neighborhood rules. They offer, except from the classification, rejection option and dynamic management of classes repository.

After gesture segmentation, the next step is gesture analysis. We intend to find a feature vector that accurately represents every sign. In addition, if these features are translational, rotational and scaling invariable, we could turn back to the gesture localization phase and re-examine the procedure. With this kind of features, one search of the image for model-matching is enough. Statistical pattern recognition theory

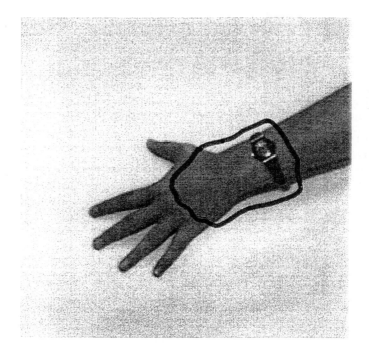

Fig. 6a

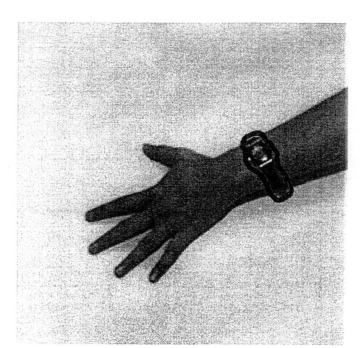

Fig. 6b

contains a number of techniques for global features' selection and performance measures.

We consider gesture segmentation and analysis as the most challenging tasks in our effort to build a gesture recognition system. The final phase, namely gesture recognition, is a rather well-posed problem. One problem, that arises from the real-time nature of the application, is the access time to the classes' database. As mentioned earlier, in applications such as sign-language understanding, the number of classes is very high. Every gesture in the sign-language vocabulary is classified as a new class. Only if the classifier has access to a specially designed database, with multiple queues and entrance points, it will retrieve a fast matching. Random-access databases offer these advantages.

References

[1] K. H. Queck, T. Mysliwiec and M. Zhao, "Finger Mouse : A Free-hand Pointing Interface", Proc Int '1 Workshop on Automatic Face and Gesture Recognition, Zurich, Switzerland, pp. 372-377, June 1995.

[2] Torige and T. Kono, "Human-Interface by Recognition of Human Gestures With Image Processing Recognition of Gesture to Specify Moving Directions" IEEE Int '1 Workshop on Robot and Human Communication, pp. 105-110, 1992.

[3] C. Downton and H. Drouet, "Image Analysis for Model-based Sign Language Coding", Progress in Image Analysis and Processing II : Proc. Sixth Int '1 Conf. Image Analysis and Processing, pp. 637-644, 1991

[4] L. Quam, "Gesture Recognition with a DataGlove", Proc. 1990 IEEE National Aerospace and Electronics Conf. , vol 2, 1990

[5] J. Sturman and D. Zeltzer, "A Survey of Glove-based Input", IEEE Computer Graphics and Applications, vol. 14, pp. 30-39, Jan 1994

[6] Cipolla and N. J. Hollinghurst, "Human-Robot Interface by Pointing with Uncalibrated Stereo Vision", Image and Vision Computing, vol. 14, pp. 171-178, Mar. 1996

[7] Darrell, I. Essa and A. Pentland, "Task-Specific Gesture Analysis in Real-Time Using Interpolated Views", IEEE Trans. Pattern Analysis and Machine Intelligence, vol. 18, no 12, pp. 1, 236-1, 242 Dec 1996

[8] Kass, A. Witkin and D. Terzopoulos, "Snakes: Active contour models", International Journal of Computer Vision, 1:321-331, 1987

[9] K. Jain, Y. Zhong and S. Lakshmanan, "Object Matching Using Deformable Templates", IEEE Pattern Analysis and Machine Intelligence, vol. 18, num. 3, 1996

[10] J. Darrell, I. A. Essa and A. P. Pentland, "Task-Specific Gesture Analysis in Real-Time Using Interpolated Views", IEEE Pattern Analysis and Machine Intelligence, vol. 18, num. 12, 1996

[11] F. Bobick and A. D. Wilson, "A State-Based Approach to the Representation and Recognition of Gesture", IEEE Pattern Analysis and Machine Intelligence, vol. 19, num. 12, 1997

[12] Lanitis, C. J. Taylor and T. F. Cootes, " Automatic Interpretation and Coding of Face Images Using Flexible Models", IEEE Pattern Analysis and Machine Intelligence, vol. 19, num. 7, 1997

[13] D. Cohen and I. Cohen, "Finite-Element Methods for Active Contour Models and Balloons for 2-D and 3-D Images", IEEE Pattern Analysis and Machine Intelligence, vol. 15, num. 11, 1993

[14] V. I. Pavlovic, R. Sharma and T. S. Huang, "Visual Interpretation of Hand Gestures for Human-Computer Interaction: A Review", IEEE Pattern Analysis and Machine Intelligence, vol. 19, num. 7, 1997

34

A Color Coordinate Normalizer Chip

I. ANDREADIS
Laboratory of Electronics
Section of Electronics and Information Systems Technology
Department of Electrical and Computer Engineering
Democritus University of Thrace
GR-671 00 Xanthi, Greece
e-mail: iandread@demokritos.cc.duth.gr

1. Introduction

The first machine vision systems used binary images of low resolution in order to handle the large amount of data contained in a television picture. Binary vision performs well when high contrast images are used and light intensities can be classified either zero or one. The shortcomings of binary vision led to the development of gray scale machine vision. Although the role of gray scale machine vision has become increasingly important for industrial applications, it has not been yet widely applied in the manufacturing industry. Reasons include difficulty to obtain repeatability in segmentation procedures, long processing times and still relatively high prices. Color adds a new dimension in machine vision and aids in building more robust and reliable systems. Limitations of possible applications of color machine vision have been associated with high cost and low processing speed of the added information. Recent progress, however, in the microelectronics industry resulted in tackling, partially, these difficulties and a limited, but increasing number, of machine vision systems which utilize color information have been reported [1-12].

The use of multispectral images considerably increases the opportunity for unique selection of features. Indeed, the use of multispectral images of the same scene has been extensively and successfully employed in remote sensing applications for analyzing geographical identifiable features (land, water, crops etc.). Color machine vision overlaps both with gray scale machine vision and colorimetry instrumentation and offers advantages over both. It will bring to the field of colorimetry instrumentation capabilities for edge detection, shape and depth analysis and other processing techniques associated with machine vision. Thus, colorimetry instrumentation will have the capability of being interfaced with industrial robots to perform various tasks, such as making adjustments and removing faulty parts from production lines.

Color requires the knowledge of three parameters and it can be represented in different ways. The available hardware for color image processing, such as color sensors and monitors are geared to RGB color space. As a result, users were forced to work with RGB color space to comply with sensor and monitor standards. RGB color space is, however, inadequate for processing images, especially in real-time [1]. Thus, it is important to transform the raw RGB data into other color spaces. Although such transformations can be easily implemented in software, they become a demanding task when processing at video rates is required. In digital signal processing there is an established need for fast and efficient hardware implementation.

This chapter presents the design and VLSI implementation of a new Application Specific Integrated Circuit (ASIC) which performs real-time conversion of the raw RGB data, obtained from a color sensor, into the rgb normalized color co-ordinates. Its frequency of operation (worst case design) is approximately 30 MHz. The high speed of operation of this ASIC is achieved by pipelining the data in a vector fashion [9]. Eight-bit color images have been used, since this resolution is adequate for encoding the composite video signal without noticeable degradation. The CADENCE VLSI CAD tool has been used to implement it. The die size dimensions for the core of the chip are 1.87 mm x 1.80 mm = 3.37mm^2, for a Double Layer Metal (DLM), 0.7 μm, N-well, CMOS technology. It is intended to be the front end of color image processing systems. Targeted tasks include real-time pattern recognition applications, such as robotics and military systems. Real-time techniques are important not only in terms of improving productivity, but also reducing operator errors associated with visual feedback delays.

2. Color Spaces

Color is one of the defining attributes of objects and it is usually represented by means of a suitable color space that consists of a co-ordinate system equipped with a distance measure. Various color spaces have been reported in an attempt to represent color perceptions by points in a space in which the distance between any two points can be taken as a measure of the magnitude of the difference between the color perceptions that are presented by the given two points. Today the development of a uniform color space is still one of the most challenging tasks in color science. The most frequently used color spaces are [4]:

(a) RGB color space. This consists of a three dimensional rectangular co-ordinate system with R, G and B axes (red, green and blue outputs of a color sensor ranging from 0 to 1V). The RGB unit vectors represent the maximum monochromatic intensity. It is a non uniform space and hardly can be related to human color perception. R, G and B are given by the following equation:

$$R,G,B \propto \int_{400}^{700} I[\lambda]O[\lambda]S_{R,G,B}[\lambda]d\lambda \tag{1}$$

where $I[\lambda]$ is the illumination spectral intensity,
$O[\lambda]$ is the object spectral reflectivity,
$S_{R, G, B}[\lambda]$ is the spectral sensitivity of the R or G or B channel of the detector and λ is the wavelength.
(b) The RGB primaries can be used to derive a new co-ordinate system via a linear or a non-linear transform. Examples of linear transformations are the YIQ and the YUV color spaces, devised by the color television industry as a way of minimizing signal bandwidth while retaining color fidelity, and the XYZ system of primaries designed to yield non negative tristimulus values for any color. Color co-ordinate systems obtained through non-linear transforms are:

(i) rgb or color fractions, or normalized color co-ordinates, or chromaticity color space. This is described in some detail in the next section.
(ii) Intensity, Hue, Saturation (IHS) color space. This presentation of colors has the advantages of the previous color space while providing interpretation of the visual result of a particular RGB stimulus.
Uniform color spaces:
(iii) L*a*b* color space. It is based on the opponent color theory and L*, a* and b are luminance, redness-greenness and yellowness-blueness, respectively. The Commission Internationale de l'Eclairage (CIE) recognized the practicality of this color space by recommending in 1976 the CIE Lab scale. This color space is extensively used in the color instrumentation and measurement industry.
(iv) L*u*v* color space has evolved from the L*a*b* system and it attempts to express mathematically, the relationship between a color and its illuminant mimicking the human perception process as far as possible.
    A characteristic common to most transformed systems is that they are more uniform than the original RGB space and that their components are substantially decorrelated.

3.   Properties of the Normalized Color Images

In this space, color co-ordinates are defined as follows: $r=R/(R+G+B)$, $g=G/(R+G+B)$ and $b=1-r-g$. Effects, such as shadows and shading variation may be reduced or eliminated, thus leading to simplified image segmentation. Figure 1 shows the profile lines of an image consisting of two objects of the same color placed at different locations within the field of view. The shadow effect appears near the edges of the object (local sharp variations-Figure 1a). As it can be noticed from Figures 1b, 1c and 1d these transient effects do not appear in the normalized images. Furthermore: (i) color fraction image values exhibit much lower σ (standard deviation) values and (ii) normalized σ does degrade with lower light intensity values but normalized mean values remain relatively stable.
    The transform is limited by the signal to noise ratio (SNR) of the three primary channels of the color sensor and it becomes unstable when signals in one or more primary channels are weak and noise is comparable to signal level. This color space

382

transform has an essential singularity (arising from the division operation), i.e. it is not defined at R=G=B=0. Near the singularity the transform becomes highly unstable as a small perturbation on one input can cause large variations of the normalized values. Consider the values RGB(1,0,0) and $R_1G_1B_1(0,1,0)$; then the corresponding normalized 8-bit values for the red image are 255 and 0, thus, illustrating this instability. In practice, this is not usually a problem; if the three primary images have sufficient dynamic ranges and the scene has enough illumination such points may be negligible.

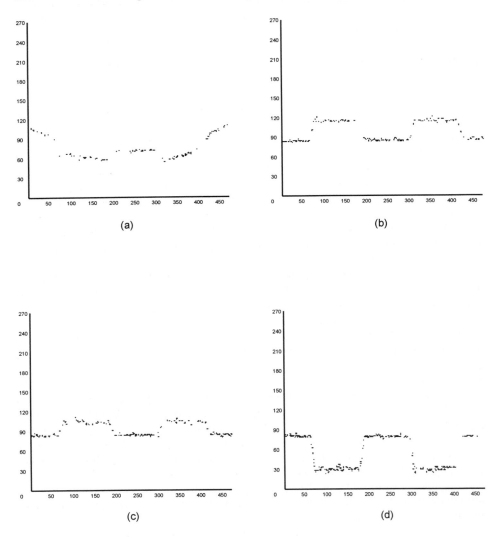

Figure 1. Image profiles: (a) Intensity image, (b) red normalized image, (c) green normalized image and (d) blue normalized image.

4. Normalizer Circuitry Description and its VLSI Implementation

The circuit diagram of the proposed color space converter is shown in Figure 2. Its inputs are RGB digital data (8-bit resolution) obtained from a color sensor and its outputs are the rgb normalized color co-ordinates. To describe the implementation of the normalizer, only one channel (b channel) will be considered as the other two will be symmetrical. Signal processing at video rates is a demanding task. The high speed of operation is achieved by pipelining the data in vector fashion. The basic function of the normalizer can be described by the function b=ROUND(255B/(R+G+B)). Coefficient 255 must be used in each normalized color co-ordinate equation in order to obtain image values in the range [0..255]. The implementation of this function can be split into two sections: (i) calculate the intensity and (ii) calculate the normalized value b. The first section can be implemented for an n-bit system using two adders (one n-bit and one (n+1)-bit). Adders are constructed with the 74283 4-bit library full adder. These devices feature carry in and carry out pins so that they can be cascaded to provide larger length adders. At each stage of the pipeline data is latched through positive edge triggered latches. The second section is more difficult to realize efficiently due to the operation of division. In this case the divider is constructed from a multiplier whose inputs are B and 255/I, where I=R+G+B. To generate 255/I a look-up table (LUT) is used to hold all the pattern pairs for all combinations of 255/I. A LUT is basically a memory array. The input to the array is applied to the address lines of the memory and the output is presented on the data lines. The input-output transform occurs simply by storing the required output data at the location in the array which corresponds to the inputs. With dramatic reductions in memory costs and short access times of memory components LUTs offer an attractive solution in this case. The size of the LUT required for an 8-bit system is 766 entries (10-bit) with an equivalent resolution of data 24 bits; 8 bits are used for the integer part and 16 bits for the decimal part. A video rate multiplier 24x8-bit wide performs the required multiplication. One multiplier has been used for all three channels (through a multiplexer) in order to minimize silicon area at the expense, of course, of speed. In the final stage rounding to the nearest integer is performed through an adder whose one input is the integer part of the previous stage operation and its second input is 0. Its carry-in input is the most significant digit of the decimal part obtained from the LUT.

The analog RGB data are digitized through three video (parallel) Analog to Digital Converters (ADCs) which must provide a sample after a rising edge in their conversion pins. The ASIC's 8-bit resolution is adequate for most industrial applications since the SNR, in all three primary channels, for the vast majority of color cameras does not exceed the SNR of an 8-bit ADC (59 dB). Typical color camera SNRs, in the three primary channels, are in the range 40-50 dB. Furthermore, the resolution of the proposed architecture can be easily scaled , e.g. to 10 bits, by simply increasing the size of the basic building blocks. In this case the required LUT will have 12 address lines and 26-bit resolution. The video multiplier will be 26x10-bit wide. The sizes of the other building blocks (latches and adders) can be easily scaled to appropriate lengths. All these size extensions are feasible in terms of silicon area. The 10-bit resolution per primary channel (70dB) exceeds the SNR of existing color sensors.

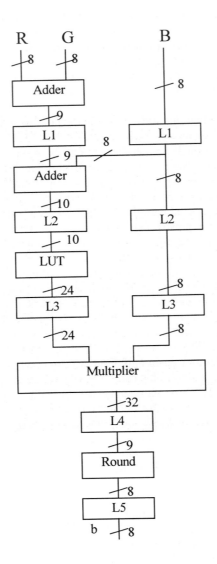

Figure 2. Circuitry of the color space converter (Blue channel).

The CADENCE VLSI CAD tool has been used to implement the chip. A block level layout of the color space converter chip, including the pads, is shown in Figure 3. The die size dimensions for the core of the chip are 1.87 mm x 1.80 mm = 3.37 mm², for a DLM, 0.7 µm, N-well, CMOS technology. The inputs to the chip are the 24-bit RGB data, the OZ tri-state output control signal, the clock, as well as the power and ground connections, whereas the outputs are the 24-bit rgb color co-ordinates. Loaded simulations revealed that its maximum speed of operation is approximately 30 MHz (worst case) and its throuput rate of operations is 5x30=150 MIPS. Its typical speed of operation is 65 MHz. Also, it can handle high resolution color images of up to 62.5 µs (duration of a video line) x 30 MHz =1875 pixels/line.

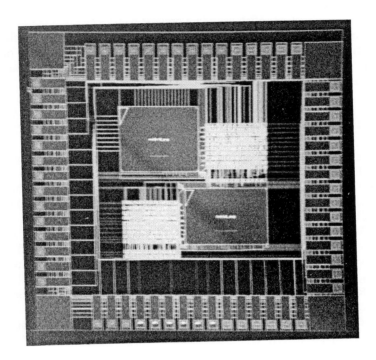

Figure 3. Block level layout of the chip.

The simulation and test language STL, a high level language (its structure is similar to PASCAL or C high level languages), has been used to examine the functionality of the chip. This supports the generation of stimuli for simulation purposes and formatted test vectors for automatic test equipment. The STL simulation output comparison capability allows to automatically compare the expected output values specified in the STL source program with the results of the simulation. Both real and computer generated data have been used to test the functionality of the chip. No errors have been detected during this process.

5. Conclusions

The design and VLSI implementation of a real-time vector pipeline color space transformer, which converts the RGB color co-ordinates to the rgb color co-ordinates has been presented in this chapter. With a DLM, 0.7 µm, N-well, CMOS technology the die size dimensions for the core of the chip are 1.87 mm x 1.80 mm = 3.37 mm^2.The chip is expected to modernize and enhance the field of colorimetry instrumentation by applying the latest microelectronics techniques. It is also intended to be used as a front end in color machine vision systems, for inspection purposes (accept/reject operations), in autonomous applications, such as robotics and military systems where image enhancement techniques are often used in a preprocessing stage in order to increase the probability of correct pattern recognition. These applications are generally characterized by data throughput requirements that can only be met by hardware capable of operating in real-time. When early and effective use of color is made it can reduce the processing burden rather than increase it as is usually assumed.

References

[1] Andreadis, I. "Colour Processing for Image Segmentation and Recognition", Ph.D. Thesis, UMIST, U.K., 1989.

[2] Bajon, J., Cattoen, M. and Liang, L., "Identification of multicoloured objects using a vision module", Proc. of the 6th Int. Conf. on Robot Vision and Sensory Controls, Paris, 1986, pp. 21-30.

[3] Barth, M., Parthasarathy, S., Wang, J., Hu, E., Hackwood, S. and Beni, G., "A color machine vision system for microelectronics: application to oxide thickness measurement", Proc. of the IEEE Int. Conf. on Robotics & Automation, San Francisco, USA, 1986, pp. 1241-1245.

[4] Hunt, R.W.G., "Measuring Color", 2nd Edition, Ellis Horwood Series in Applied Science and Industrial Technology, 1995.

[5] Lo, R.C. and Tsai, W.H., "Color image detection and matching using modified generalised Hough transform", IEE Proc. Vision, Image & Signal, 1996, Vol. 143, No. 4, pp. 201-209.

[6] Miller, R.K., "Color Machine Vision", SEAI Technical Publications, Madison, 1986.

[7] Marszalec, E. and Pietikainen, M., "Some aspects of RGB vision and its applications in industry", Int. Journal of Pattern Recognition and Artificial Intelligence, 1996, Vol. 10, No.1, pp. 55-72.

[8] Perez, F. and Roch, C., "Toward color image segmentation in analog VLSI: algorithm and hardware", Int. Journal of Computer Vision, 1994, Vol. 12, No. 1, pp. 17-42.

[9] Person, E., "A pipelined image analysis system using custom integrated circuits", IEEE Trans. Pattern Analysis and Machine Intelligence, 1988, PAMI-10, pp. 111-116.

[10] Schettini, R., "A segmentation algorithm for color images", Pattern Recognition Letters, 1993, Vol. 14, pp. 499-506.

[11] Stark, J.P.W., Mahdavieh, Y. and Tjahjardi, T., "A fast algorithm for colour region segmentation", Eurographics 82, 1982, pp. 47-56.

[12] Venetsanopoulos, A.N. and Plataniotis, K.N., "Multichannel image processing", IEEE Workshop on Nonlinear Signal & Image Processing, Chalkidiki, Greece, 1995, pp. 1-4.

PART IV

APPLICATIONS

$\vec{X}$ is the vector of unknowns given by: $\vec{X} \equiv [u_x \; u_y \; u_z \; b]^T$

$\vec{V}$ is the $(n-1)$ equation error vector given by: $\vec{V} \equiv [V_1 \; V_2 \; \cdots \; V_{n-1}]^T$

The statistics of the equation error $\vec{V}$ are calculated as being:

$$E\{V_i\} = 0$$
$$E\{V_i V_j\} = \begin{cases} 2(b - \bar{R})^2\sigma^2 + \sigma^4 & \text{for } i = j \\ (b - \bar{R})^2\sigma^2 + \frac{1}{2}\sigma^4 & \text{for } i \neq j \end{cases} \tag{8}$$

Using eq. (8) the covariance of the equation error vector $\vec{V}$ can be expressed as

$$P_V = \sigma^2(\frac{1}{2}\sigma^2 + (b - \bar{R})^2)R_N, \tag{9}$$

where R_N is a $(n-1) \times (n-1)$ matrix whose diagonal elements are 2 and off-diagonal elements are 1.

The linear regression from eq. (7) yields the minimum variance estimate $\hat{\vec{X}}$ of the vector of unknown parameters $\vec{X}$,

$$\hat{\vec{X}} = (H^T P_V^{-1} H)^{-1} H^T P_V^{-1} \vec{Z} \tag{10}$$

An interesting characteristic of the R_N matrix is that its determinant is always equal to n and it is easily inverted analytically; the inverse of R_N is required to find the inverse of the equation error covariance matrix that is explicitly given by:

$$P_V^{-1} = \frac{1}{\sigma^2(\frac{1}{2}\sigma^2 + (b - \bar{R})^2)} R_N^{-1}, \tag{11}$$

and where R_N^{-1} is the matrix inverse of R_N explicitly given by:

$$R_N^{-1} = \frac{1}{n} \begin{bmatrix} n-1 & -1 & -1 & \cdots & -1 \\ -1 & n-1 & -1 & \cdots & -1 \\ -1 & -1 & n-1 & & \vdots \\ \vdots & \vdots & & \ddots & \vdots \\ -1 & -1 & \cdots & \cdots & n-1 \end{bmatrix} \tag{12}$$

A remarkable property of the estimate in eq. (10) is the fact that it is not dependent on σ^2, the pseudorange measurement noise variance; indeed, to simplify the solution for implementation, it is noted that eq. (9) shows the equation error covariance P_V as simply R_N premultiplied by a scalar quantity. In eq. (10) the scalar premultiplier of P_V will cancel out; therefore, the minimum variance parameter estimate (position and clock bias) in eq. (10) can be rewritten in an equivalent form as:

$$\hat{\vec{X}} = (H^T R_N^{-1} H)^{-1} H^T R_N^{-1} \vec{Z} \tag{13}$$

The only inversion that needs to be performed is that of the (4×4) matrix, $(H^T R_N^{-1} H)$, which can be hardwired into the receiver's algorithm.

Furthermore, the covariance of the estimation error is given by:

$$P_X \equiv E\{(\vec{X} - \hat{\vec{X}})(\vec{X} - \hat{\vec{X}})^T\} = \sigma^2(\frac{1}{2}\sigma^2 + (\hat{b} - \bar{R})^2)(H^T R_N^{-1} H)^{-1} \qquad (14)$$

Unlike the solution estimate, the covariance P_X is dependent on σ^2, the pseudor-ange measurement noise variance; hence, σ must be known or estimated in order to compute (predict) the estimation error covariance. If a sufficiently large number of pseudorange measurements are available, the following approach can be used:

$$E\{\vec{V}^T \vec{V}\} = \sum_{i=1}^{n-1} E\{V_i^2\} = (n-1)E\{V_i^2\} = (n-1)(2(\hat{b} - \bar{R})^2 \sigma^2 + \sigma^4) \qquad (15)$$

Using the return difference from the measurement data, an expression equivalent to eq. (15) can also be obtained as follows:

$$E\{\vec{V}^T \vec{V}\} \approx (\vec{Z} - H\hat{\vec{X}})^T(\vec{Z} - H\hat{\vec{X}}) \qquad (16)$$

Evoking ergodicity, eqs. (15) and (16) are accepted as being equivalent, hence by equating eqs. (15) and (16), a quadratic equation in σ^2 is obtained; solving this quadratic equation yields the following data driven estimate of σ^2:

$$\hat{\sigma}^2 = -(\bar{R} - \hat{b})^2 + \sqrt{(\bar{R} - \hat{b})^4 + \frac{1}{n-1}(\vec{Z} - H\hat{\vec{X}})^T(\vec{Z} - H\hat{\vec{X}})} \qquad (17)$$

The expression for the estimation error covariance matrix given in eq. (14) can be rewritten in terms of the return difference data as follows:

$$P_X = \frac{1}{2(n-1)}(\vec{Z} - H\hat{\vec{X}})^T(\vec{Z} - H\hat{\vec{X}})(H^T R_N^{-1} H)^{-1} \qquad (18)$$

Experimental results showed that Eq. (17) is unreliable in low satellite avail-ability conditions typical of NAVSTAR GPS scenarios. Hence, a more rigorous derivation using weighted return differences and accounting for the estimation problem's four degrees of freedom, resulted in a much improved estimate of σ^2:

$$\hat{\sigma}^2 = -(\bar{R} - \hat{b})^2 + \sqrt{(\bar{R} - \hat{b})^4 + \frac{2}{n-5}(\vec{Z} - H\hat{\vec{X}})^T R_N^{-1}(\vec{Z} - H\hat{\vec{X}})} \qquad (19)$$

In conclusion, the derived linear regression (7), which consists of $(n-1)$ equa-tions, requires that $n-1$ be at least four to provide an initial estimate of the four parameters (user position and clock bias). This implies that a minimum satellite availability of five is required to produce the solution given in eqs. (13) and (18) for the parameter estimate and the predicted estimation error covariance, respec-tively. The solution is based on $n-1$ equations only, although n measurements are available initially, which indicates that n equations should be used to obtain the parameter estimate. This point will be revisited in Section 4 where eq. (3) is included to form an augmented set of n equations and a two stage algorithm for position and user clock bias estimation is developed.

4 Extended Kalman Filter

Motivated by the desire to obtain an estimate using an overdetermined system of n equations based on the n pseudorange measurements, we realize that the nonlinear aspect of the problem must be addressed. The closed-form solution of the $n - 1$ equations-based linear regression problem in eqs. (7) and (9) provided the parameter's mean $\vec{X}$ and the estimation error covariance P_X, viz., $\vec{X} \sim N(\hat{\vec{X}}, P_X)$, or in expanded form, $[u_x, u_y, u_z, b]^T \sim N([\hat{u}_x, \hat{u}_y, \hat{u}_z, \hat{b}]^T, P_X)$. An algorithm that employs this preliminary solution as initialization to produce an improved solution, by making use of an additional nonlinear equation, is developed. The concept behind the *Kalman update* estimation approach is similar to that of a conventional Kalman Filter. The *closed-form* solution in eqs. (13) and (18) provides a preliminary GPS solution estimate and the associated estimation error covariance matrix, (P_X). The additional n^{th} pseudorange equation (3) is now perceived as a *new*, albeit nonlinear, measurement which can be used to update the previous estimate the same way that it would be treated during the update cycle of an Extended Kalman Filter. The approach that is used entails the linearization of eq. (3) about a nominal position estimate. The linearized equation is then manipulated into the standard linear measurement form and used to update the preliminary estimate.

The *Kalman update* algorithm that is presented differs from the standard Kalman algorithms [6] in that the *new* measurement that is used to update the previous estimate is correlated with the previous estimate. The conventional Kalman Filter update equation does not allow for correlation between the new measurement and the previous estimate; hence, a novel Kalman-like update equation that can accommodate this correlation and that addresses the specific measurement situation on hand, needs to be derived.

The first step in the mathematical derivation of the *Kalman update* algorithm is the linearization of eq. (3) about a nominal user position $(u_{x_0}, u_{y_0}, u_{z_0})$ by performing a Taylor series expansion and neglecting second and higher order terms. Through equation manipulation, rearranging and redefining of terms, the following equation in the form of a linear scalar measurement model is obtained:

$$Z_n = h^T \vec{X} + w_n, \tag{20}$$

where Z_n is the scalar measurement defined as

$$Z_n \equiv R_n + \frac{(u_{x_0} - x_n)x_n + (u_{y_0} - y_n)y_n + (u_{z_0} - z_n)z_n}{\sqrt{(u_{x_0} - x_n)^2 + (u_{y_0} - y_n)^2 + (u_{z_0} - z_n)^2}},$$

h is a 4×1 vector and is defined as

$$h \equiv \begin{bmatrix} \dfrac{(u_{x_0} - x_n)}{\sqrt{(u_{x_0} - x_n)^2 + (u_{y_0} - y_n)^2 + (u_{z_0} - z_n)^2}} \\ \dfrac{(u_{y_0} - y_n)}{\sqrt{(u_{x_0} - x_n)^2 + (u_{y_0} - y_n)^2 + (u_{z_0} - z_n)^2}} \\ \dfrac{(u_{z_0} - z_n)}{\sqrt{(u_{x_0} - x_n)^2 + (u_{y_0} - y_n)^2 + (u_{z_0} - z_n)^2}} \\ 1 \end{bmatrix}$$

$\vec{X}$, we recall, is the vector of unknowns, $[u_x, u_y, u_z, b]^T$, and w_n is the measurement noise of the nth pseudorange, where $w_n \sim N(0, \sigma^2)$.

Eq. (20) is in the desired linear measurement model form that can be used to update the solution obtained from the preliminary *closed-form* algorithm in a Kalman-like update approach. Now, Z_n is actually part of the measurements that were used to obtain the *closed-form* solution and not a new measurement as would be the case in a conventional Kalman Filter application; the noise in the *new* measurement and the previously derived position estimation error therefore are correlated. This is a violation to the basic assumptions used in the derivation of the conventional Kalman Filter update equations. In order to derive the new Kalman-like update equation, it is necessary to know the relationship between the noise in the *new* measurement and the estimation error produced by the *closed-form* algorithm. The *closed-form* algorithm produced an estimate of the GPS parameters given in eq. (13) and an estimate of the estimation error covariance matrix given in eq. (18). Using the knowledge of the GPS solution estimate, the true GPS parameter vector satisfies:

$$\vec{X} \equiv \hat{\vec{X}} + \vec{W}, \tag{21}$$

where $\vec{W} \sim N(0, P_X)$. Obviously, the correlation of interest between w_n and $\vec{W}$ is defined as: $p \equiv E\{Ww_n\} \equiv E\{w_n W\}$.

To determine the relationship between $\vec{W}$ and $\vec{V}$, the linear regression in eq. (7) is multiplied from the left by $H^T R_N^{-1}$ and solved for $\vec{X}$ to obtain:

$$\vec{X} = (H^T R_N^{-1} H)^{-1} H^T R_N^{-1} \vec{Z} - (H^T R_N^{-1} H)^{-1} H^T R_N^{-1} \vec{V}$$

The first term on the right hand side of the equation is recognized from eq. (13) as $\hat{\vec{X}}$; therefore, an expression for $\vec{W}$ in terms of $\vec{V}$ is obtained:

$$\vec{W} = (H^T R_N^{-1} H)^{-1} H^T R_N^{-1} \vec{V} \tag{22}$$

Furthermore, the covariance matrix between $\vec{V}$ and w_n is determined by exploiting the noise statistics for $\vec{V}$ derived previously.

$$E\{\vec{V} w_n\} = (\bar{R} - b)\sigma^2 \begin{bmatrix} 1 \\ \vdots \\ 1 \end{bmatrix}_{(n-1) \times 1} \tag{23}$$

Using eq. (22) and the relationship in eq. (23), the required covariance between $\vec{W}$ and w_n is finally obtained:

$$p = (\bar{R} - b)\sigma^2 (H^T R_N^{-1} H)^{-1} H^T R_N^{-1} \begin{bmatrix} 1 \\ \vdots \\ 1 \end{bmatrix}_{(n-1) \times 1} \tag{24}$$

Next, an augmented linear regression is formulated by combining eq. (21) and eq. (20). The augmented linear regression is expressed as:

$$\vec{Z}_a = H_a \vec{X} + \vec{V}_a \tag{25}$$

where,

$\vec{Z}_a \equiv [\vec{X} \ Z_n]^T$ is the (5×1) augmented *measurement* vector

$H_a \equiv [I \ h^T]^T$ is the (5×4) augmented regressor

$\vec{V}_a \equiv [\vec{W} \ w_n]^T$ is the (5×1) augmented *measurement noise* vector

In order to obtain the updated estimates from the augmented linear regression in eq. (25), it is necessary to derive the covariance of the augmented noise vector $\vec{V}_a$. Since the statistics of the noise components in $\vec{V}_a$ have already been determined, the equation error covariance matrix, R_a, is given by:

$$R_a = \begin{bmatrix} P_X^- & p \\ p^T & \sigma^2 \end{bmatrix} \tag{26}$$

The updated GPS minimum variance solution estimate and the associated covariance are then given by the expressions:

$$\begin{aligned} \hat{\vec{X}}^+ &= P_X^+ H_a^T R_a^{-1} \vec{Z}_a; \ and \tag{27} \\ P_X^+ &= (H_a^T R_a^{-1} H_a)^{-1} \tag{28} \end{aligned}$$

The expressions in eqs. (27) and (28) are sufficient to obtain the required updates, but it is desirable to manipulate and reduce the equations into the more familiar and computationally efficient form of the classical Kalman filter update equations. After lengthy manipulations, and using the well known Matrix Inversion Lemma, the Kalman-like update equations in the desired form are finally obtained, viz.,

$$\begin{aligned} \hat{\vec{X}}^+ &= \hat{\vec{X}}^- + K(Z_n - h^T \hat{\vec{X}}^-); \ and \tag{29} \\ P_X^+ &= \{I - [(1 - p^T h)K + p]h^T\}Y, \tag{30} \end{aligned}$$

where the intermediate variable Y is the modified pre-update covariance matrix given by:

$$Y = P_X^- + \frac{h^T P_X^- h - 1}{(1 - p^T h)^2} p p^T + \frac{1}{1 - p^T h}(P_X^- h p^T + p h^T P_X^-) \tag{31}$$

and K is the modified Kalman filter gain given by:

$$K = \frac{1}{1 - p^T h}[\frac{1}{1 + h^T Y h} Y h - p] \tag{32}$$

It was determined experimentally that sometimes a second application of the update algorithm, eqs. (29) to (32), is required to obtain a better solution estimate. Recalling that the *new* measurement is actually the nth pseudorange equation that has been linearized about the position estimate produced by the *closed-form* algorithm, implies that how well the linearization fits the true unknown GPS parameters is dependent on how good the solution produced by the *closed-form* algorithm is to begin with. In order to alleviate this undesired dependency, after the Kalman Update algorithm has been applied once to produce an improved solution estimate, eq. (3) is once again linearized about the improved position estimate producing a better *new* linear measurement equation. This is akin to the iterated Kalman filtering algorithm used in Extended Kalman Filtering. The *Kalman update* algorithm is applied a second time using the preliminary estimate and estimation error covariance available prior to the update and produced by the linear *closed-form* algorithm, not the solution obtained as a result of the previous application of the *Kalman update*. Theoretically, this process can be continued recursively until convergence to the best possible solution is achieved; however, it was found experimentally that after the second application of the algorithm, the change in the solution estimate is insignificant; consequently, iterations are not required. The first application is strictly to obtain a suitable position estimate about which to perform a valid relinearization of the nth pseudorange equation (3) and the second application then yields the final GPS solution estimate and its covariance.

5   Experimental Results

The closed-form, linear regression algorithm-based solution developed in this chapter requires at least five pseudorange measurements to produce a stand-alone GPS solution. This is not overly restrictive since with a five degree elevation angle, there are always at least five satellites in view and at least seven satellites are in view 80 percent of the time [9]. In terms of satellite availability, the worst case scenario occurs at latitudes in the range of 35 to 55 degrees where there are at most six satellites available 20 percent of the time. Satellite availability is not dependent on user position longitude; hence, selecting a single user position in the 35 to 55 degree latitude range and assuming an elevation angle of 10 degrees will allow for simulating GPS data that is realistically indicative of worst case conditions. The experimental data was generated using GPSoft's Satellite Navigation Toolbox for Matlab [4], with an arbitrary user position of 40° N latitude, 105° W longitude, at an altitude of 300 m. The Satellite Navigation Toolbox is used to generate GPS satellite position data from which true ranges from all in view GPS satellites to the user position can be calculated. After adding an arbitrary clock bias to all the ranges, a zero mean random noise of preselected standard deviation $\sigma = 100$ m is superimposed to represent the Gaussian measurement noise. The experimental results encompass the outcome of 12 scenarios selected for greatest diversity in satellite availability and geometry. The satellite availability ranged from five to nine, which is indicative of realistic scenarios.

In addition to producing experimental results using the linear regression based closed-form algorithm and the Kalman update augmented algorithm developed in this chapter, results were also produced using the conventional ILS algorithm to provide a comparison baseline. The results discussed are the cumulative representation of 5000 Monte Carlo runs. The Gaussian pseudorange noise realization for each satellite is maintained the same between the different algorithms, for any given Monte Carlo run, in order to provide an unbiased comparison basis.

The *Kalman update* algorithm produced results comparable to the baseline ILS results. The position errors yielded by the *Kalman Update* algorithm when compared to those yielded by the ILS algorithm, all differed by less than a metre. Given the lack of confidence that can be placed on just the mean error results, the results are considered equivalent to the baseline in terms of parameter estimation errors. For the three position parameters, the experimental standard deviations are also equivalent to the baseline; however, the standard deviation associated with the error on the range equivalent user clock bias is noticeably larger than the baseline. The *Kalman update* algorithm has the capability of predicting its estimation error covariance and its performance is reasonably good: The calculated standard deviations for the three position estimation parameters where all within 14 percent of the experimentally obtained values. The performance in producing the standard deviation on the user clock bias estimation error was poorer and differed from the experimentally obtained values by as much as 44.2 percent.

Thus, a two-stage algorithm with the following attributes has been developed: 1) The algorithm is closed-form hence it can be used under any geometrical conditions without the need for externally provided initialization; 2) The algorithm has the potential to benefit from computational efficiencies due to its non-iterative nature; 3) The algorithm has the capability to predict its estimation error covariance; 4) The algorithm can produce a data driven GPS solution estimate without the knowledge of the measurement noise strength σ; 5) The performance under typical navigation scenarios, using only the NAVSTAR GPS satellite constellation, is equivalent to the performance achieved by the conventional ILS algorithm; and, 6) The horizontal positioning performance under poor geometry conditions, e.g. when ground-based planar arrays of pseudolites are used, is better than that of the conventional ILS algorithm. Moreover, there are no restrictions on the user position and an initial user position guess is not required.

6 Conclusions

A GPS solution estimate comparable to that of the conventional ILS algorithm is obtained. The strength of the *closed-form* algorithm surfaced in pseudolite ground array scenarios where the pseudolite availability is such that an excellent estimate of the pseudorange measurement noise strength, σ, could be recovered from the measurement residuals, which can then be used to calculate the estimation error covariance. The performance of the *closed-form* algorithm in estimating the horizontal user position parameters showed improvement over the ILS algorithm; furthermore, no user position restrictions were required. This may prove beneficial to test range applications where the conventional iterative algorithm is at risk of failure and this imposes restrictions on the flight test trajectory and altitude.

The results produced by the conventional ILS algorithm indicate that noise is actually treated fine in the ILS algorithm despite the fact that it does not have the means of producing an estimate of the pseudorange measurement noise strength σ. Hence, the main advantages of the work in this chapter are the potential for computational efficiency due to the closed-form nature of the algorithm and the prediction of the estimation error covariance. In general, the correct stochastic modeling presented in this chapter opens up the way for future rigorous integration of GPS and other sensors, e.g. INS. The benefits of the novel noniterative algorithm are computational efficiency, there is no need for an initial position guess, and better performance under poor geometry is realized.

References:

[1] Abel, J. and J. Chaffee. "Existence and Uniqueness of GPS Solutions," *IEEE Transactions on Aerospace and Electronic Systems*, *27*(6):952–956 (November 1991).

[2] Bancroft, S. "An Algebraic Solution of the GPS equations," *IEEE Transactions on Aerospace and Electronic Systems*, *21*(1):56–59 (November 1985).

[3] Dailey, D. J. and B. M. Bell. "A Method for GPS Positioning," *IEEE Transactions on Aerospace and Electronic Systems*, *32*(3):1148–1154 (July 1996).

[4] GPSoft, Athens, OH. *Satellite Navigation Toolbox for Matlab*, January 1997.

[5] Hoshen, J. "The GPS Equations and the Problem of Apollonius," *IEEE Transactions on Aerospace and Electronic Systems*, *32*(3):1116–1124 (July 1996).

[6] Kalman, R. E. "A New Approach to Linear Filtering and Prediction Problems," *ASME Transactions, Journal of Basic Engineering*, *82*:34–45 (1960).

[7] Krause, L. O. "A Direct Solution to GPS-Type Navigation Equations," *IEEE Transactions on Aerospace and Electronic Systems*, *23*(2):223–232 (March 1987).

[8] Siouris, George M. *Aerospace Avionics Systems: A Modern Synthesis*. San Diego: Academic Press, Inc., 1993.

[9] Spilker, J. J. Jr. "Satellite Constellation and Geometric Dilution of Precision." *GPS: Theory and Applications Volume 1*, edited by B. W. Parkinson. Washington, D.C.: American Institute of Aeronautics and Astronautics, 1996.

[10] Spilker, J. J. Jr. and B. W. Parkinson. "Overview of GPS Operation and Design." *GPS: Theory and Applications Volume 1*, edited by B. W. Parkinson. Washington, D.C.: American Institute of Aeronautics and Astronautics, 1996.

36

Digital Image Processing for Weathering Analysis and Planning of Conservation Interventions on Historic Structures and Complexes

A. Moropoulou*, M. Koui*, Ch. Kourteli*, N. Achilleopoulos*,

F. Zezza°

* National Technical University of Athens, Department of Chemical Engineering, Section of Materials Science and Engineering, Zografou Campus, 9, Iroon Polytechniou St., 15780 Zografou, Athens, Greece.

° Polytechnic of Bari, Faculty of Engineering, Institute of Applied Geology and Geotechnics, 4, Via E. Orabona, 70125 Bari, Italy.

1. Introduction

The threat to the heritage and in particular to the building materials (stone, mortars, bricks etc.) is growing due to the intensive increase of the atmospheric pollution, urbanization, industry and tourism, as well as due to inappropriate conservation treatments applied. The weathering of monuments could be considered as an interaction between the building materials and the acting environmental factors. Interest focuses on the interface between materials and the environment rather than on any intrinsic process [Baer, 1989 [1]].

Mapping techniques have been lately applied to serve conservation needs. Photogrammetrical surveys provide a general representation, of the geometric architectural and structural characteristics [Cundari, 1991 [2]], while specific mapping techniques of lithotypes and of weathering forms develop nowadays. M. Mamillan in 1991 [3] introduces ultrasonic measurements to monitor structural strength and to distinguish and evaluate zones of monuments resistance to decay accordingly. B. Fitzner in 1991 [4] combines the documenting and mapping of weathering forms of stone surface with microstructural studies as well. F. Zezza in 1989 [5], 1991 [6] worked out the capabilities of digital image processing restituting from simple photograms false colour images and ultrasonic pulses semantic for lithotypes and weathering mapping on the surfaces and in depth. The need for automatically processed documentation, inscribing in an accurate way both types of information, is apparent. Non destructive tests, like Digital Image Processing, Fiber Optics Microscopy and Ultrasound Technique, providing extended information and in short time without any risk for the monument, regard as very new methods under research as far as reliability and validity in several applications are concerned. These *non destructive techniques* [Zezza, 1996 [7]] applied in situ provide the instrumental methods to record, assess and evaluate the environmental impact on the masonries according to the physicochemical criteria (microstructure, texture and composition) [Moropoulou, 1995 [8]] which allow for the mapping of the weathering patterns (type, extent and distribution of weathering damages), along the monumental surfaces [Moropoulou, Koui, 1996 [9]].

In the present chapter, weathering mapping is performed, accompanied by non destructive techniques, based on signal and image processing applications in coordination with analysis of stone damage in order to allow for planning Conservation Interventions on Historic Structures and Architectural Complexes. An extended monument, like the Medieval Fortress of Rhodes, in a rather marine environment under different microclimatic conditions, constructed by a susceptible to salt decay highly porous stone, undergoing various physicochemical and biological degradation processes, which lead to several masonry problems at different levels, could serve as a paradigm [Theoulakis, Moropoulou, 1997 [10]]. Due to the prevailing salt decay problems, the historical archives and the survey being given, lithotype and weathering mapping assert priority.

2. Investigation Procedures, Techniques and Materials

In the present chapter five characteristic sampling points are presented (Table 1), where semantic micro-climatic conditions and various decay patterns are correlated [Theoulakis et al., 1995 [11]]. Samples were taken at four different heights all around the walls to study their textural and microstructural characteristics.

A. The following *Non Destructive* Tests were performed *on site*:

1. The applied *Digital Image Processing* technique [F. Zezza, 1991[6], 1996[7]] regards the conversion of the varying surface energy content, interpreted through the color variations of a captured image, to a false color system, which renders and distinguishes the real degradation process. The basic principal of the method is that the different levels of light reflected and diffused by the surface, correspond to the irregularity of the damaged stone surface. The final results arise from the application of the proper false color system through a process of applying various color maps and by assessing and evaluating them with physicochemical retrieval codes [Moropoulou, 1995[8]] (textural and microstructural characteristics) (fig. 1, fig. 2, fig. 5).

2. *Fiber Optics Microscopy* was applied in the field (by PICO SCOPEMAN - MORITEX), in several magnifications, x25, x50, x100, x200, to investigate the textural characteristics of the weathered stones, established by A. Moropoulou [9], M. Koui in 1996. The images were stored in a video system, and they were processed in the laboratory (fig. 3).

3. The depth of weathering of a stone surface can be evaluated by using the indirect *Ultrasonic Velocity Technique* [Zezza, 1993 [12]]. In this case the transmitter is placed on

Table 1. Selected demonstration locations - Architectural surfaces acquired by photography for general and detailed mapping.

Sampling point	Location in the moat (M) facing the sea (S)	Macroscopic observations Highly porous sandstones of various porosity characteristics
General mapping		
10 (fig. 1a-a')	Snt. Paul Gate (S)	Alveolar disease
12 (fig. 1b-b')	Intermediate Tower I, northern orientation (on the Tower) - Tongue of Spain (M)	Selective alveolar weathering of the mixed type due to air turbulent
13 (fig. 1c-c')	D'Amboise Gate	Hard carbonate and biogenic crust
Detailed mapping		
5 (fig. 2a-a')	Between the intermediate Tower X and the Sea Gate, north - east orientation facing the sea - Tongue of Castille (S)	Advanced alveolar weathering
17a (fig. 2b-b')	Masonry near Snt. John's Gate (M)	- Hard carbonate and biogenic crust - Replacement by incompatible stones and alveolation at the interface - Compact stone of dolomitic and calcitic character - Black crusts and washed out surfaces

Table 2. Microstructural data by porosimetry

Height (m)	Total Porosity (%)	Pore Radius Average (micron)	Total cumulative volume (mm³/g)	Bulk Density (g/cm²)	Total Porosity (%)	Pore Radius Average (micron)	Total cumulative volume (mm³/g)	Bulk Density (g/cm²)	Total Porosity (%)	Pore Radius Average (micron)	Total cumulative volume (mm³/g)	Bulk Density (g/cm²)
	Sampling Position 17*a				Sampling Position 13				Sampling Position 17a (new)			
0,5	19,23	26,55	88,6	2,17	26,75	30,84	136,5	1,96	11,82	24,4	60,6	1,95
1,0	9,15	26,55	45,1	2,03	18,95	27,32	89,4	2,12	12,68	20,12	53,5	2,37
1,5	22,54	26,89	109,4	2,06	15,06	27,43	64,9	2,32	18,61	25,89	83,1	2,24
2,0	8,94	25,74	36,2	2,4	-	-	-	-	-	-	-	-
	Sampling Position 12				Sampling Position 5				Sampling Position 17a (old)			
0,5	3,87	12,39	26,5	1,46	24,20	11,35	121,0	2,00	28,16	3,485	151,43	1,86
1,0	4,96	26,98	31,0	1,60	21,55	26,22	101,2	2,13	-	-	-	-
1,5	10,71	27,20	72,4	1,48	21,10	33,72	100,0	2,11	-	-	-	-
2,0	-	-		-	13,18	34,39	55,6	2,37	-	-	-	-

404

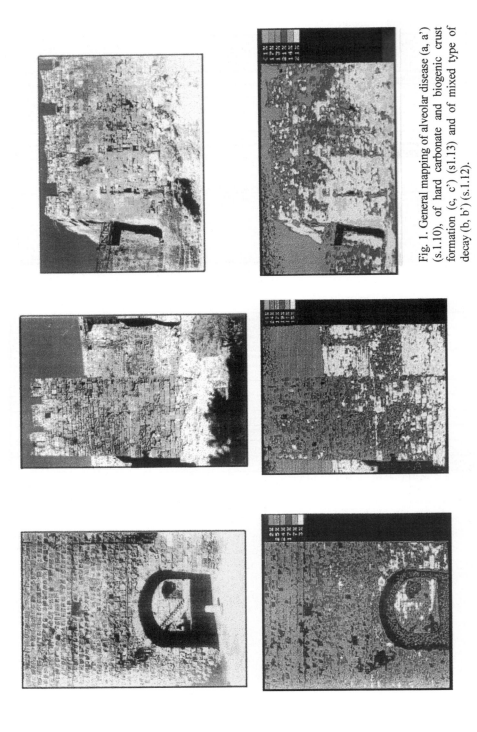

Fig. 1. General mapping of alveolar disease (a, a') (s.1.10), of hard carbonate and biogenic crust formation (c, c') (s1.13) and of mixed type of decay (b, b') (s.1.12).

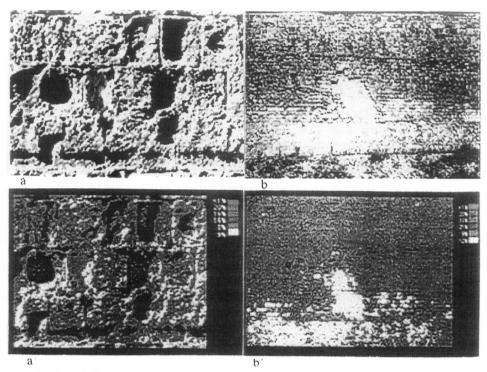

a
b
a'
b'

Fig. 2. Detailed mapping of alveolar disease (s.l. 5), varying from desagregation to pitting, cavities, interconnected cavities and full face caves (a,a'), in contrast with incompatible restoration materials and of hard carbonate and biogenic crust (s.l.17a), partial replacement by new incompatible building stones (b, b').

a
b

Fig. 3 Observations by Fiber Optics Microscopy regarding weathering mapping of masonry undergoing alveolar disease (a) (x25) and hard carbonate and biogenic crust formation (b) (x25).

a suitable point of the surface and the receiver is placed on the same surface at successive positions along a specific line. The transit time is plotted in relation to the distance between the centers of the transducers. A change of slope in the plot could indicate that the pulse velocity near the surface is much lower than it is deeper down the rock. This layer of inferior quality could arise as a result of weathering. The thickness of the weathered surface layer may be estimated as follows [Christaras, 1997 [13]]:

$$D = \frac{Xo}{2} \sqrt{\frac{Vs - Vd}{Vs + Vd}}$$

Vs: Pulse Velocity in the sound rock (Km/s)
Vd: Pulse velocity in the damaged rock (Km/s)
Xo: istance at which the change of slope occurs (mm)
D: Depth of weathering (mm)

The weathering depth is calculated, at the change of slope of the curve and the results are plotted in the form of color distribution on a section of the stone (fig. 4). A PUNDIT ultrasonic non destructive digital tester was used. Measurements are applied along the axis of the core samples and the travel time of the 25KHz source pulse is measured. Fig. 4.1 (laminar variety) and fig. 4.2 (vacuolar variety) depict the application of Integrated Computerised Analysis for stripped and alveolar weathering respectively.

B. *In the laboratory* the measurements of the microstructural characteristics were performed by a mercury *porosimeter* (FISONS Porosimeter 2000). Total Porosity, the Pore Radii distribution, the Bulk Density and the Cumulative Volume were measured. The simulation model used was the cylindrical one (Table 2).

3. Results and Discussion

3.1. ENVIRONMENTAL IMPACT ASSESSMENT - WEATHERING MAPPING

3.1.1. *Digital Image Processing (DIP)- Retrieval Codes by Fiber Optics Microscopy and Microstructural Analysis*

As far as salt decay is concerned, it suffices that the mapping technique renders the distribution of the characteristic decay patterns, i.e. the surface areas undergoing granular disintegration in the form of alveolar disease or hard carbonate crust formation. In General Mapping, as in fig.1, natural (fig.1a,b,c) and digitally processed (fig. 1a',b',c') images are presented, rendering in false colours extended masonry surfaces generally characterised by : alveolar disease (fig.1a,a') (sampling point 10), hard carbonate and biogenic crust formation (fig.1c,c') (sampling point 13) and selective alveolation due to turbulent air flow, where hard carbonate crust prevails (fig.1b,b') (sampling point 12). In Detailed Mapping, as in fig.2, natural (fig.2a,b) and digitally processed (fig.2a',b') images are presented, rendering in false colours details of masonry under alveolar disease (sampling point 5), varying from desegregation to pitting, cavities, interconnected cavities and full face caves (fig.2a,a'), and masonry presenting hard carbonate and biogenic crust (sampling point 17a), partially replaced by new structural stones of different porosity, which trigger intense alveolation and salt decay at the interface (fig.2b,b'). Materials properties like the textural and microstructural characteristics, could provide reliable classification criteria as in the case of salt decay, where alterations in pore size distribution may judge the porous stone susceptibility to decay [Theoulakis et al., 1998 [14]]. Therefore, the microstructural characteristics of the weathered stones sampled are measured (Table 2), along with their textural in situ microscopic observation (FOM, fig.3)

In the case of alveolar disease, sodium chloride crystals are growing within the pores exerting pressures which cause a breaking of grain contacts. The granular disintegration in process results in the exposition of a more rough surface (fig.3a) to the light entailing reflection towards higher grey levels. The prevailing alveolar pattern is discernible by false colours towards black in the sequence of higher damage levels (fig. 2 a, a'). Image classification to the level of more or less severe alveolar damage, according to IAEG in 1979 [15], is possible. This is in contrast to the hard crust formation that due to its coherent micro-crystalline calcitic surface (fig.3b) exposed to the light, attains reflection towards the lightest shades of grey and is distinguished by false colors towards yellow (fig. 2 b, b'), hindering however a potential advanced damage level due to the internal relaxation zone that it develops. Biological attack to the hard carbonate crust entails disaggregation of the external calcitic surface, resulting to higher grey levels, not as high though as in the case of alveolar disease. The substrate, i.e. whether it concerns hard crust or mere the porous building stone, determines the restitution of biological crust by grey-ochre or maroon tonalities.

The same contrasts distinguishes compact (new) to dissaggregated porous stones (old) and in generally incompatible to historic restoration materials (like cement mortar joints. The experimental conservation intervention, which took place recently with freshly cut and less porous stones from the Stegna Quarry, is rendered clearly by yellow-ochre grey-light maroon (≈21%), in the same way that the compact hard calcitic crust does. This is in contrast to the more susceptible, more porous and friable stone, which reflects light towards higher shades of grey, in the same way that the already weathered by granular disintegration stone does. It is also observed that a condensation of darker colours at the interface between new / old stones indicating intensified alveolar disease in the old stones around, occurs.

Hence, the *DIP* results show the characteristic distribution pattern of the weathering forms, when the microstructural and textural characteristics of weathered stone are used as interpretation criteria.

3.1.2. Integrated Computerised Analysis for Weathering (ICAW)

Ultra sound measurements accompanied by Image Processing of the weathered stones provide insights for the forms and the depth of decay layers. Stripped and alveolar decay patterns regarded as laminar or vacuolar calcarenite varieties (monumental samples) [Zezza 1989 [5], 1991 [6], 1996 [7]] are discerned and the thickness of the weathered layers are estimated and represented (fig.4).

As regard the laminar variety of calcarenite the differential deterioration, which reveals in the form of laminas in relief and of excaved laminas (fig. 4.1) and which proceeds from the exterior to the interior, is due to the textural characteristics of the laminas having different packing and dimensions of the clasts; in this sense the coarse grained laminas and the fine-medium grained laminas are characterised by different porosity values. The reconstruction of the buried structure beneath the decayed surface has made it possible to establish that two different *decay layers* are present in the considered block (fig.4.1e): the first one, discontinuous and irregular, reaches à

408

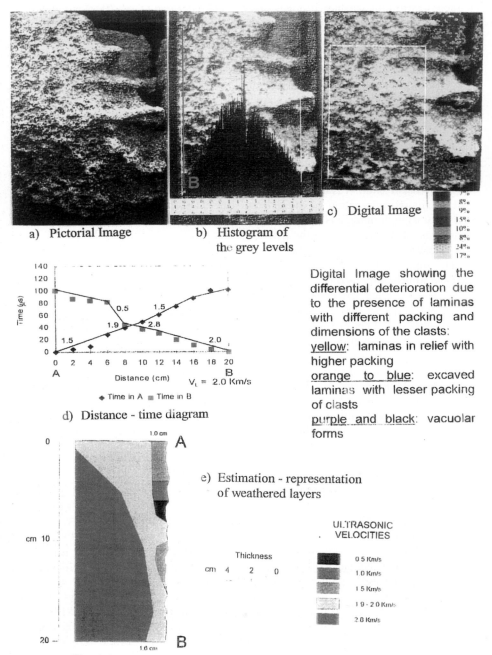

a) Pictorial Image

b) Histogram of
the grey levels

c) Digital Image

d) Distance - time diagram

e) Estimation - representation
of weathered layers

Digital Image showing the
differential deterioration due
to the presence of laminas
with different packing and
dimensions of the clasts:
<u>yellow</u>: laminas in relief with
higher packing
<u>orange to blue</u>: excaved
laminas with lesser packing
of clasts
<u>purple and black</u>: vacuolar
forms

ULTRASONIC
VELOCITIES

Thickness

cm   4   2   0

0.5 Km/s
1.0 Km/s
1.5 Km/s
1.9 - 2.0 Km/s
2.0 Km/s

Fig. 4.1. Integrated Computerised Analysis for Weathering
Lithology : calcarenite (laminar variety)

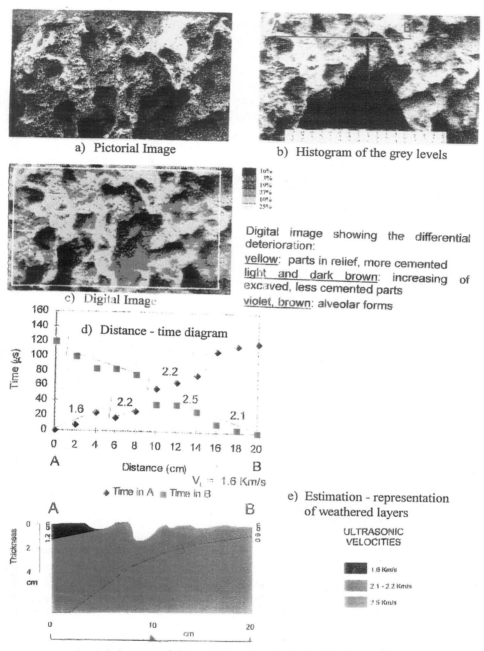

a) Pictorial Image

b) Histogram of the grey levels

c) Digital Image

Digital image showing the differential deterioration:

yellow: parts in relief, more cemented
light and dark brown: increasing of excaved, less cemented parts
violet, brown: alveolar forms

d) Distance - time diagram

Time (µs)

Distance (cm)

A   B

$V_1 = 1.6$ Km/s

◆ Time in A   ▪ Time in B

e) Estimation - representation of weathered layers

ULTRASONIC VELOCITIES

1.6 Km/s
2.1 - 2.2 Km/s
2.5 Km/s

A   B

Thickness

cm

Fig. 4.2. Integrated Computerised Analysis for Weathering
Lithology : calcarenite (vacuolar variety)

thickness of 1.0 cm; it is characterised by the presence of structure with dikes having different velocity values (0.5 km/s,1.0 Km/s and 1.5 Km/s) in relation to the presence of laminas with different packing of the clasts. The second decay layer (velocity 1.9 - 2.0 Km/s) reaches thicknesses which, from 1.6 cm (or less) in the final part of the section (in B), deepen progressively towards the beginning (in A) in relation to the presence of the edge of the block for which the weathering obviously starts from both the external surfaces. The relatively integral stone material is characterised by a velocity of 2.8 Km/s.

For vacuolar variety of calcarenite the differential deterioration shows in the form of more cemented parts in relief and less cemented parts excavated to the extent as to originate alveolar forms (fig. 4.2). In this sense the textural characteristics reveal the presence of fossil cavities and alternating parts with greater and lesser packing of clasts in relation to a different degree of diagenesis. Porosity is of both inter- and intra-granular types. The buried structure (fig. 4.2 e) is similar to the laminar variety in that there are two distinct decay layers (respectively 1.6 Km/s and 2.1- 2.2 Km/s) on the relatively integral material below (2.5 Km/s). The first of these is present only in correspondence with the initial part of the section, that is at the edge of the block, where it reaches a thickness of 1.2 cm. The second layer is of 0.9 cm at the final part of the section (in B) and deepens gradually towards the initial part for the above mentioned reason.

3.2. GIS MANAGEMENT OF DATA

A G.I.S. application was developed by using Arc/Info software for the display, query and generally management, of the data. More specifically, the user has the option to view the available base maps, zoom in/out, pan, etc. A raster image of the old city serves as the main orientation base map (fig. 5). A vector map of the Wall contains more details. Side views of the Wall in vector format provide even more detail for various locations of the Wall. Measurements of distance and area can be performed directly on-screen on the displayed base-maps [Maguire, 1991 [16]]. The test points are displayed as marker symbols on-top of the Wall map at their correct spatial location. Alternatively, the user can query the data base to identify,for example, for which test points a specific measurement /data is available and have the results of the query displayed graphically on the screen. This way the user could, for example, identify all the test points for which Digital Image Processing and Integrated Computerised Analysis for Weathering thermographic images exist, and display all these images simultaneous on windows on the screen. Such a visualisation of the data allows for quick comparison and evaluation of the specific method for the different test points, or in the second approach of the different imaging techniques for a single point [Benhardsen, 1992 [17]]. This in turn allows for better strategic planning intervention. The user can select to view the base-maps, along with the test points-locations, information on available data for each test point or finally the full data for a specific test point. Fig. 5 visualises all the characteristic weathering and materials mapping by Digital Image Processing and Integrated Computerised Analysis for Weathering over the walls and permits conservation planning and environmental management.

411

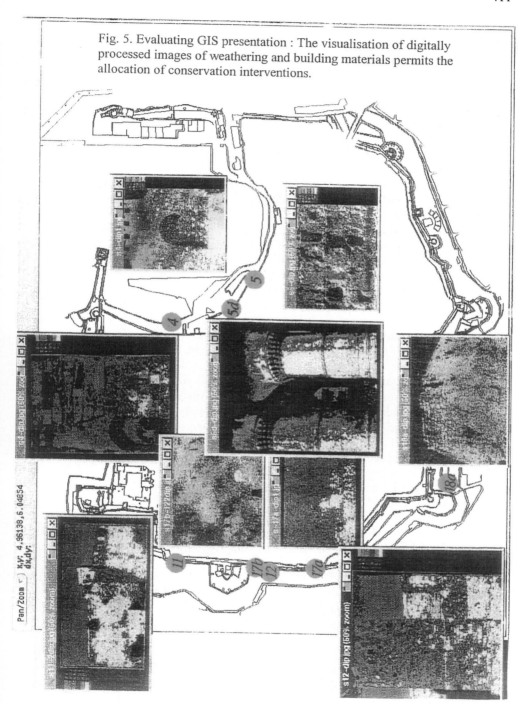

Fig. 5. Evaluating GIS presentation : The visualisation of digitally processed images of weathering and building materials permits the allocation of conservation interventions.

3.3. EVALUATING MAPPING INFORMATION FOR CONSERVATION PLANNING

The visualisation of digitally processed images of weathering and building materials permits the allocation of conservation interventions. As it results from the materials decay analysis, the conservation planning regards as *necessary conservation interventions* :

- *Conservation substitution for old stones by compatible new ones.* Mapping permits the allocation of incompatible substitions and in coordination with porosity measurements or in situ assessment (FOM-ICAW), allows for the choice of the proper quarry for compatible new stones excavation. The criterion to meet is compatibility among old and new stones concerning compactness of cementing material and pore systems. Lithotypes mapping along the walls permits the proper choice per surface area according to the lithotype by which it is constructed.

- *Reconstruction and filling of washed out joint mortars by compatible restoration mortars.* Mapping permits reallocation of the most damaged areas to be reconstructed as well as the washed out joint mortars which under the exerted stresses produce masonry pathology (fissuring, cralls, etc.). The criterion to meet is compatibility of restoration mortars, with the porous stone and the historic ones. The advanced caveneous alveolar disease of the porous building stones (fig 1a') due to the incompatible cement mortar of the Italian restoration (in the middle-war period) permits to discern incompatible mortar joints in order to intervene properly.

- *Consolidation of disaggregated porous stones due to advanced alveolar disease.* Mapping permits the allocation of masonry surfaces areas damaged by alveolar disease due to salt crystallisation, where consolidating materials should be induced by proper techniques into the masonry. Integrated Computerised Analysis might be used to assess and evaluate consolidation pilot intervention on the masonry according to the results of the relevant consolidation study. From the existing analysis of the decayed stones it results that a level of 2,5-2,8 Km/sec of ultrasonic velocities should be attained in order to achieve the compactness of the intact core. The decayed layers estimated at a thickness of more or less <1 cm indicate the level of the depth of intervention. The in between intermediary area of ultrasonic velocities of more or less 2 Km/sec provides the acceptability levels of the intervention.

- *Integration of stone cavities.* Mapping permits as well the allocation of very corroded stones where big cavities have to be filled. In that case artificial stone could be produced by proper, very dense mortar,comprised by crushed porous stone as aggregate.

- *Mechanical connection of bearing building stones* units in characteristic architectural elements (arches, gates, etc.) by compatible metal joints to the very aggressive, corrosive marine environment. In that case iron, even steel bars should be excluded due to the high salinity of the atmosphere while titanium joints are suggested, when compatible.

- *Cleaning* of facades and especially marble decorations from the black crust is indicated by the relevant allocation and mapping. Proper materials and techniques have to be suggested by a relevant pilot project. Biological crust should be cleaned only in the case of buildings under restoration, whereas hard carbonate and biogenic crust should be left intact on the masonry. Any intervention on masonry crusts would produce a retreat in the facade due to the ablation of the outer surface of the crust, which is weakened underneath its hard exposed end.

4. Conclusions

The Digital Image Processing on historic architectural surfaces results on the characteristic distribution pattern of the weathering forms, when the microstructural and textural characteristics of weathered stone are used as interpretation criteria. Image classification to the damage level, according to IAEG, is possible in the case of alveolar decay. The same processing distinguishes in general restoration materials as compatible or incompatible to their original ones. Ultra sound measurements accompanied by Image Processing of the weathered stones (Integrated Computerised Analysis for Weathering) provide insights for the form and the thickness of the decay layers.

The visualisation of digitally processed images of weathering and building materials permits the allocation of the proper conservation interventions. The management of data by GIS allows for conservation planning regarding specifically : conservation substitution for old stones by compatible new ones, reconstruction and filling of washed out joint mortars by compatible restoration mortars, consolidation of disaggregated porous stones due to advanced alveolar disease, integration of stone cavities, mechanical connection of bearing building stones, cleaning of facades.

Hence, the combined use of Digital Image Processing, and Integrated Computerised Analysis for Weathering develops a reliable, as far as physicochemical processes of decay are concerned, integrated automatic risk mapping method, regarding the weathering of architectural surfaces and the evaluation of compatible conservation interventions. The coordinated methods should be further searched and standardised in various environments and per various materials, in order to become a tool in the conservation practice.

5. References

1. Baer, N., Sabbioni, C., Sors, A., (1991), "Science, Technology and European Cultural Heritage", Proc. of the European Symposium, Bologna, 1989, Butterworth-Heinemann Ltd..
2. Cundari, C., (1991), "Verso un sistema informativo dei beni culturali. Il complesso monumentale di Monteoliveto in Napoli. Disegnare idee immagini", Universita degli Studi di Roma "La Sapienza", n 3, Roma, p. 57-68.
3. Mamillan, M., (1991), "Methodes d' evaluation des degradations des monuments en pierre", Colloque Int.: Deterioration des materiaux de construction, La Rochelle, France, p. 12-14.
4. Fitzner, B., (1991), "Mapping, mearurements and microstructure analysis-combined investigations for the characterization of deteriorated natural stones", Weathering and air pollution, Community of Mediterranean Universities, Venezia, Milano.
5. Zezza, F., (1989), "Computerized analysis of stone decay in monuments", Proc. 1st Int. Symp. on the Conservation of Monuments in the Mediterranean Basin, Bari.
6. Zezza F., (1991), "Digital image processing of weathered stone in polluted atmosphere", Weathering and air pollution, Community of Mediterranean Universities, Venezia, Milano.
7. Zezza F., (1996), "Decay Patterns of Weathered stones in Marine Environment", E.C. Research Workshop : Origin, mechanisms, and salts on degradation of monuments in marine and continental environment, Bari, Italy.

8. Moropoulou A., Koui, M., Theoulakis P., Kourteli Ch., Zezza F., (1995), "Digital Image Processing for the Environmental Impact Assessment on Architectural Surfaces", J. Environmental Chemistry and Technology, No1, p. 23-32.

9. Moropoulou, A., Koui, M., Tsiourva, Th., Kourteli, Ch., Papasotiriou, D., (1996), "Macro - and micro- non destructive tests for environmental impact assessment on architectural surfaces", Materials Issues in Art and Archaeology IV, Materials Research Society.

10. Theoulakis P., Moropoulou A. (1997), "Porous stone decay by salt crystallization - The case of the building material of the Medieval City of Rhodes", Proc., 4th Int. Symp. : Conservation of Monuments in the Mediterranean Basin, Rhodes, Publ. Technical Chamber of Greece, Vol. 4, p. 647-673.

11. Moropoulou, A., Theoulakis, P., Chrysophakis, T., (1995), "Correlation between stone weathering and environmental factors", Atmospheric Environment, Vol. 29, No 8, p.895-903.

12. Zezza F., (1993), "Evaluation criteria of the effectiveness of treatments by non destructive analysis", Proc. 2nd course of CUM University School of Monument Conservation, Heraklion, p.198-207.

13. Christaras B., (1997), "Estimation of damage at the surface of stones using non destructive techniques", Structural Studies, Repairs and Maintenance of Historical Buildings, Advances in Architectural Series, Computational Mechanics Publications, Vol.3, p.121-128.

14. Theoulakis P., Moropoulou A. (1998), "Porous stone decay mechanism by salt crystallization", J. Construction and Building Materials.

15. I.A.E.G., (1979), "Classification of rocks and soils for engineering geological mapping. Part I, Rock and soil materials", Bull, IAEG, 19, p. 364-371.

16. Maguire D., Goodchild M., Rhind D., (1991), "Geographical Information Systems, Principles and Applications".

17. Benhardsen T., (1992). "Geographical Information Systems".

37

A Thinning-Based Method for Extracting Peri-Urban Road Network from Panchromatic Images

V. KARATHANASSI, C. IOSSIFIDIS, AND D. ROKOS
National Technical University of Athens
Laboratory of Remote Sensing
Heroon Polytechniou 9, Zographos, 15780, Athens, Greece
Tel: +30-1-7722593, Fax: +30-1-7722594
E-mail: karathan@ survey.ntua.gr

1. INTRODUCTION

The extraction of linear elements from remotely sensed imagery is a quite complex process. The extraction, for example, of the road network requires also its recognition which is not an easy operation, because road segments are not represented by a constant range of digital numbers in the input images, the shapes of the roads are ill-defined, and pixels with the same digital number often represent land cover types which are not related to roads. That happens because in remotely sensed images, certain types of land cover, e.g. buildings, have similar spectral characteristics with roads and, thus, are indistinguishable, at a pre-processing stage, from the roads.

Consequently, line detection and image representation / description algorithms, give noisy results, in the presence of: a) line segments which do not belong to the road network, like the edges of the buildings, b) discontinuous linear features, and c) spikes and irregularities.

Road tracking and linking algorithms [1,2] have been developed, since they overcome these defects in semi-automatic methods, by using seed points on roads: a) as starting points, b) near roads discontinuities, and c) near local direction changes.

Another approach to cope with this problem is the extraction of cartographic primitives. In the relevant research the linear elements of maps or images, are considered as primitives and extracted at a first step by digitisation or lower level techniques. Emphasis is given then into their interpretation. Examples include: a) the building of a rule based system for interpreting linear map features [3]; b) the road network extraction by interpreting two types of primitives: crossroads and road segments [4]; and c) the implementation of a system for the extraction of a forest road network by using both procedural and knowledge-based modules. The later defines the road network by eliminating extraneous elements derived from the image [5]. Although the results obtained by the rule-based systems are encouraging, the complexity of the knowledge required if the system should be expanded to incorporate more linear objects and features, cause serious inconvenience for their broad use.

A thinning based method has been developed by the authors [6] that enables users to extract the road network of peri-urban areas from panchromatic SPOT images, 10m resolution. This method is based on three key factors. First, it uses the fact that geographical features which have similar spectral characteristics with the road network but are not part of it, normally, had different size and shape from the features of the transportation network. Thus, there is a stage in the method that eliminates from further processing the clusters of pixels that have sizes and shapes that are different from those that are typical of road network features. Second, it processes several binary images produced by the original panchromatic image, where every one of them represents a subset of the road network with different completeness and refinement. Third it uses the digital numbers of the pixels in the original panchromatic image to expand the linear features that have been extracted in the early stages of the processing. Such an expansion fills the gaps that have resulted from the thinning and linking algorithms of the method. This combination of excluding pixels that represent information unrelated to the road network and including pixels that are likely to be part of the network had as a consequence the substantial improvement of the performance of the extraction algorithms. As a result, the road network that is obtained from this process has a high degree of completeness. The road network was detected in a sufficient accuracy.

In this effort the developed algorithms were applied on very high resolution simulated data similar to these expected from the upcoming remote sensing microsatellites and the performance of these algorithms is examined. The simulated data were derived by scanning panchromatic air photos (scale 1:20000) and they had result pixel size of 5 m which is similar to the resolution expected by the sensors to be launched in the near future.

The chapter is organised into four sections. In the next section, we cite the main issues of the developed method in article [6]. In the third section we present some experimental results and we compare them with the results obtained by the processing of the SPOT Panchromatic images. Finally, in the last section we give some conclusions.

2. THE PROPOSED METHOD

The method consists of five major algorithms. Figure 1 shows graphically the proposed procedure. Every algorithm characterises a step of the method.

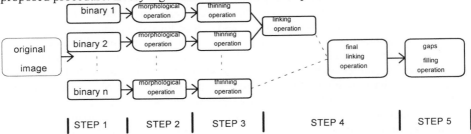

Figure 1: The proposed method

The first is a Thresholding Algorithm which produces several binary images of likely road features. The second is a **Morphological Algorithm** which excludes all pixels that were classified as road pixels by the first algorithm but are unlikely to represent road features due to characteristics of shape and size of pixel clusters that include them. The third algorithm is a **Thinning Algorithm** which refines the road segments by determining their axes based on the morphological corrected images. The fourth algorithm is a **Linking Algorithm** which links parts of the refined road segments found on the different binary images. Finally, the fifth algorithm is a **Gap Filling Algorithm** which completes gaps of the network by the use of the maximum local value of the original image.

2.1 THE THRESHOLDING ALGORITHM

This is a very simple algorithm that is used to select a set of pixels that satisfy a user-specified condition. The algorithm gets as input the digital panchromatic image and a threshold value, and generates a binary image that contains values from the set {0, 255}.

2.2 THE MORPHOLOGICAL ALGORITHM

The binary images that result from the application of a threshold value, contain a classification error which is caused by the inclusion of pixels that represent land cover types other than roads into the set of pixels that represent roads. Examples of such land cover types are certain kinds of agricultural cultivations, built areas, and barren land that has specific moisture. This happens because the grey-level values of the erroneously classified pixels are similar to those of the pixels that represent parts of the road network and, thus, the corresponding pixels cannot be differentiated using only radiometric information.

The Morphological Algorithm is based on the fact that roads are linear features and, thus, their width is substantially smaller than their length. Other land cover types, such as agricultural cultivations, urban built areas and barren land, have a substantially different pattern in terms of shape, size, and ratio of width to length. This difference provides the means to separate clusters of pixels which represent roads from those that represent other land cover types.

The algorithm takes as input a binary image which has resulted from the original panchromatic image by applying a threshold value and deletes all clusters of "active" pixels that have greater width than the width of a specific structure (e.g. crops, building, storehouse).

2.3 THE THINNING ALGORITHM

The main goal in developing a thinning algorithm is to determine the "skeleton" of features shown on a binary image as accurately and as completely as possible. This goal can be achieved by algorithms that satisfy the following five conditions [7]:

- Parts of the feature that are connected in the image must be connected in the "skeleton" of the feature (connectivity preservation).
- The "thinned" result must be minimally 8-connected.
- The endline locations of the feature must be maintained as much as possible (no excessive erosion).
- The "skeletons" that result from the thinning procedure must approximate the medial lines (or axes) of the feature (medial line approximation), and

- Extraneous spurs in the resulting "skeleton" must be avoided (boundary noise immunity).

The thinning algorithm that is used in this method is based on the above five conditions. It works iteratively and, in each iteration, it removes "active" pixels that are on the outer boundary layer of features that are thinned. The removal of "active" pixels is made using eight (8) templates (Figure 2). The algorithm checks if a given "active" pixel in the binary image matches any of these templates and, if it does, then it removes that pixel from the set of "active" pixels. Specifically, templates (a)-(d) are used to remove pixels that are at the top, left, bottom, and right edges of a feature while templates (e)-(h) are used to remove pixels that are on top-right, top-left, bottom-left, and bottom-right edges of such a feature.

Thus, in a given iteration, the outer layer of a feature that is thinned is removed. Successive iterations "peel" a feature and leave only the pixels that comprise its "skeleton".

	0	0
x	255	x
x	x	x

(a)

0	x	x
0	255	x
0	x	x

(b)

x	x	x
x	255	x
0	0	0

(c)

x	x	0
x	255	0
x	x	0

(d)

x	0	0
x	255	0
x	x	x

(e)

0	0	x
0	255	x
x	x	x

(f)

x	x	x
0	255	x
0	0	x

(g)

x	x	x
x	255	0
x	0	0

(h)

[**Note**: An "x" denotes the fact that the corresponding pixel can take any of the two values that are permissible in the binary image (0 or 255). However, two or more pixels that are denoted with an "x" in each template must take the value of 255 in order for the thinning algorithm to operate properly.]

Figure 2: Thinning templates

The above procedure preserves most of the conditions specified above for the correct "skeletonization" of a feature. Specifically, the requirement that at least two of the pixels denoted with an "x" in each template (Figure 2) must have the value of "255" in order for the central pixel to be deleted, preserves, in most cases, the local connectivity condition (condition 1). Also, the same requirement satisfies the minimal endline location erosion and minimal connectivity conditions (condition 2 and 3). Finally, the use of each template in each iteration minimises biases in the determining the medial line of an feature because features are "peeled" gradually from their outer boundaries to their interior (condition 4)(Figure 3, scheme 2).

One of the limitations of the Thinning Algorithm is that, in cases where there is a transition from one part of linear feature which is at least two pixels wide, to another which is at least one pixel wide. It creates gaps in the "skeleton" of the relevant feature (violation of the first condition) (Figure 3, scheme 1). We have not attempted to fix this limitation at this stage because it will be fixed by the Linking and Gap Filling algorithms.

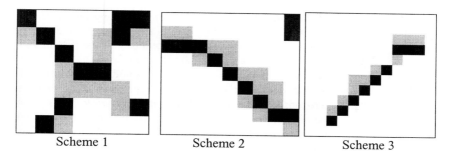

Scheme 1	Scheme 2	Scheme 3

Figure 3: Examples of the thinning algorithm application
Scheme 1:   Violation of the first condition
Scheme 2:   Medial line approximation
Scheme 3:   Features without noisy boundaries

As far as the elimination of extraneous spurs that may result from the Thinning Algorithm is concerned (condition 5), we have tried to avoid them by using as input, the output of the morphological algorithm as well as many binary images (Figure 3, scheme 3). This way, we have avoided use of cleaning templates to reduce noise in the boundaries of features, as it is often done in research and applications [7].

2.4 THE LINKING ALGORITHM

The segments of the road system that resulted from the application of the Thinning Algorithm on the different binary images can not be unified using an "add images" operation because such an operation would violate the unit-width requirement (condition 2). This violation would occur when parts of the road segment are not aligned perfectly with each other in two binary images. Consequently, after the "add images" operation, it is likely that the resulting binary image would contain segments that are more than one pixel wide.

The Linking Algorithm uses as input two images that have resulted from the Thinning Algorithm. One of them, the most comprehensive, serves as the basis which must be completed (Figure 4a). The other serves as an auxiliary which would provide information for completing the first one (Figure 4b).

The algorithm scans the two images simultaneously using a 3x3 window in each of them. In each step of the iteration, the algorithm examines whether the central pixel of the window in the first image is an endline pixel. This is the case when that pixel is "active" and one of its adjacent eight (8) pixels is "active" too (Figure 4a) or two consecutive adjacent pixels are "active" (Figure 4b).

255	0	0
0	255	0
0	0	0

255	255	0
0	255	0
0	0	0

(a)   (b)

Figure 4: Sample templates defining endlines of road axes

420

Then the algorithm checks the corresponding window in the second image. If one or more non-central pixels in that window are "active" and, in addition, are not adjacent to each other nor adjacent to pixels that are "active" in the boundary of the first image, then those non-central pixels are turned into "active" in the first image.

The algorithm operates iteratively. In each iteration, if its conditions are satisfied, it adds new "active" pixels at the endpoints of the road segments. This process continues until no further extensions can be made on the road network in the first image. It must be noted that the preconditions of the Linking Algorithm guarantee that the extensions made in the road network in the first image are connected to at least one segment of the road network (connectivity preservation). Also, they preserve the unit-width requirement, they reduce the problem of excessive "erosion" at the endlines of the road segments, and satisfy the medial line approximation requirement.

It must be noted that the algorithm does not transfer in the first image parts of the network which appear only in the second image and are not linked to any parts of the network of the first. This limitation can be handled, to some degree, by choosing as base image the one that contains as many non-connected pieces of the road network as possible. Experiments on the panchromatic SPOT image of Attica Greece showed that the binary image resulted from the lower threshold plus 10 brightness values upwards, is always the most recommended.

Figure 5 demonstrates how the Linking Algorithm works. Scheme 1 and 2 show the base and auxiliary images respectively that are input into the algorithm. Those images have resulted from the Thinning algorithm applied on the different thresholded binary images. Scheme 3 shows the output of the algorithm.

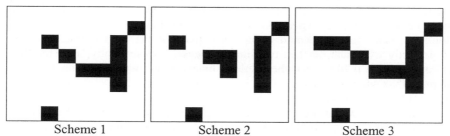

Scheme 1   Scheme 2   Scheme 3

Figure 5: Example of the Linking Algorithm result after one iteration
Scheme 1:Base image (result of the Thinning Algorithm)
Scheme 2:Auxiliary image (also result of the Thinning Algorithm)
Scheme 3:Output image

2.5 THE GAP FILLING ALGORITHM

This algorithm completes the road network which was obtained from the Linking Algorithm by adding segments that do not appear on any image produced by the Thresholding Algorithm. The Gap Filling algorithm is based on the fact that in peri-urban areas the grey-level values of the road segments are substantially higher than the values of the adjacent land cover types. This difference in the grey-level values enables us to distinguish visually roads from the other land cover types in panchromatic images. The information which is portrayed by this difference, however, cannot be detected by the thresholding algorithm because, often, the absolute grey-level values of the roads in the panchromatic image are smaller than the specified threshold value and, thus, those roads are not represented in the binary images. Even when we use low threshold values in the Thresholding Algorithm, still, we would miss certain road segments as a result of the morphological algorithm.

The Gap Filling Algorithm works iteratively and fills small gaps in the roads that have resulted from the Linking Algorithm by taking into consideration the grey-level values of the pixels of the original panchromatic image. Specifically, the algorithm determines whether a pixel is an endline pixel of a road segment using the same procedure as the one described in the Linking Algorithm and, if it is, the algorithm examines the eight (8) peripheral pixels of the 3x3 window which is centred on the endline pixel to determine which one has the highest grey-level value in the original panchromatic image. If that pixel is not "active" nor adjacent to an "active" pixel in the binary image then the algorithm turns it to an "active" pixel and continues to process the next endline pixel. Otherwise, the algorithm searches to find the peripheral pixel of the same 3x3 window that has the second highest value. If that pixel is not "active" nor adjacent to an "active" then the algorithm turns that pixel to an "active" one and continues processing from the next endline pixel. Otherwise, it continues to examine the pixel with the third highest value. If that pixel is not "active" it turns it to an "active" and continues processing from the next endline pixel of the image.

In the case in which there is a gap between two segments endlines that are separated by one pixel, the algorithm fills the gap by turning into "active", the intermediate pixel.

The Gap Filling Algorithm is effective in filling up, pixel by pixel, gaps of the road system without creating extraneous spurs or violating the thinning requirements. Algorithm includes "ring correction" when, at a given endline location, the pixels with the high grey-level values form a small scale circle. The basic weakness of the algorithm is its inability to fill gaps when certain segments of the road not being detected by the Thresholding and Linking Algorithms do not match any endline pixel in the binary image which is input in the Gap Filling Algorithm.

Figures 6 demonstrate partially how the Gap Filling Algorithm works. Pixels shown in black colour indicate pixels that are part of the road network which resulted from the Linking Algorithm, and pixels that are dark grey are pixels which were turned to "active" by the Gap Filling Algorithm. Pixels shown in lighter tones of the grey come from the original grey-level value which is also represented.

3. EXPERIMENTAL RESULTS. METHOD EVALUATION

The proposed method was applied on a number of scanned air-photos in order to optimise and automate the batch procedure described in figure 1. The value of the lower threshold is the only input required by the user every time the method is applied. Eight more thresholds are calculated automatically by increasing the input value by a step of 10. This occurs because the analysis of the scanned aerial photography is higher. So, road segments are presented by an increased number of digital values. Because of the increased resolution of the air-photos, the Gap algorithm didn't give similar results as on SPOT images. Indeed, the road network as well as other objects, are represented with many highlighted pixels due to the 5 meter resolution. Consequently, the gaps are not correctly completed when the Gap algorithm is performed.

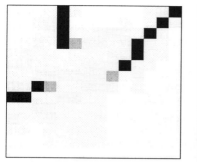 

Figure 6: Examples of how the Gap Filling Algorithm works

In this section three typical examples are presented. These represent areas which are currently under development and are found in the peri-urban area of Athens, Greece. They are characterised by a dense irregular road network with complicated patterns. Other land cover classes like buildings and vegetation are also represent. Pictures 1,3,5 show portions of the air-photos. Pictures 2,4,6 show the final results of the applied method overlaid on the air-photos.

Picture 1: The scanned air-photo   **Picture 2**: Extracted roads.

It has been observed that the majority of the road network has been detected correctly. Few linear features which can be interpreted in the air-photos as roads were not detected. This occurred because at the resolution of 5m some large roads were represent by a wide array of pixels. Consequently, the morphological algorithm has eliminated them. Furthermore, the effects of the shadows, which were intensively present on air-photos, may have resulted in missing small sections of the road network. We observe also that small segments of the detected roads do not correspond to the road network but to other land uses classes, such as parking areas, buildings, etc. A set of rules would remove them, but not without an effect on the detected road network since it has been rather arbitrarily developed, and not according to an urban planning design.

In order to estimate the effectiveness of the presented method in determining the road network and its accuracy, we extracted by photointerpretation the axes of the road network and we compared the results of the above examples, with digitised road network as those were interpreted from the air-photos.

Picture 3: The scanned air-photo

Picture 4: Extracted roads.

Picture 5: The scanned air-photo

Picture 6: Extracted roads.

The percentage of the road network which was correctly detected using our method was 83% for the first example, 85.5% for the second, and 82,5% for the third one. The erroneously detected road network was under 5% for each one of the three examples.

For comparison purposes, we give an example with the results of the method when applied on the SPOT Panchromatic image of Attica captured on 22/04/89 (Picture 7-12). In this case, the correctly detected road network was over 92%.

The differences in the application of the developed procedures between the 10m and 5 m resolution datasets are the following: effectiveness of the Morphological algorithm as well as the Gap algorithm in these higher resolution images and the lack of shadows which were present on the air-photos.

424

The differences of the batch procedure, when data sets with different resolution (10m and 5 m) are used, are the following:

- In the 10m resolution images the method is not hardly depended on the value of the user specified lower threshold since our experiment showed that the expected road network was not sensitive to the specification of the lower threshold in the range of 10 brightness values. This is due to the applied morphological correction. Unlikely, in the 5m resolution images, the method requires a well defined low threshold in order to extract many details of the road network.

- Four thresholds are required for the processing of the 10m resolution data. If 5m resolution data are used nine thresholds are required. This is due to the wider range of digital values with which the road network is represented in the second dataset.

Picture 7:A portion of SPOT image

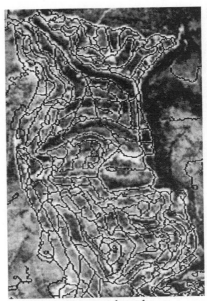

Picture 8: Extracted roads

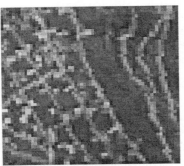

Picture 9:A portion of SPOT image

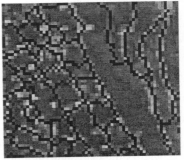

Picture 10: Extracted roads

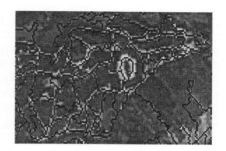

Picture 11:A portion of SPOT image   **Picture 12**: Extracted roads

4. CONCLUSIONS

In this chapter, a method developed by the authors [6] was applied for the extraction of the road network in peri-urban areas using scanned panchromatic air-photos, to simulate the very high spatial resolution remotely sensed images of the microsatellites to be launched in the near future. A batch procedure is executed, which combines in a particular way the results of the developed algorithms.(thresholding, morphological, thinning, linking; and gap filling). The method is almost automatic since the lower threshold is the only parameter required by the user. Concerning its characteristics we can conclude the following:

- It handles successfully road direction changes even if these are significant.
- It can successfully extract roads with significant width and without shadows.
- It efficiently excludes buildings and other land cover classes with similar brightness values.

Although the accuracy of the method as applied to higher resolution data is lower when compared with those resulted by the use of SPOT panchromatic data, the results seems promising for updating maps at the scale of 1:20000. Vectorization of the results must follow for this purpose.

It is recommended that more sophisticated thinning algorithms need to be developed and tested, in order to overpass some of the constrains introduced by the higher resolution images.

References

1. TON J.,JAIN A.K.,ENSLIN W.R. and HUDSON W.D.,1989, Automatic road identification and labelling in Landsat 4 TM images. *Photogrammetria*, 43(5), 257-276.
2. GRUEN A. and LI H., 1995, Road extraction from aerial and satellite images by dynamic programming. *ISPRS Journal of Photogrammetry and Remote Sensing,* 50(4), 11-20.
3. SCHENK T., and OFER ZILDERSTEIN, 1990, Experiments with a Rule-Based System for Interpreting Linear Map Features. *Photogrammetric Engineering and Remote Sensing, 56,* 911-917.
4. CLEYNENBREUGEL V.J.,FIERENS F.,SUETENS P.,AND OOSTERLINCK A.,1988, Knowledge-Based Road Network Extraction on SPOT Satellite Images.

Proceedings of the Fourth International Conference on Pattern Recognition, Cambridge, U.K., March 1988, (New York: *Lecture notes in Computer Science No301, Springer Verlag),* pp.352-359.

5. DOMENIKIOTIS C., LODWICK G.D., WRIGHT G.L.,1995, Intelligent Interpretation of SPOT Data for Extraction of a Forest Road Network. *Journal of the Australian Institute of Cartographers: Cartography*, 24,47-57.

6. KARATHANASSI V., IOSSIFIDIS C., ROKOS D., 1998, A Thinning-Based Method for Recognizing and Extracting Peri-Urban Road Network from SPOT Panchromatic Images. International Journal of Remote Sensing. (to be published)

7. ROLAND T. CHIN and HONG-KHOON WAN,1987, A One - Pass Thinning Algorithm and its Parallel Implementation. *Computer Vision, Graphics, and Image Processing,* 40, 30-40.

38

Integration of Multiple Software Tools for the Design and Analysis of Structures

J.R.García-Lázaro, J.F. Bienvenido
Dpto. Lenguajes y Computación
University of Almería
E-04120, Almería (SPAIN)

1. Introduction

There is, actually, a normalization process on Europe, leaded by the EU institutions, that includes the elaboration of norms for the greenhouse structures, conditioning the financing for the farmers to the presentation of adequate projects. In the Province of Almería, taking advantage of its favorable weather conditions and with the development of a proper type of low-cost greenhouse, is situated the largest concentration of greenhouses in the world. Now, this normalization process and the liberalization of markets threaten these advances, becoming necessary to face these challenges by means of technological advances adapted to our environment.

In order to obtain a new range of greenhouses adapted to our conditions and facilitate the elaboration of building projects, we detected the existence of software tools fitted to resolve partially the process of design. There were general tools for structural analysis, drawing, and cost evaluation, but no one resolved the full specific problem, being necessary to manage a different description of each structure with each different tool. Another problem was the fact that, due to the actual conditions of building these structures, the technicians of the building companies were not skilled with these different tools. Our proposal was to elaborate an integrated tool that facilitates the definition of the proposed structures using a user oriented windows interface and integrating some proved general tools. The tool (complemented with a simulation one) allows the definition of customized structures, generating automatically project plans, budgets and structural analysis reports. It manages automatically specific commercial tools for structural analysis (SAP) and drawing plans (AutoCad or CorelDraw), and integrates our budget generator.

In order to improve the development of greenhouse projects, DAMOCIA project [5], which objective was the "Computer-Aided Design for the Construction of Automated Greenhouses", included a special workline oriented to the implementation of a design tool, that would be adapted to the special characteristics of our zone. The result of this workline has been DAMOCIA-Design. This tool generates greenhouse projects (including their plans, budgets and structural analysis reports) and formal definitions of the designed greenhouses (that can be used by a simulation tool). In other workline was developed a simulation tool, DAMOCIA-Sim [4] [14], which characterizes actually greenhouses as passive radiation captors, permitting to compare different proposals. Next versions will include other parameters as temperature or humidity.

In this chapter we will present, first, how we modelize the design problem as a mass customization problem [11][15][16], defining their required characteristics. Once the problem is established, we will resume the applied techniques, taking special care of the integration of multiple external tools in the resolution of different subproblems and the use of formal representations of the designed object (greenhouse structure). Using the proposed schema, we propose to integrate two sorts of intelligent modules, counselors, which propose specifications, and evaluators, that criticize different proposals and suggest modifications.

2. Problem definition

Initially we started studying the state of the greenhouse design in our zone and outside, finding out these facts:

- In our zone there is a *great diversity* of structures. As a matter of fact, there is a strong *temporal evolution* from the original "parral" (flat roof) structures to more advanced, automated and diversified structures. Different external climatic conditions (e.g. wind conditions originated by our orography), founding possibilities of farmers, product diversity (with multiple varieties), crop periods, final markets (with different requirements) and investment redemption periods generate this extremely diversification of structures [11]. This great diversity is extended with the application of diverse control possibilities and the incorporation of special subsystems. [9]
- Existing global design proposals that have been developed for other zones (mainly for cold winter regions), are linked with special structures. They consider construction modules that are assembled repetitively (as fixed diameter tunnel structures used in France), offering short customization possibilities.
- Great renovation ratio. All the structures are partially or totally renovated each 3-4 years, existing a fixed market of about 40-80 M$ per year (just in our province). This renovation is due to difficult external conditions as strong winds, high insolation and humidity, and cultural methods (e.g. "enarenado" or sand cropping).
- *Convenience of integrating multiple external tools* into the design process. These proven commercial tools simplify the design process, applying them in order to get

specific results. The same task can be executed using diverse alternative tools. Actually, building technicians and agronomists use no standard tools.
- *Multiple usable description.* Greenhouse structures can be described from different points of view. Farmers describe them compactly, specifying the model and some general conditions (user view). Technicians describe them in a strict and detailed form, using plans, budgets, etc.

With all these considerations in mind (model variability, temporal evolution, convenience of integrating commercial tools, multiple usable descriptions and acceptable market size), we propose to handle the problem as a special case of mass customization [16]. Now, we are going to present a general model of the design process that we used in the development of DAMOCIA-Design.

The design process can be modeled as a set of activities, that refines successively the definition of the designed object, finishing with its full and final specification, as it is shown in figure 1. We start with an initial description (included into the problem domain), evolving to more complex and refined descriptions (included into the work domain), until we reach the final result, which is a complete description of the designed object (included in the solution domain).

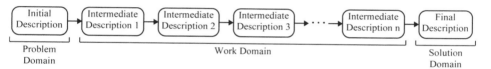

Figure 1. Design process as a refined sequence of descriptions.

In this design process, we can detect mainly two types of activities: *specification* and *translation* activities. Translation activities convert a given specification of the designed object to a more detailed one. Specification activities look for some new information in order to execute some translation process, they generate some complement data required by translation tasks. There is a special specification task (Initial Specification Task) that can be considered as a specification task, it extracts from the user the preliminary definition of the required object. But, it can be evaluated as a translator from the user point of view of the object to its initial formalized description. Figure 2 shows this model of the design process.

Complement data, which are formalized by the specification tasks, are given by the user or are extracted from a general design repository. In the first case, the process associated to the specification task is part of the user-interface or it is associated to it. The design repository stores massive information about the design process, possible subelements with their characteristics, and design heuristics (used by the specification tasks). There is a main control sequence that links the different translation tasks (1). Depending on the design conditions the specification tasks can be executed at the beginning of the design process (as it occurs in our simplest case of greenhouse design)

430

or can be interpolate between the translation tasks (2). Sometimes a translation task uses directly data stored into the repository; in this case the specification task is minimal, being integrated into the translation task. Really, the sequence of tasks it is no linear, it can be structured as a tree structure of tasks, generating each branch a different part of the global description (e.g. plans and budgets).

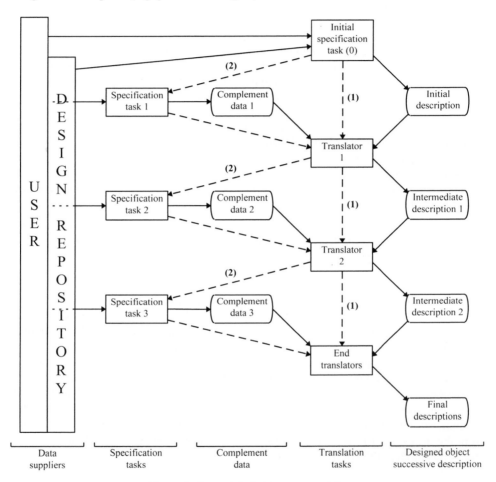

Figure 2. General design process model.

With this schema, we have modeled the design process using as backbone a sequence of translation tasks that refines the designed object description, requiring sometimes specification task to prepare sets of data. This sequence of translation tasks can be implemented as classical transformation processes, without an intelligent behaviour. They only require a formal (and strict) description of the designed objects. The implementation of these tasks can be done integrating preexisting tools in order to obtain some partial results. Possible use of different commercial tools in order to reach the same task objective supposes its expansion to different implemented processes [3].

In order to expand the capacities of the design tool, we proposed to add some intelligent processes [10] [13] classified in two levels. The first level, called *counselor level*, supposes the substitution of simple specification processes (that take data directly from the user or the design repository), by more active agents that generate part of the complement data taking account of repository experience data and using changeable blocks of rules [2][3][8]. In this case, the new element defines complement data using knowledge based-techniques. We have proposed diverse modules that use multicriteria selection algorithms with continuous evaluation of weights. The specification processes can be broken on sequence of subspecification subprocesses, generating each one part of the complement data (using separate unintelligent and intelligent procedures).

The second level of intelligence, called *evaluation level*, corresponds with new processes that evaluate the results of translation processes. These evaluation processes use mainly heuristics and alternative blocks of rules, generating conclusions about the global design process and its actual results. They can demand to active a backtracking process, modifying some complement data of the return point, or they can classify diverse alternative solutions, giving some sort of a multidimensional score to each solution.

3. Resolution procedure

Now, we are going to present how we have implemented the proposed design model, specifying, first, the applied techniques and a development proposal. Finally, we will describe their use designing greenhouses for mild winter regions. The techniques applied were:

- *Different level declarative descriptions*. High level definitions describe structures in a compact user oriented format. Low level definitions describe them using their most basic (and general) elements. The different declarative languages are formalized with BNF grammars [12]. This multilevel declarative descriptions approach isolates the effects of modifications into the high level definition.
- *A two level distributed software architecture*. In the first level the whole process is broken into separate processes, that act, mainly, as translators between the diverse description languages or final results. The second level of distribution corresponds with the internal structure of these translators, distributing the treatment of the different elements of the declarative descriptions. We propose to use twice (recursively) our distributed architecture DACAS [7] in the implementation of the global systems.
- *Simple, unique and separate interface*. The whole design process is presented to the user by a unique user interface, which manages the different distributed elements of the tool, including external tools. [1]

- **Multiple tool integration**. Some designing tasks can be executed using existing commercial tools. Sometimes there are diverse alternative tools for the same task, in these cases we can implemented multiple options, facilitating the user to obtain results using tools that he/she owned, or taking account of their simplicity or price. Figure 3 shows the general schema used integrating the diverse external tools.

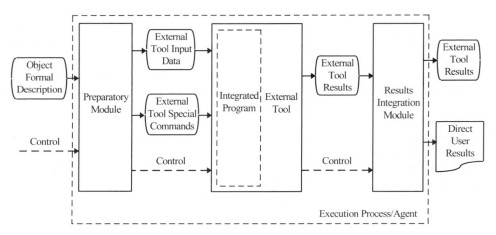

Figure 3. General structure of an external tool manager.

The "Preparatory module" extracts information of the actual designed object description and generates a file, "External tool input data", that contains data with the format required by the external tool (whatever possible we use standard formats). This module calls the external tool, passing it the demands it has received from de central control block via an "External tool special commands" file. The external tool executes demanded operations using a internal program, "Integrated program", that acts automatically when it starts execution (this program will be written in the special control language of the tool, when it exists, as AutoCAD's AutoLISP or Excel's macros). Finally, when the external tool finishes its execution, control goes to the "Result integration module", which transforms the generated data into the general tool format and generates direct results (when it is required).

Our proposal includes a series of steps in order to develop design systems adjusted to the proposed model:

1. Definition of the diverse description levels, developing their formal description languages.

2. Analysis of the diverse translation tasks. Specification of the diverse translators and selection of their implementation mode (specific external tools or customized code).

3. Implementation of translators. They will use complement data obtained directly from the user-interface or supplied via text files.

4. Detailed definition and implementation of the simple specification processes. Implementation of classic data acquisition and validation modules.

5. Specification and implementation of counselor modules. It is possible to implement alternative modules with different expertise.

6. Specification and implementation of simple evaluation modules. These will analyze independently the results of the different translation and specification modules.

7. Specification and implementation of general evaluation modules. These will evaluate great sectors of the design process, proposing backtracking and modifying control.

In our case, the design of greenhouse structures, we have defined the design description levels shown in figure 4.

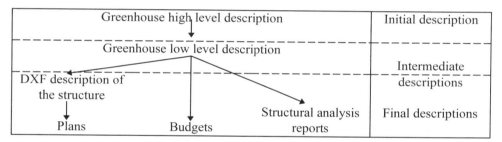

Figure 4. DAMOCIA-Design description levels.

The high level description corresponds with the user point of view and includes the global elements of the design as plot characteristics, structure, foundations, plastics, etc. Figure 5 presents a sample of the high level definition BNF grammar. The low level description refers the independent subelements of the structure as individual pilots, cables, foundation blocks, tubes, etc. [6] Later, we use this general description, generating specialized partial descriptions/ results as sets of plans, budgets documents and sets of structural analysis reports.

Figure 6 shows the general architecture of DAMOCIA-Design, detailing the second level decomposition of the high to low level description translator. General control blocks that use specific behaviour definitions as proposed in our DACAS architecture manage the different processes. The execution processes of the first level of distribution correspond to design description translators. Those of the second level corresponds to processes that manage different aspects of the high level description (as foundations or covers).

<High level definition> ::= <Plot> <Greenhouse type> <Ground plan> <Structure> <Pilots> <Cable ties> <Foundations>
                <Openings><Wire mesh> <Mosquito net> <Drainpipe> <Plastic>
<Plot> ::= <Straight sides plot> | <Curve sides plot>
<Straight sides plot> ::= 0 0 <Angle> <Number> <Latitude> <Longitude> <Point> <Points> <Color> <Layer> <Line type>
<Curve sides plot> ::= 0 1 <Angle> <Number> <Latitude> <Longitude> <Point> <C_Points> <Color> <Layer> <Line type>
<Greenhouse type> ::= 0 2 <Integer>
<Ground plan> ::= <Rectangular plan> | <Polygonal plan>
<Rectangular plan> ::= 1 1 <BD_sel> <BD_sel> <BD_sel> <BD_sel> <BD_sel> <Length> <Length> <Orientation>
<Polygonal plan> ::= 1 2 <BD_sel> <BD_sel> <BD_sel> <BD_sel> <BD_sel> <Number> <Points>
<Structure> ::= <Plane_str.> | <INACRAL_str> | <Symmetric_str> | <INVERNAVE_str> | <Tunnel_str> | <INAMED_str>
<Plane_str> ::= 2 1 <Height><Number> <Number> <Number> <Number>
<INACRAL_str> :: = 2 2 <Height> <Height> <Number> <Side> <Number> <Number> <Number> <Number> <Number>
<Symmetric_str> ::= 2 3 <Height> <Height> <Height> <Side> <Number> <Number> <Number> <Number> <Number>
<INAMED_str> ::= 2 4 <Height> <Height> <Side> <Number> <Number> <Number> <Length-height> <Length-Height>
<Tunnel_str> ::= 2 5 <Height> <Height> <Side> <Number> <Number> <Number>
<INAMED_str> ::= 2 6 <Height> <Height> <Height> <Side> <Number> <Number> <Number> <Number>
<Pilots> ::= <Plane_pil.> | <INACRAL_pil.> | <Symmetric_pil.> | <INVERNAVE_pil.> | <Tunnel_pil.> |
              <INAMED_pil.>
<Plane_pil.> ::= 3 1 <BD_sel> <BD_sel> <BD_sel> <BD_sel> <BD_sel>
<INACRAL_pil.> ::= 3 2 <BD_sel> <BD_sel> <BD_sel> <BD_sel> <BD_sel> <BD_sel> <BD_sel> <BD_sel> <BD_sel>
<BD_sel><BD_sel> <BD_sel> <BD_sel> <BD_sel> <BD_sel> <BD_sel> <BD_sel> <BD_sel>
<Symmetric_pil.> ::= 3 3 <BD_sel> <BD_sel> <BD_sel> <BD_sel> <BD_sel> <BD_sel> <BD_sel> <BD_sel> <BD_sel>
<BD_sel><BD_sel> <BD_sel> <BD_sel> <BD_sel> <BD_sel> <BD_sel>
<INAMED_pil.> ::= 3 4 <BD_sel> <BD_sel> <BD_sel> <BD_sel> <BD_sel> <BD_sel> <BD_sel> <BD_sel>

. . . .

Figure 5. Sample of the high level definition BNF grammar.

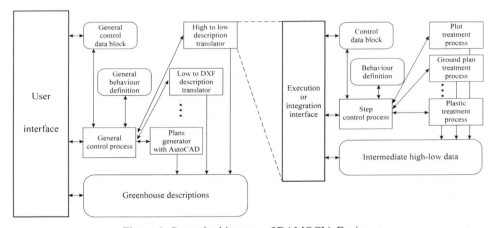

Figure 6. General rchitecture of DAMOCIA.Design.

Some of the user interface windows of the tools are shown in figure 7, including the main window and a structure selection one. Main window presents the different typologies managed by the tool (seven typologies by the moment), the state (defined or not) of the different constructive elements (plot, ground plan, structure, foundations, pilots, ten-cables, openings, wire meshes, mosquitoes nets, drainpipes and cover characteristics). It lists, too, the remaining blocks of data to fill up (tree of windows related with the constructive elements). Using the options defined in the general Windows menu, we can ask for the desired results as those shown on figure 8.

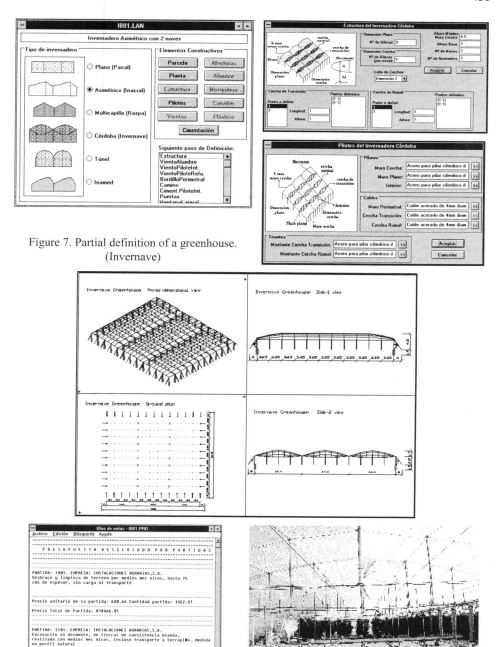

Figure 7. Partial definition of a greenhouse.
(Invernave)

Figure 8. Diverse results of DAMOCIA-Design.

As intelligent modules we can mention, as example, these:

- An evaluator of the most adequate orientation of structures. It objective is to increase or reduce insolation (depending of the crop period), given a typology (taking account of its effectivity and price) and a plot (making the most of the available surface). It is a specification level module.

- A supplier selector, that takes account of the different prices of the elemental pieces, their delivery time and the advantages of reducing the number of suppliers. It gives a global proposal of suppliers. It is a specification level module that works in two modes, as a unique proposal selector or as a supplier database reducing agent.

- An evaluator of structure parameters, that analyzes the user-proposed parameters. It studies pilot distances, materials, etc., taking account of the greenhouse use. As example, it recommends to extend distance between pilots in order to use machinery.

- A wind resistance evaluator. It uses structural analysis data and specific climatic conditions (by land sector) in other to fit the structure resistance. It recommends reduction or extension of the pilot width (changing price and radiation efficiency).

- An evaluator of sets of structures. It analyzes main data of several possible structures, evaluating their behaviours. It generates comparatives using several criteria (cost and multiple performance parameters), summing up the analysis in an ordered sequence. It can propose to evaluate other alternatives. There are several sets of rules and weights that are generated by different experts.

4. Conclusions, next works and acknowledgements

As conclusions we can point:

- Design of mild winter region greenhouses can be considered as a mass customization problem. It is possible to develop specifics project using distributed techniques.

- It is possible to assemble design tools using formal declarative descriptions of the designed objects. The kernel of the design process is a series of translators.

- Use of PSMs techniques in design (with distributed architectures) facilitates the required customization, offering a huge adaptability.

- We include intelligent design elements into de global process, adding *counselor modules*, which specify design parameters, and *evaluation modules*, that analyze and compare proposals. These modules use mainly knowledge based systems, and fuzzy and multicriteria decision techniques.

- We have included easily external proven tools, integrating them into the whole system. User has not to use them directly. Multiple options facilitate reuse of existing software.

- DAMOCIA-Design integrates multiple (and diverse) tools and techniques uniformly using a user-friendly unique interface.

Further works are mainly:

- Inclusion of new subsystems into the design, as heating and fertirrigation systems. We are working now in this sense in next version of the tool.

- Extension of the typologies managed by the design tool.

- Development of an automatic developer of multiple design proposals, evaluating them.

- Logical formalization of translators, using logic programming. Building a library of formal object transformations.

DAMOCIA-Design is implemented and working effectively for commercial and educational purposes. This work was accomplished as part of the DAMOCIA project, financed by the EU (P7510 PACE) and the Spanish Ministry of Industry (PATI PC-191). Structural analysis of the structures has be done using SAP at the Polytechnical University of Valencia. More information can be obtained through the World Wide Web URLs hhtt://www.prosoma.lu/cgi-bin/show.py?opt=flat&id=662&page=result and http://www.ualm.es/Universidad/Depar/Spiaam/spiaam.htm.

5. Bibliography

1. Avouris, N.M.: *User interface design for DAI applications: an overview.* Distributed Artificial Intelligence: Theory and Praxis. Kluwer Academics Publishers, 1.992.
2. Benjamins, V.R.: *Problems-Solving Methods in Cyberspace.*
3. Benjamins, V.R.: *Problem Solving Methods for diagnosis.* Doctoral Thesis. University of Amsterdam. 1993.
4. Bienvenido, J.F. et al.: *DAMOCIA-Sim, a generic tool for radiation simulation into mild winter region greenhouses.* Proceedings of the First European Conference for Information Technology in Agriculture. Copenhagen, Denmark, 1997.
5. Bienvenido, J.F. et al.: *Application of modern information technologies on the design and simulation on Mediterranean greenhouses, an experience.* Mediterranean Colloquium on Protected Cultivation, Agadir, Morocco, 1.996.

6. Bienvenido, J.F.; et al. *DAMOCIA: Computer-Aided Design for the Construction of Automated Greenhouses. AgEng'96,* 1:376-377, Madrid, Spain, 1.996.

7. Bienvenido, J.F. et al.: *DACAS: a Distributed Architecture for Changeable and Adaptable Simulation,* Proceedings of the EIS'98, Tenerife, Spain, 1.998

8. Breuker, J.: *Problems in Indexing Problem Solving Methods.* Proceedings of the Workshop Problem-Solving Methods for Knowledge-Bases Systems. Nagoya (Japan), 1997.

9. Chandrasekaran, B.: Models versus rules, deep versus compiled, content versus form. IEEE-Expert, 6(2), 1991.

10. Cuena, J.: *Knowledge architectures for real-time decision support.* In Second Generation Expert Systems, JM. David, JP. Krivine and R. Simmons (eds.), Springer-Verlag. 1993.

11. Das, D. et al.: *Generating redesign suggestions to reduce setup cost: A step towards automated redesign.* Computer Aided Design, 28 (10), 1996.

12. Garcia, J.R. et al.: *A Design Methodology Base on Multilevel Declarative Definitions and a Distributed Architecture. Application to the Design of Greenhouses for Mild Winter Regions.* In 2nd Annual Conference on Industrial Engineering Applications and Practice. Vol. I. San Diego, USA, 1.997.

13. Guida, G. and Zanella, M.: *Knowledge-based design using the multi-modeling approach.* In Second Generation Expert Systems, JM. David, JP. Krivine and R. Simmons (eds.), Springer-Verlag, 1993.

14. Guirado, R. et al.: *Multimodel Simulation of the Canopy effect on an Inner Radiation Model for Parameterized Greenhouses.* International Conference Engineering of Decision Support Systems in Bio-Industries. Montpellier, France. 1998.

15. Gupta, S.K. & Nau, D.S.: *A systematic approach for analyzing the manufacturability of machined parts.* Computer Aided Design, 27(5), 1995.

16. Tseng, M.M. & Jiao, J.: *A Framework of Design for Mass Customization.* Proceeding of the 2nd Annual International Conference on Industrial Engineering Applications and Practice. Vol. I. San Diego, California. USA. 1997.

39

Application of Distributed Techniques in a Set of Tools to Climatic Sensor Signal Validation and Data Analysis : DAMOCIA-VAL and DAMOCIA-EXP

F. Rodríguez, A. Corral, F. Bienvenido
Dpto. Lenguajes y Computación
University of Almería
E-04120, Almería (SPAIN)

1. Introduction

The main income sources of the province of Almería (Spain) are related with the crop growing in greenhouses (more than 35,000 hectares). They are low-cost greenhouses of medium yield, normally passive, that takes advantages of favorable outside climatic conditions. In order to improve the quality of the horticultural products, it is necessary to enhance the greenhouse climatic conditions using a better control. The established techniques used in other zones (North and Central Europe) with fully automated greenhouses are not directly applicable in our conditions, because we want to regulate the microclimatic conditions in a more passive way. First, we used a direct digital control with classical control techniques, starting now to apply advanced and more intelligent methods as multivariable, adaptive, model predictive and robust control. These techniques need knowledge about the climate behaviour, via mathematical models of the passive greenhouses. Now, we are involved in the second phase of our ideal control system development, working on a climatic model of the Mediterranean greenhouse.

In order to achieve these tasks figures the *DAMOCIA* project, *"Computer-Aided Design for the Construction of Automated Greenhouses"* [4], financed by the *European Union* within the framework of the *ESPRIT projects (Special Action P7510 PACE)* and the *Spanish Ministry of Industry (PATI PC191)*. Its overall objective is the computer aided design of customized, controlled greenhouses and the development of marketable prototypes. In its *modellization* workline, we developed a tool for simulating the radiation behaviour into the greenhouses (base of other variable behaviour) [2] [11] [12] [13]. This tool has two kinds of models: mathematical models based on ideas by

Bot [6], Critten [7] and our own ones; and experimental models that use sensor measurements and interpolation methods. The radiation simulator [4] [10] uses as input a formal definition of the greenhouse, which is generated by the *DAMOCIA-DESIGN* tool [3], and the desired simulation profile. It uses finite-elements in two levels, greenhouse surfaces and volumes. In the *acquisition and control* workline [16], we developed a system to control the greenhouse prototypes and test the results of the climatic models. When they are verified, we use them enhancing the control algorithms.

The experiences were realized using seven different greenhouse structures, representative of our zone. We took measurements of temperature, humidity, global radiation, P.A.R. radiation, balance radiation, wind speed and wind direction. We developed a flexible system based on the definition of a measurement grid in the greenhouses, allowing up to 56 measure points per greenhouse of 1200 m^2, plus an external meteorological station. The sample rate was six samples per minute and the system stored 7 Mbytes of data daily (990 Mbytes of data per growing season). Although the system contained signal conditioning hardware modules, generally it was necessary software preprocessing of the sensor measurements. So we developed an automatic data validation software tool, *DAMOCIA-VAL*, that contains twenty-four filters to detect and correct automatically the wrong sensor readings. Furthermore, a database of abnormal behaviours is created to check their repetition and find patterns that permit a later automatic correction. On the other hand, it was necessary to develop an exploitation software tool, *DAMOCIA-EXP*. This facilitates agronomic engineers to browse the validated data, allowing an easy manipulation of the very large volume of stored data and the generation of tables, charts, simulations, etc. These tools, implemented using distributed techniques, incorporate some intelligent supervisor and counselor modules [8].

In this chapter, we will describe the techniques we used in the implementation of the validation and exploitation tools and their results. The application of these results, in order to develop the intelligent control systems, requires a deeper analysis of the actual data, in which we are involved now.

2. Methodology

DAMOCIA-VAL and *DAMOCIA-EXP* were developed using the following techniques:

2.1 DISTRIBUTED MULTIAGENT ARCHITECTURE (*DACAS*)

The different operations performed with the data are executed by 'independent' modules, which communicate via 'interchange data structures'. All these modules are managed by a special module, the *Control Block*. We have implemented both tools using *DACAS* (*Distributed Architecture for Changeable and Adaptable Simulation*) [5] whose general architecture is shown in Figure 1. Its elements are:

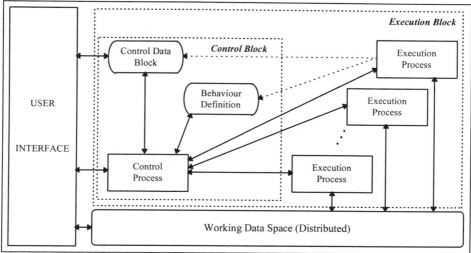

Figure 1.- DACAS general architecture.

- *Behaviour Definition.* It is formed by a declarative representation of the general control program. It uses a language that includes capabilities of sequential, alternative, parallel and cyclic execution of the different processes. (See figure 2). [15].

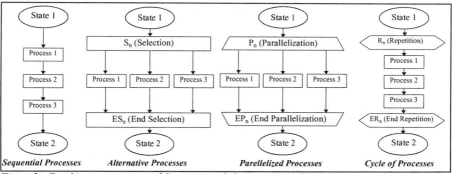

Figure 2.- Graphic representation of the processes behaviour possibilities.

- *Control Process.* It instantiates a Behaviour Definition applying the control parameters defined into the Control Data Block. It evaluates the sequence of execution of the different processes and their performance.
- *Control Data Block.* It contains the control parameters used instantiating the plan stored into the Behaviour Definition. These data are supplied by the User Interface and modified by the Control Process and the Evaluation Processes.
- *Working Data Space.* It stores all the domain specific data. It is made up of Interchange Data Structures (*IDS*), which are data blocks (with standard formats) that are interchanged between the different application processes.

- *Execution Processes.* They perform a domain specific task using a set of input data stored (as IDSs) into the Working Data Space. They can be described as simple input-process-output structures.
- *Evaluation Processes.* They are execution processes that can modify the control parameters (Control Data Block) or the plan (Behaviour Definition).
- *User Interface.* It supplies the input data and shows the results.

This architectures presents two ways of real-time control adaptability: *Reactive Planning*, the Evaluation Processes modify the control parameters (*Control Data Block*) using a sequence of constraints obtained of the execution of certain processes; and *Deliberative Planning*, an Evaluation Process changes the plan itself, modifying the Behaviour Definition (it changes, as example, the process sequence, the execution criteria, the available alternatives or the selection criteria) [14]. The execution and evaluation processes can be composed by sets of subprocesses, in a lower level of decomposition. These subprocesses can be integrated using the same DACAS architecture (appearing as multilevel distributed environment).

Its main advantages are: easy new agents adding (including the new process and changing the Behaviour Definition), generic, open, intelligent control (reactive y deliberative), generic Control Process and incorporation of specific expert systems, classic modules and external tools. Its main disadvantages are complex coherence conditions of the Evaluation Processes, centralization of the Control Block, IPO structure of the Execution Processes and limitation of the communication between active processes. All these disadvantages will be compensated with next architecture versions.

2.2. MULTILEVEL INDEPENDENT INTERFACE

The interface is independent of the rest of the application. Its main functions are providing the input data, executing the control block and showing the execution evolution and primary results [1]. The control block acts as an agent manager with a high level of complexity. The agents perform the selected operations. All these elements can be executed in different platforms. In basis of the kind of user, the interface offers three different user levels:

- *Basic.* The user can only execute a fixed set of experiments, using mainly menus.
- *Definition.* The user can define new experiments using all the available operations; he/she adds new options to the basic user menus.
- *Development.* It develops new calculation agents.

This interface presents a high integration level. Internally, the user interface can be structured as a unique block or a distributed subsystem. The possible options are described, internally, by declarative lists of alternatives.

2.3. DISTRIBUTED DATA STORAGE

Owing to the large volume of data, their different applications and their retrospective use we have implemented a distributed data storage system [9], with the possibility of using different devices. Data are stored on autonomous, independent, and self-described CD-ROMs. Each one includes these files:

–   An identification or label file, that indicates the physical device number and the beginning and ending date of the stored data.
–   A description file, which stores the organization of its data files.
–   Experimental files. The database is composed by a set of files with a standard format (text separated by tabs).

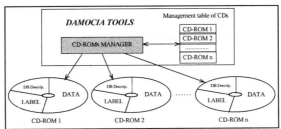

Figure 3.- Management of the CD-ROMs.

In order to use efficiently the data devices, we have implemented a *CD-ROMs Manager*, that interface the applications with the database. It selects and demands automatically the devices required to do a given analysis. It obtains the distribution and organization of data into the CD-ROMs from a management table, where it is stored the content of the description files, as it is showed in Figure 3. This storage technique presents some advantages: data availability (different researchers can use the different storage devices), general safety (use of several CD-ROMs), system efficiency (faster access than that with a single large device). On the other hand, anytime we change the organization of a determined device, it must be restructured its description file and the management tables of the instances of the CD-ROM Manager.

3. Specific control architecture of *DAMOCIA-VAL* and *DAMOCIA-EXP* tools

DAMOCIA-VAL validates the climatic sensor readings. It contains a set of twenty four independent agents, that do some specific actions over the data, as detecting time intervals without measurements and correcting them using linear interpolations, detecting time periods with abnormal behaviours of the signals, sample interpolation between standard deviation bands, etc.

444

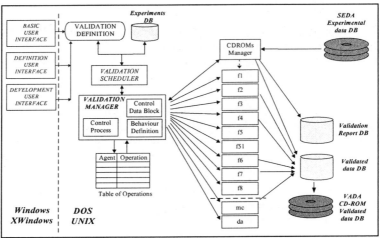

Figure 4.- DAMOCIA-VAL architecture.

The user defines an experiment *(Validation Definition)* selecting the experimental data files, the filters to apply and their input parameters. The experimental data are prepared by the *CD-ROMs Manager*, which acts as an evaluation process. The *Validation Scheduler* uses the *Validation Definition* and the *Experiments DB* (that contains information about previous analysis) to generate the Control Data Block. This *Validation Scheduler* acts as an evaluation process. The user interfaces use the *Experiments DB* to activate prefixed validation schemes or obtaining new schemes by modification of the preexisting ones. The *Validation Manager*, that corresponds with the specific Control Block, uses a *Table of Operations* file to find the agents. The selected modules generate the *Validated Data DB* and incidences files.

DAMOCIA-EXP allows an easy manipulation of the experimental or validated data, facilitating their analysis. It generates tables, charts, simulations, reports, etc. It includes a set of modules that do statistical operations, graphic representations, etc. Its structure and working are similar to that of *DAMOCIA-VAL*. The user selects the time period, the climatic variables, the desired operations and their input parameters. After that, the *CD-ROMs Manager* prepares input data, the *Exploitation Scheduler* instantiates the *Behaviour Definition* and the *Exploitation Manager* launches the agents that generate results and error files.

The user establishes the *Validation* or *Exploitation Definitions*, composed by the agent execution sequences, their input parameters and constraints. The domain constraints (stored in *Interchanged Data Structures*) include an experiment execution maximum time, storage devices available capacities, required configuration files, available computers, etc. The input parameters of the agents are part of the control parameters defined in the *Control Data Block*. The *Validation* and *Exploitation Managers,* that include the *DACAS Control Process,* execute the defined experiment, applying the corresponding parameters to the established operations sequence.

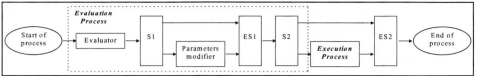

Figure 5.- Agent structure.

Figure 5 shows how the agents can be divided in two subprocesses. First, there is an *Evaluation Process*, which evaluates the domain constraints, modifying the control parameters and the plan (when required). Second, if the constraints are fulfilled, it is executed the real *Execution Process*. A development user can substitute any pure execution process by this structure. In order to clarify this *Reactivity Control*, we are going to describe the *Behaviour Definition* of *DAMOCIA-VAL*, considering only five error managing modules. Figure 6 shows *Behaviour Definitions* plan.

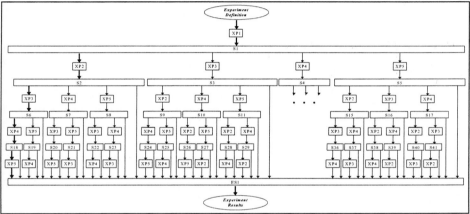

Figure 6.- Behaviour Definition diagram.

We have highlighted, in the previous figure the trace of a possible experiment, that consists of the following operations: normalization of the time format of the data files (XP1); detection of intervals without measurements due to a power supply interruption and interpolation using linear methods (XP2); detection of intervals with signal saturation and correction using linear interpolation (XP3); sample interpolation using different error bands in each interval of day (XP4); and sample interpolation between standard deviation bands (XP5). Generally, an *Evaluation Process* (EV), as that described previously in Figure 5, can be executed before each *Execution Process (XP)*. These EVs can modify the control parameters of the agents based on the domain constraints. Figure 7 shows the initial execution sequence of this experiment.

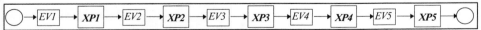

Figure 7.- Initial execution sequence of the defined experiment.

When the user defines the experiments, he establishes the control constraints. Before the experiment is executed, the initial instance of the *Control Data Block* is:

446

	Input directory and file masks	Output directory	Time	Daily Time intervals	Length interval	Saturation values
XP1	d:\mar_98*.*	c:\valida\...\salid_sh\	1			
XP2	c:\valida\...*.*	c:\valida\...\salid_01\			1000	0.0
XP3	c:\valida\...*.*	c:\valida\...\salid_07\				
XP4	c:\valida\...*.*	c:\valida\...\salid_02\		500		
XP5	c:\valida\...*.*	c:\valida\...\salid_04\				

Table 1.- Initial instance of the Control Data Block.

On the other hand, the experiment presents these domain constraints: 480 minutes of maximum execution time; an available capacity of <a, b, c, d, f, g, h> = <1, 1, 75.3, 0, 450, 550, 255> Mbytes in the storage devices; damocia.cfg, config.cfg and config.lim as configuration files; and four available computers <lab1, lab2, lab3, lab4>. Once process XP3 is executed, consuming 130 minutes, the state of domain constraints is:

Time	Storage devices							Configuration files			Computers			
	a	b	c	d	f	g	h							
250	1	1	10.3	0	450	500	225	damocia.cfg	config.cfg	config.lim	1	2	3	4

Table 2.- State of the domain constraint.

The *Evaluation Process EV4* evaluates the required resources to store the data generated by the *Execution Process XP4*, as example 30 Mbytes. The *Control Data Block* indicates the output directory of XP4 process is unit *c:*, but it offers only 10.3 Mbytes. Therefore, *EV4* searches a device that satisfies this constraint, modifying the control parameters of the *Control Data Block* and the domain constraints. Now, *XP4* will be executed using unit *f:* as storage device. The IDSs that interrelate this agent with the next ones will be in the selected unit.

Time	Storage devices							Configuration files			Computers			
	a	b	c	d	f	g	h							
175	1	1	10.3	0	420	500	225	damocia.cfg	config.cfg	config.lim	1	2	3	4

Table 3.- Final state of the domain constraint.

The tool generates new validated data files, incorporating them to the *Validated Data Database*. Also, it generates reports and incidences files. Next figure shows the original signal of a radiation sensor and the validated signal eliminating the noise and interpolating time intervals and examples of report files.

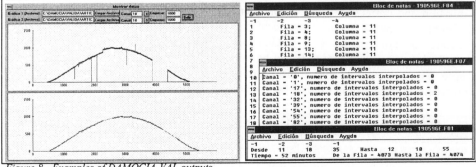

Figure 8.- Examples of DAMOCIA-VAL outputs.

4. Results

4.1. A TOOL SET FOR CLIMATIC DATA AND SENSOR SIGNAL ANALYSIS

We have developed a signal treatment toolset, named *DAMOCIA-AVE*, that includes *DAMOCIA-VAL* and *DAMOCIA-EXP*. These tools, that validate and analyze large sets of experimental data, are complemented by *DAMOCIA-AYC* [16], a signal acquisition tool. Although the toolset was designed for the treatment of climatic variables, its is easily generalizable to any other application environment where it is necessary to take any kind of measurements, validate them, and analyze the data.

4.2. AN EXPERIMENTAL CLIMATIC DATABASE, *DAMOCIA-DB*

We store all the generated data and annotate the incidents detected during the treatment processes. Figure 9 shows the generated databases (stored on CDs), the temporal and special databases (stored on hard disk) and the treatment processes [16]. The global database is composed, really, by three main data collections:
- *SEDA*. Compressed sensor data. It is a log of the acquisition process.
- *VADA*. Validated database generated by *DAMOCIA-VAL*.
- *REDA*. It stores the results of exploitation experiments. Their definitions are stored in the *Experiment Database*. The *Exploitation action database* stores the definition of the intermediate operations.

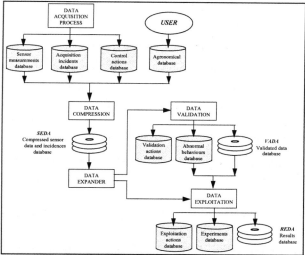

Figure 9 .- DAMOCIA-DB database.

The validation tool marks time periods with abnormal behaviour, storing them in a database. We use *DAMOCIA-EXP* to analyze these behaviours and characterize them. Therefore, we will design and implement new DAMOCIA-VAL modules to detect and correct automatically these errors.

The system has three databases containing accessory information that help us to characterize correctly these abnormal behaviours: an *Acquisition incidences database*, a *Control actions database* and an *Agronomic database* (that stores information introduced by the agronomic engineers related with the growing crop, agrochemical products used, fruit production, achieved manual actions, fertilizers composition, etc.).

4.3. VERIFICATION OF THE PROPOSED THEORETICAL MODELS

With the objective of verifying the climatic variables mathematical models [2] [12], centered now on the modellization of the radiation, we use *DAMOCIA-EXP* to analyze the data obtained by the radiation sensors located in different zones out and into the greenhouses. We check the proposed models using two ways:

- *Radiation surface maps*. A result of the theoretical models is a series of radiation maps that shows the incident radiation values on each zone of the greenhouses inner surfaces, for determined temporal sequences. In order to verify the radiation distribution of the model, we compare the experimental and theoretical maps.
- *Radiation values in determined locations*. Other result of the theoretical models is the daily evolution of the radiation in determined locations. We compare mathematical and real daily evolutions in a location where there is a radiation sensors. Figure 10 shows the comparison of the radiation on September the 14th 1996 using the outdoor radiation sensor, and the results of one of our own external radiation models.

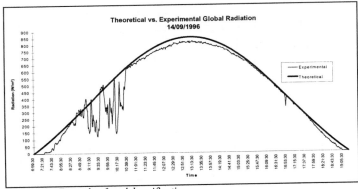

Figure 10.- Example of model verification.

Once we have analyzed these results, we modify the theoretical models, changing some equation parameters. Now, we are verifying the radiation models, adjusting them to our zone and most representative structures.

5. Conclusions and further developments

We can point the following conclusions from our developed work:

- The use of distributed software architecture in the implementation of data and signal analysis tools is convenient. It eases adding new analysis modules, integration of external tools and parallelization of tasks.
- In order to store a large volume of data it is convenient to distribute the database on different devices. CD-ROM devices offer large storage capacity, high security, standard use and low price.
- The implantation of some agents as knowledge based systems (using AI techniques) facilitates the administration of this large amount of data and its flexible and extendible analysis. From our own experience we propose two levels of intelligence: resources management and specific analysis.
- *DAMOCIA-VAL* and *DAMOCIA-EXP* are part of a set of software tools that constitutes a decision support system in order to:
  - Design control systems in two levels: *installation level*, and *algorithm level*.
  - Determinate the optimum climate set-point
  - Verify climatic models and define crop models.
  - Obtain automatically reports of experimental data.

Once finished *DAMOCIA* project, we are working, now, in a new project, *CAMED* (*Control and Management of the Mediterranean Greenhouses Agrosystems*), financed by the S*panish Ministry of* Industry (CDTI/CICYT, 970068). Some of the planned further developments related are: enhancement of *DAMOCIA-VAL* and *DAMOCIA-EXP* adding new modules, improving of the control algorithms with intelligent and multivariable techniques, improving the computing performance, distributing the different processes between several processors, extending the scope of the simulation (temperature, humidity, condensation, air flow, etc.), adding more accurate mathematical and experimental models, and storage of the experimental data using distributed k-d trees in order to improve the access time.

6. Bibliography

1) Avouris, N.M.: "User interface design for DAI applications: an overview". In Distributed Artificial Intelligence: Theory and Praxis. Kluwer Academics Publishers, 1992.
2) Bienvenido, J.F. et al.: "DAMOCIA-Sim, a generic tool for radiation simulation into mild winter region greenhouses". Proceedings of the First European Conference for Information Technology in Agriculture, *EFITA '97*. Copenhagen, Denmark, 1997.
3) Bienvenido, J.F.; et al.: "Application of modern information technologies on the design and simulation on Mediterranean greenhouses, an experience". Mediterranean Colloquium on Protected Cultivation, Agadir, Morocco, 1996.
4) Bienvenido, J.F.; et al.: "DAMOCIA: Computer-Aided Design for the Construction of Automated Greenhouses". *AgEng'96*, 1:376-377, Madrid, Spain, 1996.
5) Bienvenido, J.F.; et al.: "DACAS: a Distributed Architecture for Changeable and Adaptable Simulation". Proceedings of the *EIS'98*, Tenerife, Spain, 1.998.

6) Bot, G.P.A.: "Greenhouse Climate From Physical Processes to a Dynamic Model". Wageningen. 1983.

7) Critten, D.L.: "Light Transmission Through Structureless Multispan Greenhouses Roofs of Gothic Arch Cross Section". J. Agr. Eng. Res, N° 41, Pag. 319-325, 1988.

8) Cuena, J.: "Knowledge architectures for real-time decision support". In Second Generation Expert Systems, J.M. David, J.P. Krivine and R. Simmons (eds.), Springer-Verlag, 1993.

9) Duijnhouwer, R.; Dekkers, W.A.: "EPROS, A system to store, retrieve and analyze data of field trials". *EFITA '97*, 63-67, Copenhagen, Denmark, 1997.

10) Garcia, J.R. et al.: "A Design Methodology Base on Multilevel Declarative Definitions and a Distributed Architecture. Application to the Design of Greenhouses for Mild Winter Regions". In 2nd Annual Conference on Industrial Engineering Applications and Practice. Vol. I. San Diego, USA, 1997.

11) Guida, G. and Zanella, M.: "Knowledge-based design using the multi-modeling approach". In Second Generation Expert Systems, J.M. David, J.P. Krivine and R. Simmons (eds.), Springer-Verlag, 1993.

12) Guirado, R. et al.: "Multimodel Simulation of the Canopy Effect on an inner Radiation Model for Parameterized Greenhouses". In International Conference Engineering of Decision Support Systems in Bio-Industries, *BIO-DECISION'98*, Montpellier, France, 1998.

13) Kirchner, T.B.: "Distributed processing applied to ecological modeling". In Simulation Practice and Theory, Vol. 5-1, 1997.

14) Kurihara, S., Aoyagi, S. and Onai, R.: "Adaptative Selection of Reactive/Deliberate Planning for the Dynamic Environment". In Multi-Agent Rationality, M. Boman and W. Van de Welde (eds.), Springer-Verlag, 1997.

15) Lalis, I. and Menhart, P.: "Object oriented toolset for sequential and distributed simulation". In Proc. European Simulation Multiconference *ESM'95*, Praha, 1995.

16) Rodríguez, F. et al.: "A distributed set of tools for experimental data treatment in agroresearch: DAMOCIA-AVE / DAMOCIA-DB". In International Conference Engineering of Decision Support Systems in Bio-Industries, *BIO-DECISION 98*, Montpellier, France, 1998.

40

Current Techniques in Distributed Sensor Networks and Connection with AI and GIS : An Overview

SPYROS G. TZAFESTAS, and LOUKAS N. LALIOTIS

Computer Science Division
Electrical and Computer Engineering Department
National Technical University of Athens
Zographou 15773, Athens, Greece

1. Introduction

1.1 WHAT IS A DISTRIBUTED - SENSOR NETWORK (DSN)

A distributed-sensor network (DSN) is a group of sensors connected by a communication network to a set of information processing elements. These elements process the measurements and communicate with each other to extract features. A DSN consists of many sensors that can pool their information to achieve a better overall estimate. The estimation structure consists of a central processor (or station) and two local processors. An extension of the results to the case of m local stations is straightforward. The situation in which the local measurement data at each local station are synchronised with each other may also considered. Although many techniques have been developed for the DSN system, more improved and powerful techniques are needed in areas such **GPS** *(Global Position Systems)* detection resource allocation, communication control...etc [1]
In this chapter we will penetrate mostly to the most popular application of DSN that is GPS

1.2 DIFFICULTIES IN USING DSN SYSTEMS

There are several difficulties involved in the blending (fusion) of multiple information sets. First, to identify the object responsible for each individual measurement is unknown so that *there is uncertainty on how to associate information sets from one sensor which are obtained at one time and location*

452

to those of another point in time and location. This data association it further complicates the situation by the following facts.

(a) The target may not be detected by some sensors, due to the variations of signals and the sensor characteristics.

(b) Dense false measurements may be present and are not distinguishable from the true target measurements.

(c) Target models may not be known exactly and they may vary in time.

*The two major problems in distributed sensor networks are **detection** and **estimation**.* In practical estimation problems, difficulties result from the uncertainties in measurement origins and system models. In this research these two important problems will be studied in a distributed framework

Due to the uncertainty of the measurement origins, data association is one of the most important problems of estimation. The problem is how to associate the right measurement to the right target at the right time in the low detection and cluttered environment. This problem becomes substantially more complicated in a DSN system where measurements may come from different detection and measurement models. [1]

2. Report on potential and problems on DSN/GPS linkage

2.1 THE GPS TECHNOLOGY

GPS is one of the most significant inventions of our century, a great positioning tool. It consists of 24 satellites (Fig.1) which orbit the earth in 12 hours,

Figure 1 GPS consists of 24 satellites

keeping a distance from the earth of at least 20,000 km. The satellites move in 12-hour circular orbits some 20,000 km above the earth and are tracked by a global network monitored from a control center. They are arranged in 6 orbital planes in a structure which assures that at any time three satellites should be locked to capture two-dimensional position, and four should be locked to obtain the three- dimensional position.[4],[9]

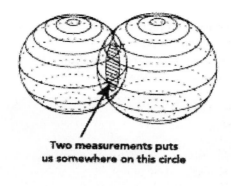

Two measurements puts
us somewhere on this circle

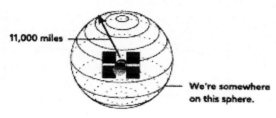

11,000 miles

We're somewhere
on this sphere.

Three measurements puts us
at one of two points

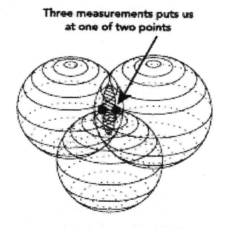

Figure 2 Schematic view on how GPS is working

Figure 2 shows how GPS is working, *using satellites as sensors* for real time positioning.

For accurate positioning we need four satellites to determine more accurately a possible position. This allows GPS receivers to solve four equations and determine *latitude, longitude and height* of any point with respectful accuracy.

In brief, GPS consists of 3 parts: *the space segment, the user segment, and the control segment.* (Figure 3)

(a) *The space segment* consists of 24 satellites, each in its own orbit 11,000 nautical miles above the Earth.

(b) *The user segment* consists of receivers, which you can hold in your hand or mount in a car, aeroplane, and so on.

(c) *The control segment* consists of ground stations (five of them, located around the world) that make sure the satellites are working properly. [8]

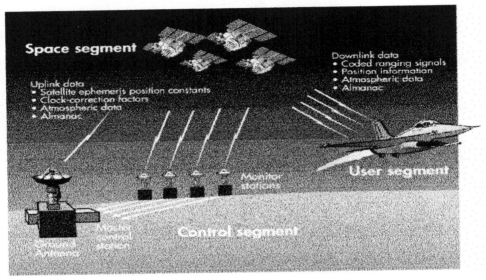

Figure 3 Parts of GPS

2.2 APPLYING DSN TECHNOLOGY FOR REAL TIME POSITIONING (GPS)

2.2.1 General

GPS is most attractive to the *engineer* who is often looking for relative position accuracy in terms of centimeters.

Surveyors use GPS to mark legal boundaries and to collect data for GIS with electronic data collectors. The accuracy requirements range from millimeters to several hundred meters depending on the capabilities of the GPS and by factors such as the quality and number of receivers used, the number of satellites contacted and the time taken to make the fix. Also the use of multiple receivers is of special interest to surveyors because it can greatly increase positioning accuracy and many times to overcome the potential problem of selective availability [10]

The *Gulf War* proved the capabilities of GPS to the public as well as the military. GPS receivers were used on ships, tanks, helicopters etc.. to provide nearly accurate positional information. Hand-held receivers helped prevent troops from being lost in the featureless desert and helped identify which troop locations were friendly and which were enemy.

In applications such as *power and traffic* control systems, the measurement data are collected by a network of sensors distributed over a large geographic region. *These systems are inherently distributed sensor networks.* [11]

For example *vehicle tracking* is one of the fastest-growing GPS applications. GPS-equipped fleet vehicles, public transportation systems, delivery trucks, and courier services use receivers to monitor their locations at all times.

Automobile manufacturers are offering *moving-map displays guided by GPS* receivers as an option on new vehicles. The displays can be removed and taken into a home to plan a trip. Several rental car companies are demonstrating GPS-equipped vehicles that give directions to drivers on display screens and through synthesized voice instructions.

GPS is also helping to save lives. Many *police, fire,* and *emergency medical* service units are using GPS receivers to determine the police car, fire truck, or ambulance nearest to an emergency, enabling the quickest possible response in life-or-death situations.

Mapping and surveying companies use GPS extensively to *determine population distribution patterns and possible sources of crop diseases.* GPS-equipped balloons are monitoring holes in the ozone layer over the polar regions, and air quality is being monitored using GPS receivers.

Archaeologists and explorers are using the system. Anyone equipped with a GPS receiver can use it as a reference point to find another location. [7], [11],[12].

2.2.2 The GIS approach

Geographic information comprises data about the surface, subsurface, 43 atmosphere of the earth and any other data that can be represented in a map, interpretations and explanations applied to those data, and an organizational framework for understanding the information. Data may be acquired through direct measurement, remote sensors, defined through survey or legal description; interpreted from data analysis or simulation. Geographic information obtains meaning from both spatial and non-spatial aspects of the data, i.e. "where" and "what"

Usually the most important geographic data are fields or arrays of point measurements observations, and interpretations. Such data are used to maintain historical records, to populate model parameters, to construct interpolated surfaces, and to compute distributions. The spatial location of these data points may be arrayed in a regular grid, randomly distributed, located at specific points within a network, or associated with a constant value of a measured attribute. [13]

2.2.3 GIS – GPS (DSN) connectionism [5], [13],[14]

A very operational advantage of GPS is the portability of equipment which enables the user to find the *exact location* of any point, as well as collect data for any purpose of his job. In addition, the technology itself is likely to develop further in order to obtain quick and more accurate measurements. Development is now being in progress using technologies such as CAD, GIS and GPS technologies together in order for GPS locations to be located on the calibrated map base rather than the screen by displaying the latitude-longitude locations. Furthermore GPS systems have been structured to be placed into the cars showing *a map of the area* and the position of the car in it. GPS systems will soon be very common for walkers and ramblers at low prices.

As GPS technology becomes more commonplace, geospatial data will be collected at an ever increasing rate. Real-time airborne GPS techniques have already been used in aerial surveys to collect high-resolution aerial photographs with accurate positioning information in order to be used in GIS applications. This technology is moving towards the use of GPS on the air-craft where ground control points can be located in an accuracy up to five meters.

All project tasks, where GPS technology involved, are categorised according to the required accuracies, which will determine the appropriate equipment cost :

(a) *100 meter accuracy* (single-receiver SPS {standard positioning service} projects, Low-cost)

(b) *1-10 meter accuracy* (differential SPS code Positioning Medium-cost)

(c) *20 meter accuracy* (single-receiver PPS {precise positioning service} projects, High-cost)

(d) *1 mm to 1 cm accuracy* (differential carrier phase surveys, High-cost) [9]

Topics that should be addressed in GIS -GPS linkage also include:
Description of GPS satellite system, geodetic principles, coordinate systems, geodetic datum's, height systems, GPS observable, receiver technology, least squares adjustment, static GPS, kinematic GPS, pseudo kinematic GPS, high-precision networks, standards and specifications, planning a survey, field procedures, data processing and analysis, adjustment and classification, and data-management issues.

Other topics where GPS and GIS have been used with valuable savings in cost and accuracy are: In Mapping archaeological sites, Forestry service, cars , navigation, air planes (especially in the Gulf War), etc.

GPS will very soon have many civilian users, most of whom do not belong to the surveying profession: environmental specialists, GIS data collectors, navigators, geophysicists, and civil, transportation, and utility engineers.

2.2.4 Problems in using GPS technology

The positional accuracy achieved with GPS receivers depends on a number of things:

One factor is the service supported by the military custodians of the system who provide a *standard positioning service (SPS)* and a *precise positioning service (PPS)*. This dual service is called selective availability. Military PPS receivers may be much more accurate but cost more and are difficult to obtain.

The accuracy of degraded SPS positioning, which is planned for general public use, falls in the range of 100 meters (about the length of a city block) for most places and times and the accuracy of PPS falls in the range of 20 meters accuracy.

Also a serious disadvantage in GPS surveying is also the risk that signals open to the public will be terminated in times of national security concern.

An other handicap is that most of the times in surveying use is made of at least two receivers at a time and typically expect centimetre position accuracy which always refers to the three-dimensional relative position between co-observing receivers.

One of the biggest handicap in using GPS is the requirement that at least *four satellites* be available at any given time. It is surprising how many trees, buildings, bridges, etc., there are to cause signal blockage.

In addition in *"kinematic" GPS surveying* much effort is required in planning the path of an antenna to avoid dropping below the required number of four satellites. Of course, one can go back to the last determined position and choose an alternative route, but this is an emergency solution .

In the *"Static"* approach the observations are accumulated over time until there is sufficient strength to estimate the coordinates and ambiguities at the same time in the least-squares sense. The solution generally can be more accurate according to the time of observation and according to the ability to observe the satellites with the better geometry.[3],[10]

3 Current applications using DSN Technology

3.1 A WATERING SOLUTION FOR THE RIVER IJSEL   [6]

The north sea floods of 1953 more than 2000 people lost their lives and in the decades that followed, a national effort was mobilised to make the country's sea defences impregnable.

In this quest for the definite model in the Ijssel delta they realized that the topographic data would be adequate. A flood modelling package developed ten years before relied on photogrametric data. The procedure for modelling 'breaklines' (the points at which dykes and river banks burst) was expensive and time-consuming. The greater point density and convenient digital format of a DTM would offer faster and more reliable prediction of water levels.

Hills, buildings and other topographic objects displace dramatically different volumes of water at different levels of immersion, causing the course of a flood to change by a second. In these circumstances, an inaccuracy of a few centimetres can undermine the best-laid flood prevention plans. Using traditional methods would have taken an army of surveyors a matter of years.

" Laser altimetry" was the route chosen to create the DTM.

The team carried out aerial laser scanning at a rate of eight to nine points per square metre. The data was processed using an aerial corridor mapping system. The team could be confident that the least one of the scanning points would always reach the ground.

A specially equipped helicopter was used to create this DTM (Digital Terrain Model). Four GPS receivers were augmented by two video cameras, one pointing forward, the other directly down. The video footage would provide visual information for any map feature that operators could not readily identify from the DTM, as well as being synchronized with the GPS and laser data.

The flight crew relied on a radio link for differential GPS corrections to help calculate the aircraft's attitude. A minimum of six satellites were kept in view (for better accuracy) at all times and all the receivers, whether on the ground or in the air. Pitch and roll was measured, using an on – board gyroscope. The data from the gyro was recorded along with that from the laser scanner, onto the removable hard disk of a PC.

Scanning 40 times a second the data collected by the laser contained 200 range measurements. With a width of 60 degrees its coverage was approximately equal to the helicopter altitude

Clear still weather was also important. Any disturbance to foliage can create a false reading, while raindrops diffract the laser beam and produce holes in the data. Lakes and rivers likewise appear as black holes, and photogrammetrists can determine the heights of rivers and lakes from their banks and shores. Air temperature, cloud cover and wind speed and direction were all logged at reference stations along the route of the flight.

The coordinates of no fewer than 20 million points was in Universal Transverse Mercator (UTM) projection. Laser intensity was also recorded. This differentiates ground cover.

Should they have any lingering doubts, operators can click on the feature, and the system will automatically access the time-coded video footage. DTM program model was used for visualizing and editing the data in 3D.

To link the laser data to the helicopter's position the team produced vector offsets from the GPS base stations on the ground to the primary navigation receiver on the helicopter. Using a "least square" technique, they reconstructed the flightpath of the helicopter in 3D.

The data from the four on –board GPS receivers was processed to give the aircraft heading-again, to the half second. They then time matched the position and altitude information with the gyro and laser data, to give accurate X, Y and Z coordinates of the laser returns as they were received up to 8000 times a second.

With DTM data to add to its existing sources of information about river depths and water levels, we will be able to forecast water levels up to two days in advance, and model the effects of new buildings or the implications of dredging the riverbed.

Thus in the future there will be a possibility to control flood damage with building restrictions, and flood control works.

3.2 THE CASE FOR EMERGENCY VEHICLE (TRAFFIC PREEMPTION) [15]

Without preemption, drivers of emergency vehicles (fire engines, police cars, ambulances, buses, and other agency service vehicles) often find themselves stuck in congested traffic or are forced into dangerous traffic manoeuvres. These include entering opposing traffic lanes and running red lights. The latter is particularly dangerous, since a motorist with the green may not see the approaching emergency vehicle until it is too late. In case of an accident, the motorist with the green is technically in the right. Thus in these kind of vehicles what the driver really wants is the service of traffic preemption.

Using GPS, DSN technology plus GIS software, one can achieve traffic preemption. Thus a single vehicle can preempt a whole series of intersections, and choose the alternative route (Fig.4) so that the driver has nothing but green lights along the emergency route.

In more details the system functioning has as following:

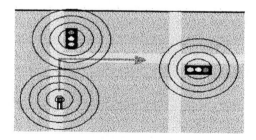

Figure 4 Choosing alternative routes

The GPS system works like this (Fig.5): A set of 24 Global Positioning System satellites send signals to emergency vehicles and traffic lights, allowing each to calculate exactly where it is. Then, as the truck approaches the liqht, it sends a signal that interrupts the normal green-yellow-and-red cycle. The light turns green, allowing the emergency vehicle to cruise through it without stopping.

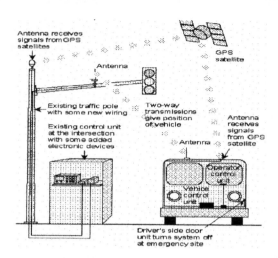

Figure 5 Traffic preemption system overview

The preemption range for each approach to an intersection (GPS and radio antennas) is programmabled based on distance or time, whichever provides the earlier preemption call, taking in account other mapped data such as real-time position, speed, and direction data and turn signal indications, which shall be sensed by the preemption equipment in the vehicle. The preemption system shall make provision for all approaches per intersection. Neurones technology can be used at that phase so the system can learn and decide in more accurate way.

A preemption call shall be issued to the traffic controller at the intersection when the ETA of the approaching vehicle is equal to, or less than, a pre-programmed lead-time for the onset of preemption.

A preemption call to the traffic controller at the intersection shall be held when a first vehicle has passed through the intersection, and a second vehicle is about to enter the ETA-based preemption window. This will prevent the controller from cycling to opposing traffic phases, which would need to be terminated by a second preemption event. The maximum gap between vehicles shall be programmable as time in seconds or as distance in feet or meters.

Precise vehicle position, direction and speed shall be derived from GPS data, which reflects absolute latitude and longitude.

To improve on the accuracy of raw GPS position information, which includes Selective Availability (SA) errors, each intersection shall be used as a GPS base station with a fixed, known position. Each intersection shall calculate its own position to an accuracy of 1 meter (3.3 ft). Differential GPS corrections shall be calculated and transmitted from each intersection to the approaching vehicles using digitally-coded two-way radio. With the differential corrections, 95% of reported vehicle positions shall be within 5 meters of true position.

The system is clever and automatically drops the preemption call when the vehicle is parked at the emergency site, thus eliminating the possibility of unwanted preemption of intersections downstream on the preemption approach. The intersection equipment shall also be able to use the GPS data from the vehicle to detect lack of motion and automatically drop the preemption call if the vehicle has not moved for a specified number of seconds.

Logged data include the start and stop times of each preemption event, the vehicle I.D. number, vehicle priority level, agency I.D. number, approach number, preemption output channel number, and reason for terminating the preemption call.

3.3 ATHENS REAL TIME TRAFFIC CONGESTION MAP [16]

In this application raw data covering traffic volumes and occupancies are selected by sensors and arrive to the NTUA CSST (Control Strategy Selection Tool) communication handler every 90 seconds. These data are compiled to a file, which is passed on to the a central PC. All data are batch processed, with a combination of specific tools and algorithms. The process produces, among others, statistical data and conclusions concerning the overall traffic in Athens For the observer all numerical data are compiled into graphical representation, that makes better sense, for more meaningful comprehension.

The first step is to graphically represent the Athens road network. To do this, a CAD model of Athens is used. A set of points in a Cartesian (x,y) space is used to designate each and every node of the road network. Each node represents a physical node of the Athens road network. A road in the network is thus defined by the nodes it connects. Two files are then used: the first containing the node coordinates and the second containing all links between nodes. These two files define completely the network on which all information is gathered and will be presented.

The analysis of the processed data yields a quantitative estimation of the traffic between nodes in the network. In order to depict the freshly collected data, a file containing this numeric information is generated every quarter of an hour. The magnitude of a traffic volume in each link is represented by a number, which is translated to a colour using the tables of numbers.

Processing all the information contained in the data files, GIF images are generated which are included in Web Pages in the NTUA Faculty of Civil Engineering Web Server giving access to information on the traffic situation in Athens to any Internet user. (Fig.6).

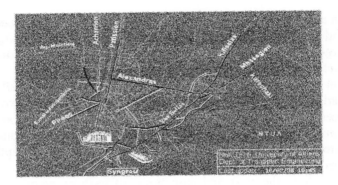

Figure 6 Athens real time traffic map

4. General applications, in research, using the DSN technology in Video Surveillance and Monitoring methods [2]

4.1 GENERAL

The technical focus is a how a forest of stationary sensors, can detect any kind of difference from a spatially distributed set of sensors. Detecting the difference is critical in nowadays.This is a difficult task which is very different from the tasks of change detection and target identification. Much of the success has been based on the elegant use of geometric models. In fact, the fundamental question has been changed: what we need is *activity recognition. This section will be discussed in the next pages.*

4.2 TECHNICAL APPROACH

In the approach two major technological advances were necessary:
One: *techniques for seamlessly fusing visual information observed at different times and from different locations.*
Two: *a framework to construct activity models so activity can be reliably and efficiently detected.*
   An activity model should provide the mechanism both for fusing and interpreting the many available sources of sensory information. To coordinate their surveillance and monitoring, the system must *self-calibrate both in space and in activity,* i.e., the forest of sensors must automatically determine where each camera is with respect to the others, and then must coordinate the observation of activities between the viewpoints, using sensors with optimal views, and to coordinate monitoring tasks, i.e. to perform activity calibration.

   In more details such systems are able to:

(a) construct rough site models, i.e., it must use observed static and dynamic cues to block out obstacles, to identify open space, etc.

(b) perform generic detection of objects of interest, i.e., it must have methods for detecting objects such us people and vehicles.

(c) perform detection based on dynamic properties as well as spatial properties, that is, detectors must be trained so as to match dynamic patterns of motion as well as spatial shape.

(d) learn using AI (Artificial Intelligence) algorithms to model coordinated patterns of activity amongst large numbers of primitive elements, so as to perform activity recognition.

4.3 APPLICATIONS IN RESEARCH

4.3.1 *Primitive detection of moving objects amongst a forest of sensors.*

The tracking system uses an adaptive backgrounding method to model the appearance of the scene without any moving objects present.

This model approximates the recent RGB values of each pixel. The particular colour pixels which are more consistent and persistent are chosen as the background model.

The pixels in the current image, which are not within two standard deviations of the background pixel model, are assumed to be produced by a moving object. Connected regions of these pixels are used to approximate the position and size of the objects present in each frame.

A form of multiple hypothesis tracking is used to determine which regions correspond from frame to frame and to filter regions which are not persistent. The end result is the ability to continuously track multiple discrete objects in a cluttered and changing environment.

The result of this stage is a robust tracking system that can track multiple objects in real time, and acquire statistical data about each tracked object.

An example of tracking patterns are shown below:

The *left* image in Figure 7 shows the observed area, the *right* image shows the patterns of tracking, with colour encoding direction and intensity encoding speed, and the third image (Fig.8) below them is an overlay of the two.

In each case, lanes of vehicle traffic as well as speed direction are easily identified.

4.3.2 *Refinement of simple site models using tracking of moving objects.*

With the ability to calibrate multiple cameras, it will be possible to construct spatial models of the environment.

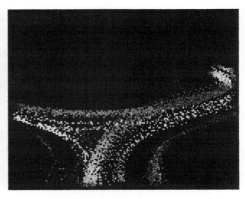

Figure 7 Observed area   Patterns of tracking, with colour encoding direction

Figure 8 Overlay of the two images in Figure 7

This method initially models the space as completely occupied. As objects move in the environment, the regions of space between the camera and the visible parts of the objects is removed from the model. This eventually creates an efficient model of the areas where visible objects can move and where they are occluded.

In figure 9 is shown an example of the system reconstructing a site. On the left is an example for the processing, showing the base image, the current image, and the extracted moving object. At the bottom right is the current map built from tracking this person.

5. Conclusions

DSN enhanced with GPS and GIS, as well as, other sensor capabilities, can be used to support useful operations in a variety of applications. Accurate X,Y and Z coordinates of any point on the ground can be found using laser altimetry in order to create the DTM (Digital Terrain Model). With DTM data one will be

The result from this stage is a method for constructing rough site models by tracking moving objects through the scene.

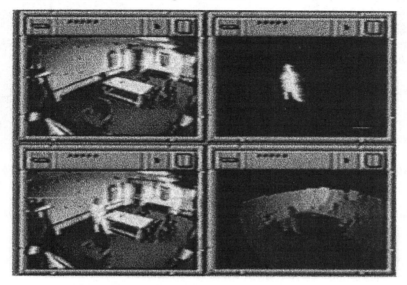

Figure 9 Example of a system reconstructing a site

able to forecast water levels, and model the effects of new buildings, in order to prevent flood damage with building restrictions, and flood control works.

Using sensors new methods in real time monitoring of a traffic magnitude, and, traffic preemption for emergency vehicles are being developed.

Video surveillance and Monitoring methods are now used to detect any kind of difference from a spatially distributed set of sensors and any detection of moving objects. DSN - GPS combination can help to determine population distribution patterns and possible sources of crop diseases. In addition using Geographic information systems one can obtain meaning from both spatial and non-spatial aspects of the data, i.e. "where" and "what" i.e. Police, fire, emergency medical service units are using GPS receivers to determine the police car, fire truck, or ambulance nearest to an emergency. In nowadays automobile manufacturers are offering moving-map displays guided by GPS receivers as an option on new vehicles.

The issues discussed in this chapter show clearly the increased capabilities of DSN technologies and how useful can be in many sections of everyday life.

References

[1] S.G. Tzafestas, K.Watanabe (1993) *Stochastic Large scale engineering systems Marcel Dekker*, New York

[2] E. Grimson, P. Viola, Massachusetts Institute of Technology Department: Artificial Intelligence Lab, (1997), *A Forest of Sensors*, http:// www. ai. mit. edu/ projects/ darpa/ vsam/

[3] Geo-reference and Map Electronic Sensor Measurements From External Devices (1997) *http://www.georesearch.com/xds.htm*

[4] E.Livieratos, A.Fotiou (1996) *Ellipsoid geodesy & geodetic Networks* (in Greek)

[5] S. J. Wormley, Iowa State University, (1998), *DGPS–Differential Global Positioning System http: //www .cnde.iastate.edu/staff/swormley /gps /dgps.html*

[6] A. Schofield GIS Europe (February 1988) *A watering solution for the river Ijsel*, 26-28

[7] R. J. Collier (1992) *GPS Uses in Everyday Life The Aerospace Corporation, http: // www. aero.org / publications / GPSPRIMER / EvryDyUse. html*

[8] The Aerospace Corporation, (1998) *GPS Elements http:// www. aero.org/publications / GPSPRIMER/ GPSElements.html*

[9] P.H. Dana (1994) Department of Geography, University of Texas at Austin. Global Positioning System Overview, *http://www. utexas.edu/ depts/ grg/ gcraft/notes/gps/ gps.html*

[10] Alfred Leick (1992) Department of Spatial Information, University of Maine, Delineating Theory for GPS Surveying http: //www. spatial. maine. edu/~leick/pub7.htm

[11] Trimble Navigation Limited, (1997), Why do we need Differential GPS? http://www.trimble.com/gps/diffGPS/aa_dg2.htm

[12] Trimble Navigation Limited, (1997), *How does Differential GPS work? http://www.trimble.com/gps/diffGPS/aa_dg3.htm*

[13] P.A Borough. (1988). *Principles of GIS* ,137

[14] J.Star, J. Estes, (1984) *Geographical Information Systems -An Introduction*

[15] Midwest Traffic Products Inc (1997) Traffic Signal Preemption for Emergency and Transit Vehicles, *http://mtp-gps.com/index.html*

[16] G.Mouzakitis & A. Stathopoulos (1997), *Athens real time traffic Congestion*, http:// www. transport. ntua. gr/ map/ aboutmap.html

41

A New Technique for Non-Invasive Assessment of Aortic Pressure Modulations during Treadmill Running

A. Qasem, A. Avolio, F. Camacho, T. Stephan, G. Frangakis

Graduate School of Biomedical Engineering
The University of New South Wales, Sydney, 2052, Australia
email:a.avolio@unsw.edu.au

1. INTRODUCTION

The interaction between ventricular ejection and body movement during running in humans causes blood pressure to be modulated with a frequency equal to the difference between heart rate and step rate (1). The beating effect may be due to the transmission of ground impact forces through the blood column (mainly the aortic trunk) and the change in cardiac ejection due to altered venous return. However, when heart rate and step rate are equal, the beating effect vanishes and pulse pressure is maximal when cardiac ejection and foot strike are in phase, and minimal when out of phase (1,2). It is not known if this occurs spontaneously with inherent control such that blood pressure is optimised to cause minimal cardiovascular energy expenditure. However, the physical characteristics of humans indicate that, in endurance running (eg marathon), values of heart rate and step rate are quite close, raising the possibility of a type of coupling or entrainment arising from a resonance between the action of the heart and the vertical movement of the body.

Blood pressure modulation during running has been observed in peripheral pressure signals (1,2). However, the arterial pressure pulse is amplified in its travel from the heart to the periphery due to the elastic non-uniformity of the arterial vasculature and the presence of wave reflection (3,4). The transfer characteristics are frequency dependent (5), hence pulse amplification is strongly heart rate dependent. In exercise, therefore, when substantial changes in heart rate occur, measurement of peripheral pulse changes do not necessarily indicate changes in the loading characteristics of the heart which are dependent on aortic pressure (6,7).

The aim of this study was *(i)* to develop a new technique to investigate changes in the human blood pressure pulse measured non-invasively during running on a treadmill at different speeds and *(ii)* to determine the corresponding changes in the aortic pulse pressure derived from the peripheral pulse using mathematical transformation techniques.

467

2. INTERACTION BETWEEN HEART RATE AND STEP RATE

The vertical movement of the body and the ground impact produce an effect similar to a ballistocardiogram (8) where the momentum of blood ejected from the ventricle can be registered as a movement of the body (in the supine position). These forces are assumed to be sinusoidal waves, the heart wave $H(t)$ being due to the arterial pressure, and the body wave $B(t)$ due to the vertical movement and ground impact force (9). The result of their interaction can be determined as the algebraic sum of $H(t)$ and $B(t)$.

For $H(t) = A_1 sin(\omega_1 t + \theta_1)$ and $B(t) = A_2 sin(\omega_2 t + \theta_2)$, where A = peak to peak amplitude, ω = frequency, θ = phase delay, the output will depend on the frequency and the phase delay.

For $a = \omega_1 t + \theta_1$ and $b = \omega_2 t + \theta_2$, beating occurs when $\omega_1 \neq \omega_2$. The output $y(t)$ is

$$y(t) = (A_1 + A_2)cos((a-b)/2)sin((a+b)/2) + (A_1 - A_2)sin((a-b)/2) \, cos((a+b)/2)$$

and the beating frequency is $\omega_1 - \omega_2$.

For in-phase coupling, $\omega_1 = \omega_2 = \omega$ and $\theta_1 = \theta_2 = \theta$,

$$y_1(t) = (A_1 + A_2)sin(\omega t + \theta).$$

For out-of-phase coupling, $\omega_1 = \omega_2 = \omega$ and $\theta_1 = \theta_2 - \pi$,

$$y_2(t) = (A_2 - A_1)sin(\omega t - \theta_2)$$

The amplitude of $y_2(t)$ is less than that of $y_1(t)$, indicating that the pressure pulse would decrease for the case when ground foot strike is out of phase with the heart beat.

3. METHODS

3.1 Estimation of central aortic pressure

Aortic pressure was estimated from the peripheral pressure signal using the non-invasive SphygmoCor system (PWV Medical Pty Ltd, Sydney). This device utilises an on-line computerised mathematical transfer function for the adult human arm developed previously from invasive data (10). Fig. 1 shows the frequency characteristics of the transfer function between the ascending aorta and the radial artery. Modulus increases as a function of frequency to a resonant peak at approximately 4 Hz. Since most of the energy of the arterial pulse is in the lower harmonics, usually below 4 Hz (4), this explains the general amplification of the propagating pressure pulse (Fig. 2). Since the finger is close to the radial artery, and the waveforms are essentially similar, this transfer function was applied to the finger pulse in the reverse direction to derive the aortic pulse.

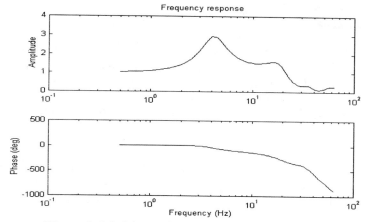

Figure 1. Modulus and phase of transfer function between central (aortic) pressure and peripheral (radial) pressure.

Figure 2. Amplification of the peripheral pulse (radial) pulse compared to the aortic pulse.

3.2 Measurements

3.2.1 Peripheral pressure

Peripheral arterial pressure was measured non-invasively in the middle finger of the left hand of eight volunteer healthy subjects running on a treadmill using and finger cuff device (Finapres, Ohmeda). Measurements were made with the arm extended and the hand fixed at heart level as well as the arm moving freely synchronous with the body. This was done to determine the effect of hand movement *per se* on the modulated pressure signal. Heart rate and pulse interval were determined from the ECG signal and step rate determined from an accelerometer attached to the left leg. Running speeds

470

were 9,12,15 and 18 km/hr with the treadmill platform horizontal. The speed of 18km/hr was used to determine maximal heart rate which was used for normalisation.

3.2.2   Beating amplitude

The beating amplitude was estimated using the envelope of the measured peripheral and derived aortic pressure signal. A peak detection algorithm was used to determine the beating amplitude shown in Fig. 3.

Figure 3. Beating amplitude of the peripheral (finger) and central (aortic) pressure signals. A biphasic pattern occurs at the point of minimal pulsations.

3.2.3   Biofeedback

Not all subjects were able to achieve identical values of heart rate and step rate at the designated speeds on the treadmill. (Theoretically, there should be a combination of speed and work load [angle of inclination] to produce similar values of heart rate and step rate for each subject. However, this combination was not explored in all subjects in this study). A tendency for coupling was found in one subject with similar heart rate and step rate at around 12 km/hr. If heart rate and step rate are in the same vicinity, the subject is able to modify the stride length on the treadmill platform turning at constant speed to maintain coupling for some minutes. In this subject the electrocardiogram (ECG) was used to produce an auditory signal which was used as feedback through a controlled time delay unit. The aim was to modify the pressure pulse amplitude by altering the time delay so as to produce 'in-phase' and 'out-of phase' interaction between the action of the heart and movement of the body.

4.   RESULTS

The beating effect was seen in the peripheral pressure signal both with the hand fixed at heart level and hand moving freely with the body. However the effect of the movement of the hand was to increase the pulse amplitude due to artefact since the swing of the arm was synchronous with body movement. The comparison of beating amplitude between hand fixed and hand free is shown in Fig. 4.

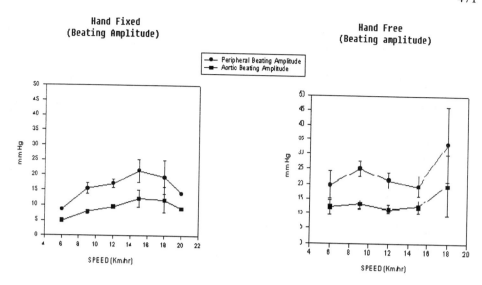

Fig. 4. Beating amplitude for peripheral and aortic pressure for hand fixed at heart level and hand free to move with movement of the body during running.

Figure 5 illustrates the modulation of measured peripheral and derived aortic pressure during steady state running (beating) and during coupling (no beating) between heart rate and step rate brought about by audible biofeedback of the heart signal to the subject. This enabled adjustment of stride length to maintain coupling between heart rate and step rate. A biphasic pattern, similar to arterial counterpulsation (4,9), is seen in the out-of-phase coupling.

A comparison of beating and pulse amplitudes for peripheral and derived aortic pressure is given in Table for running speeds 9,12,15 km/hr. Heart rate values for 18 km/hr for each subject was used to normalise heart rate for each speed.

Table

Speed (Km/hr)	HR* %	PBA (mmHg)	AoBA (mmHg)	Ratio (PBA/AoBA)	Ratio (PP/AoP)
9	78±3.3	15.5±1.7	7.9±0.6	1.94±0.64	1.87±0.03
12	88±2.6	17.1±1.4	9.6±0.6	1.81±0.18	1.97±0.03
15	96±1.4	21.4±3.9	12.2±2.8	1.87±0.2	2.05±0.04

HR*:% of Heart Rate at 18 km/hr; PBA: peripheral Beat Amplitude; AoBA: Aortic Beat Amplitude; PP: peripheral pulse amplitude; AoP: Aortic Pulse Amplitude.

472

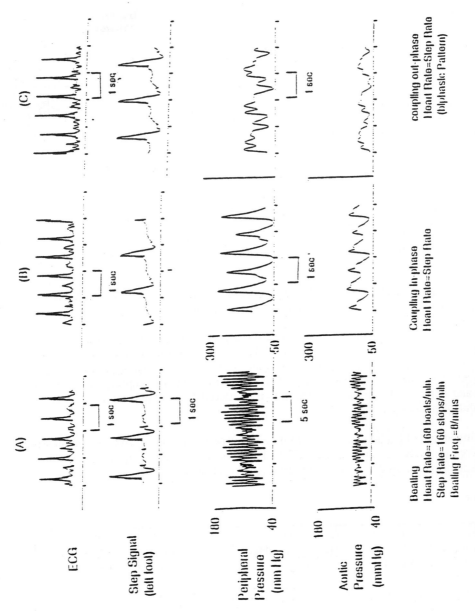

Fig. 5 (A) Beating of arterial pressure when heart rate and step rate are not equal. (B) In-phase coupling between heart rate and step rate. (C) Out-of-phase coupling between heart rate and step rate.

5. DISCUSSION

Results of this investigation show that the interaction between the arterial pressure generated by ventricular ejection and body movement can be characterised by non-invasive measurements of arterial blood pressure. The finding of beating in the non-invasive pressure signal agrees with observations made with invasive catheters (1). During movement the calibration procedure of the Finapres device was disabled to obtain continuous recording. Although the true level of mean arterial pressure could therefore not be obtained, the relative changes were quite reliable as can be observed from the marked changes in pulse pressure obtained during coupling of heart rate and step rate in and out of phase.

The use of a mathematical transformation technique for the aortic pressure pulse has been shown to be useful in characterising the relative changes in the modulations observed in the peripheral pulse. This takes into account the heart rate dependency of the amplification characteristics of the propagating pressure pulse (4,5,10) and so improves the estimation of the changes in work load on the heart.

The technique proposed here, including the provision of a feedback signal of cardiac origin to adjust step rate, can be used both for assessment and training of endurance athletes. It has been found that significant efficiency can be gained by running at a speed of around 3 steps per second (11). This corresponds to a frequency of 180/min, which is around the maximal heart rate for adults. This suggests that if internal feedback mechanisms exist to maintain coupling between heart rate and step rate, improved cardiovascular performance could be achieved by simple mechanical maneuvers, such as accurate timing of foot strike with respect to the phase of the cardiac cycle.

In addition to athletic performance, this technique could be used to assess the relative phase relationship of foot strike and cardiac cycle in cardiac patients undergoing routine exercise tests on the treadmill. By monitoring the peripheral pulse and the phase relationship, detrimental increase on cardiac load could be avoided by provision of external feedback signals to modify the step rate or stride length.

6. CONCLUSIONS

A computerised on-line mathematical technique can be used for non-invasive estimation of aortic pressure from the peripheral pulse during running. With increased heart rate, the peripheral pulse can be more than twice the central aortic pulse. The beating seen in the peripheral pulse is also present in the aortic pulse, but with a lower amplitude. Beating and pulse amplitudes are amplified to the same degree. Non-invasive estimation of the central aortic pressure may provide a more accurate assessment of the pressure-related effects of running on cardiac load. This may have important implications for assessment of patients with heart disease using treadmill exercise protocols.

7. REFERENCES

1. Palatini P, Mos L, Mormino P. di Marco A, Munar A, Fazio G, Giuliano G, Pessina AC, dal Palu C. Blood pressure changes during running in humans: the "beat" phenomenon. *J Appl Physiol* 1989;67:52-59.

2. O'Rourke M, Avolio A. Stelliou V, Young J, Gallagher D. The rhythm of running: can the heart join in? *Aust NZ J Med* 23:703-710,1993.

3. Rowell LR, Brengelmann GL, Blackmon JR, Bruce RA Murrey JA. Disparities between aortic and peripheral pulse pressures induced by upright exercise and vasomotor changes in man. *Circulation* 1968;37:954-964.

4. Nichols WW and O'Rourke MF. McDonald's blood flow in arteries. 4th edition, Arnold, London, Baltimore, 1998.

5. Chen CH, Nevo E, Fetics B, Pak PH,Yin FCP, Maughan L, Kass DA. Estimation of central aortic pressure waveform by mathematical transformation of radial tonometry pressure - Validation of generalised transfer function. *Circulation*, 1997;95:1827-1836.

6. Elzinga G, Westerhof N. How to quantify pump function of the heart. *Circulation Research* 1979;44:303-308.

7. Suga H, Sagawa K, Demer L. Determinants of instantaneous pressure in canine left ventricle. *Circulation Research*, 1980;46:256-263.

8. Starr I. The relation of the balistocardiogram to cardiac function. Am J Cardiol, 1958 :737-747

9. O'Rourke M, Avolio A. Improved cardiovascular performance with optimal entrainment between heart rate and step rate during running in humans. *Coronary Arteries Disease*, 1992,3:863-869.

10. Karamanoglu M, O'Rourke MF, Avolio AP, Kelly RP. An analysis of the relationship between central aortic and peripheral upper limb pressure waves in man. *European Heart J.* 14:160-167, 1993

11. Kaneko M., Mechanics and energetics in running with special reference to efficiency. *J. Biomechanics* ,1990, 23(Suppl 1): 57-63.

42

The Role of Planning in Scheduling Patient Tests in Hospital Laboratories

C.C. Marinagi, C.D. Spyropoulos, S. Kokkotos,
C. Halatsis***
Software and Knowledge Engineering Laboratory
Institute of Informatics and Telecommunications, N.C.S.R.
"Demokritos",
e-mails: [costass | katerina]@iit.demokritos.gr*
*National Documentation Center,
e-mail: skokko@iris.ekt.org.gr*
***Department of Informatics, University of Athens,
e-mail: halatsis@di.uoa.gr*

1. Introduction

Recently, many researchers have been interested in applying AI planning technology to practical real-world problems. An interesting problem where planning technology can be applied is the problem of scheduling patient tests in hospital laboratories. Doctors prescribe tests to be performed in order to assist the diagnosis. Hospital laboratories that perform tests, must cooperate in order to maximize the utilization of their equipment and minimize patient waiting time. The actual timing of the tests prescribed for a particular patient, depends on several factors that require both planning and scheduling technology.

Until now, approaches that cope with this problem use pure scheduling techniques [1,2]. Among them, there are approaches that consider scheduling tests in a single laboratory [2] and approaches that support multi-laboratory test scheduling by assigning different schedulers to different laboratories [1]. In [3], a dynamic distributed scheduling approach has been proposed. In [4] we made a first attempt to integrate planning and scheduling technology to solve problems of this domain.

In the present chapter a more thorough approach is given. We first examine the need to

¹ This work was developed during the project PENED 561: CHRONOBASI (TEDRAS), funded by the European Commission (EC) and the Greek General Secretary for Research and Technology of the Ministry of Development.

incorporate and integrate AI planning technology within the problem of scheduling patient tests in hospital laboratories. We also suggest appropriate techniques that should be applied in order to represent and reason in such a domain. Then we describe an enhanced version of a planning system, called TRL-Planner [5,6], that fits the requirements posed by the domain. Finally, an improved version of the dynamic distributed planning/scheduling paradigm presented in [4] is given, which supports incremental scheduling as well.

More specifically, this chapter is structured as follows. In section 2 we define the problem of scheduling patient tests in hospital laboratories, we present related work and give the motivations of our approach. In section 3 we give the requirements that must be satisfied by a planning system for representing and reasoning in the domain, we describe the TRL-Planner and demonstrate how the domain is represented. In section 4 we describe the improved version of the patient test planning/scheduling system and in section 5 we discuss conclusions.

2. Scheduling Patient Tests in Hospital Laboratories

2.1 THE PROBLEM

Hospital laboratories perform patient tests that have been prescribed by doctors. Laboratory personnel perform the tests, either on patients (e.g. X rays) or on samples (e.g. blood tests), using devices from the existing laboratory equipment. Each laboratory is able to perform a pre-defined set of tests. Tests results are returned to doctors. Usually the laboratories are responsible for scheduling their own tests. When the communication between laboratories is loose, the schedules are also loosely integrated and patients have to wait between tests. Sometimes incompatibilities between tests are not considered from the beginning and patients have to visit the hospital more than once. Moreover, since laboratory equipment is very expensive to stay idle, its high utilization should always be a target. Concluding, scheduling of patient tests in hospital laboratories should enable better patients service and better utilization of hospital equipment.

2.2 RELATED APPROACHES

In the domain of patient tests scheduling in hospital laboratories only a few efforts exist. These are based solely on scheduling techniques [1,2,3]. The approach described in [1] deals with the problem from a hospital-wide perspective, instead of dealing with tests in a single laboratory as in [2]. In [1] a distributed scheduling approach is followed. Instead of having a single scheduler to decide for every aspect of the system, like in centralised scheduling, this approach assigns a scheduler to each laboratory or equipment group. This equipment-wise distribution usually results in better resource utilization, at the cost of extensive overhead in communication between schedulers and usually it may unexpectedly delay to produce a schedule.

In [3] the idea of dynamic distributed scheduling based on patient-wise distribution is introduced. In patient-wise distribution every request for tests is handled as an independent entity since it involves only one patient. Schedulers are dynamically generated every time a new request is entered into the system. Each scheduler is assigned to a request and "dies" when the request has been served. This prototype scheduling system takes into account the constraints among tests imposed by various medical protocols. However the whole system is made quite complicated even under the concept of patient-wise distribution. The need to integrate a planning system is clear.

A first approach, presented in [4], uses an integrated planning/scheduling prototype. In that version, a new plan needs to be created for every request, even though same requests are frequently repeated. This means that in case of a major destructive event, all previously scheduled requests have to be planned and scheduled from scratch. Thus no incremental scheduling is supported.

2.3 MOTIVATIONS FOR A NEW APPROACH

The tests, which doctors usually prescribe for their patients, require laboratories to follow specific medical protocols during their implementation. These protocols concern the procedure and possible steps that have to be followed during the tests, correlation between steps, incompatibilities among tests, durations of particular actions and so on. In such a domain, a system which is solely based on scheduling techniques can not deal with the problem efficiently. It needs the support and integration of a planning system, which is able to organize the tests, consider the medical protocols and deal with all resource types appearing in hospital laboratories.

Identical medical cases require the same set of tests to be performed. To save time it is desirable that such requests would not be planned again, if their solution plans exist in a repository. A planning system can be provided with possible requests of tests performance and produce plans to be stored in a plan repository. This also facilitates the application of incremental scheduling.

With this work, we intend to present an improved version of our planning/ scheduling system that will support all resource types involved in the domain, as well as incremental scheduling. This system is expected to result in increased performance.

3. Planning in the Domain of Scheduling Patient Tests

3.1 REQUIREMENTS POSED BY THE DOMAIN

A planning system that could be applied in the domain of scheduling patient tests in hospital laboratories must meet the following requirements.

Such a system could provide a uniform formalism for representing activities. Heuristic knowledge should be easily encoded using the formalism and be utilized appropriately.

This ability is useful when new information has to be included, for instance, when a laboratory obtains new equipment able to perform new kind of tests for which new medical rules hold.

Another requirement should be the ability to represent and reason about partially ordered actions so that parallel plans can be generated. For instance, after taking a patient's blood, the blood test is being performed, while at the same time the patient may be involved into another test.

The particular domain, as well as most real-world domains, require some form of hierarchical task network planning (HTN). The main advantage of this planning technique is the flexibility that provides to the domain writer to specify particular courses of actions. Actions can be decomposed to lower level actions until a ground level is reached where only primitive actions exist. Such a technique is useful in order to represent patient tests, since tests can be analyzed into steps and steps can then be analyzed into a set of actions related with the involved resources.

Moreover, a planning system applied in a scheduling domain will be useful only if it is able to represent and reason about time and resources. Temporal and resource constraints must be handled, since the domain requires a schedule to be produced each time a new request is entered into the system.

Although some existing planning systems, like O-PLAN2 [8,9], SIPE-2 [10] and IxTeT [11] that integrate planning and scheduling technology, might be able to be used in the particular domain, we have used a particular planning system, called TRL-Planner, which satisfies the above mentioned requirements.

3.2 TRL-PLANNER

TRL-Planner [5,6] is a general purpose temporal planning system which produces hierarchical, partially ordered plans and supports user interaction. It has already been used as a core module to the real-world problem of planning cargo handling operations for chemical carriers. TRL-Planner is built upon the TRLi temporal reasoning system [7] which is an extension of the Horn clause logic programming language (e.g. Prolog).

TRL-Planner enables the representation of different types of activities using a uniform representation schema where goals, activities, preconditions, effects and resources may be labeled by TRLi's temporal references. Temporal references are *temporal points* T, *temporal intervals* <T1,T2>, *temporal instances* [T1,T2], and *uncertain temporal intervals* <[T1,T2],[T3,T4]>. A duration and a set of temporal constraints between time points appearing as variables in the schema are defined for each activity.

Two types of activities are necessary in the domain, called *macro actions* and *primitive actions*. Macro actions can be further decomposed to subactions while primitive actions can not. We adopt two types of preconditions, called *assumptions* and *subgoals*. Assumptions are filter conditions that must be true in order for the particular activity to be selected. Subgoals are first checked to be true. If they are not, they can be achieved by selecting an appropriate activity. The effects are distinguished into *primary effects* and *general effects*. Primary are these effects that satisfy subgoals of other activities. General effects, are these that are produced as side effects when the activity is performed.

TRL-Planner supports hierarchical task network planning combined with reasoning about partially ordered actions. In general, the plan is generated backwards starting from goals and is guided by temporal inferences and constraint solving techniques. TRL-Planner is currently able to represent and reason about unsharable and sharable reusable resources as well as consumable resources. All resources are currently represented using a separate 'Resource' statement in the representation schemes. This is an approach followed by most planning systems that are able to handle these resource types [8,9,10,11].

3.3 REPRESENTATION IN TRL-PLANNER

Three types of resource appear in hospital laboratories which are represented and handled in the proposed approach. :

Unsharable reusable : All persons (patients, doctors, nurses, technicians), and laboratory devices which can perform only one action at a time such as X-rays and ultrasound devices.
Sharable reusable : Laboratory devices which have a maximum capacity defining the number of tests that the device can perform simultaneously. For example, an imunoassy analyser which can elaborate a number of blood samples per time.
Consumable : injections, tubes, catheters, films, shading liquids used for X-rays, liquids used in blood tests, etc.

Information concerning the specifications and allocations of each resource is stored in a common database and consist the initial state of the world for each request to be solved. For example the following information specifies a resource name, type, allocation type and initial available capacity/quantity :

resource(philips_Integris_3000,   x_rays,   unsharable_reusable,   1).
resource(smelter_1523/NR291,   film_developer,   unsharable_reusable,   1).
resource(chiron_ACS180plus,   imunoassy_analyser, sharable_reusable,   60).
resource(magnevist,   shading_liquid,   consumable,   150).

Information determining resource allocations include the resource name, the allocated quantity and the identification number of the test that allocates the resource. Temporal references determine the allocation intervals. For example:

<9:00, 10:00>   : reserved(philips_Integris_3000,   1,   test_id123)
<9:00, 10:00>   : reserved(drSmith,   1,   test_id123)
<12:00,12:10>   : consume(magnevist,   8,   test_id345)

Doctors examine patients and issue *requests*, prescribing tests to be performed. A request consists of a set of goals that ask tests for one patient to be completed until particular due dates/times. Let us assume a goal included in a request where Dr. Fox asks for an angiography test to be performed for Mr. Adams. The test should be completed at some time between 10:30 and 15:00:

[10:30,15:00] : angiography_test_completed(test31, mrAdams, drFox,
PersonnelList, ResourceList)

A *test* is decomposed into an ordered set of *steps*, with predefined temporal relations between subsequent steps. In TRL-Planner, tests are encoded as 'macro actions'. The steps of each particular test are the sub-actions into which the test is decomposed. Temporal constraints are used to define the order or even the temporal distance between steps. These are included in the Subactions Temporal Constraints list. The goal that requires an angiography test to be completed will call the 'angiography_test' macro action, which has a primary effect matching with this goal. In Figure 1, the analysis of the angiography test into three levels is illustrated : a) the tests level, b) the steps level and c) the resources level. In the following, we give the representation schema of the angiography test :

Name	= $<T1,T2>$: angiography_test(Test_id, Patient,Doctor, [Nurse,Technician1,Technician2], [Shading_liq,Film,Device1,Device2])
Duration	= D1
Type	= macroaction (
	Subactions:
	$<T1,T3>$:pour_shading_liquid(Test_id,Patient,Nurse,Shading_liq)
	$<T4,T5>$:x_ray(Test_id, Patient, Technician1, Device1,Film)
	$<T6,T7>$:film_development(Test_id, Technician2, Device2)
	$<T8,T2>$:check_results(Test_id, Doctor)
	Subactions_TempConstrs : T3≤T4, T4≤T3+10, T5≤T6, T7≤T8)
Assumptions	= ∅
Subgoals	= $<Ts1,Ts2>$: patient's_case_history_taken(Patient, Doctor)
PrimaryEffect	= T2 : angiography_test_completed(Test_id, Patient, Doctor, [Nurse,Technician1,Technician2], [Shading_liq,Film,Device1,Device2])
GeneralEffects	= ∅
Resources	= ∅
TempConstrs	= Ts1≤T1

For every resource involved in a *step* three actions are defined. For patients these actions are *preparation, busy* and *rest*. During *preparation* the patient may undress or drink some liquid. During *busy* the patient is subjected to a step of a test, while during *rest* the patient may dress or just rest on a bed. For doctors, nurses and technicians only a *busy* action is defined. For equipment the *initialization, busy* and *reset* actions are defined. During *initialization* the equipment is prepared to be used, during *busy* it is used and during *reset* it is returned to its initial condition. Finally, for consumable resources only a *consumption* action is defined during which the resource is consumed.

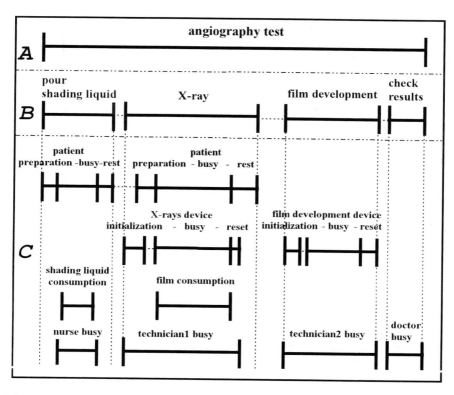

Figure 1. Analysis of the angiography test in three levels : *A*)tests level *B*)steps level *C*)resources level

In TRL-Planner, *steps* of tests are also encoded as 'macro actions'. Each step is analyzed into a predefined set of sub-actions, depending upon the resources, which are allocated. The following 'film_development' step involves an unsharable reusable resource, the film development device.

Name	= <T1,T2> : film_development(Test_id, Technician, Device)
Duration	= D1
Type	= macroaction (Subactions:
	<T1,T3>: film_development_equipment_initialization(Test_id, Device)
	<T4,T5>: film_development_equipment_busy(Test_id, Device)
	<T6,T7>: film_development_equipment_reset(Test_id, Device)
	<T1,T2>: film_development_personnel_busy(Test_id, Technician)
	Sub-actions_TemporalConstraints : T3≤T4, T4≤T3+5, T5≤T6, T7≤T2)
Assumptions	= ∅
Subgoals	= ∅
PrimaryEffects	= T2 : film_development_completed(Test_id,Technician,Device)

GeneralEffects = ∅
Resources = <T1,T7> : reserve(Device, 1, Test_id)
                         <T1,T2> : reserve(Technician, 1, Test_id)
TempConstrs = ∅

The sub-actions into which steps are decomposed are primitive actions. Their duration is usually an integer which corresponds to minutes, but it can also have an uncertain value. For example, the film development device smelter 1523/NR291 develops film in 2 minutes, so the duration of the primitive action film_development_equipment_ busy(Test_id, smelter_1523/NR291) is 2 minutes.

4. The Patient Test Planning and Scheduling System

An improved version of the design architecture of the test planning/scheduling system proposed in [5], is pictured in Figure 2. We will describe the main modules and explain their operation focusing on the upgraded parts. According to the dynamic distribution planning/scheduling approach the planning and scheduling load is distributed to several computer machines. We try to minimize the communication among computer machines.

Users enter their requests, which are stored into the *common database*. This database also contains information about the allocation of resources to tests, the full timetables of equipment, patients and personnel, as well as static information concerning equipment, patient and personnel data.

The *common knowledge base* records scheduling methods, like best fit, worst fit, etc. as well as heuristic medical rules which are imposed by doctors. It also contains activity representation schemes, which describe activities such as tests and steps of tests.

The *plan repository* contains plans generated using the TRL-Planner. These plans may be generated off-line as solutions to possible requests and may involve single tests or combinations of tests. In case that an unplanned request is entered, TRL-Planner produces a set of new plans that solve the request and augment the plan repository with them.

The *TRL-Planner/Scheduler* uses four sub-modules : the communication module, the local database, the local knowledge base, and the local plan base. The communication module handles all communication between the TRL-Planner/Scheduler and the common database, the common knowledge base and the plan repository. The local database and knowledge base contain information copied from the common database and knowledge base, which concerns the particular request. The local plan base contains related plans copied from the plan repository that solve the request. Therefore communication overhead is reduced.

The *organizer* is the central part of the system which acts as central control, dispatcher and communication channel. The organizer repeats a particular procedure each time a new request is entered into the system. More specifically the organizer:

• Recognizes the arrival of the request to the common database and specifies the computer platform with the less load, onto which it generates a new TRL-Planner/Scheduler.

• Passes to this TRL-Planner/Scheduler the relevant information about the request from the common database and knowledge base and a set of solution plans which might exist in the plan repository.

In case that the plan repository contains no appropriate plan, the TRL-Planner produces a set of new ones, stores them to the local plan base and at the same time updates the plan repository through the organizer. During plan generation, information contained in local database and knowledge base is considered.

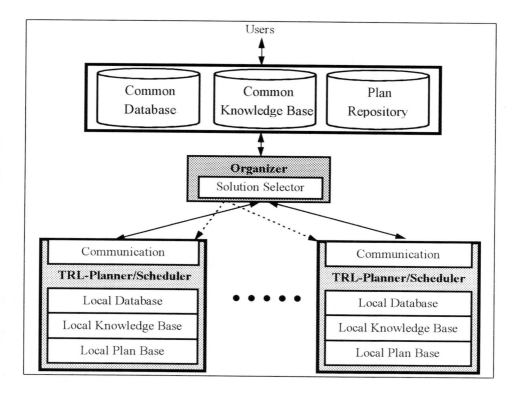

Figure 2. Design architecture of the test planning and scheduling system to support incremental scheduling

Not all the plans included in the local plan base are useful. In order to select only these that may solve the request, the local database is consulted to give information about resource availabilities. Plans involving resource types not currently available are rejected. There may exist different plans for different resource types as their analysis at the level of steps may differ. Among the rest of the candidate plans the scheduling component chooses one and considers scheduling methods in order to produce a set of schedules. The number of the generated schedules is predetermined and depends on the system performance.

• Activates its *solution selector* which rejects invalid schedules. Invalid schedules may be produced because the TRL-Planner/Scheduler is isolated from the common database after its activation, for reasons of communication overhead minimization. As a result, the resource availabilities may have changed in the meanwhile. The reason under such change may be either the destruction of an available resource or its allocation by another request that has been scheduled in the meanwhile. Among valid schedules, the best one is chosen according to the objective function described in [4]. The rest of the valid schedules are kept as alternatives in the common database in case incremental scheduling will need to be applied.

In case no schedules are resulted or all produced schedules proved to be invalid, then the mechanism of incremental scheduling is activated. That is, the organizer retrieves from the common database the already scheduled requests that overlay with the current request in resource allocations. If such a request has greater priority than the priority of the current request, then the current request has to be scheduled again. If such a request has less priority than the priority of the current request, its solution is cancelled. If any of its alternative schedules can be concerted with the current request, then it is accepted, else this request will have to be scheduled again. For every request the organizer allows a standard number of attempts to be rescheduled. If all attempts fail, the request can not be solved and is postponed to be scheduled another day/time.

• Sends the best schedule to the common database where it is stored as the accepted solution to the current request and it discontinues the operation of the current TRL-Planner/Scheduler.

• Removes from the common database all schedules as well as their alternatives when their execution has been fulfilled.

Additionally, the organizer detects the occurrence of any unexpected events that may cause the destruction of a resource that has already been scheduled to be allocated by one or more requests. In such a case the organizer chooses among the accepted solutions of the requests contained in the common database, those ones which are influenced by the destructive event. For each of the affected solutions the organizer generates a new request and starts the procedure described above in order to satisfy the request. However, this time the solution plan is certainly contained in the plan repository and the time needed for rescheduling is reduced.

The above described architecture has been implemented into an improved version of the planning/scheduling system. An earlier version of this system has been developed which is presented in [4]. We are currently testing the system and monitoring its performance against the original system, which does not use a plan repository and incremental scheduling.

5. Concluding Remarks

In this chapter, we have reported the role of planning to the task of scheduling patient tests in hospital laboratories. We have followed an integrated planning and scheduling approach, where a particular planning system it used, called TRL-Planner, which meets the requirements posed by this real-world domain. The architecture of the planning and scheduling system is based on the patient-wise dynamic distributed planning/scheduling paradigm, according to which the system dynamically creates planner/schedulers over several machines and assigns them to requests in order to reduce their communication overhead.

TRL-Planner is initially used off-line to generate plans which are stored in a plan repository. These plans are possible solutions to particular problems of the domain. Each planning problem is a request of tests that have to be performed on a patient or on samples taken from a patient. TRL-Planner is also used when a new unplanned request is arrived in order to provide a set of alternative plans and update the plan repository. After selecting a plan, the time windows available are considered in order to produce a schedule ready for execution. Incremental scheduling is supported when resource allocations have changed in the meanwhile of the planning/scheduling process or when unexpected events kick out of order some resources. Then the schedule is adapted to the new situation without rejecting those parts that are not influenced by the change.

Finally, we believe that the use of a plan repository to the task of scheduling patient tests gives us the opportunity to eliminate the required time for performing incremental scheduling. Additionally, it actually eliminates the need for plan modification as the plan repository is updated each time an unplanned request is entered. As the system is being actively operated, the plan repository will contain plans for almost any possible request, which are most useful combinations of tests. For the future we intend to finalize our prototype system and thoroughly test it.

References

1. Kumar A. D., Ow P. S., Prietula M. J. "Organizational Simulation and Information Systems Design: An Operations Level Example", Management Science, Vol. 39, No. 2, pp. 218-240, 1993.
2. Sullivan W. G., Blair E. L. "Predicting Workload Requirements for Scheduled Health Care Services with an Application to Radiology Departments", Socio-

Economic Planning Science, Vol. 13, pp. 35-39, 1979.

3. Kokkotos S., Ioannidis E., Spyropoulos C.D., "A System for Efficient Scheduling of Patient Tests in Hospitals", accepted for publication in Medical Informatics, 1997.

4. Spyropoulos C.D., Kokkotos S. and Marinagi C.C., "Planning and Scheduling Patient Tests in Hospital Laboratories", presented at the 6th Conference on Artificial Intelligence in Medicine Europe (AIME'97), Grenoble, France, March 1997, appears also in Lecture Notes of Artificial Intelligence (LNAI), no 1211, "Artificial Intelligence in Medicine Europe", eds E. Keravnou et al, Springer, 1997, pp 307-318.

5. Marinagi C.C., Panayiotopoulos T., Spyropoulos C.D. "Planning through the TRLi temporal reasoning system", presented at the 10th International Conference on Applications of Artificial Intelligence in Engineering (AIENG'95), Udine, Italy, July 1995, appears also in "Applications of Artificial Intelligence in Engineering X", eds G. Rzevski, R.A.Adey, C.Tasso, Computational Mechanics Publications, Boston, pp. 19-27, 1995.

6. Marinagi C.C., Panayiotopoulos T., Vouros G.A., Spyropoulos C.D. "Advisor: a Knowledge-based Planning System", International Journal of Expert Systems Research and Applications, Special Issue on Knowledge Based Planning, Vol. 9, No. 3, pp. 319-353, 1996.

7. Panayiotopoulos, T. and Gergatsoulis, M., "A Prolog like temporal reasoning system", in Proceedings of the 13th IASTED International Conference on Applied Informatics, Innsbruck, Austria, Feb.21-23, 1995, IASTED-ACTA PRESS, pp.123-126.

8. Currie, A. Tate, "O-PLAN: the open system architecture", Artificial Intelligence 52, pp.49-86, 1991.

9. Drabble B., Tate A. "The Use of Optimistic and Pessimistic Resource Profiles to Inform Search in an Activity based Planner", 2nd International Conference on AI Planning Systems, AIPS-94, Chicago, IL, pp. 243-248, 1994.

10. Wilkins D.E. "Can AI Planners Solve Practical Problems", Computational Intelligence, Vol. 6, pp. 232-246, 1990.

11. Laborie P., Ghallab M. "Planning with Sharable Resource Constraints", 14th International Joint Conference on AI - IJCAI'95, pp.1643-1649, 1995.

The following diagrams show the steps of the MRA decomposition process.

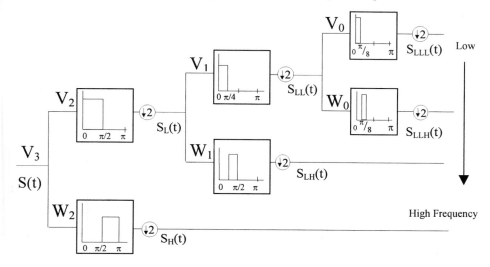

Figure 3. MRA decomposition chart for 8 data points.

where

$$S_L(t) = \sum_{k=0}^{3} c_{2k}\phi(2^2t-k), \qquad c_{2k} = \int_0^1 S_L(t)\phi(2^2t-k)dt \quad \text{(decomposition level 1)}$$

$$S_{LL}(t) = \sum_{k=0}^{1} c_{1k}\phi(2t-k), \qquad c_{1k} = \int_0^1 S_{LL}(t)\phi(2t-k)dt \quad \text{(decomposition level 2)}$$

$$S_{LLL}(t) = c_{00}\phi(t), \qquad c_{00} = \int_0^1 S_{LLL}(t)\phi(t)dt \quad \text{(decomposition level 3)}$$

and

$$S_H(t) = \sum_{k=0}^{3} d_{2k}\psi(2^2t-k), \qquad d_{2k} = \int_0^1 S_H(t)\psi(2^2t-k)dt \quad \text{(decomposition level 1)}$$

$$S_{LH}(t) = \sum_{k=0}^{1} d_{2k}\psi(2t-k), \qquad d_{2k} = \int_0^1 S_{LH}(t)\psi(2t-k)dt \quad \text{(decomposition level 2)}$$

$$S_{LLH}(t) = d_{00}\psi(t), \qquad d_{00} = \int_0^1 S_{LLH}(t)\psi(t)dt \quad \text{(decomposition level 3)}$$

2.2.2. MRA Reconstruction

The MRA reconstruction process is the inverse process of the decomposition. The end result of the 3-step decomposition is the recovery of a signal $S(t)$ and is given by the following expression.

$$S(t) = S_{LLL}(t) + S_{LLH}(t) + S_{LH}(t) + S_H(t)$$

where $S_{LLL}(t) \in V_0$, while $S_{LLH}(t) \in W_0$, $S_{LH}(t) \in W_1$, and $S_H(t) \in W_2$.

492

Using the expressions developed earlier in the decomposition section, the above equation can be written as follows

$$S(t) = c_{00}\phi(t) + d_{00}\psi(t) + \sum_{k=0}^{1} d_{1k}\psi(2t-k) + \sum_{k=0}^{3} d_{2k}\psi(2^2t-k)$$

where

$$c_{00} = \int_{0}^{1} S_{LLL}(t)\phi(t)dt, \qquad d_{jk} = \int_{0}^{1} f(t)\psi(2^{j}t-k)dt$$

where $f(t)$ is $S_{LLH}(t)$, $S_{LH}(t)$, or $S_{H}(t)$ depending on the resolution level.

The following diagram represents the steps of the MRA reconstruction process.

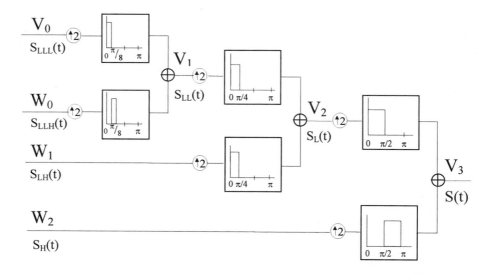

Figure 4. MRA decomposition chart for 8 data points.

2.3. Entropy

Entropy, a thermodynamic concept, is a measure of the degree of randomness of a set of random variables. If we consider a set of n random variables

$$s_1 = s(t_1), \ s_2 = s(t_2), \ s_3 = s(t_3), \ \dots, \ s_n = s(t_n)$$

with probabilities

$$p_1 = p(s_1), \ p_2 = p(s_2), \ p_3 = p(s_3), \ \dots, \ p_n = p(s_n)$$

then the entropy is defined as

$$H = -\sum_{i=1}^{n} p_i \log_2 p_i$$

The entropy can range from 0 to $\log_2 n$. More specifically, if $p_1 = 1$, $p_2 = \ldots = p_n = 0$, then, $H = 0$, and if $p_1 = p_2 = p_3 = \ldots = p_n = 1/n$, then H takes its maximum value, which is $\log_2 n$.

In this study, we examine the characteristics of the random SEMG signal as a function of its entropy. (In our case, p_i was calculated from the zero crossings of the signal.)

3.   Review of fatigue analysis

3.1. Frequency analysis

Frequency analysis of SEMG signals provides important signal characteristics such as the identification of certain frequencies of interest and the energy distribution of the signal. More specifically, the frequency power spectrum has been used extensively in muscle fatigue identification studies during a variety of job conditions [2]. It has been demonstrated that with sustained muscle contraction the high frequency components of the signal decrease while the low frequency component increase. This change results in a shift of the mean/median frequency of the power spectrum towards lower frequency (Fig. 5).

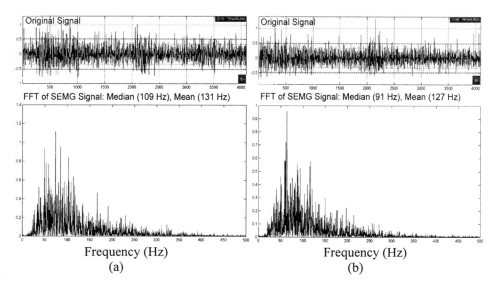

Figure 5.   Mean and median frequency shifts (a) at the beginning, and (b) at the end of a fatigue test.

It was observed that the mean frequency is more sensitive to noise than the median frequency. However, researchers [3] have found that although some test subjects show median frequency shift, no significant difference was found across subjects, and therefore, questioned the validity of such fatigue test. They claim that the frequency shift may be due to load or muscle length changes [4, 5] occurring during isometric (static) contraction [6].

3.2. Zero Crossings

The number of times the signal crosses the time axis is known as zero crossings as it is shown in Fig. 6.

Figure 6. The zero crossings.

It appears that this may be related to muscle contraction force and that the number of zero crossings increases as the muscle activity increases. However, it was observed that at high level of muscular activity, the number of zero crossings did not increase. Some investigators studied shoulder and neck muscle fatigue based on zero crossings and claimed that their results are encouraging [7].

4. **Experimental Method**

4.1. Experiment [1]

Four subjects with no known history of neuromuscular disease volunteered to participate in the fatigue experiment. SEMG signals were recorded from the erector spinae muscle bilaterally at L3 level 3cm from the midline. Each subject performed a lifting/lowering-twisting task at the NIOSH 1991 lifting guideline [8] LI = 1.0, using a 45 N weight, once every 5 seconds, over a 200 second period. The load was grasped by the hands and was moved from a height of 56 cm above the floor to a height of 96.5 cm above the floor, through an arc of 90 degrees (+/- 45 degrees) with feet pivoted in place. Erector spinae muscle electrical activity outcome was recorded for the first four and the last four seconds of a 60-second, Sorenson-like [9, 10] fatigue test (Fig. 7) after the 200-second lifting/lowering-twisting (LLT) task. SEMG data were recorded at 1000 points per second.

Each subject rested for 5 minutes between each trial of the outcome-task-outcome sequence and was evaluated under the following testing conditions:

(1) standing still wearing no support and performing no LLT task
(2) wearing no support and performing LLT task
(3) wearing high back support tightly, bridging pelvis and ribcage, and performing LLT task
(4) wearing lumbar support loosely around pelvis, and performing LLT task
(5) wearing sacroilliac belt tightly, and performing LLT task
(6) wearing lumbar support tightly around pelvis, and performing LLT task
(7) wearing no support and performing LLT task

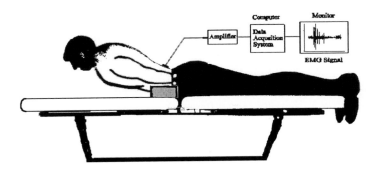

Figure 7. Modified Sorenson test for fatigue testing.

4.2. Fatigue Identification based on Wavelet/Entropy Analysis

In this restricted study we test the hypothesis that the entropy of the surface EMG signal is different at the beginning and at the of a fatigue test. Furthermore, we hypothesize that high entropy means high fatigue and vice versa. The study of the signal was done by applying a wavelet based technique known as multiresolution analysis (MRA) using Daubechies wavelet [11-14] of 16th order. The surface EMG signal was decomposed into eleven levels (corresponding to the 4096 data points) and partially reconstructed by summing levels from 8 to10 only (which corresponds to major power spectrum EMG activity). Subsequently, the entropy of the reconstructed signal was computed for all the test cases involved.

5. Results

The results have shown (Fig. 8) that the entropy is higher in the final four seconds for all subjects and for all seven cases examined which is consistent with the results obtained by the frequency method, where the median frequency shifts toward lower frequencies indicating more fatigue.

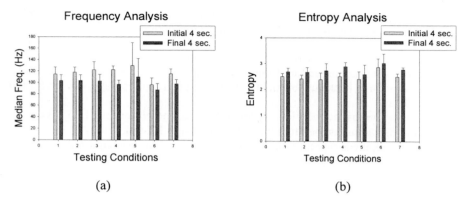

Figure 8. Summarized results from (a) frequency and (b) entropy analysis.

6. Conclusion

Fig. 9 shows: (a) the average median frequency shift and (b) the average entropy increase (final(4 sec.) – initial(4 sec.) difference).

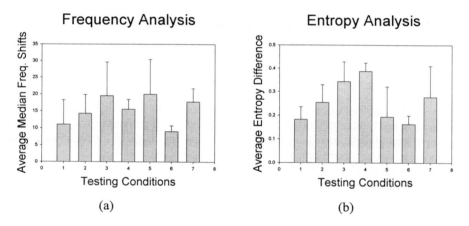

Figure 9. Comparison of frequency and entropy based fatigue analysis.

Next, we discuss the significance of the findings for each testing condition.

Testing condition 1: Shows low fatigue level as expected.
Testing condition 2: Shows increased fatigue level as expected.
Testing condition 3: Shows increased fatigue level, contrary to what is expected.
                      Consequently, back support did not reduce the fatigue level.

Testing condition 4: Shows increased fatigue level as expected, contrary to the result obtained from frequency analysis.

Testing condition 5: Shows decreased fatigue level as expected, contrary to the result obtained from frequency analysis.

Testing condition 6: Produced less fatigue (best result) in both frequency and entropy based analysis indicating possible best low back support.

Testing condition 7: Shows consistent increase in fatigue as expected in both analysis.

Clearly, as it can be seen in Fig. 9, the test condition 6 (lumbar support worn snugly around the pelvis), produced less fatigue in both frequency and entropy based analysis. This is significant in the sense that if a statistically large number of test cases show the same trend, then a new entropy based muscle fatigue test could be developed. This could have significant influence in not only the design of new lumbar support but also the manner the support is worn.

References

1. Wilder D, Lee J, Pope M, *et al. Erector spinae muscle fatigue and circumferential compression of the trunk and pelvis.* in *International Society for the Study of the Lumbar Spine.* Brussels, Belgium. 1998.

2. DeLuca C, *Myoelectrical manifestations of localized muscular fatigue in humans.* CRC Critical Review in Biomedical Engineering, 11: p. 251-279, 1985.

3. Matthijsse P, Hendrich K, and Rijnsburger W, *Ankle angle effects on endurance time, median frequency and mean power of gastroenemius EMG power spectum: A comparison between individual and group analysis.* Ergonomics, 30: p. 1149-1159, 1987.

4. Gander R and Hudgins B, *Power spectral density of the surface myoelectric signal of the biceps brachii as a function of static load.* Electromyogr Clin Neurophysiol, 25: p. 469-474, 1985.

5. Christensen H, Monaco ML, and Dahl K, *Processing of electrical activity in human muscle during a gradual increase in force.* Electroencephalogr Clin Neurophysiol, 58: p. 230-239, 1984.

6. Bazzy A, Karten J, and Haddad G, *Increase in electromyogram low-frequency power in nonfatigued contracting skeletal muscle.* J Appl Physiol, 61: p. 1012-1017, 1986.

7. Hagg G, Suurkula J, and Liew M, *A worksite method for shoulder muscle fatigue measurements using EMG, test sontractions and zero crossing technique.* Ergonomics, 30: p. 1541-1551, !987.

8. Waters T, Putz A, Garg A, *et al., Revised NIOSH equation for the design and evaluation of manual lifting tasks.* Ergonomics, 36(7): p. 749-776, 1993.

9. Biering-Sorenson F, *Problems in measuring of isometric endurance of the back muscles.* Back and muscle research, ed. T. Bendix, *et al.* Copenhagen: Dansk Fysiurgisk Selskab. 41-52, 1983

10. Biering-Sorenson F, *Physical measurements as risk indicators for low-back trouble over a one-year period.* Spine, 9(2): p. 106-19, 1984.

11. Daubechies I, *Ten Lectures on wavelets*. Philadelphia: Society for Industrial and Applied Mathematics, 1992
12. Akansu A and Haddad R, *Multiresolution signal decomposition*. San Diego: Academic Press, 1992
13. Vaidyanathan P, *Multirate systems and filter banks*. Englewood Cliffs: Prentice Hall, 1993
14. Vetterli M and Kovacevic J, *Wavelets and Subband coding*. Englewood Cliffs: Prentice Hall, 1995

44

European Thematic Network for Intelligent Forecasting

B.E. BITZER
University of Paderborn
Steingraben 21; D-59494 Soest
Germany

1. Introduction

The European Union funds the thematic network for intelligent forecasting systems (IFS) by the Brite-EuRam program.

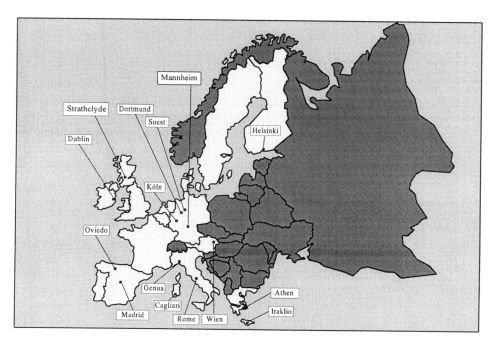

Figure 1. European members of the IFS thematic network

This program is in the field of industrial and material technologies. The network brings together industry and universities all over Europe for cooperation in forecasting systems of refineries and power systems. This chapter describes the general tasks, objectives and worksteps associated with the subprojects of forecasting in refineries and power systems.

Besides this first results of the cooperation project between the University of Paderborn and the local distributor of the German town Dortmund are presented.

2. Thematic Networks

Thematic networks will bring together research carried out by universities, manufacturers, and end-users on a paricular technological research objective. The objectives of the project featured in this article are intelligent forecasting systems for refineries and power systems. So the network includes end users as utilities and refineries and also producers and research institutes from universities and industry. Countries all over Europe, from Crete and Rome in the south to Dublin and Helsinki in the north, work together in this project (Figure 1). The project is funded by the Brite-EuRam program of the EU-Commission. The BriteEuRam program defines a work program on industrial and material technologies and is the acronym for Basic Research in Industrial Technologies for Europe-European Research on Advanced Materials.

3. IFS Tasks

In general, the IFS network covers the following research fields in the Brite-EuRam program [1]:

- Developing new, quality-oriented, intelligent and flexible production systems based on applications of artificial intelligence and advanced automation
- Optimizing processes and identifying more efficient process configurations, particularly using simulation and modeling
- Monitoring and predicting online the behavior of complicated structures and systems, such as utility distributions.

In the first phase of the project, this complex task was reduced to forecasting systems.

For the project part of power systems, it means the forecasting of electrical load. Good prediction of future load demands is necessary when making an optimal production plan for the power plants of power systems. In the last few years, several techniques for short- and long-term load forecasting have been discussed, such as kalman filters, regression algorithms, and neural networks. These methods try to provide accurate predictions, but mostly do not consider the performance and the complexity required.

There is a need for a system that can easily be adapted to new trends without great changes in the structure. Conventional methods such als kalman filters and regression algorithms have trouble identifying a model that adapts to the problem in the best way.

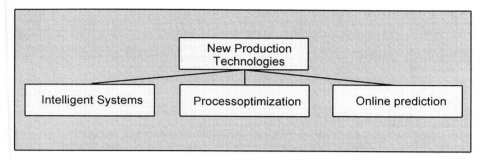

Figure 2. Brite EuRam research field

Most of the time spent on the realization of this method is on the development of the model. Some methods do not adapt well to changes in the structure and can involve the necessity of a new development.

4. Objectives and Worksteps

Most work in the field of intelligent systems is done by basic research. Universities have had many research projects in the field of neural networks and expert systems, but only some applications are realized today, especially in the field of power systems and refineries.

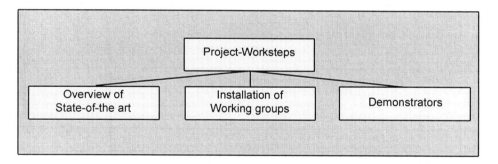

Figure 3. Project worksteps

The main problem is still the exchange of experiences between the fundamental research in universities and the more practical experiences of the industry with its lack of developing applications with intelligent systems like neural networks, expert systems, or fuzzy logic.

The general aim of the thematic network is the realization of an information exchange between universities, research centers, and industry to guarantee the knowledge transfer

502

of basic research from universities/research centers and experiences from industry applications.

This knowledge transfer includes the exchange of personnel between the institutes and the training of industry personnel. Therefore, the network has to install the support for this knowledge transfer, including:
- Current overview of the state of the art and the main research activities in the field of intelligent forecasting systems
- Installation of common working groups between universities and industry
- Workshops for personnel training
- Project-wide demonstrations to show the advantages of intelligent forecasting systems.

5. The Network Structure

The 20 active members work within three task groups:

- Energy Time Series
- Fault & State forecasting
- Modelling & Control Strategies (Fig. 4)

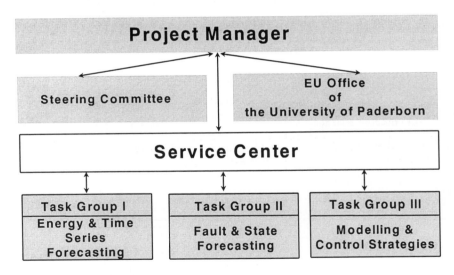

Figure 4. The IFS Structure

The results of these Task Groups are presented at national and international workshops. The Service Center has the central role of the thematic network. It organizes the communication and knowledge transfer between the network members and supports the organization of workshops and meeting.

There is also an information board in the internet (see www.uni-paderborn.de/~IFS) with the following services:

- Structure, aims and schedule of IFS
- IFS activities and subprojects
- Research activities, research groups in intelligent forecasting
- Commercial and available Softwareproducts
- Companies and services
- Conferences, workshops and seminars

Because of the network size a strong project management was installed by one central project manager, project office for the administration support and steering committee from universities and industries to bring together the scientific and industrial aspects and needs in this network.

6. National Subprojects

The main task of the thematic network is to organize the communication and knowledge transfer between all active and interested members. The research and industrial development results come from national subprojects which are funded from national programmes or industrial budgets. The following list gives an overview about these subprojects for the three task groups [3-10].

Table 1: **Subprojects of the Task Group I**

„Energy & Time Series Forecasting"

No.	Subprojects Members	Title
P 11	Technical Educational Institute, Crete University of Paderborn Public Power Coperation Crete	Online load forecasting systems for autonomous supplies
P12	Dublin City University ESB, National Grid	Intelligent time series modelling techniques with application to electrical load forecasting on short, medium and long time scales
P13	Technical Educational Institute, Crete Load Dispatch Center, Public Power Coperation, Crete University of Paderborn	ES System Load Data Bank Review: Structure Singular Points Update

No.	Subprojects Members	Title
P14	University of Strathclyde UK utilities	Predictive condition monitoring data analysis: data mining and knowledge discovery for forecasting techniques
P15	Technische Universität Wien Siemens AG, Austria	Power System Load Forecasting - rapid prototyping of neural network prediction models by standard softwaretools -

Table 2:   **Subprojects of the Task Group II**

„Fault & State Forecasting"

No.	Subprojects Members	Title
P 21	National Technical University of Athens Hellenic Aspropyrgos Refinery S.A	Fault Forecasting and detection
P 22	University of Paderborn DEA Mineraloel Werk UK Wesseling	Forecasting of state parameters in refineries
P 23	INITEC, Madrid	Integrated Tool for Fault Detection and Control forecasting System
P24	Forschungsgemeinschaft für Hochspannungs- und Hochstromtechnik e.V.	Forecasting in maintenance systems for power systems

Table 3:   **Subprojects of the Task Group III**

„Modelling & Control"

No.	Subprojects Members	Title
P 31	Eniricerche Italy Università di Genova Università di Gagliari	Technology Monitoring of IFS industrial applications
P 32	University of Gagliari University of Genova Centro Ricerche ENI SARAS	Methodologies for modeling and control in the Chemical Process Industries, CPI
P 33	University of Oviedo Hydroelectrica del cantabrico S.A.	Global modeling of 2 multi-product powerfacility for the optimization of consumption

No.	Subprojects Members	Title
P34	Program S.r.L, Rome University of Genova	Forecasting systems for early, detecting chemical reaction parameters far away

A detailed description of these subprojects will be given on the IFS-internet pages (adress see above).

7. Horizontal Themes

All the network members work together on topics which are relevant to several task groups.

These horizontal themes were defined as robustness, exactness and validity of forecasting systems. Besides this the preparation, analyzing and preprocessing of the system data and the online behaviour were also chozen as important subjects. All these horizontal themes are listed in detail:

- robustness & online-capability of forecasting systems
- exactness & validity of the prediction
- data preparing for predictions (techniques and validation)
- adaptivity on new forecast problems, rapid prototyping techniques for forecasting
- generalization capabilities of demonstrators, integration in existing EMS
- effectiveness of intelligent tools for real world

8. An Application Example

8.1 The Situation

As mentioned above the research results of the network come from the national subprojects. As an example of one of these subprojects the cooperation of the University Paderborn with a German utility company is described.

This cooperation is in the field of short-term load forecasting and is part of the IFS task group I. The following sentences describe the situation and motivation for this national subproject.

In January 1995 the power distribution of the German town Dortmund changed from the regional power distributor VEW to the local distributor who is now responsible for the gas, water and power delivery. The power generation is still realized from VEW. The calculation and costs of the energy import is fixed on a contract that is based on the average of the three largest peak loads in a year. The local distributor disposes of small own energy capacities for a short time. They are able to reduce load peaks if the energy demand is known in time. Up to now, load forecasting is done empirically and experiences with different load forecasting techniques are missing. A computer operated Supervisory Control and Data Acquisition (SCADA) will be installed, including load forecasting elements based on regression methods.

8.2 The Project

The cooperation project has the aim to examine the practical use, advantages, disadvantages and differences of regression methods and neural networks for a 24 hour load forecasting in Dortmund. Therefore 15 minutes - load data from two years are available as well as weather information like temperature, wind speed and clouds from two weather stations in this region.

In a first step an overview of the data analysis will be given to determine the influence of historical load data and temperature on the load demand for the next 24 hour. Based on this information the inputs for both methods will be set. In a next step the results of a 24 hour load forecasting with neural networks and a multiple regression for three exemplary months will be shown and the results will be discussed. Improvements, advantages and disadvantages in the use of one of these methods will be presented.

8.3 Neural network structure

For the given application several types and structures of neural networks are discussed. A feedforward network trained with an rprogalgorithm achieved good results. As shown in figure 5 the day was splitted into three periods. This improves the prediction. The minimal and maximal temperature influence each period, as opposed to the day light, which only improves the load prediction for the second period.

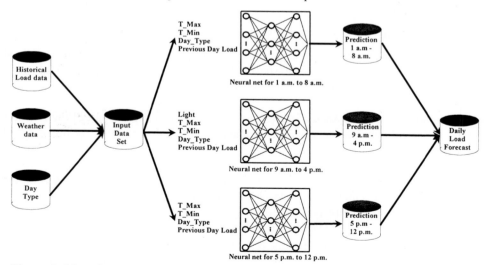

Figure 5. Neural network structure

8.4 Results

The first results of the subproject show that neural networks are suitable for daily and peak load forecasting.

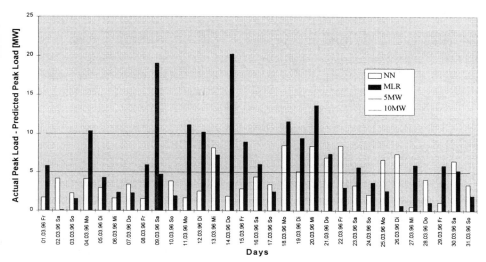

Figure 6. Peak load error for neural network and multiple linear regression

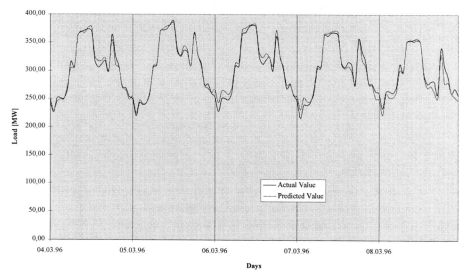

Figure 7. Forecast and daily load demand

508

The comparison of the neural network method (NN) with a multiple linear regression (MLR) is shown in figure 6. The NN method is always better in the peak load forecasting, if the MLR method has relevant errors (> 10 MW).
Only on one saturday the NN method has an error higher than 10 MW, but this is not so important on a weekend.

Network with different input structures and three periods as shown in figure 5 improve the load prediction.

Most peak load errors are under the 5 MW limit. This was defined above as a very good result. Also the RMS-error is less then 5% for the important week days (see figure 7) as it was wished above.

References:

[1]   European Commission: Brite/Euram - Workprogramme; Brüssel, 1994.

[2]   Bitzer, B.: European Project: Intelligent forecasting systems, IEEE computer application in power CAP, 4/97, Columbus, Ohio 1997.

[3]   Bitzer, B.; Rößer, F.: Intelligent load forecasting for the electrical power system on Crete, UPEC '97 - Universities Power Eng. Conf., UMIST-Univ. of Manchester, 1997.

[4]   Bitzer, B.; Rößer, F.: „Energy Control of a Refinery by Artificial Neural Networks" Isth IMACS World Congress 1997, Vol. 4, p. 191 - 196.

[5]   Ringwood J.V., Murray, F.T.: Improvement of electricity consumption forecasts using temperature inputs; Simulation Practice and Theory, Vol: 2 Iss: 3 p. 121-139, 1994.

[6]   Seyed-Masoud-Moghaddas-Tafreschi, Mueller, H.: Medium-term forecast of network load for the operational planning per year in the electric supply with help by concepts of neural networks and the fuzzy set theory, VEO Journal, Vol. 7-8, p. 133-140, 1995.

[7]   Asar, A.U., McDonald, J.R.: A specification of neural network applications in the load forecasting problem; IEEE Transactions on Control Systems Technology, p. 135-141, Vol.2, 1994.

[8]   Baratti,R., Vacca, G., Servida, A.: Neural Network modelling of distillation columns; Process Control & Information Systems, June 1995.

[9]   Tzafestas, S., Verbruggen, H.B.: Artificial Intelligence in Industrial Decision Making, Control and Automation, Kluwer Academic Publishers, 1995.

[10]   Baratti, R., Corti, S., Servida, A.: A feedforward neural controller strategy for destillation columns, EANN 96, London, June 17-19, 1996.

45

Review of ANN Based Load Forecasting Research for the EPS of Crete

T.M. Papazoglou
E.E. Dept., Technological Educational Institute Iraklio
P.O. Box 1365, Iraklio 711 10, Crete, Greece

1. Introduction

Short Term Load Forecasting (STLF) is very important for the autonomous Electric Power System (EPS) of Crete. This is especially so, because up to now the System's reserves of installed power have been chronically very low, due to strong public resistance to additions of new thermal power generating plants on the island, as a result the system is often marginally stable and load shedding is practiced during periods of peak demand (mostly at summer). Also, because of peak-to-minimum load ratios as high as 3 and load factors around 55%, the load demand exhibits relatively fast variations. An additional difficulty is the fast growth of power demand, averaging about 7% p.a. [19], as well as the fast growing wind generation on the island. Keeping the balance between generation and power demand, in view of the peculiarities of generation units, reduces the likelihood of operational disorders which could jeopardize the safety of the system, and, at the same time, makes it possible to supply electric power in the most economical way while satisfying the technical constraints. Accurate load forecasting has crucial influence on security as well as on efficient operation of the EPS and is useful for system control, unit scheduling and maintenance.

Thus far, STLF for the Cretan EPS is done empirically. Measurements of loads, ambient temperature, etc., are logged every minute into a database. Load curves are drawn on 24-hour basis and experienced system-operators predict the forthcoming load demand taking into account judiciously chosen recent-past load curves. It has been observed, for example, that special events on the island play an important role in shaping the electric demand.

Recently, an on-line Supervisory Control and Data Acquisition (SCADA) system has been installed and is successfully operating. Thus, it is possible to add Energy Management System (EMS)-functions, such as intelligent forecasting, to aid the system operators. Past efforts to test on-line Neural Network (NN)-based software tools for load forecasting did not show satisfactory user-friendliness. System operators have found them, up to now, more time consuming than practically useful. However, recently improved versions show promise for substantial upgrades.

For comparison, the EPS of Cyprus is 70% larger than that of Crete. It operates with a comfortable power reserve margin of about 30%. Even so, accurate load forecasting for this autonomous system contributes to security and efficient operation of the EPS and is useful for system control, unit scheduling and maintenance. In the past, a STLF-software tool based on Kalman Filter, has been used for peak load predictions. Recently, a NN-based software Load Predictor was installed and is undergoing training and tuning.

2. Review of Techniques

In the past, statistical forecasting techniques and later Kalman Filters have been used for load predictions [1,2,3,4,5]. More recently, there has been wide interest in developing Intelligent Forecasting Systems (IFS), and in using Artificial Neural Networks (ANN) for electric power load forecasting [6,7,8,...,20]. Different approaches have been adopted in which both feedforward [9,17,18,19,20] and recurrent [14,15,16] nets were tried. In some cases modular nets (one subnet for each hour to be forecast) have been used [14]. Such a network is claimed to be of high predictive accuracy for a 24-hour span, provided that the number of variables in the input vector is large enough. Also, expert systems [21], [22] and fuzzy logic [23] have been applied.

It should be noted that conventional methods like Kalman Filters and regression algorithms present the difficulty of identifying the model which adapts best to the particular problem. On the other hand, ANN are adaptable, self-learning and more robust against disturbances. Comparison of the forecasting accuracy between the above mentioned approaches has been reported in the past [24]. The author and co-workers [17] have compared Kalman Filters to ANN for STLF for the EPS of Crete and have found ANN to be indeed superior with respect to robustness against disturbance, accuracy of prediction and adaptivity to changing data sets. In the same work [17], it was shown that Multilayer Perceptron (MLP) networks performed better than Radial Basic Function (RBF) networks. In another paper [20], the author and co-workers have compared MLP networks favorably to Multiple Linear Regression (MLR) and Multiple Non-linear Regression (MNR).

3. Network Architectures and Inputs

The author and co-workers [17,18,19,20] have tried different NN structures, sizes and learning parameters. Multilayer Feedforward Networks (MLF) were found suitable for predictions as well as for pattern recognition. Extensive testing resulted in two alternative NN architectures, depending on whether or not seasonal trends were classified. The first, a MLF network with one hidden layer and with transfer functions: linear for the input and output layers and sigmoid for the hidden layer. Two different sizes were used for this MLF net: (55,10,1) and (17,15,1). In the first case, the input vector composition was richer, as follows:

- the expected temperatures of current day (min, max)
- the average temperatures of last three days
- the load of same hour of previous day
- the last 48 measurements of loads
- the load of same hour of previous week.

In the second case, the input vector composition was more selective, as follows:

- the expected temperatures of the current day (min, max)
- the average temperatures of the last three days
- the last six measurements of loads
- the load of same hour, and preceding two loads of the previous day
- the load of same hour and preceding two loads of previous week.

The second MLF architecture had two hidden layers: one with gaussian activation function and the other with logistic activation function, and size (9,10,10,1) neurons. Nine element inputs were used, corresponding to values characterizing the month and day of year, the expected minimum and maximum temperatures, the expected weather condition (cloudiness), and the load of previous four hours.

Other NN architectures have also been successfully tried for STLF models applied in the case of the Cretan EPS. In [25] the STLF model is based on a three layer feedforward NN with size (64,48,24).

4. Acceptance Criteria for Load Forecasting

The author with co-workers is currently studying, in the context of an European Union BRITE-EURAM II Program-sponsored IFS Thematic Network, the question of acceptable criteria for load forecasting with respect to the particular case of the EPS of Crete, along the following directions.

4.1 Acceptable Accuracy of Prediction

Defining the acceptable accuracy of prediction for system loads is an important issue. At least five different measures of accuracy of the ANN based load forecasts have been applied by various approaches:

- Mean Absolute Deviation (MAD)
- Mean Absolute Percentage Error (MAPE)
- Root Mean Square deviation (RMS)
- Average Error (AE)
- Maximum Error (ME).

The maximum acceptable error for a load forecast is actually system-dependent. It depends on:

- The point in time

- The composition of power generation in the EPS and its characteristics, e.g. the intermittent property of wind power generation [26]
- Consequences on Power Quality and Electric-Consumer economics.

It is therefore of value for a given EPS, with respect to its particular characteristics, to determine what are the Meaningful Accuracy Levels (MAL) and which are the Maximum Tolerable Errors (MATE) for load forecasting.

Also, the acceptable accuracy tolerances depend on the time horizon of the forecast, especially for the EPS of Crete where the generating units have different characteristics depending on their ratings and the system is marginally stable.

4.2 User-friendly Load Forecasting Software Tools

A lot of attention has to be given to assuring the user-friendliness of a load forecasting software tool. To this end one must assure the compatibility with the operating SCADA system, must establish the necessary and sufficient data base continuously updated. The tool should require minimal effort from the system operators and provide information of assured practical value. Also, timewise, forecasting should be compatible with the actual system operation scheduling horizon.

4.3 Databases

Defining the necessary and sufficient database is crucial for intelligent load forecasting. It is a task that should include:

- Pointing the important input variables (environmental such as: temperature, humidity, luminosity, wind chill factor, etc., as well as societal such as: special events, etc.)
- Study of the mechanisms of factors affecting load demand (or shaping load demand)
- Collect the relevant characteristics of the EPS, e.g. technical constraints of generating plant, etc.
- Utilization and formulation of the operators-experience
- Isolation and rejection of spurious data (such as: faulty measurements, system faults, unrecorded measurements, etc.)
- Batch-files for training of NN.

5. Conclusions

An attempt was made to review the research done, mainly by the author and his co-workers, in the last five years on ANN-based load forecasting with emphasis to the EPS of Crete. This research eventually led to the formation of an European Union BRITE-EURAM II Program-sponsored IFS Thematic Network for Power Systems and Refineries.

Load forecasting in the Cretan EPS is complicated by the System's characteristics. It is autonomous with fast growing load demand and wind power penetration, etc.. The importance of reliable STLF for the EMS of the autonomous EPS of Crete is shown. The techniques used were reviewed. The different network architectures and inputs that were tested aiming at realizing the best possible estimations are compared. The criteria of acceptance of load forecasting, including the question of acceptable accuracy of prediction, the user-friendliness of the load predictor tools and the appropriate databases, were presented. Research work is currently extended to the study of critical parameters for optimum load forecasting.

6. Acknowledgments

The author wishes to express his sincere thanks to the personnel of the Load Dispatch Center for Crete of the Public Power Corporation for fruitful discussions and useful cooperation.

7. References

[1] Box G.E.P. and Jenkins G.M. *Time Series Analysis Forecasting and Control*, Revised Edition. Holden-Day. Oakland, 1976.

[2] Hubele N. and Chuen-Sheng C. «Identification of seasonal short-term load forecasting», IEEE Trans. on Power Systems, Vol. 5, No. 1, 1990.

[3] Papalexopoulos A.D. and Hestenberg T.C. «A regression based approach to Short Term Load Forecasting», IEEE Trans. on Power Systems, Vol. 5, No. 4, pp. 1535-1547, 1990.

[4] Strasser H., et al. «Short term load forecast using multiple regression analysis or adaptive regression analysis», Proceedings of PSCC. Graz, 1990.

[5] Wang X. And McDonald J.R. *Modern Power System Planning*, McGraw-Hill Int.. London, 1994.

[6] Dillon T., et al. «An adaptive neural approach in load forecasting of power systems», First International Forum on Applications of Neural Nets to Power Systems. Seattle, WA. pp. 17-21. July 23-26, 1991.

[7] Dillon T., et al. «Short Term Load Forecasting Using an Adaptive Neural Network», Int. Journal EPES, Vol. 13, No. 4, pp. 186-192, April, 1991.

[8] Park D., et al. «Electric load forecasting using an artificial neural network», IEEE Trans. on Power Systems, Vol. 6, 2:442-448, May, 1991.

[9] Ho K., et al. «Short term load forecasting using a multilayer neural net with an adaptive learning algorithm», IEEE Trans. on Power Systems, Vol. 7, 1:141-149, Feb. 1992.

[10] Lee K., et al. «Short term load forecasting using artificial neural nets», IEEE Trans. on Power Systems, Vol. 7, 1:124-132, 1992.

[11] Lu C., et al. «Neural Network based short term load forecasting», IEEE Trans. on Power Systems, Vol. 8, 1:336-342, 1993.

[12] Khotanzad A., et al. «Forecasting Power System peak loads by an adaptive neural network», Intelligent Engineering Systems through Artificial Neural Networks. 3:891-896. Edited by Dagli C., et al. ASME Press, 1993.

[13] Peng T., et al. «Advancement in the application of neural networks for short term load forecasting», IEEE Trans. on Power Systems, Vol. 8, 3:1195-1202, 1993.

[14] Khotanzad A., et al. «An adaptive and modular recurrent neural network-based power system load forecaster», Proceedings IEEE Int. Conf. ICNN'95, Perth, Western Australia, 2:1032-1036, 1995.

[15] Czernichow T., et al. «Improving recurrent net load forecasting», Proceedings IEEE ICNN'95, Perth, 2:899-904, 1995.

[16] Mandal J. K., et al. «Application of recurrent neural networks for short-term load forecasting in Electric Power System», Proceedings IEEE Int. Conf. ICNN'95, Perth, 5:2694-2698, 1995.

[17] Papazoglou T., et al. «Load forecasting for the Power System of Crete. A comparative analysis», Proceedings of UPEC'95, Vol. 2, pp. 435-438, University of Greenwich, Sept. 1995.

[18] Bitzer B., Roesser F., Papazoglou T. «24-hour load forecasting for the EMS of Crete», Proceedings of UPEC'96, Vol. 1, pp. 346-349, Techn. Educational Institute, Iraklio-Crete, Sept. 1996.

[19] Bitzer B., Papazoglou T.M., Roesser F. «Intelligent Load Forecasting for the Electrical Power System of Crete», Proceedings of UPEC'97, Vol. 2, pp. 891-894, UMIST, Manchester, Sept. 1997.

[20] Papazoglou T.M., et al. «ANN based Load Forecasting for the Energy Management System of a Mediterranean Island», Proceedings of UPEC'98 under publication, Napier University, Sept. 1998.

[21] Ho K., et al. «Short term load forecasting for Taiwan Power System using a knowledge-based expert system», IEEE Trans. on Power Systems, Vol. 5, No. 4, Feb. 1990.

[22] Rahman S., Hazim O. «A generalized knowledge-based short-term load-forecasting technique», IEEE Trans. on Power Systems, Vol. 8, No. 2, Feb. 1992.

[23] Dash P.K., et al. «Peak load forecasting using a fuzzy neural network», Electric Power System Research 32, pp. 19-23, 1995.

[24] Brace M.C., et al. «Comparison of the forecasting accuracy of Neural Networks with other established techniques», IEEE Proceedings, First International Forum of Applications of Neural Networks to Power Systems, Seattle, WA, July 23-26, 1991, pp. 31-35.

[25] Kiartzis S.J., et al «DAPHNE a Neural Network based Short-term Load Forecasting Program. Application to an Autonomous Power System», Proceedings EURISCON'98 under publication, Edited by S. Tzafestas, Athens 1998.

[26] Papazoglou T.M. «Quality of Power Supply and Wind Power Penetration Prospects for the EPS of Crete», VDE ETG-Conference Proceedings on Quality of Power Supply, Vol. 70, pp. 55-60, VDE-VERLAG GmbH, 1997.

46

Critical Factors in Load Forecasting for the Autonomous Electric Power System of Crete

D.I. Stratakis, T.M. Papazoglou
Technological Educational Institute, Iraklio.
Stavromenos 71500, Iraklio, Crete, Greece.

1. Introduction.

The Load Demand (LD) problem has to do with financial problems and problems about planning the ways of electric power production and distribution. This problem is particular for any power system and has important difficulties in the case of Autonomous Energy Power Systems such as that of Crete [1]. For this reason we start this chapter with a brief description of the characteristics of the Energy Power System (EPS) of Crete and the problems that it faces. Then we deal with the designation of accuracy levels that a certain forecast program must perform for the EPS of Crete as well as for the way of the data base formation and its demands. A great effort is made to designate the way of the optimum operation of the EPS of Crete.

2. Analysis of the EPS of Crete.

The recent figures of the EPS of Crete (up to early 1998) are presented in table 1. System development in energy production and maximum power demand from 1964 to 1997 is shown in figure 1 [2],[3],[4].

According to these records we come to the following conclusions:
(a) EPS of Crete consists of 26 power units. They have small net capacity of power production and this in relation to the power requirements especially in load peak periods results in their constant incorporation and release from the production system (except the base units which are in constant operation).
(b) Temperature increase and other factors (e.g. the pollution of the cooling system of the units) results in the decrease in the units capacity which can lead even to the failure of the system or in extensive load shedding especially in summer.
(c) There are many stability and power quality problems because of the implementation of the Wind Turbines in the EPS of Crete.
(d) There is a continually increasing load demand estimated about 7% for the following years. Here we must say that the variations in power demand throughout a year are due to the different type of consumers, especially household, commercial and agricultural consumers whose LD is not stable.
(e) There is a great financial problem concerning the cost of electric power production because of the peculiarities of the units (fuel cost, maintenance cost, years of operation e.t.c.).

All the above mentioned in combination with the great fluctuation of demand during 24 hours cause many problems in the safety of the operation of the system, in the quality of the produced power and in the operational cost of the system. In the following paragraphs we shall try to focus our attention on the major problems that must be faced.

Table 1. Figures of the EPS of Crete up to early 1998.

Serial number	Kind of Unit	Name of Unit	Installation Year	Installed Power (MW)	Maximum Win. Load (MW)	Maximum Summer load (MW)	Minimum load for stable operation (MW)	Spec. demand consumption 1997 Kg/KWh	Spec. demand consumption 1997 Lt/KWh	Availability 1997 (%)
1	Stream Electric (SE) Units	No1 Lin	1965	6,2	5,9	5,9	4	0,369		93,26
2		No2 Lin	1971	15,0	14,3	14,3	8	0,325		81,38
3		No3 Lin	1971	15,0	14,3	14,3	8	0,325		87
4		No4 Lin	1977	25,0	23,5	23,5	14	0,290		87,67
5		No5 Lin	1981	25,0	23,5	23,5	18	0,296		90,39
6		No6 Lin	1981	25,0	23,5	23,5	18	0,288		92,34
Total power of SE Units				**111,2**	**105,0**	**105,0**	**74**			
7	Diesel (D) Units	No1 Lin	1989	12,3	11,8	11,8	3	0,198	0,228	91,28
8		No2 Lin	1990	12,3	11,8	11,8	3	0,202	0,234	91,37
9		No3 Lin	1990	12,3	11,8	11,8	3	0,198	0,228	90,18
10		No4 Lin	1990	12,3	11,8	11,8	3	0,198	0,228	91,07
Total power of Dies. Units				**49,2**	**47,2**	**47,2**	**12**			
11	Gas Turbine (GT) Units	No1 Lin	1973	16,2	15,0	14,0	3		0,501	91,28
12		No2 Lin	1974	16,2	15,0	14,0	3		0,491	91,37
13		No1 Ch	1969	16,2	15,0	14,0	3		0,577	89,19
14		No4 Ch	1985	24,0	20,0	18,0	3		0,457	68,43
15		No5 Ch	1987	36,0	30,0	28,0	5		0,483	75,48
16		No11 Ch	1998	59,0	59,0	57,0	8		;	;
17		No12 Ch	1998	59,0	59,0	57,0	8		;	;
Total power of GT Units				**226,6**	**213**	**202**	**33**			
18	CC Unit	No6 Ch	1992	45,5	40	37,0	8		0,443	11,26
19		No7 Ch	1992	45,5	40	35,0	8		0,456	86,09
20		No1 Ch	1994	42,4	39	36,0	19		0,285	57,67
Total power of CC Units				**133,4**	**119**	**108**	**35**			
21	Wind Turb (WT) Un.	Toplou	1993	6,6	6,6	6,6	*		-	~
22		Rocas	1998	10,2	10,2	10,2	*		-	~
23		Sitia	-	0,5	0,5	0,5	*			~
24		Anogia	-	0,15	0,15	0,15	*		-	~
Total power of WT Units				**17,45**	**17,45**	**17,45**	*****			
25	HE Un	Almiros	-	0,3	0,3	0,3	*		-	~
26		Agia	-	0,3	0,3	0,3	*		-	~
Total power of HE Units				**0,6**	**0,6**	**0,6**	*****		**-**	**~**

(CC = Combined Cycle, HE = Hydroelectric, Lin = Linoperamata, Ch = Chania, symbols * and ~ mean that the corresponding value is unknown because of wind velocity and symbol ; means that the corresponding value is unknown because the unit was put in operation recently).

From the above table we can see that the total installed power of the EPS of Crete is 538,45 MW, the maximum winter load of the units is 502,25 MW and the maximum summer load of the units is 480,25 MW.

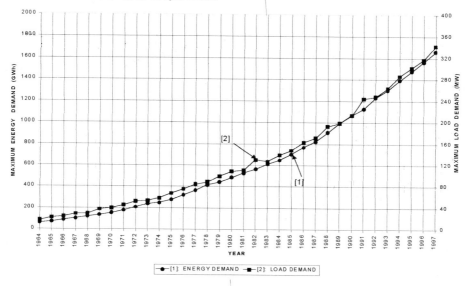

Fig 1. System development in energy production and in peak power demand.

3. Accuracy requirements and utility of the load forecast.

According to the up to now mentioned it is an urgent need of creation of a load forecast model for the EPS of Crete which may aim at giving the necessary information to the operators of the system for the estimation of future development of LD within a certain time horizon. The accuracy of that model has an important role in the technical and financial management of the system.

The accuracy limits depend on the time horizon of the forecast. As a conclusion for the EPS of Crete we can say that the model accuracy for half an hour time horizon must be extremely high, to say 99%, so that the system operators are able to:
(a) React properly in sudden incidents (e.g. short circuits, failure of a unit or failure of a power switch e.t.c.),
(b) Decide for the incorporation or release of a unit in or from the system.

A limit to the above time horizon is half an hour because: i) in the EPS of Crete it is usual that the LD in half an hour time horizon exceeds the capacity of power production of many units (it is sometimes more than 25 MW), ii) the slowest peak unit needs 12 – 13 minutes to be inserted in the system and this action must have the time needed to be repeated two times.

On the condition that the previous accuracy demand is justified the accuracy demands for long term load forecast (LTLF) become more flexible. For the 24 hours time horizon the forecast accuracy has to be more than 97%, so that partial service of certain peak units (which usually needs some hours to be completed) can be programmed during a day (since for the given period of time there are the necessary reserves to cover the power demand of the system). Lower forecast accuracy, but not less than 96% may have a LTLF (of some months) so that the entire shut down off of a unit must be designated for its annual service.

As a conclusion we can say that the accuracy demand of 99% for the next half an hour especially in the case of load demand increase in the direction of a local peak in the load curve must be maintained no matter what happens. In accordance with the above demand which, the system operator must have the necessary support to decide for the optimum management of a unit or optimum unit incorporation or release to the system in the basis of economic criteria, technical restrictions, quality of the produced electric power and environmental effects. If the above mentioned come true then there is a point in the existence of the forecast model.

4. Data Base requirements.

The statistic nature of 24 hours load curve exerts an important influence in the formation of the necessary data base for the EPS of Crete. The data base must include the necessary information for the forecast of the future behavior and development of the system. The accuracy demand of half an hour period mentioned in the previous paragraph inserts the first important demand of that data base: to be continually updated and constantly available. This means that the data must be inserted in the base at the moment of their collection, and must be directly available for further processing because, all the more reason, the direct future state of demand has to do with the present demand to a very high degree.

The insertion of the data in the base of course assumes full control of their correctness and probably their modification if needed so that to be correct. This can be achieved by the comparison of the incoming data with [5]:

(a) Data that have just been inserted,
(b) Typically expected data which refer to the historical past of the given period of time examined (e.g. by comparison the incoming data with another day's data with similar characteristics at the time of examination).

The above demand leads to the following:

(a) The volume of the data base must be not larger than the exactly necessary, which is determined by the accuracy levels we have mentioned before, so that the time of access and processing of the data may be the shortest possible. For the above reason the factors that influence the load curve and that lead to the determination of the important and unimportant variables which will determine the form of the data of the base, must be examined furthermore. This is discussed in the next paragraph.
(b) The use of the data base must be easy as: i) except the receiving data that will update the base, the operator must have the opportunity of inserting (when necessary) more crucial data in the base in a very easy way, ii) must give the

possibility of direct illustration in a graphical and quantitative way of important information for the past state of the system at certain periods of time.

5. Determination of important and unimportant variables that must be taken into account for the data base formation.

The determination of important and unimportant variables that must be taken into account for the data base formation is of crucial importance for the better solution of the problem of load forecast, because there are many variables which contribute to the formation of LD curve and have different influence in different systems (e.g. autonomous, connected, small and large geographic area systems e.t.c.).

For the autonomous system of Crete a first grouping of the variables which affect LD in order of importance according to the study of load curves since 1996 to May 1998, is the following [6]:

(a) Hourly fluctuation of load demand,
(b) 24 hours, weekly and seasonal fluctuation of load demand,
(c) Weather conditions,
(d) Special days,
(e) Random factors,
(f) Economic factors.

From the above categories of variables the most important must be chosen (demand of minimum necessary volume of data base) which will be subject to further processing, without meaning that information needn't be kept for the rest variables (for the reason of a deeper knowledge and understanding of the mechanisms by which they influence the load curve).

To be more specific let's examine the LD curves of two typical weekdays (fig 2). As we can mention from the curves in fig 2 the 24 hour LD can be divided into five separate areas with separation limits the local extremes (two local maxima and two local minima). The conclusions of the load curves study are the following:

(a) In the area number 1 the fluctuation of the LD is not important. The main factors that affect that fluctuation are the municipal lighting and the working hours of entertainment centers (night clubs) that varies from season to season.

(b) In the area number 2 the fluctuation of LD is more important and the increase is due to household and commercial consumers (the industrial consumers have smaller and more stable effect, because they have more certain characteristics and because they are fewer than the above mentioned).

(c) The decrease in the LD which appears in the area number 3 is because most of the works in the households are finished and because of the midday closing of the shops.

(d) The bigger increase in the area number 4 is due to commercial consumers, night cooking, municipal lighting, television programs and starting of the night clubs.

(e) The decrease in the area number 5 is due to the gradual decrease of the LD of the household consumers and to the rapid decrease in the LD of the commercial consumers which takes place about 21:00 until 21:30 a.m. The rest development of the load curve is towards the area number 1 of the next 24 hours curve with some variations of course.

Here we have to say that the most important effect in the formation of the 24 hour load curve for the EPS of Crete has the perfecture of Iraklion with the 53% on the total LD, then the perfecture of Chania with 23,5% and then the perfecture of Rethimnon and Lasithi with the rest 23,5% of the total LD. Thus the data from the perfecture of Iraclion must have the primary role in the load curve formation.

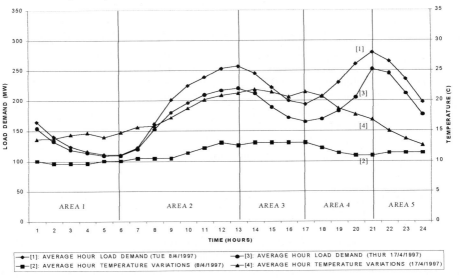

Fig 2. Typical weekdays load curves for the EPS of Crete.

Weekly fluctuation of LD can be separated in four categories: Fluctuation of LD from day to day for weekdays, for weekends especially for Sundays, for holidays within the week and three days holiday. For the first we can say that the variations are due to working hours of the shops (they are closed every Monday, Wednesday and Saturday afternoons except the large super markets which are closed only on Sunday). For the second we can say that the decrease in LD is due to commercial and industrial consumers. For the third we can say that local holidays have an unimportant effect except the local holiday of Saint Minas in Iraclion (Nov. 11th) which must be examined like the other standard holidays of the whole year. For the last category we can say that there is a successive decrease (from day to day) in the LD with a little increase in the last day's evening in relation to the previous day.

For the seasonal fluctuation of LD except the temperature which has a great effect we can conclusively say the following:

(a) Touristic season has a great effect. A constant increase in LD appears from the middle of March to the middle of July and a decrease up to the middle of October. The duration of the touristic period is not stable every year and we can not fully estimate its contribution because it is affected by other factors, that is political, economic, international coincidences, improvement of the given services, promotion and advertisement. Judging however from the rate of increase in the hotel construction especially in the northern road axis of Crete we can say that under normal conditions there must be a constant increase in LD because of the tourism. Also we must say that the most sudden changes in the load curve take place during the typical beginning and the typical end of the touristic period.

(b) Oil factories. From the middle of November the LD because of the oil factories activity shows a rapid increase from the beginning of the olive harvest, with a decrease from the middle of February ending at the middle of March.

(c) Average rate of the annual demand increase. The average rate is estimated about 7% for the following years. This increase is not linear all the year round, but it is greater towards the end of Spring and the beginning of Summer. This increase depends on the total number of new connections, the duration of the touristic period, the percentage of the tourists from the total number of the inhabitants, and the use of new electric devices into the houses.

(d) Summer time. This insert an discontinuity in the load curve which however is limited mostly in the first day and less in the two following days after the change of conventional time. Then the form of the LD curve returns to its usual form.

(e) Irrigations during summer. The last years they affect the increase of the LD especially after long periods of drought and we must take them seriously into account.

(f) The beginning and the end of the school period affects the LD of household consumers only during the first days after it, and its effect is not very important.

Weather conditions. The effect of the weather conditions on LD is also very important. The formation of a model of weather conditions which will have a direct relation to the demand is very difficult if not impossible because of the random appearance of the weather phenomena. For the EPS of Crete the most important climate variables are the following:

(a) Temperature. Its effect is very important mainly during summer and winter. During winter the increase of the temperature leads to a relative decrease in LD mainly during the day. During summer the increase of temperature leads to an important increase in LD due to the always increasing use of air conditions, machines which are easy to be bought and easy to be installed. This reason in combination with the touristic period leads to the transfer of the peak LD in the summer months. (The problems arised here are many and concern mainly the production system of Crete and this will be discussed however in the next paragraph).

(b) Direction and speed of the wind apparently has to do with the temperature change, so it is an important variable for the forecast program. Conclusively we can say that south winds in the winter lead to the decrease in LD which depends on their speed. On the contrary south winds in summer lead to significant increase in LD which is the greater as their speed is the higher. The exact opposite happens with the north winds (in winter and summer) whereas the effect of intermediate directions winds is less important than the above mentioned cases.

(c) Sunlight and cloudiness are indirectly related to the temperature and the winds, and they have an important effect due to the consumers psychological reactions.

Less important climatic variables are the humidity and atmospheric pressure which also affect the temperature change as well as the direction and speed of the winds. It is considered however that the study of temperature covers the effect of the above magnitudes. Finally the intensive weather conditions (showers, rainstorms, or intense snowfall and most of all that of great heat during summer months, must be placed among the climatic important variables.

Standard special days. Are the fourteen standard holidays and St Minas celebration as well as the three days holiday of the great religious celebrations and the new year's day. Their research is of particular interest because of the demand decrease appearance and which is mainly due to commercial consumers, and less to the industrial and household consumers. Some load curves of special days are shown in fig 3, where the typical load curve of 8/4/1997 (Tuesday) and the load curve of day of maximum load (Thursday 7/8/1997) demand is also shown for comparison reasons.

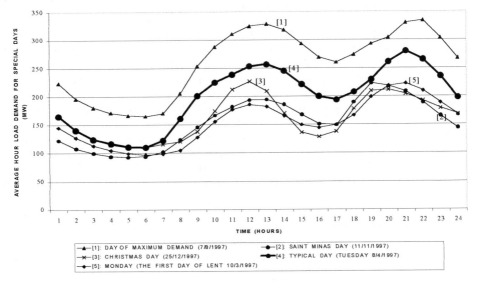

Fig 3. Load demand curves for different special days.

Random factors. It is not possible to estimate the time of their effect. Such factors in order of importance are:

(a) The T.V. programs transmission of significant interest. This leads to LD increase that by the time going becomes less important by the reason of the general increase in the LD in the EPS of Crete. Here we must say that in pre-election periods in the past the LD was increasing during the T.V. transmission of a politician's speech. This with the power reserves which Public Power Corporation (P.P.C.) must keep for such situations (e.g. for big pre-election concentrations) must be examined further.

(b) The Strikes. We talk about strikes in the private sector (General Confederation of Workers of Greece), in the public sector (Association of Public Servants) and general strikes. The two former don't have any important effect in the EPS of Crete because: i) the participation percentage is very low in the strikes of private sector, ii) the demand percentage of the industry for the EPS of Crete is directly small (of 10%) and iii) the LD for the public services is small. General strikes therefore have no significant influence within the accuracy limits put on the EPS of Crete.

(c) Visits of Very Important Persons: political, artistic or religious celebrities don't exert influence on LD but do so on power production due to amount of power reserves that probably has to be maintained by P.P.C.

(d) Holding of international congresses and special artistic events should be included to the above categories. In this case the use of power reserves is the important factor and mostly concerns the production system.

Financial Factors. For the time being the consumers of the EPS of Crete practice little economy in the use of electric power. It should be said that the rise of the standard of living leads rather to an increase in LD because of the use of new technology electric devices which are sometimes very consumptive, and because of a change in the way of living. Economy concerns mostly the power production and is discussed later.

Finishing this paragraph we can say that the most important variables which must be taken into account for the construction of the on-line data base are: hourly demand, weekly demand (day to day change of demand), seasonal demand, temperature, direction and speed of the wind and special occasion demand. According to the above mentioned the load forecast of half an hour time horizon is of greater value, which, in combination with the optimum management of the production system will lead to significant economic profits for the enterprise and probably for the consumers, as the decrease in the price of 1 KWh for the production system of Crete will have an effect in the decrease of the sale price in the whole area of Greece provided that the present economic status will be maintained (Nowadays production of a KWh in Crete especially in load peak periods costs more than the production cost of a KWh in the rest power system of Greece).

6. Relation between load forecast and power production.

The exact forecast of electric demand on the conditions put, by itself can't lead to a significant decrease in the production cost of electric power. But the use of the forecast for the optimum management of the production system [7] under the technical restrictions which it is subject to, will lead to significant economic benefits. The justification of the above requirements must lead to:

(a) Requirements of availability and easiness in use of the forecast results at the crucial time needed. This means easiness of reading (recording) and understanding the presentation of the forecast results in an on-line base so that they can be used directly by the system operator for the final decision (for which his experience has until now the major role).

(b) Directness of results presentation at the minimum possible time and in an on-line basis since, as it is referred, a greater importance for the EPS of Crete is given to the accuracy of forecast for half an hour time horizon.

(c) The operators interference possibility for the insertion of more data because of random factors which concern however the production system mainly (e.g. short-circuits, failure of power switches, failure of power transformers, total or partial units maintenance, reserves requirements e.t.c.).

The above conditions lead us to the following figure of access with the operators of the system:

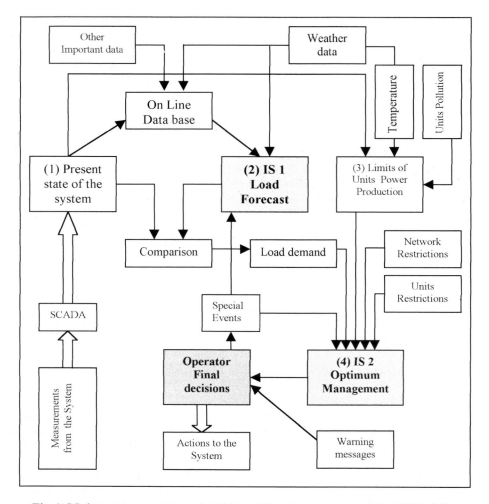

Fig 4. Major components needed for on-line management of the EPS of Crete.

In reference to the above figure we can say that except the operator's experience, four more, important stages are needed. On the stage 1 the present condition of the system based on measurements which come from the SCADA system should be presented. These actual power requirements will update directly the data base and together with the rest data of the base (the important variables as determined) throughout an intelligent system (IS 1) will lead to the forecast of load demand more important being the forecast for the immediate time horizon. There should be a comparison between data from stage 1 and forecast data from stage 2 so that the increase or decrease in LD requirements may be presented directly. Stage 2 should be able to present the demand curve graphically and its data (demand of active and inactive power) quantitatively. Those data must be potential changeable and not static as the inputs are potential variables changing with the present state of the power system. The operator as well should have the possibility of a long term forecast within the accuracy limits put for a long term planning.

The most important part of the whole process may be the stage 3 and mainly the stage 4. The stage 3 must be updated for the present state of the production units which mostly depends on the temperature of the environment and other technical restrictions (e.g. units cooling system pollution). This section should display graphically and quantitatively the maximum limit in power production of the already operating units but in actual values.

In the stage 4 a second intelligent system (IS 2) is really required with the following inputs:

(a) The actual capacity of the units in the load production,
(b) The requirements for the covering of the forecast demand,
(c) The perhaps actual existing problems of the system,
(d) The technical restrictions of the production units (e.g. number of starts and shut downs of a unit in combination with the maintenance and repair cost, environmental consequences of the use of a unit which are related directly to people's reactions but also to economic factors (as, for the recovery of accidental pollution much more money must be spent), the actual economic cost of the operation or no operation of a certain unit (which is due to fuel price, transport cost, storing and preprocessing cost).
(e) Special events. The operator as well should have the possibility of inserting more data (e.g.: unit switch off for maintenance, or partial maintenance), so that the operator can come to some conclusions for the behavior of the system which will have direct relation to the time planning. On the other hand in this way he will be able to check the capability of the system in remaining stable to random disturbances.

We believe that this stage may be the most important in the whole forecast process, the estimation of the state of the power system and the final decision for the power system development, and a particular attention must be given to its implementation. The output of stage 4 (in the case the above requirements are justified) should be the optimum units management and really it must be the suggestion to the operator of the system for the final decision. Of course the profound research of power problem of the EPS of Crete is a time consuming procedure from the aspect of collecting data, their recording in data bases, and of finding the proper way of their processing.

7.   Conclusions.

Our effort was concentrated on the accurate determination of the LD problem of the EPS of Crete, the finding out of the major variables that affect it and the way with which we must handle the whole forecast problem and optimum system management.

It is clear that for the autonomous Power System of Crete, the problem of the accurate demand forecast in combination with the optimum economical operation of the system is difficult to be solved. The efforts should be focused on the creation of an on-line forecast model with 1% error in a half an hour time horizon, which should have as inputs the most important variables so that the size of the data base and the time of access and processing its data may be minimized.

For the confining of the data base volume the necessary variables have already been suggested (historical seasonal and weekly data, day to day correlation coefficients [6], temperature, time of the day, direction and speed of the wind, random events). On the above suggested conditions we believe that the on line updating of the data base and the on-line demand forecast is feasible. We mustn't forget however the existence of the second intelligent system because only then the accurate load forecast has meaning and is more useful for the consumers as well as for the Public Power Corporation.

Acknowledgements.

We want to express our thanks to the dispatching center personnel of P.P.C. of Iraclion of Crete for their offering the necessary figures for the present survey, as well as to Mr John Sfakianakis from the climatic center of Iraclion for his remarks for the climatic conditions of Crete.

References.

[1] **Katsigiannakis G.,** "The Energy Problem Structure in Crete", Two Days Conference for the Energy Development of Crete, Technical Chamber of Greece, (Iracklio 1-2 Mar 1989). (In Greek)

[2] **P.P.C.,** "System of Crete 1997", Iracklion 23-2-1998. (In Greek)

[3] **P.P.C.,** "Development of the Electric Power System of Crete", Athens 1987 (In Greek)

[4] **P.P.C.,** "The Greek Electric System, 1995", Athens 1996. (In Greek)

[5] **Gross, G., and Galiana, F.,** "Short-Term Load Forecasting", Proceedings of the IEEE, vol. 75, no. 12, December 1987, 1558-1572.

[6] **Bitzer, B., Rosser, F. and Papazoglou, T.,** "24 Hours Load Forecasting for the Energy Management System on Crete – Online Training and Forecasting of Load Data", Proceedings of the 31st Universities Power Engineering Conference 1996, vol.1, September 1996, 346-349.

[7] **Bruce, F.W. and Toshiaki, S.,** "Artificial Intelligence in Power System Operations", Proceedings of the IEEE, vol. 75, no. 12, December 1987, 1678-1685.

47

Reliability Cost Assessment in Composite Power Systems using the Monte Carlo Simulation Approach

N.C. Koskolos, S.M. Megalokonomos, E.N. Dialynas
National Technical University of Athens
Department of Electrical and Computer Engineering
42 28ᵗʰ October street, Athens 10682 GREECE
dialynas@power.ece.ntua.gr

1. Introduction

The ability of a power system to provide an adequate and secure supply of electrical energy at any point in time is referred to as the reliability of the system and supply interruptions, regardless of their cause, constitute a reduction in this reliability. An optimal operation plan should achieve minimum operation and interruption cost [1-4]. Both, system operating and customer interruption costs depend on system states and on the type of customers at each load bus. The reliability cost / reliability worth assessment of Composite Generation and Transmission System is seriously affected by time of occurrence, duration and frequency of interruptions. Therefore, it is necessary to simulate a considerable number of system contingency states involving various load levels in order to obtain the expected values of these costs [1,5]. The chapter describes a computational technique for evaluating the costs of interruption and, hence, the reliability worth in composite generation and transmission power systems considering the time varying loads and the type of customers at each load bus, in relation with the above factors of interruptions. This technique was developed using the sequential Monte Carlo simulation approach and using the linear programming method for achieving a minimum operating and interruption cost for system customers. The basic features of the technique are the recognition of different interruption costs at different buses for different customers and the consideration of the duration of system contingencies. The computational method has been implemented efficiently into an interactive computer program which was written in FORTRAN77. Various case studies were conducted using the IEEE reliability test system (RTS) [6,7] and the obtained results are presented and discussed.

2. Basic Concepts of Power System Outage Cost

The primary function of a modern electric power system is to supply the customer's requirements at a reasonable level of reliability and quality. The precise specifications of this reasonable level of reliability is a very difficult problem of balancing the need for continuity of power supply and the cost involved. Although, the majority of outages are caused by factors that are not related to load demand levels, the load demand affects seriously the characteristics of outages (e.g., frequency, duration, unserved energy, etc.) and, consequently, their associated customer economic costs. The customer economic costs due to outages resulting load curtailment can be estimated using the customer survey method where customers are surveyed to estimate their economic losses and to create a customer damage function for each customer group [2,8]. The composite customer damage function for a particular service area defines the total customer costs for that area as a function of the interruption duration and the mix of the respective customers. The broadest application of a customer damage function is its use to relate the composite customer losses to the socioeconomic worth of electric service reliability for an entire utility service area. Reliability worth can be expressed in terms of the expected customer interruption cost which can be estimated by multiplying the expected energy not supplied to customers due to power interruptions by a suitable cost index. This index is the Interrupted Energy Assessment Rate (IEAR) which relates the customer losses caused by electric power interruptions to the worth of electric service reliability and it is expressed in $/kWh [9].

3. Monte-Carlo Simulation Approach

The Monte-Carlo simulation approach is a stochastic simulation procedure and can be used for estimating the reliability indices of an electrical power system by simulating its actual behavior [10-13]. The problem is treated as a series of real experiments conducted in simulated time steps of one hour. A series of system scenarios is obtained by hourly random drawings on the status of each system component and determining the hourly load demand (sequential simulation). The desired reliability indices are calculated for each hour with the process repeated for the remaining hours in the year. The annual reliability indices are calculated from the year's accumulation of data generated by the simulation process. The year continues to be simulated with new sets of random events until obtaining statistical convergence of the desired indices. The main features of the sequential simulation approach being developed at NTUA are the following [14]:

1) The pseudo-random numbers are generated applying the multiplicative congruent method. The antithetic sampling technique has been also used for variance reduction.
2) The classical two-state Markovian model is generally used to represent the system component operation while actual or equivalent generating units may be represented by a multiple state model in order to recognize their derated states. Common-cause transmission line outages are also considered applying appropriate models.
3) The generation system includes thermal and hydroelectric plants. A thermal plant

consists of a number of single generating units while a hydroelectric plant also includes a reservoir. The source of energy in hydroelectric plants is the inflow of the water, which is stored in the reservoirs located along the respective rivers. A hydro-chain can also exist consisting of plants located on the same river.

4) A generation rescheduling technique is applied after the occurrence of a generation unit outage for compensating its loss.

5) The network branch flows are obtained for any given hour of the simulation period using a DC load flow model.

6) Branch overloading is alleviated by rescheduling generation and/or load curtailment at appropriate system bus-bars by applying two alternative algorithms.

7) Various operating and water management policies are simulated which affect significantly the system reliability performance when there are limitations in the energy produced by the hydroelectric plants.

Three sets of indices are calculated for each system under study. The first set forms system indices which reflect the adequacy of the entire system. The second set forms system interruption indices which reflect the characteristics of the interruptions being occurred in the system. The third one is also calculated for each system hydroelectric plant providing valuable information about their operation. The following indices are calculated and have the respective acronyms:

(a) System indices
- Loss of load expectation (LOLE) in hours/year
- Loss of energy expectation (LOEE) in MWh/year
- Expected demand not supplied (EDNS) in MW/year
- Frequency of loss of load (FLOL) in occ./year
- Energy index of reliability (EIR)
- Average energy produced by all the hydroelectric plants (AEHG) in GWh/year
- Fraction of total energy demand supplied by hydroelectric generation (FEHG) in %

(b) System interruption indices
- Energy not supplied per interruption (ENSINT) in MWh/interruption
- Demand not supplied per interruption (DNSINT) in MW/interruption
- Expected duration per interruption (DOI) in hours/interruption

(c) Hydroelectric plant indices
- Average water used to produce electric energy (AWE) in Mm^3/year
- Average energy produced from hydroelectric plant (EAWE) in GWh/year
- Average water spilled (AWS) in Mm^3/year
- Average energy production lost due to water spillage (EAWS) in kWh/year
- Average water pumped (AWP) in Mm^3/year
- Average energy used to pump water (EAWP) in kWh/year

Figures of all the above indices as functions of the number of sampling years can be also produced in order to identify their variations very easily. Furthermore, frequency histograms and distributions of these reliability indices can be produced which provide some interesting and important insights into the random behavior of the system.

4. Computational Method for the Evaluation of Outage Cost

Three alternative computational methods have been developed applying the features of the Monte-Carlo simulation approaches that were described previously. Their main objective is to calculate a set of indices which quantify the reliability cost of composite generation and transmission systems. The basic features of the first method (method A) are the following:

a) An interruption sequence is defined as a successive occurrence of failure states, resulting in the same amount of load curtailment in each failure state at the load points in question.

b) The interruption cost to customers at bus i for an interruption sequence m having duration d, can be obtained using the composite customer damage function of the jth user sector $c_j(d)$ that is connected at bus i. The cost of interruption (ECOST) at bus i for an interruption sequence of duration d is calculated as:

$$ECOST_{i,\,m} = \sum_{j=1}^{ID} L_i W_{ij}\, c_j(d) \ \text{(in k\$/yr)} \tag{1}$$

where ID is the number of sectors present at bus i, W_{ij} is the fraction of the load bus i taken by the jth user sector of that bus, which is constant for every hour of the year, and L_i is the load being curtailed at bus i.

The expected interruption cost at bus i is calculated as:

$$ECOST_i = \sum_{m=1}^{k} ECOST_{im} / N \ \text{(in k\$/yr)} \tag{2}$$

where k is the total number of interruption sequences being experienced during the total number of simulation years (N).

c) The system interruption cost can be calculated as follows:

$$ECOST = \sum_{i=1}^{n} ECOST_i \ \text{(in k\$/yr)} \tag{3}$$

where n is the total number of system load buses.

d) The interrupted energy assessment rate at bus i, can be calculated as follows :

$$IEAR_i = \frac{ECOST_i}{LOEE_i} \ \text{(in \$/kWh)} \tag{4}$$

where $LOEE_i$ is the Loss of Energy Expectation Index of bus i.

e) The system index IEAR can be obtained as follows :

$$IEAR = \sum_{i=1}^{n} IEAR_i\, q_i \ \text{in (k\$/kWh)} \tag{5}$$

where q_i is the fraction of system load supplied to load bus i.

The second method (method B) estimates the outage cost using the same procedure as was described previously for method A. There is only one difference concerning the definition of an interruption sequence. In method A, that is defined as a successive occurrence of failure states, without reference to the amount of load being curtailed during each failure state. In method B, the load being curtailed is obtained as the ratio of the unserved energy during an interruption to the duration of the interruption.

The third method (Method C) estimates the outage cost utilizing the sector contribution to each load bus which varies according to system load duration curve. This curve is divided into a number of steps that is formed by aggregating the hourly system load curve and the fraction W_{ij} is varied for each step of the load duration curve. The cost of interruption at bus i for an interruption sequence m of duration d is calculated as:

$$\text{ECOST}_{i, m} = \sum_{j=1}^{\text{ID}} L_i \, c_j(d) \sum_f W_{ijf} \, t_f \, / d \quad \text{(in k\$/yr)} \tag{6}$$

where t_f is the sum of hourly intervals belonging in the fth step of load duration curve.

Additionally, a minimization model of operating and interruption cost has been developed applying the features of linear programming method. The minimization model is used for each hourly interval experiencing line overloads and its intention is to alleviate them by rescheduling generation outputs to avoid load curtailment, if possible, or, to minimize total load curtailment, if unavoidable. Additionally the objective of this model is to minimize the sum of the operating and interruption costs while satisfying the power balance, the DC load flow relationships and the limits on line power flows and generator outputs. The objective function (OF) of the minimization model is the following:

$$\text{minimize} \ \ OF = \sum_{i=1}^{\text{NG}} \sum_{j=1}^{\text{NGi}} b_{ij} P_{ij} + \sum_{i=1}^{\text{NLB}} \sum_{j=1}^{\text{ID}} c_{ij} LC_{ij} \tag{7}$$

while the following model equations and inequalities must be satisfied:

$$T_n = \sum_{i=1}^{\text{NB}} A_{ni} \left(\sum_{j=1}^{\text{NGi}} P_{ij} + \sum_{j=1}^{\text{ID}} LC_{ij} - \sum_{j=1}^{\text{ID}} L_{ij} \right) \tag{8}$$

$$\sum_{i=1}^{\text{NG}} \sum_{j=1}^{\text{NGi}} P_{ij} + \sum_{i=1}^{\text{NLB}} \sum_{j=1}^{\text{ID}} LC_{ij} = \sum_{i=1}^{\text{NLB}} \sum_{j=1}^{\text{ID}} L_{ij} \tag{9}$$

$$P_{ij}^{\min} \le P_{ij} < P_{ij}^{\max} , (i = 1,..., \text{NG}, j = 1,...., \text{NGi}) \tag{10}$$

$$0 \le LC_{ij} < L_{ij}, (i = 1,..., \text{NLB}, j = 1,...., \text{ID}) \tag{11}$$

$$|T_n| \le T_n^{\max}, (n = 1,...., \text{NL}) \tag{12}$$

532

where:

b_{ij} = generation cost (in \$/kWh) of the jth generator at bus i

P_{ij} = generation of the jth generator at bus i

P_{ij}^{min}, P_{ij}^{max} = lower and upper bounds of P_{ij}

c_{ij} = damage cost (in \$/kW) of the jth customer sector at bus i for an interruption hourly duration

L_{ij} = load of the jth customer sector at bus i

LC_{ij} = load curtailment of the jth customer sector at bus i

T_n = power flow on line n

T_n^{max} = capacity of line n

A_{ni} = element of the relation matrix between line flows and power injections at buses

NG = number of generator buses

NGi = number of generators at bus i

NLB = number of load buses

ID = number of customer sectors

NB = total number of system buses

NL = total number of system lines.

The above described computational methods have been implemented efficiently into an interactive computer program **SIRLCT** which was written in FORTRAN77 and tested on a PC-IBM compatible 586 microcomputer. The basic principles of interactive computation were employed which permit a natural, flexible and convenient way of communication between the user and computer.

5. Analysis of a Typical Power System

The developed computational method and the respective computer program **SIRLCT** have been applied for conducting reliability and cost assessment studies on the IEEE-RTS [6,7]. This system consists of 24 buses, 36 transmission lines, 9 stations and 32 generating units. The minimum and the maximum ratings of the generating units are 12 MW and 400 MW respectively. The total installed generation capacity is 3405 MW and the system peak load demand is 2850 MW which means that the system operating reserve is 555 MW. The full set of the system reliability and cost indices have been evaluated and are presented in Table 1. The following base case system study and five alternative case studies were considered:

Case 1: base case study considering the hourly load demand model and method A is applied for the outage cost calculation.

Case 2: the constant peak load demand model is considered.

Case 3: as in case 1 but method B is applied for the outage cost calculation.

Case 4: as in case 1 but method C is applied for the outage cost calculation.

Case 5: as in case 1 but the minimization model is applied.

Case 6: as in case 3 but the minimization model is applied.

Table 1: System Reliability and Cost indices

Case Study Indices	1	2	3	4	5	6
LOLE	20.875	740.520	20.875	20.875	17.989	17.989
LOEE	2480.20	128468	2480.20	2480.20	2093.92	2093.92
EDNS	73.6944	165.727	73.6944	73.6944	67.9669	67.9669
FLOL	4.5995	19.0369	4.5995	4.5995	4.0740	4.0740
EIR	0.999838	0.99484	0.999838	0.999838	0.999863	0.999863
ENSINT	539.233	6748.34	539.233	539.233	513.972	513.972
DNSINT	77.0147	115.165	77.0147	77.0147	74.1876	74.1876
DOI	4.5393	38.8992	4.5393	4.5393	4.41556	4.41556
ECOST	9087.2	648828	9133.6	15988.7	8917.8	8787.3
IEAR	3.3238	5.2776	3.5763	5.4777	3.3273	3.5205

Additionally, Table 2 presents the reliability indices for all system buses being obtained applying methods A and B while Table 3 presents the bus and system interruption cost indices being obtained applying all methods A, B, C. Finally, Figure 1 shows the variation of the interruption cost index in particular load buses according to the three methods being obtained (A, B and C).

Table 2: Reliability indices of load buses

Index	FLOL (occ./year)		DOI (hrs/int.)		ENSINT MWh/int.)	
Bus No.	Method A	Method B	Method A	Method B	Method A	Method B
1	5.3840	1.5985	1.2124	4.0835	4.9501	16.6728
2	6.0025	1.7465	1.1738	4.0344	4.3966	15.1104
3	0.0180	0.0065	1.4167	3.9231	15.6543	43.3503
4	0.2340	0.0820	1.2372	3.5305	6.3209	18.0376
5	0.2025	0.0705	1.3086	3.7589	6.8682	19.7279
6	0.0260	0.0130	1.2885	2.5769	6.7758	13.5517
7	16.850	4.6160	1.2294	4.4875	7.5592	27.5935
8	12.598	3.5415	1.2350	4.3931	8.5294	30.3414
9	11.934	3.3600	1.2327	4.3780	8.5419	30.3376
10	11.932	3.3590	1.2326	4.3784	9.4857	33.6941
13	13.209	3.6300	1.2202	4.4399	46.4417	168.988
14	0.8975	0.2790	1.1989	3.8566	13.0104	41.8523
15	13.206	3.6280	1.2199	4.4405	40.7411	148.293
16	2.7555	0.7625	1.1629	4.2026	11.6167	41.9799
18	12.650	3.4430	1.2227	4.4923	54.4539	200.070
19	0.0010	0.0010	1.5000	1.5000	6.4243	6.4243
20	3.7505	0.9790	1.2409	4.2722	26.6817	91.8596

534

Table 3: Interruption Cost indices of system load buses

Index	Method A		Method B		Method C	
Bus No.	ECOST (k$/yr)	IEAR (k$/kWh)	ECOST (k$/yr)	IEAR (k$/kWh)	ECOST (k$/yr)	IEAR (k$/kWh)
1	124.614	4.6757	142.302	5.3394	175.054	6.5683
2	54.466	2.0639	101.133	3.8322	95.939	3.6354
3	1.054	3.7409	1.329	4.7162	1.586	5.6290
4	3.758	2.5407	5.516	3.7295	6.145	4.1546
5	5.985	4.3034	6.806	4.8936	9.096	6.5398
6	0.733	4.1596	0.733	4.1619	1.307	7.4182
7	513.854	4.0343	550.429	4.3214	714.788	5.6118
8	354.397	3.2981	459.558	4.2768	731.901	6.8113
9	237.191	2.3269	211.734	2.0772	470.740	4.6181
10	353.676	3.1249	364.766	3.2230	508.734	4.4950
13	2678.87	4.3671	2871.41	4.6809	4777.78	7.7887
14	35.376	3.0296	33.344	2.8556	45.971	3.9369
15	1727.72	3.2113	1585.42	2.9468	3143.89	5.8436
16	78.618	2.4561	97.704	3.0523	102.350	3.1975
18	2734.25	3.9694	2443.22	3.5468	4942.88	7.1756
19	0.013	1.9885	0.013	4.8052	0.016	2.4563
20	182.591	2.0304	258.148	2.8705	260.787	2.8965
System	9087.166	3.3238	9133.569	3.5763	15988.664	5.4777

A considerable number of comments can be drawn from the obtained results but the most important ones are the following:
- The application of the peak load demand model increases the system reliability and interruption cost indices drastically as indicated by comparing the results obtained for cases 1 and 2. This means that the load variation during the study period affects significantly the results of the reliability assessment and needs to be handled as an important parameter of such studies. However, the computer running time, using the hourly load demand model, is expected to increase significantly.
- The application of method B gives a higher estimate of the interruption cost index compared with that obtained when applying method A as its shown by the results obtained for case 3 because this method gives failure states which have smaller frequency but greater duration. This conclusion can also be shown by comparing the bus indices FLOL and DOI in Table 2 for methods A and B.
- The application of method C gives highest estimates of the interruption cost index since the sector contribution to each load bus varies according to the steps of the system load duration curve being applied.
- The application of the minimization model decreases the reliability and interruption cost indices as its indicated by comparing the results obtained for cases 1 and 5 or 3 and 6. This means that the minimization model avoids load curtailment if possible or minimizes the total load curtailment if unavoidable and curtails bus load according to the interruption cost of the customer sectors in the bus.

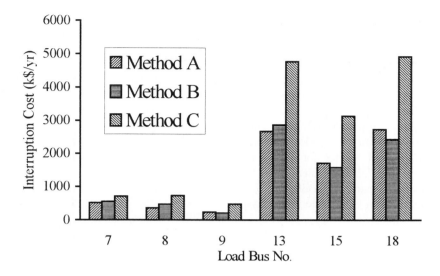

Figure 1. Interruption cost index of load buses

6. Conclusions

The chapter describes the main concepts and features of a computational method that was developed at NTUA for evaluating the costs of interruption and, hence, the reliability worth in composite generation and transmission power systems considering the time varying loads and the type of customers at each load bus, in relation with these factors of interruptions. This method was developed using the sequential Monte-Carlo simulation approach and the linear programming method for achieving a minimum operating and interruption cost for system customers. Different techniques are presented for estimating the cost of interruption at system load buses according to the type of load model being used. The technique being used for calculating the interruption cost is significantly influenced by the definition of interruption sequences and the curtailment strategies being used. Three different techniques for calculating the interruption cost (Methods A, B and C) were described and discussed. The choice of method to be applied depends on the availability of data regarding the individual bus loads. The chapter also presents the system reliability and cost indices being obtained for each system bus and the overall system from the analysis of various case studies for the IEEE-RTS. Finally, the reliability assessment studies being conducted demonstrate the increased information that can be gained about the reliability performance of the system.

Acknowledgments

The research work described in this chapter was fully financed by the General Secretariat of Research and Technology of the Ministry of Development in Greece. The authors gratefully acknowledge this financial support under the research programme of PENED 95.

References

1.   Billinton, R., Allan, R.N., "Reliability Evaluation of Power Systems", Longmans, London, 1984..
2.   Billinton, R., Allan, R.N., "Reliability Assessment of Large Electric Power Systems", Kluwer Academic Publisher, Accord Station, 1988.
3.   Wacker, G., Wojczynski, E., Bollinton, R., "Comprehensive Bibliography of Electrical Service Interruption Cost", IEEE Transactions, Vol. PAS-102, 1983, pp. 1831-1837.
4.   Sanghvi, Arun, P., "Measurement and Application of Customer Interruption Cost/Value of Service for Cost - Benefit Reliability Evaluation: Some Commonly Raised Issues", IEEE Transactions, Vol. PWRS-8, 1993, pp. 761-771.
5.   Sankarakrishnan, A., Billinton, R., "Effective Technique for Reliability Worth Assessment in Composite Power System Networks Using Monte Carlo Simulation", IEEE Summer Meeting, paper 95SM 512-4-PWRS, Portland, 1995.
6.   IEEE APM Subcommittee Report: "The IEEE-Reliability Test System", IEEE Transactions, Vol. PAS-98, 1979, pp. 2047-2054.
7.   IEEE APM Subcommittee Report: "The IEEE-Reliability Test System - 1996", IEEE/PES Winter Meeting, paper 96WM 329-9PWRS, Baltimore, 1996.
8.   Tollefson, G., Billinton, R., Wacker, G., Chan, E., Aweya, J., "A Canadian Customer Survey to Assess Power System Reliability Worth", IEEE Transactions, Vol. PWRS-9, 1994, pp. 443-450.
9.   Oteng-Adjei, J., Billinton, R., "Evaluation of Interrupted Energy Assessment Rates in Composite Systems", IEEE Transactions, Vol. PWRS-4, 1989, pp.83-93.
10. Ross, M.S., "A Course in Simulation", Maxwell Macmillan, New York, 1991.
11. Salvaderi, L., Allan, R.N., Billinton, R., Endrenyi, J., Mc Gillis, D., Manning, P., Ringlee, R., "State of the Art of Composite-System Reliability Evaluation", 1990 CIGRE Session, Paris, Paper 38-104.
12. Anders, G.J., Endrenyi, J., Pereira, M.V.F., Pinto, L., Oliveira, C., Cunha, S., "Fast Monte - Carlo Simulation Techniques for Power System Reliability Studies", 1990 CIGRE Session, Paris, Paper 38-205.
13. Salvaderi, L., Billinton, R., "A Comparison Between Two Fundamentally Different Approaches to Composite Reliability Evaluation", IEEE Transactions, Vol. PAS-104, 1985, pp.3486-3492.
14. Dialynas, E.N., Koskolos, N.C, "Comparison of Contingency Enumeration and Monte - Carlo Simulation Approaches Applied to the Reliability Evaluation of Composite Power Systems", European Journal of Diagnosis and Safety in Automation, Vol.5, 1995, pp.25-48.

INDEX

International Series on
MICROPROCESSOR-BASED AND INTELLIGENT SYSTEMS ENGINEERING

Editor: Professor S. G. Tzafestas, *National Technical University, Athens, Greece*

KLUWER ACADEMIC PUBLISHERS – DORDRECHT / BOSTON / LONDON